Solitons and the Inverse Scattering Transform

SIAM Studies in Applied Mathematics

WILLIAM F. AMES, Managing Editor

This series of monographs focuses on mathematics and its applications to problems of current concern to industry, government, and society. These monographs will be of interest to applied mathematicians, numerical analysts, statisticians, engineers, and scientists who have an active need to learn useful methodology for problem solving.

The first three titles in this series are: *Lie-Bäcklund Transformations in Applications*, by Robert L. Anderson and Nail H. Ibragimov; *Methods and Applications of Interval Analysis*, by Ramon E. Moore; and *Ill-Posed Problems for Integrodifferential Equations in Mechanics and Electromagnetic Theory*, by Frederick Bloom.

Mark J. Ablowitz and Harvey Segur

Solitons and the Inverse Scattering Transform

siam *Philadelphia/1981*

Copyright 1981 by Society for Industrial and Applied Mathematics. All rights reserved.

Library of Congress Catalog Card Number: 81-50600
ISBN: 0-89871-174-6

Photography, Figure 4.7b, courtesy of T. Toedtemeier; Figure 4.10, courtesy of NASA; Figure 4.13, courtesy of R. W. Terhune.

to Carol and Enid

Contents

Prologue		ix
Chapter 1: The Inverse Scattering Transform on the Infinite Interval		1
1.1.	Introduction	1
1.2.	Second order eigenvalue problems and related solvable partial differential equations	8
1.3.	Derivation of a linear integral equation and inverse scattering on the infinite interval	15
1.4.	Time dependence and special solutions	28
1.5.	General evolution operator	42
1.6.	Conservation laws and complete integrability	52
1.7.	Long-time behavior of the solutions	67
	Exercises	84
Chapter 2: IST in Other Settings		93
2.1.	Higher order eigenvalue problems and multidimensional scattering problems	93
2.2.	Discrete problems	114
2.3.	Periodic boundary conditions for the Korteweg-deVries equation	134
	Exercises	148
Chapter 3: Other Perspectives		151
	Overview	151
3.1.	Bäcklund transformations	153
3.2.	Pseudopotentials and prolongation structures	161
3.3.	Direct methods for finding soliton solutions—Hirota's method	171

3.4. Rational solutions of nonlinear evolution equations 191
3.5. N-body problems and nonlinear evolution equations 203
3.6. Direct approaches with the linear integral equation 217
3.7. Painlevé transcendents 233
3.8. Perturbations and transverse stability of solitons and solitary waves 250
Exercises 261

Chapter 4: Applications 275

4.1. KdV problems and their cousins 276
4.2. Three-wave interactions 300
4.3. The nonlinear Schrödinger equation and generalizations . . 313
4.4. Equations of the sine-Gordon type 327
4.5. Quantum field theory 339
Exercises 342

Appendix: Linear Problems 351

A.1. Fourier transforms 351
A.2. Failure of the Fourier transform method 373
Exercises 384

Bibliography 393

Index 415

Prologue

The basic theme of this book can be stated quite simply: *Certain nonlinear problems have a surprisingly simple underlying structure, and can be solved by essentially linear methods.* Typically, these problems are in the form of *evolution equations*, which describe how some variable (or set of variables) evolves in time from a given initial state. The equations may take a variety of forms, including partial differential equations, differential-difference (discrete space, continuous time), partial difference (discrete time and space), integro-differential, as well as coupled ordinary differential equations (of finite order). What is surprising is that even though these problems are nonlinear, one may obtain the general solution that evolves from arbitrary initial data (within an appropriate class) without approximation. It is perhaps equally surprising that some of these exactly solvable problems arise naturally as models of physical phenomena. These applications have helped to generate interest in the subject.

Several viewpoints about these exactly solvable problems are common. One of them identifies the general solution of an appropriate initial value problem as *the* objective of the analysis. This solution is obtained by the Inverse Scattering Transform (IST), which is described in detail in Chapters 1 and 2. It can be viewed as a generalization of the Fourier transform, by which linear problems may be solved.

The problems in question have such a rich structure that they may be considered from several other viewpoints, which may be rather unrelated to IST. Some of these other perspectives are examined in Chapter 3. Many of these are more useful if one is primarily interested in special solutions, such as solitons, rather than in the general solution of an initial value problem. A number of physical applications are discussed in detail in Chapter 4.

The value of IST is that one treats nonlinear problems by essentially linear methods. This value is marginal, of course, unless one is already familiar with the methods and results of linear theory. Because of the fundamental role played by linear theory, we have included an extensive appendix which deals with linear problems and their solutions. These serve as useful guides against which to compare the corresponding solutions of the nonlinear problems that are the subject of this book.

Before we plunge in, here is our opinion regarding the order in which the book should be read. The Appendix contains material that is preliminary, although not necessarily trivial. It will be most useful to those unfamiliar with Fourier transform methods if it is read first. Because it is introductory, a substantial set of fairly straightforward exercises is included at the end of the Appendix.

Chapter 1 is fundamental. Later chapters often build on the material in this chapter, and refer back to it. We recommend that all of Chapter 1 be read.

Many avenues are available after Chapter 1. Chapters 2, 3 and 4 depend on Chapter 1, but not particularly on each other. They may be read in any order desired. To a lesser extent, the sections within each chapter may be considered independent of each other as well. This permits the reader with a specialized interest to gain access to his/her material relatively quickly.

Finally, a word about the exercises. These cover a range of difficulty, from merely filling in some missing steps, to research problems whose answers, to our knowledge, are open. Usually the wording of the problems identifies to the reader the ones that are open.

Chapter 1

The Inverse Scattering Transform on the Infinite Interval

1.1. Introduction. In 1965 Zabusky and Kruskal discovered that the pulselike solitary wave solution to the Korteweg–deVries (KdV) equation had a property which was previously unknown: namely, that this solution interacted "elastically" with another such solution. They termed these solutions *solitons*. Shortly after this discovery, Gardner, Greene, Kruskal and Miura (1967), (1974) pioneered a new method of mathematical physics. Specifically, they gave a method of solution for the KdV equation by making use of the ideas of direct and inverse scattering. Lax (1968) considerably generalized these ideas, and Zakharov and Shabat (1972) showed that the method indeed worked for another physically significant nonlinear evolution equation, namely, the nonlinear Schrödinger equation. Using these ideas Ablowitz, Kaup, Newell and Segur (1973b) and (1974) developed a method to find a rather wide class of nonlinear evolution equations solvable by these techniques. They termed the procedure the Inverse Scattering Transform (IST).

This monograph is devoted to this subject: i.e., to solitons and IST. There have been numerous developments in this area, which has aroused considerable interest among mathematicians, physicists and engineers. We hope that by capturing many of the main ideas and putting them into one location, we will be helpful to both beginners and the "pros" in the field. The main difficulty in doing this comes from the vigor with which the field has and is (at this time) continuing to develop.

Some review articles[1] and some collected works[2] on the subject are available. At the time of writing, there are not any other monographs extant on this topic, but we expect that this state of affairs will undoubtedly change quickly.[3]

---

[1] See, for example, Scott, Chu and McLaughlin (1973), Miura (1976), Ablowitz (1978) and Makhankov (1979).

[2] See, for example, Newell (1974a), Miura (1974), Moser (1975b), Calogero (1978a) and Lonngren and Scott (1978).

[3] In fact, by the time galley proofs for this monograph were received, both Zakharov, Manakov, Novikov and Pitayevsky (1980) and Lamb (1980) had appeared.

The study of solitary waves began with the observations by J. Scott Russell (1838), (1844) over a century ago. Russell, an experimentalist, first observed a solitary wave while riding on horseback beside a narrow barge channel. When the boat he was observing stopped, Russell noted that it set forth

> a large solitary elevation, a rounded, smooth and well defined heap of water, which continued its course along the channel apparently without change of form or diminution of speed ⋯. Its height gradually diminished, and after a chase of one or two miles I lost it in the windings of the channel. Such, in the month of August 1834, was my first chance interview with that singular and beautiful phenomenon. (Russell (1838)).

This observation inspired Russell to initiate an extensive experimental investigation of water waves. He divided all of water waves into two classes, the "great primary wave of translation" (which would eventually be called a solitary wave), and all other waves, which "belong to the *second or oscillatory order of waves*;" the latter waves "are not of the first order" (Russell (1838)). Clearly he regarded the solitary waves as being of primary importance and concentrated most of his attention on them. Among his many results, we should note particularly the following.

1. Solitary waves, which are long (shallow water) waves of permanent form, *exist*. This is undoubtedly his most important result.

2. The speed of propagation of a solitary wave in a channel of uniform depth is given by

$$v = \sqrt{g(h+\eta)},$$

η "being the height of the crest of the wave above the plane of repose of the fluid, h the depth throughout the fluid in repose, and g the measure of gravity" (Russell (1844)). Considering the accuracy of the experimental equipment available to him, this result is somewhat remarkable.

Russell found that no mathematical theory available at the time predicted a solitary wave, but noted that

> it was not to be supposed that after its existence had been discovered and its phenomena determined, endeavors would not be made to reconcile it with previously existing theory, or in other words, to show how it ought to have been predicted from the known general equations of fluid motion. In other words, it now remained to the mathematician to predict the discovery after it had happened; i.e., to give an *a priori* demonstration *a posteriori*. (Russell (1844)).

Russell seems to have been particularly contemptuous of Airy, who published a theory of long waves of small but finite amplitude in his *Tides and Waves* (1845). This theory is summarized in Lamb (1932, §§ 175 and 187), who states "when the elevation η is not small compared with the mean depth, h, waves, even in an uniform-canal of rectangular section, are no longer propagated without change of type." Thus, Airy concluded that solitary waves of permanent form do not exist! He also found an approximate formula for the wave speed,

$$v = \sqrt{gh}\left(1 + \frac{1}{2}\frac{\eta}{h}\right),$$

which agrees with Russell's result to first order in (η/h). From this, Airy decided that "we think ourselves fully entitled to conclude from these experiments [i.e., Russell's] that the theory [Airy's] is entirely supported." Russell described Airy's conclusion as "completely the opposite of that to which we should be led on the same grounds."

This controversy raged on for another 50 years before it was finally resolved by Korteweg and de Vries (1895). They derived an equation (now known as the Korteweg–deVries, or KdV equation), which governs moderately small, shallow-water waves. Their equation had permanent wave solutions, including solitary waves.

Boussinesq (1871), (1872), also derived a nonlinear evolution equation governing such long waves. Both Boussinesq (1871), (1872) and Rayleigh (1876) obtained solitary wave solutions.

As Miura (1976) points out, despite this early work, apparently no new applications of the equation derived by Korteweg and deVries were discovered until 1960. Then, while studying collision-free hydromagnetic waves, Gardner and Morikawa (1960) also derived the Korteweg–deVries equation.

The physical problem which motivated the recent discoveries related to the KdV equation was the Fermi–Pasta–Ulam (FPU) problem (1955). In 1914 Debye suggested that the finiteness of the thermal conductivity of an anharmonic lattice is due to its nonlinearity. This led Fermi, Pasta and Ulam to undertake a numerical study of a one-dimensional anharmonic lattice. They felt that, due to the nonlinear coupling, any smooth initial state would eventually relax to an equipartition of energy among the various degrees of freedom of the system.

The model they considered consisted of identical masses connected to their nearest neighbors by nonlinear springs with the force law $F(\Delta) = -K(\Delta + \alpha\Delta^2)$. The equations of motion are

(1.1.1)
$$\frac{m}{K}y_{i,tt} = (y_{i+1} + y_{i-1} - 2y_i) + \alpha[(y_{i+1} - y_i)^2 - (y_i - y_{i-1})^2],$$
$$i = 1, 2, \cdots, N-1, \quad y_0 = y_N = 0,$$

with a typical initial condition of $y_i(0) = \sin i\pi/N$, $y_{it}(0) = 0$ (typically N was taken to be 64). Here y_i measures the displacement of the ith mass from equilibrium.

According to Fermi, Pasta and Ulam (1955),

> the results of our computations show features which were, from the beginning, surprising to us. Instead of a gradual, continuous flow of energy from the first mode to the higher modes, ··· the energy is exchanged, essentially, among only a certain few. ··· There seems to be little if any, tendency towards equipartition of energy among all degrees of freedom at a given time. In other words, the systems certainly do not show mixing.

In order to understand this phenomenon, Kruskal and Zabusky (1963) considered a continuum model. Calling the length between springs h, $t' = \omega t$ ($\omega = \sqrt{K/m}$), $x' = x/h$ with $x = ih$, and expanding $y_{i\pm1}$ in Taylor series, reduces (1.1.1) to (dropping the primes)

$$(1.1.2) \qquad y_{tt} = y_{xx} + \varepsilon y_x y_{xx} + \frac{h^2}{12} y_{xxxx} + O(\varepsilon h^2, h^4),$$

where $\varepsilon = 2\alpha h$. A further reduction is possible if we look for an asymptotic solution of the form (unidirectional waves)

$$y \sim \phi(X, T), \quad X = x - t, \quad T = \frac{\varepsilon t}{2},$$

whereupon (1.1.2) gives

$$(1.1.3) \qquad \phi_{XT} + \phi_X \phi_{XX} + \delta^2 \phi_{XXXX} + O\left(h^2, \frac{h^4}{\varepsilon}\right) = 0,$$

where $\delta^2 = h^2/12\varepsilon$. By setting $u = \phi_X$, (1.1.3) is reduced to an equation directly related to that originally discovered by Korteweg and deVries (1895):

$$(1.1.4) \qquad u_T + uu_X + \delta^2 u_{XXX} = 0.$$

Kruskal and Zabusky computed (1.1.4) typically with sinusoidal initial conditions. With δ^2 taken small, the slope of the initial function steepens until the third derivative terms become important. At this stage the solution develops an oscillatory structure of a definite form. The oscillations interact in a very definite and surprising way, which we will discuss presently. The process of trying to understand this phenomena is what led to our present understanding of the properties and solutions of the KdV equation. (Interestingly Lax and Levermore (1979) have reinvestigated (1.1.4) with δ^2 small.)

Hereafter we shall work with the KdV equation in the following form:

$$(1.1.5) \qquad K(u) = u_t + 6uu_x + u_{xxx} = 0.$$

Equation (1.1.5) is equivalent to (1.1.4) upon a scale change (note that any constant coefficient may be put in front of each of the three terms by suitably scaling the independent and dependent variables).

It was clear to Kruskal and Zabusky (and was well known) that KdV had a special permanent wave solution, the solitary wave,

(1.1.6) $$u = 2k^2 \operatorname{sech}^2 k(x - 4k^2 t - x_0),$$

where k and x_0 are constant. Note that the velocity of this wave, $4k^2$, is proportional (by a factor of 2 here) to the amplitude, $2k^2$. What was not clear to previous researchers, and what is so surprising, is the way these waves interact with each other elastically. Indeed, in trying to understand the nature of the oscillations discussed above, Zabusky and Kruskal discovered the following. Suppose that at time $t = 0$, two such waves as (1.1.6) are given, well separated and with the smallest to the right. Then after a sufficient time the waves overlap and interact (the bigger one catches up). Following the process still longer, the bigger one separates from the smaller, and eventually (asymptotically) regains its initial shape and hence velocity. The only effect of the interaction is a phase shift; i.e., the center of each wave is at a different position than where it would have been if it had been traveling alone (see Fig. 1.1).

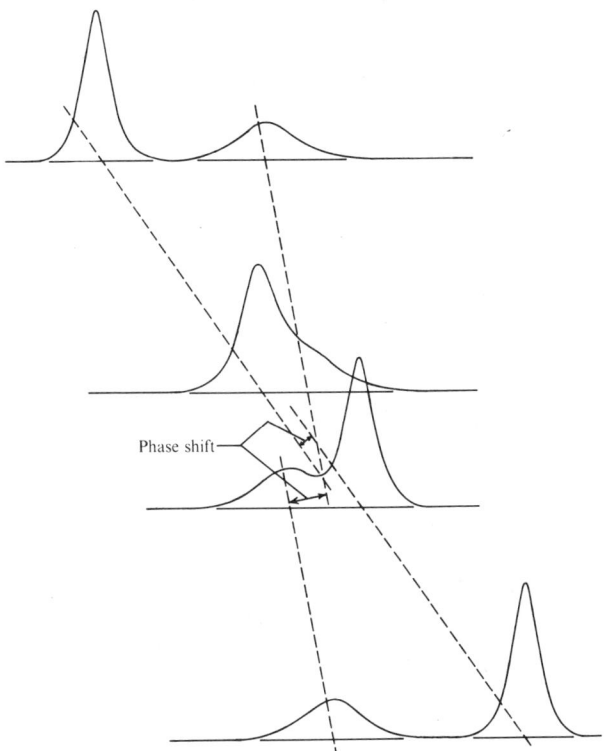

FIG. 1.1. *A typical interaction of two solitons at succeeding times.*

Because of the analogy with particles, Zabusky and Kruskal referred to these special waves as *solitons*. We shall follow their lead and refer to any localized nonlinear wave which interacts with another (arbitrary) local disturbance and always regains asymptotically its exact initial shape and velocity (allowing for a possible phase shift) as a soliton.

We refer to waves which interact inelastically as solitary waves. We also note that there are many different working definitions in the literature of what is, and what is not, a soliton. For our purpose the above definition is adequate.

In attempts to understand the initial onset of the oscillations in the numerical calculation of (1.1.4), the question of a "reversible" shock arose. A shock requires jump conditions; hence the question of jump conditions and conservation laws arose naturally. (A conservation law is an equation of the form $\partial T/\partial t + \partial F/\partial x = 0$, where T is called the density and F the flux, by analogy with fluid flow.) Early on, four conservation laws were obtained, Miura subsequently discovered a few more (Miura (1976)), and it was conjectured that there were an infinite number.

After studying these conservation laws, and those associated with a completely new evolution equation (which is commonly called the modified KdV equation or mKdV)

$$(1.1.7) \qquad M(v) = v_t - 6v^2 v_x + v_{xxx} = 0,$$

Miura (1968) discovered the following transformation. If v is a solution of (1.1.7), then

$$(1.1.8) \qquad u = -(v^2 + v_x)$$

is a solution of (1.1.5). Specifically,

$$(1.1.9) \qquad K(u) = -\left(2v + \frac{\partial}{\partial x}\right) M(v).$$

Because of the operator in the right-hand side of (1.1.9), the transformation is single-valued in one direction only.

It was the transformation (1.1.9) that led to the other important results related to the KdV equation. Originally, (1.1.9) was the basis of a proof that the KdV equation indeed had an infinite number of conserved quantities, (Miura, Gardner and Kruskal (1968)). The basic idea is as follows. Since the KdV equation is Galilean invariant, the transformation

(1.1.10a)
$$x' = x + \frac{6}{\varepsilon^2} t, \qquad t' = t,$$

$$u(x, t) = u'(x', t') - \frac{1}{\varepsilon^2}$$

leaves the KdV equation invariant, whereas setting

(1.1.10b) $$v(x, t) = -\varepsilon w(x', t') + \frac{1}{\varepsilon}$$

transforms the mKdV equation into

(1.1.10c) $$w_{t'} + \frac{\partial}{\partial x'}(3w^2 + 2\varepsilon^2 w^3 + w_{x'x'}) = 0.$$

Clearly $\int_{-\infty}^{\infty} w \, dx'$ is a conserved quantity of (1.1.10c). Similarly, from the Miura transformation (1.1.8), (1.1.10a, b) yield

(1.1.10d) $$u' = 2w + \varepsilon w_{x'} - \varepsilon^2 w^2.$$

Thinking of $\varepsilon \ll 1$, we may solve (1.1.10d) recursively for w as a function of u', i.e.,

(1.1.10e) $$w = w_0 + \varepsilon w_1 + \varepsilon^2 w_2 + \cdots = \frac{u'}{2} - \frac{\varepsilon}{4} u'_{x'} + \frac{\varepsilon^2}{4}\left(\frac{u'_{x'x'}}{2} + u'^2\right) + \cdots.$$

Hence, (1.1.10e) allows us to obtain an infinite number of conserved quantities. Later, in § 1.6, we shall give alternative proofs of the fact that KdV, mKdV, etc., have an infinite number of conserved quantities (or densities). Moreover, it can be shown that the even ones are nontrivial (i.e., not perfect derivatives).

The most significant result of all, however, was the development of a new method in mathematical physics, the Inverse Scattering Transform (IST). It too was motivated by (1.1.8). Note that (1.1.8) may be viewed as a Ricatti equation for v in terms of u; the well-known transformation $v = \Psi_x/\Psi$ linearizes (1.1.8), yielding

$$\Psi_{xx} + u\Psi = 0.$$

Since the KdV equation is Galilean-invariant, and to be as general as possible, Miura, Gardner and Kruskal (1968) considered

(1.1.11) $$\Psi_{xx} + (\lambda + u)\Psi = 0.$$

It turns out that this equation provides an implicit linearization of the KdV equation. Indeed, (1.1.11) is not an insignificant equation itself. It is the time-independent Schrödinger equation of quantum mechanics.

Gardner, Greene, Kruskal and Miura (1967), (1974) first discovered the method of solution of KdV by employing (1.1.11). Although we deviate from their original procedure, the ideas are of course similar. We postulate an associated time evolution equation,

(1.1.12) $$\Psi_t = A\Psi + B\Psi_x,$$

where A, B are scalar functions independent of Ψ (note that this is the most general local, linear form of time dependence). We find that, if the KdV equation (1.1.5) is satisfied and if we choose

(1.1.13) $$A = u_x, \quad B = 4\lambda - 2u,$$

then the eigenvalues are invariant in time, i.e., $\lambda_t = 0$. In fact, the reader can verify that forcing the compatibility condition $\Psi_{txx} = \Psi_{xxt}$ yields

(1.1.14) $$[K(u) + \lambda_t]\Psi = 0.$$

Hence if $K(u) = 0$, then $\lambda_t = 0$. In § 1.2 we shall give a deductive procedure for finding A, B. We will show that there are infinitely many equations associated with (1.1.11) in this way, with different A, B.

In subsequent sections we shall discuss in detail how the results (1.1.11)–(1.1.14) can be used to reconstruct potentials $u(x, t)$, given $u(x, t = 0)$. The method is somewhat sophisticated, and applies to a number of physically interesting evolution equations. The results in this field apply to a variety of physical problems, as discussed in Chapter 4. Moreover the mathematics used is also quite broad, ranging from classical analysis to differential geometry to algebra and to algebraic geometry (see also Chapter 3).

1.2. Second order eigenvalue problems and related solvable partial differential equations. As mentioned briefly in § 1.1, the inverse scattering transform (IST) was first developed and applied to the Korteweg–deVries (KdV) equation and its higher order analogues by Gardner, Greene, Kruskal and Miura (1967), (1974). At that time and shortly thereafter it was by no means clear if the method would apply to other physically significant nonlinear evolution equations. However, Zakharov and Shabat (1972) showed that the method was not a fluke. Using a technique first introduced by Lax (1968) they showed that the nonlinear Schrödinger equation

(1.2.1) $$iq_t = q_{xx} + \kappa q^2 q^*, \quad \kappa > 0$$

is related to a certain linear scattering problem. Applying direct and inverse scattering ideas, they were able to solve (1.2.1) given initial values $q(x, 0)$ that decayed sufficiently rapidly as $|x| \to \infty$. Shortly thereafter, Wadati (1972), using these ideas, gave a method of solution for the modified Korteweg–deVries (mKdV) equation

(1.2.2) $$q_t + 6q^2 q_x + q_{xxx} = 0,$$

and Ablowitz, Kaup, Newell and Segur (1973a) did the same for the "sine-Gordon" equation

(1.2.3) $$u_{xt} = \sin u.$$

These results already showed the power and versatility of IST in order to solve certain physically interesting nonlinear PDE's.

Then Ablowitz, Kaup, Newell and Segur (1973b), (1974) developed procedures which, given a suitable scattering problem, allow one to derive the nonlinear evolution equations solvable by IST with that scattering problem. For example, it turns out that the KdV, modified KdV, nonlinear Schrödinger, and sine-Gordon equations can all be shown to be related to one master eigenvalue problem.

We begin by briefly considering the essential ideas behind Lax's (1968) approach. Consider two operators L, M, where L is the operator of the spectral problem and M is the operator of an associated time evolution equation

(1.2.4a) $$Lv = \lambda v,$$

(1.2.4b) $$v_t = Mv.$$

Associated with KdV is the Schrödinger scattering problem (1.1.11). Hence, in this case $L = \partial_x^2 + u(x, t)$.

Taking the time derivative of (1.2.4a) and assuming $\lambda_t = 0$, we have $L_t v + Lv_t = \lambda v_t$. Substitution of (1.2.4b) yields a condition which is necessary for (1.2.4a,b) to be compatible:

(1.2.4c) $$L_t + [L, M] = 0,$$

where $[L, M] = LM - ML$ (the commutator of L and M). Equation (1.2.4c) contains a nonlinear evolution equation if L and M are *correctly* chosen. Given L, Lax (1968) indicates how to construct an associated M so as to make (1.2.4c) nontrivial.

The difficulties with this method are that (a) one must "guess" a suitable L and then find an M in order to satisfy (1.2.4a, b) and (b) it is often awkward to work with differential operators (e.g., sine-Gordon (1.2.3)). As an alternative Ablowitz, Kaup, Newell and Segur (1974) proposed a technique which, very generally, can be formulated as follows. Consider two linear equations

(1.2.5a) $$\mathbf{v}_x = X\mathbf{v},$$

(1.2.5b) $$\mathbf{v}_t = T\mathbf{v},$$

where $\mathbf{v}$ is an n-dimensional vector and X, T are $n \times n$ matrices. Then cross differentiation (i.e., taking $\partial/\partial t$ (1.2.5a), $\partial/\partial x$ (1.2.5b) and setting them equal) yields

(1.2.6) $$X_t - T_x + [X, T] = 0.$$

This is, in essence, the equivalent of (1.2.4c). It turns out that, given X, there is a simple deductive procedure to find a T such that (1.2.6) contains a nonlinear evolution equation. In order for (1.2.6) to be effective, the associated operator X should have a parameter which plays the role of an eigenvalue, say ζ, and obeys $\zeta_t = 0$. Moreover, a complete solution to the associated nonlinear evolution equation on the infinite interval can be found when the associated scattering problem is such that analytic inverse scattering can be effectively carried out. (Even though there are numerous nonlinear evolution equations which satisfy (1.2.6), a complete scattering theory of many of the associated equations (1.2.5a) has, at this time, not yet been successfully developed.)

As an example let us first consider the case of a 2×2 eigenvalue problem. A modification of the scattering problem of Zakharov and Shabat (1972) is given by

(1.2.7a)
$$v_{1x} = -i\zeta v_1 + qv_2,$$
$$v_{2x} = i\zeta v_2 + rv_1,$$

and the most general linear time dependence which is local is

(1.2.7b)
$$v_{1t} = Av_1 + Bv_2,$$
$$v_{2t} = Cv_1 + Dv_2,$$

where A, B, C, D are scalar functions, independent of $\mathbf{v}$. (1.2.7a, b) play the role of (1.2.5a, b) and X, T are given by the right-hand sides of (1.2.7a, b) respectively. Note that if there were x-derivatives on the right-hand sides of (1.2.7b) they could be reduced out by use of (1.2.7a). Furthermore, when $r = -1$ (1.2.7a) may be reduced to the Schrödinger scattering problem

$$v_{2xx} + (\zeta^2 + q)v_2 = 0$$

(in this case ζ^2 plays the role of the parameter λ in (1.2.4a)).

It is interesting to note that when $r = -1$ or $r = \pm q^*$ (or $r = \pm q$ if q is real), physically significant nonlinear evolution equations are consequences of the formalism. Moreover when $r = -1$, i.e., when we have the Schrödinger equation, the question of inverse scattering has been considered by numerous authors. On the infinite line a review of this work can be found in Faddeev (1963), and more recently in Deift and Trubowitz (1979). Similarly, the system (1.2.7a) is sometimes referred to as a Dirac system of differential equations (note: the inverse scattering problem of a special case of (1.2.7a) was considered by Gasymov and Levitan (1966)).

In what follows in this section we shall describe a simple technique which allows us to find nonlinear evolution equations in the form (1.2.6) for the special case of 2×2 systems. In subsequent sections we shall consider the question of carrying out the direct and inverse scattering, as well as considering higher order systems.

Compatibility of (1.2.7a) and (1.2.7b) is that a certain set of conditions on $A, \cdots, D$ must be satisfied. Requiring that $(v_{ix})_t = (v_{it})_x$, $i = 1, 2$ (i.e., cross differentiation of (1.2.7a) and (1.2.7b) and assuming the eigenvalues are invariant in time $(\partial \zeta / \partial t = 0)$), we readily find that the equations for $A, \cdots, D$

satisfy

$$A_x = qC - rB,$$
$$B_x + 2i\zeta B = q_t - (A - D)q,$$
$$C_x - 2i\zeta C = r_t + (A - D)r,$$
$$(-D)_x = qC - rB.$$

Without loss of generality we can take $D = -A$ in what follows. Thus we have

(1.2.8a) $$A_x = qC - rB,$$
(1.2.8b) $$B_x + 2i\zeta B = q_t - 2Aq,$$
(1.2.8c) $$C_x - 2i\zeta C = r_t + 2Ar.$$

At this point we wish to solve the set of equations (1.2.8) for A, B, C. Doing this ensures that (1.2.7a, b) are compatible. In general we find that this can be done if still another condition is satisfied; this latter condition is the evolution equation. The evolution equation results from solving (1.2.8). There are various methods that are feasible; here we will describe an expansion procedure. In § 1.5 a general evolution equation (via an operator method) is derived. Alternatively, Zakharov and Shabat (1974) have developed a technique starting from a postulated linear integral equation. We shall discuss this latter technique in § 3.6.

Since ζ, the eigenvalue, is a free parameter (it might be small), we try for an exact truncated power series solution to (1.2.8) in powers of ζ. A simple expansion which yields an interesting nonlinear evolution equation is

(1.2.9)
$$A = A_2\zeta^2 + A_1\zeta + A_0,$$
$$B = B_2\zeta^2 + B_1\zeta + B_0,$$
$$C = C_2\zeta^2 + C_1\zeta + C_0.$$

Substitute (1.2.9) into (1.2.8) and equate coefficients of powers of ζ. The coefficients of ζ^3 [(1.2.8b) and (1.2.8c)] immediately yield $B_2 = C_2 = 0$. For ζ^2, (1.2.8a) gives $A_2 = a_2 = $ const.; (1.2.8b) gives $B_1 = ia_2q$; (1.2.8c) gives $C_1 = ia_2r$. Next, the coefficients of ζ yield the following. Via (1.2.8a), $A_1 = a_1 = $ const. For simplicity we take $a_1 = 0$ (if $a_1 \neq 0$ a more general evolution equation is obtained). Then (1.2.8b) gives $B_0 = -a_2q_x/2$; (1.2.8c) gives $C_0 = a_2r_x/2$. Finally, for the coefficients of ζ^0, (1.2.8a) gives $A_0 = a_2qr/2 + a_0$; again, $a_0 = $ const, and we take $a_0 = 0$. Then (1.2.8b) and (1.2.8c) respectively yield, at ζ^0,

(1.2.10a) $$-\tfrac{1}{2}a_2q_{xx} = q_t - a_2q^2r,$$
(1.2.10b) $$\tfrac{1}{2}a_2r_{xx} = r_t + a_2qr^2.$$

This is a coupled pair of nonlinear evolution equations, which are reminiscent of the nonlinear Schrödinger equation (1.2.1). Indeed the nonlinear Schrödinger equation results if we let $r = \mp q^*$. Then (1.2.10a) and (1.2.10b) are compatible if and only if $a_2 = i\alpha$, α real. Furthermore, if we take $\alpha = 2$ we find the equation

(1.2.11) $$iq_t = q_{xx} \pm 2q^2 q^*.$$

If we take the + sign special soliton solutions can be found, whereas with the − sign no solitons exist for potentials decaying sufficiently rapidly at ∞ (since the spectral operator in (1.2.7a) is Hermitian).

In summary, if we postulate the eigenvalue problem (1.2.7a) and the associated time dependence (1.2.7b), then a compatibility condition is (1.2.8). In this example, taking the expansion (1.2.9) and inserting it into (1.2.8), we deductively and systematically find, with $r = \mp q^*$,

(1.2.12)
$$A = +2i\zeta^2 \pm iqq^*,$$
$$B = 2q\zeta + iq_x,$$
$$C = \mp 2q^*\zeta \pm iq_x^*,$$

which satisfies (1.2.8) so long as the nonlinear Schrödinger equation (1.2.11) holds for $q(x, t)$.

This procedure can be carried out for any polynomial expansion in ζ. We shall simply quote the results for the most significant cases. The interested reader can verify these results using the above ideas. In the case of polynomials in ζ to the third power we find

$$A = a_3\zeta^3 + a_2\zeta^2 + \frac{1}{2}(a_3 qr + a_1)\zeta + \frac{1}{2}a_2 qr - \frac{ia_3}{4}(qr_x - q_x r) + a_0,$$

(1.2.13) $$B = ia_3 q\zeta^2 + \left(ia_2 q - \frac{1}{2}a_3 q_x\right)\zeta + \left(ia_1 q + \frac{i}{2}a_3 q^2 r - \frac{1}{2}a_2 q_x - \frac{i}{4}a_3 q_{xx}\right),$$

$$C = ia_3 r\zeta^2 + \left(ia_2 r + \frac{1}{2}a_3 r_x\right)\zeta + \left(ia_1 r + \frac{i}{2}a_3 qr^2 + \frac{1}{2}a_2 r_x - \frac{i}{4}a_3 r_{xx}\right),$$

and the evolution equations

(1.2.14a) $$q_t + \frac{i}{4}a_3(q_{xxx} - 6qrq_x) + \frac{1}{2}a_2(q_{xx} - 2q^2 r) - ia_1 q_x - 2a_0 q = 0,$$

(1.2.14b) $$r_t + \frac{i}{4}a_3(r_{xxx} - 6qrr_x) - \frac{1}{2}a_2(r_{xx} - 2qr^2) - ia_1 r_x + 2a_0 r = 0.$$

Evolution equations of physical interest are obtained as special cases. Taking $a_0 = a_1 = a_2 = 0$, $a_3 = -4i$ and $r = -1$, we have

(1.2.15) $$q_t + 6qq_x + q_{xxx} = 0 \quad \text{(KdV)}.$$

If $r = \mp q$,

(1.2.16) $\qquad q_t \pm 6q^2 q_x + q_{xxx} = 0 \qquad$ (mKdV).

Note that if we take $a_0 = a_1 = a_3 = 0$, $a_2 = -2i$ and $r = \mp q^*$, we obtain (1.2.11).

In the same way that we found the evolution equations corresponding to the expansion of A, B, C in positive powers of ζ, we may also find equations corresponding to expansion in inverse powers of ζ (or both). For example, taking

$$A = \frac{a(x, t)}{\zeta}, \quad B = \frac{b(x, t)}{\zeta}, \quad C = \frac{c(x, t)}{\zeta}$$

yields

(1.2.17) $\qquad a_x = \tfrac{1}{2}(qr)_t, \qquad q_{xt} = -4iaq, \qquad r_{xt} = -4iar.$

Special cases are

(1.2.18)
$$a = \left(\frac{i}{4}\right) \cos u, \quad b = c = \left(\frac{i}{4}\right) \sin u, \quad q = -r = -\frac{u_x}{2},$$
$$u_{xt} = \sin u \qquad \text{(sine-Gordon)}$$

and

(1.2.19)
$$a = \left(\frac{i}{4}\right) \cosh u, \quad b = -c = \left(\frac{i}{4}\right) \sinh u, \quad q = r = \frac{u_x}{2},$$
$$u_{xt} = \sinh u \qquad \text{(sinh-Gordon)}.$$

The above are only a few of the evolution equations obtainable by this expansion procedure.

When $r = -1$, an alternative formulation is to use the Schrödinger scattering problem and appropriate associated time dependence, namely,

(1.2.20a) $\qquad v_{xx} + (\lambda + q)v = 0,$

(1.2.20b) $\qquad v_t = Av + Bv_x.$

The above-described procedure amounts to taking the time derivative of (1.2.20a) and setting it equal to the second space derivative of (1.2.20b). Equating coefficients of v, v_x yields the compatibility conditions

(1.2.21a) $\qquad A_{xx} - 2B_x(\lambda + q) - Bq_x = -q_t,$

(1.2.21b) $\qquad B_{xx} + 2A_x = 0,$

which are the analogues of (1.2.8). Expanding A, B in various powers of λ (when $r = -1$ in (1.2.7a), $\lambda = \zeta^2$) yields nonlinear evolution equations. For example, expanding A, B as

$$A = A_1 \lambda + A_0, \qquad B = B_1 \lambda + B_0$$

and equating coefficients of λ gives

$$B_1 = b_1 = \text{const.}, \qquad A_1 = a_1 = \text{const.},$$

$$B_0 = -\frac{b_1}{2}q, \qquad A_0 = \frac{b_1}{4}q_x + a_0 \qquad (a_0 = \text{const.}),$$

an additional restriction being the nonlinear evolution equation. Taking $b_1 = 4$, $a_1 = 0 = a_0$, we have the KdV equation

(1.2.22) $$q_t + 6qq_x + q_{xxx} = 0,$$

with the time dependence of the eigenfunctions given by

(1.2.23) $$v_t = q_x v + (4\lambda - 2q)v_x.$$

By solving for λv from (1.2.20a), we may see that (1.2.23) has the alternative forms

(1.2.24a) $$v_t + 4v_{xxx} + 6qv_x + 3q_x v = 0$$

or

(1.2.24b) $$v_t + v_{xxx} + 3(q - \lambda)v_x = 0.$$

This is the typical form of the time dependence of the eigenfunctions generated by Lax's (1968) approach.

Other nonlinear evolution equations are obtained by taking different expansions from that above. This procedure gives the same result as does Lax (1.2.4), but of course it is essentially algebraic in nature.

An interesting variation of this approach was given recently by Kaup and Newell (1978a) and Wadati, Konno, and Ichikawa (1979). If we replace (1.2.7) with

(1.2.25a) $$\begin{aligned} v_{1x} &= -i\zeta v_1 + \zeta q v_2, \\ v_{2x} &= i\zeta v_2 + \zeta r v_1 \end{aligned}$$

and

(1.2.25b) $$\begin{aligned} v_{1t} &= Av_1 + Bv_2, \\ v_{2t} &= Cv_1 - Av_2, \end{aligned}$$

then their compatibility requires that

(1.2.26) $$\begin{aligned} A_x &= \zeta(qC - rB), \\ B_x + 2i\zeta B &= \zeta(q_t - 2qA), \\ C_x - 2i\zeta C &= \zeta(r_t + 2rA) \end{aligned}$$

instead of (1.2.8). As before, finite expansions of A, B, C in powers of ζ yield a variety of nonlinear evolution equations compatible with (1.2.25). As an example, if we take $r = -1$, $q = u - 1$,

$$A = -4iu^{-1/2}\zeta^3 - u^{-3/2}u_x\zeta^2,$$
$$B = 4u^{-1/2}(u-1)\zeta^3 + 2iu^{-3/2}u_x\zeta^2 - (u^{-3/2}u_x)_x\zeta,$$
$$C = -4u^{-1/2}\zeta^3,$$

we obtain Dym's equation (Kruskal (1975))

(1.2.27) $$u_t = 2(u^{-1/2})_{xxx}.$$

1.3. Derivation of a linear integral equation and inverse scattering on the infinite interval. In this section we will study the direct and inverse scattering problem associated with the system (1.2.7a). We shall attempt to illustrate the basic ideas without a preponderance of mathematical formulae and theorems. Indeed, even for the classical Schrödinger scattering problem the rigorous theory is very substantial (see, for example, Agranovich and Marchenko (1963), Faddeev (1963), Newton (1966), Deift and Trubowitz (1979)), and the scattering problem (1.2.7a) presents new difficulties.

In what follows we shall assume that q and r vanish rapidly as $|x| \to \infty$. We note that this is a very important assumption, since a scattering theory with different boundary conditions gives markedly different results. With this in mind, define the eigenfunctions ϕ, $\bar{\phi}$, ψ, $\bar{\psi}$ with the following boundary conditions on $\zeta = \xi$ (note that $\zeta = \xi + i\eta$ is the eigenvalue):

(1.3.1) $$\left. \begin{array}{c} \phi \sim \begin{pmatrix} 1 \\ 0 \end{pmatrix} e^{-i\xi x} \\ \bar{\phi} \sim \begin{pmatrix} 0 \\ -1 \end{pmatrix} e^{i\xi x} \end{array} \right\} \text{ as } x \to -\infty, \quad \left. \begin{array}{c} \psi \sim \begin{pmatrix} 0 \\ 1 \end{pmatrix} e^{i\xi x} \\ \bar{\psi} \sim \begin{pmatrix} 1 \\ 0 \end{pmatrix} e^{-i\xi x} \end{array} \right\} \text{ as } x \to +\infty.$$

(Note that $\bar{\phi}$ is not the complex conjugate of ϕ. We use ϕ^* for complex conjugate.) These solutions of (1.2.7a) are defined at a fixed time (say $t = 0$), and all of the scattering theory (direct and inverse) developed in this section is at the same fixed time. We will see in the next section how to obtain the time-dependent eigenfunctions that satisfy both (1.2.7a, b) from these time-independent ones. Hereafter in this section we omit the time dependence in our notation. Now if $u(x, \xi)$ $(u(x, \xi)$ is a 2×1 column vector with components $u_i(x, \xi)$, $i = 1, 2)$ and $v(x, \xi)$ are two solutions of (1.2.7a), we have

(1.3.2a) $$\frac{d}{dx} W(u, v) = 0,$$

where $W(u, v)$, the Wronskian of u, v, is

(1.3.2b) $$W(u, v) = u_1 v_2 - u_2 v_1.$$

From (1.3.1) we see that $W(\phi, \bar{\phi}) = -1$ and $W(\psi, \bar{\psi}) = 1$. The solutions $\psi, \bar{\psi}$ are linearly independent, so we may write

(1.3.3a) $$\phi = a(\xi)\bar{\psi} + b(\xi)\psi,$$

(1.3.3b) $$\bar{\phi} = -\bar{a}(\xi)\psi + \bar{b}(\xi)\bar{\psi}.$$

(The minus sign is for convenience.) We also note that the scattering matrix is usually defined as

(1.3.3c) $$S = \begin{pmatrix} a & b \\ \bar{b} & -\bar{a} \end{pmatrix}.$$

Using (1.3.3) and $W(\phi, \bar{\phi}) = -1$, we obtain

(1.3.4) $$a(\xi)\bar{a}(\xi) + b(\xi)\bar{b}(\xi) = 1.$$

Furthermore, we will establish the following analytic properties of the scattering data (as functions of the complex variable ζ). So long as $q, r \in L_1$ (i.e., absolutely integrable) then $e^{i\zeta x}\phi$, $e^{-i\zeta x}\psi$ are analytic in the upper half plane ($\eta > 0$) and $e^{-i\zeta x}\bar{\phi}$, $e^{i\zeta x}\bar{\psi}$ are analytic in the lower half plane ($\eta < 0$). These facts immediately imply that

$$a = W(\phi, \psi) = \phi_1 \psi_2 - \psi_1 \phi_2$$

is analytic in the upper half plane, and

$$\bar{a} = W(\bar{\phi}, \bar{\psi})$$

is analytic in the lower half plane. In general, $b = -W(\phi, \bar{\psi})$, $\bar{b} = W(\bar{\phi}, \psi)$ need not be analytic anywhere.

To establish these properties one usually converts the scattering problem into an integral equation. For example, (1.2.7a) for ϕ obeys

(1.3.5a) $$\phi_1(x, \zeta) e^{i\zeta x} = 1 + \int_{-\infty}^{x} dy \int_{-\infty}^{y} dz \, q(y) r(z) e^{2i\zeta(y-z)} \phi_1(z, \zeta)$$

or

(1.3.5b) $$\phi_1(x, \zeta) e^{i\zeta x} = 1 + \int_{-\infty}^{x} M(x, y, \zeta) \phi_1(y, \zeta) e^{i\zeta y} \, dy,$$

(1.3.5c) $$\phi_2(x, \zeta) e^{i\zeta x} = \int_{-\infty}^{x} e^{2i\zeta(x-y)} r(y) \phi_1(y, \zeta) e^{i\zeta y} \, dy,$$

where

(1.3.5d) $$M(x, y, \zeta) = r(y) \int_{y}^{x} e^{2i\zeta(z-y)} q(z) \, dz.$$

Since q, r vanish rapidly enough as $|x| \to \infty$, so that $q, r \in L_1$ and

$$R_0(x) \equiv \int_{-\infty}^{x} |r(y)| \, dy, \quad Q_0(x) \equiv \int_{-\infty}^{x} |q(y)| \, dy$$

exist, then the Neumann series of the Volterra integral equations can be shown to converge absolutely in the upper half plane ($\eta > 0$). Specifically, from (1.3.5a),

$$|\phi_1(x, \zeta) e^{i\zeta x}| \leq 1 + \int_{-\infty}^{x} dy \int_{-\infty}^{y} dz \, |q(y)| |r(z)|$$

(1.3.6)
$$+ \int_{-\infty}^{x} dy \int_{-\infty}^{y} dz \, |q(y)| |r(z)|$$

$$\int_{-\infty}^{z} dy_1 \int_{-\infty}^{y_1} dz_1 \, |q(y_1)| |r(z_1)| + \cdots ,$$

so that one simple estimate gives

$$|\phi_1(x, \zeta) e^{i\zeta x}| \leq 1 + Q_0(x) R_0(x) + \frac{1}{(2!)^2} Q_0^2(x) R_0^2(x)$$

(1.3.7)
$$+ \frac{1}{(3!)^2} Q_0^3(x) R_0^3(x) + \cdots$$

$$= I_0(2\sqrt{Q_0(x) R_0(x)})$$

and we see that the Neumann series for the integral equation (1.3.5) is absolutely convergent for $\eta > 0$. This fact immediately implies that $\phi_j e^{i\zeta x}$ is bounded for $\eta > 0$. Analyticity of the corresponding functions $e^{i\zeta x}\phi_i(x, \zeta)$, for $\eta > 0$, is established by repeating this procedure on the integral equations obtained by differentiating (1.3.5) with respect to ζ.

Simply requiring $q, r \in L_1$ does not yield analyticity on the real axis ($\eta = 0$); more stringent conditions must be placed upon r and q to do better. Using the basic ideas above, one can show that if

$$|r(x)| \leq C e^{-2K|x|}, \quad |q(x)| \leq C e^{-2K|x|},$$

where C, K ($K > 0$) are constants, then $e^{i\zeta x}\phi(x, \zeta)$, $e^{-i\zeta x}\psi(x, \zeta)$ and $a(\zeta)$ are analytic for all $\eta > -K$, and $e^{-i\zeta x}\bar{\phi}(x, \zeta)$, $e^{i\zeta x}\bar{\psi}(x, \zeta)$ and $\bar{a}(\zeta)$ are analytic for all $\eta < K$. Moreover, $b(\zeta)$ and $\bar{b}(\zeta)$ are analytic in the strip $K > \eta > -K$.

Having r, q vanishing faster than any exponential as $|x| \to \infty$ implies that the respective functions discussed above are entire functions of ζ. The very special case of compact support is particularly easy to understand, since in this case (1.3.5) are Volterra integral equations on a finite interval. Such equations always have absolutely convergent Neumann series solutions (see, for example Pogorzelski (1966)).

Returning to either (1.3.5) or (1.2.7a), we may compute the asymptotic expansions for large ζ. From (1.3.5) we simply integrate by parts to find for ζ

in the upper half plane:

(1.3.8a) $$\phi_1 e^{i\zeta x} = 1 - \frac{1}{2i\zeta} \int_{-\infty}^{x} r(y)q(y)\, dy + O\left(\frac{1}{\zeta^2}\right),$$

(1.3.8b) $$\phi_2 e^{i\zeta x} = -\frac{1}{2i\zeta} r(x) + O\left(\frac{1}{\zeta^2}\right).$$

(Alternatively, from (1.2.7a) we may find (1.3.8) via WKB analysis.) Similarly, the other eigenfunctions have the expansions for $\psi(\zeta)$ (ζ in the upper half plane)

(1.3.9)
$$\psi_1 e^{-i\zeta x} = \frac{1}{2i\zeta} q(x) + O\left(\frac{1}{\zeta^2}\right),$$
$$\psi_2 e^{-i\zeta x} = 1 - \frac{1}{2i\zeta} \int_{x}^{\infty} r(y)q(y)\, dy + O\left(\frac{1}{\zeta^2}\right),$$

while for (ζ in the lower half plane) $\bar{\phi}, \bar{\psi}$,

(1.3.10)
$$\bar{\phi}_1 e^{-i\zeta x} = -\frac{1}{2i\zeta} q(x) + O\left(\frac{1}{\zeta^2}\right),$$
$$\bar{\phi}_2 e^{-i\zeta x} = -1 - \frac{1}{2i\zeta} \int_{-\infty}^{x} q(y)r(y)\, dy + O\left(\frac{1}{\zeta^2}\right),$$

(1.3.11)
$$\bar{\psi}_1 e^{i\zeta x} = 1 + \frac{1}{2i\zeta} \int_{x}^{\infty} q(y)r(y)\, dy + O\left(\frac{1}{\zeta^2}\right),$$
$$\bar{\psi}_2 e^{i\zeta x} = -\frac{1}{2i\zeta} r(x) + O\left(\frac{1}{\zeta^2}\right).$$

Thus in each respective half plane we have, as $|\zeta| \to \infty$,

(1.3.12a) $$a(\zeta) = 1 - \frac{1}{2i\zeta} \int_{-\infty}^{\infty} q(y)r(y)\, dy + O\left(\frac{1}{\zeta^2}\right),$$

(1.3.12b) $$\bar{a}(\zeta) = 1 + \frac{1}{2i\zeta} \int_{-\infty}^{\infty} q(y)r(y)\, dy + O\left(\frac{1}{\zeta^2}\right).$$

So long as q, r are not too "small" (conditions which we shall discuss later in this section), the scattering problem (1.2.7a) can possess discrete eigenvalues (bound states). These occur whenever $a(\zeta)$ has a zero in the upper ($\eta > 0$) half plane or whenever $\bar{a}(\zeta)$ has a zero in the lower half plane ($\eta < 0$). We shall designate the zeros of $a(\zeta)$ by ζ_k, $k = 1, 2, \cdots, N$, where N is the number of bound states. Then at $\zeta = \zeta_k$, ϕ is proportional to ψ (recall $a = W(\phi, \psi)$) or,

(1.3.13a) $$\phi = C_k \psi.$$

Similarly, whenever $\bar{a}$ is zero in the lower half plane, i.e., at $\zeta = \bar{\zeta}_k, k = 1, \cdots, \bar{N}$, we have bound states and

(1.3.13b) $$\bar{\phi} = \bar{C}_k \bar{\psi}.$$

As discussed earlier, if r, q decay sufficiently rapidly as $|x| \to \infty$ then $a, b, \bar{a}, \bar{b}$ are entire functions. In this case we can extend $b, \bar{b}$ to find $C_k = b(\zeta_k)$, $\bar{C}_k = \bar{b}(\bar{\zeta}_k)$. So, for convenience, here we shall assume that r, q decay fast enough such that:

$$\int_{-\infty}^{\infty} |x|^n \left\{ \left| \begin{matrix} q(x) \\ r(x) \end{matrix} \right| \right\} dx < \infty,$$

for all n. In this case $a(\zeta), \bar{a}(\zeta)$ are analytic on the real axis as well as in the upper and lower half planes respectively. This assures us that $a(\zeta)$ has only finitely many zeros for Im $(\zeta) \geq 0$ (i.e., $a(\zeta)$ is analytic for Im $(\zeta) \geq 0$, $a(\zeta) \to 1$ as $|\zeta| \to \infty$; hence all zeros of $a(\zeta)$ are isolated and lie in a bounded region).

As an eigenvalue problem, (1.2.7a) with (1.3.1) differs from the Schrödinger problem in several respects: (i) the zeros of $a(\zeta)$ (i.e., the eigenvalues) are not necessarily restricted to the imaginary axis; (ii) $a(\zeta)$ may have multiple zeros; (iii) $a(\zeta)$ may vanish on Im$(\zeta) = 0$. The last are not proper eigenvalues as they have no square-integrable eigenfunction (see also Ablowitz, Kaup, Newell and Segur (1974)).

The important (physically significant) cases occur when r is proportional either to q^* or q. In the case where $r = \pm q^*$, from (1.2.7a) we have symmetry relations

(1.3.14a)
$$\bar{\psi}(x, \zeta) = \begin{pmatrix} \psi_2^*(x, \zeta^*) \\ \pm \psi_1^*(x, \zeta^*) \end{pmatrix},$$
$$\bar{\phi}(x, \zeta) = \begin{pmatrix} \mp \phi_2^*(x, \zeta^*) \\ -\phi_1^*(x, \zeta^*) \end{pmatrix},$$

which imply (from $a = W(\phi, \psi)$, etc.) that

(1.3.14b) $$\bar{a}(\zeta) = a^*(\zeta^*), \quad \bar{b}(\zeta) = \mp b^*(\zeta^*),$$

and consequently

(1.3.14c) $$\bar{N} = N, \quad \bar{\zeta}_k = \zeta_k^*, \quad \bar{C}_k = \mp C_k^*.$$

Similarly, when $r = \pm q$ we have

(1.3.15a) $$\bar{\psi}(x, \zeta) = \begin{pmatrix} \psi_2(x, -\zeta) \\ \pm \psi_1(x, -\zeta) \end{pmatrix}, \quad \bar{\phi}(x, \zeta) = \begin{pmatrix} \mp \phi_2(x, -\zeta) \\ -\phi_1(x, -\zeta) \end{pmatrix},$$

which imply

(1.3.15b) $$\bar{a} = a(-\zeta), \quad \bar{b}(\zeta) = \mp b(-\zeta)$$

and consequently

(1.3.15c) $\qquad \bar{N} = N, \quad \bar{\zeta}_k = -\zeta_k, \quad \bar{C}_k = \mp C_k.$

Hence if $r = \pm q$ and q is real then all of the above symmetry conditions hold. This implies that when ζ_k is an eigenvalue then so is $-\zeta_k^*$. This means eigenvalues either are on the imaginary axis or are paired. The above relationships have important implications for the mKdV, nonlinear Schrödinger and sine-Gordon evolution equations.

We also note in passing: (i) When $r = +q^*$ the eigenvalue problem (1.2.7a) is Hermitian. In this case, for $q \to 0$ sufficiently rapidly as $|x| \to \infty$, there are no eigenvalues with Im $\zeta > 0$. (ii) Estimates can be given to assure that there will be no bound states when $r = -q^*$. For example, since $a = W(\phi, \psi)$ we have

(1.3.16a) $\qquad a(\zeta) = \lim_{x \to \infty} \phi_1(x, \zeta) e^{i\zeta x},$

and from (1.3.5a)

(1.3.16b) $\qquad |a(\zeta) - 1| \leq 1 + \int_{-\infty}^{\infty} dy \int_{-\infty}^{y} dz \, |q(y)| |r(z)| |\phi_1(z, \zeta) e^{i\zeta z}| - 1.$

Using (1.3.7) we have

(1.3.16c) $\qquad |a(\zeta) - 1| \leq I_0(2\sqrt{Q_0(\infty) R_0(\infty)}) - 1$

($I_0(x) \geq 1$ if $x \geq 0$). Thus if

(1.3.16d) $\qquad I_0(2\sqrt{Q_0(\infty) R_0(\infty)}) \leq 2$

or

(1.3.16e) $\qquad Q_0(\infty) R_0(\infty) < 0.817$

there are no bound states to (1.2.7a) (note that if $r = -q^*$, $Q_0 = R_0$).

The above discussion is related to what is commonly referred to as the direct scattering problem. Next we shall investigate the inverse scattering problem.

We shall derive our inverse scattering formulae by making the assumption that the scattering data $(a, \bar{a}, b, \bar{b})$ are entire functions. For this to hold, it is sufficient to assume q, r decay faster than any exponential as $|x| \to \infty$. This very restrictive assumption can be relaxed, but the simple derivation we give here must then be modified.

First we assume the following integral representations for the eigenfunctions ψ and $\bar{\psi}$:

(1.3.17a) $\qquad \psi = \begin{pmatrix} 0 \\ 1 \end{pmatrix} e^{i\zeta x} + \int_x^{\infty} K(x, s) e^{i\zeta s} ds,$

(1.3.17b) $\qquad \bar{\psi} = \begin{pmatrix} 1 \\ 0 \end{pmatrix} e^{-i\zeta x} + \int_x^{\infty} \bar{K}(x, s) e^{-i\zeta s} ds,$

where $\zeta = \xi + i\eta$ has $\eta \geq 0$ and $K, \bar{K}$ are two component vectors, e.g., $K(x, s) = \binom{K_1(x,s)}{K_2(x,s)}$. The integral terms involving $K, \bar{K}$ represent the difference between the boundary values at $x = \infty$ and the true eigenfunctions. To build in the appropriate boundary condition it is natural to assume that the integral terms run from x to ∞. The crucial new step is, however, that the kernels $K, \bar{K}$ are independent of the eigenvalue ζ. This was noted by Gel'fand and Levitan (1955) in their original work.

To prove this we need only to substitute (1.3.17) into the eigenvalue problem (1.2.7a). For example, doing this with (1.3.17a) yields

(1.3.18a)
$$\int_x^\infty e^{i\zeta s}[(\partial_x - \partial_s)K_1(x, s) - q(x)K_2(x, s)]\, ds$$
$$-[q(x) + 2K_1(x, x)]e^{i\zeta x} + \lim_{s\to\infty} [K_1(x, s)\, e^{i\zeta s}] = 0,$$

(1.3.18b)
$$\int_x^\infty e^{i\zeta s}[(\partial_x + \partial_s)K_2(x, s) - r(x)K_1(x, s)]\, ds$$
$$-\lim_{s\to\infty} [K_2(x, s)\, e^{i\zeta s}] = 0.$$

It is necessary and sufficient to have

(1.3.19)
$$(\partial_x - \partial_s)K_1(x, s) - q(x)K_2(x, s) = 0,$$
$$(\partial_x + \partial_s)K_2(x, s) - r(x)K_1(x, s) = 0,$$

subject to the boundary conditions

(1.3.20)
$$K_1(x, x) = -\tfrac{1}{2}q(x),$$
$$\lim_{s\to\infty} K(x, s) = 0.$$

To see that a solution of (1.3.19) exists subject to the boundary conditions (1.3.20), introduce the coordinates

$$\mu = \tfrac{1}{2}(x+s), \qquad \nu = \tfrac{1}{2}(x-s).$$

Upon transforming to these coordinates, (1.3.19–20) become

$$\partial_\nu K_1(\mu, \nu) - q(\mu + \nu)K_2(\mu, \nu) = 0,$$
$$\partial_\mu K_2(\mu, \nu) - r(\mu + \nu)K_1(\mu, \nu) = 0,$$
$$K_1(\mu, 0) = -\tfrac{1}{2}q(\mu),$$
$$\lim_{\mu\to\nu} K(\mu, \nu) = 0.$$

(Note from the above we have $K_2(\mu, 0) = \tfrac{1}{2}\int_\mu^\infty r(\mu')q(\mu')\, d\mu'$.) From the theory of characteristics (these equations are what is commonly referred to as a

Goursat problem), the solution exists and is unique. Similarly, one can show that $\bar{K}$ exists and is unique.

Next we derive the linear integral equations (Gel'fand–Levitan–Marchenko integral equation) of inverse scattering. Consider ζ on a contour C in the complex ζ plane starting at $\zeta = -\infty + i0^+$, passing *over all* zeros of $a(\zeta)$ and ending at $\zeta = +\infty + i0^+$. Since we have strong decay on q, r, (1.3.3a) may be extended into the upper half plane so that

(1.3.21) $$\frac{\phi(x, \zeta)}{a(\zeta)} = \bar{\psi}(x, \zeta) + \frac{b(\zeta)}{a(\zeta)} \psi(x, \zeta).$$

Substitute (1.3.17) into (1.3.21) to find

(1.3.22) $$\frac{\phi}{a} = \binom{1}{0} e^{-i\zeta x} + \int_x^\infty \bar{K}(x, \zeta) e^{-i\zeta s} ds$$
$$+ \frac{b}{a}(\zeta)\left(\binom{0}{1} e^{i\zeta x} + \int_x^\infty K(x, s) e^{i\zeta s} ds\right).$$

Operate on this equation with $(1/2\pi) \int_C d\zeta\, e^{i\zeta y}$ for $y > x$, use $\delta(x) = (1/2\pi) \int_C e^{i\zeta x} d\zeta$ ($\delta(x)$ is the Dirac delta function), interchange integrals and obtain

(1.3.23a) $$I = \bar{K}(x, y) + \binom{0}{1} F(x+y) + \int_x^\infty K(x, s) F(s+y)\, ds,$$

where

(1.3.23b) $$F(x) \equiv \frac{1}{2\pi} \int_C \frac{b}{a}(\zeta) e^{i\zeta x} d\zeta,$$

(1.3.23c) $$I \equiv \frac{1}{2\pi} \int_C \frac{\phi(x, s)}{a(s)} e^{i\zeta y} d\zeta.$$

Since $\phi e^{i\zeta x}$ is analytic in the upper half plane, $y > x$, and the contour C passes over all the zeros of a, we have that $I = 0$. Hence we have

(1.3.24) $$\bar{K}(x, y) + \binom{0}{1} F(x+y) + \int_x^\infty K(x, s) F(s+y)\, ds = 0.$$

Performing the equivalent operations on the analytic extension of (1.3.3b) in the lower half plane yields

(1.3.25a) $$K(x, y) - \binom{1}{0} \bar{F}(x+y) - \int_x^\infty \bar{K}(x, s) \bar{F}(s+y)\, ds = 0,$$

where

(1.3.25b) $$\bar{F}(x) = \frac{1}{2\pi} \int_{\bar{c}} \frac{\bar{b}}{\bar{a}} e^{i\zeta x} d\zeta.$$

THE INVERSE SCATTERING TRANSFORM ON THE INFINITE INTERVAL 23

and $\bar{C}$ is a contour like C but passing below all zeros of $\bar{a}(\zeta)$. A special case of these formulae is obtained by assuming that $a(\zeta)$ does not vanish on the real axis ($\zeta = \xi$, $\eta = 0$) and $a(\zeta)$ has isolated simple zeros (multiple roots are obtained as a limiting case of coalescing simple zeros). Contour integration in (1.3.23b) and (1.3.25b) gives

(1.3.26a) $$F(x) = \frac{1}{2\pi} \int_{-\infty}^{\infty} \frac{b}{a}(\xi) e^{i\xi x} d\xi - i \sum_{j=1}^{N} C_j e^{i\zeta_j x},$$

(1.3.26b) $$\bar{F}(x) = \frac{1}{2\pi} \int_{-\infty}^{\infty} \frac{\bar{b}}{\bar{a}}(\xi) e^{-i\xi x} d\xi + i \sum_{j=1}^{\bar{N}} \bar{C}_j e^{-i\bar{\zeta}_j x},$$

where

$$C_j = \frac{b(\zeta_j)}{a'(\zeta_j)}, \quad \bar{C}_j = \frac{\bar{b}(\bar{\zeta}_j)}{\bar{a}'(\bar{\zeta}_j)}.$$

In the case of slower decay, where a, $\bar{a}$, b, $\bar{b}$ cannot be extended, (1.3.26a, b) still hold, but here the normalization constants C_j, $\bar{C}_j$ are found from the proportionality of the eigenfunctions $\phi_j = \phi(x, \zeta_j)$, ψ_j. For example, $\phi_j = \tilde{C}_j \psi_j$ and $C_j = \tilde{C}_j/a'_j$, etc. (Note the slight change in notation from (1.3.13).) The integral equations (1.3.24, 25) can be put into a convenient single matrix equation by defining

(1.3.27a) $$\mathcal{K} = \begin{pmatrix} \bar{K}_1 & K_1 \\ \bar{K}_2 & K_2 \end{pmatrix}, \quad \mathcal{F} = \begin{pmatrix} 0 & -\bar{F} \\ F & 0 \end{pmatrix},$$

whereby we have

(1.3.27b) $$\mathcal{K}(x, y) + \mathcal{F}(x, y) + \int_{x}^{\infty} \mathcal{K}(x, s) \mathcal{F}(s + y) \, ds = 0.$$

As pointed out earlier, in the special (but physically significant) case $r = \pm q^*$ there are numerous symmetry relationships. Using (1.3.14, 15) we have

(1.3.28a) $$\bar{F}(x) = \mp F^*(x),$$

(1.3.28b) $$\bar{K}(x, y) = \begin{pmatrix} K_2^*(x, y) \\ \pm K_1^*(x, y) \end{pmatrix}.$$

The integral equations (1.3.27b) with the above symmetry conditions reduce to

(1.3.29a) $$K_1(x, y) \mp F^*(x + y) \pm \int_{x}^{\infty} \int_{x}^{\infty} K_1(x, z) F(z + s) F^*(s + y) \, ds \, dz = 0.$$

(When $r = \mp q$, q real, then $F(x)$ and $K(x, z)$ are real.)

Finally using (1.3.20), we obtain the potential

(1.3.29b) $$q(x) = -2K_1(x, x).$$

In the more general case of (1.3.24) the potential $r(x)$ satisfies

$$r(x) = -2\bar{K}_2(x, x)$$

and is obtained in the same way as (1.3.29b) (see (1.3.18), etc.).

The question of existence and uniqueness of the linear integral equations (1.3.24) is usually examined by use of the Fredholm alternatives (see, for example, Pogorzelski (1966)). For example, the restriction $r = -q^*(x)$ is sufficient to guarantee that the solutions of (1.3.24) exist and are unique.

To show this, consider the homogeneous equations corresponding to (1.3.24) ($y > x$):

(1.3.30a) $$h_1(y) + \int_x^\infty h_2(s) F(s+y) \, ds = 0,$$

(1.3.30b) $$h_2(y) - \int_x^\infty h_1(s) \bar{F}(s+y) \, ds = 0.$$

Suppose $h(y) = \binom{h_1}{h_2}$ is a solution of (1.3.30) which vanishes identically for $y < x$. By the Fredholm alternatives, it is sufficient to show that $h(y) \equiv 0$. Multiply (1.3.30) by $[h_1^*, h_2^*]$, integrate in y and use

$$\int_x^\infty |h_j(y)|^2 \, dy = \int_{-\infty}^\infty |h_j(y)|^2 \, dy.$$

One obtains

(1.3.30c)

$$\int_{-\infty}^\infty \left\{ |h_1|^2 + |h_2|^2 + \int_{-\infty}^\infty [h_2(s) h_1^*(y) F(s+y) - h_1(s) h_2^*(y) \bar{F}(s+y)] \, ds \right\} dy = 0.$$

If $r = -q^*$ then the symmetry condition (1.3.28) allows this latter equation to be written

$$\int_{-\infty}^\infty \left\{ |h_1|^2 + |h_2|^2 + 2i \, \text{Im} \int_{-\infty}^\infty h_1^*(y) h_2(s) F(s+y) \, ds \right\} dy = 0.$$

The real and imaginary parts must both vanish, from which it follows that

$$h(y) \equiv 0,$$

and the solution of (1.3.24) exists and is unique.

Second, if

$$r(x) = +q^*(x),$$

the problem is formally self-adjoint, the spectrum lies on the real axis and
$$\bar{F}(s+y) = -F^*(s+y).$$
In this case, (1.3.30c) becomes

(1.3.30d) $$\int_{-\infty}^{\infty} \left\{ |h_1|^2 + |h_2|^2 + \text{Re} \int_{-\infty}^{\infty} h_1^*(y) h_2(s) F(s+y) \, ds \right\} dy = 0.$$

Since we have $r = +q^*$ then $|a|^2 - |b|^2 = 1$ on the real axis, so $|a(\zeta)| > 0$. Moreover, the problem is self-adjoint, hence
$$|a(\zeta)| > 0, \quad \eta \geq 0.$$
This implies that there are no discrete eigenvalues and therefore $F(z) = (1/2\pi) \int_{-\infty}^{\infty} (b/a)(\xi) e^{i\xi z} \, d\xi$. The Fourier transform of $h_j(y)$ is $\hat{h}_j(\xi) = \int_{-\infty}^{\infty} h_j(y) e^{-i\xi y} \, dy$, which satisfies Parseval's relation
$$\int_{-\infty}^{\infty} |h_j|^2 \, dy = \frac{1}{2\pi} \int_{-\infty}^{\infty} |\hat{h}_j|^2 \, d\xi.$$

Substituting these results into (1.3.30d) and reversing the order of integration yields
$$\int_{-\infty}^{\infty} \left\{ |\hat{h}_1(-\xi)|^2 + |\hat{h}_2^*(\xi)|^2 + 2\,\text{Re}\left[\frac{b}{a}(\xi) \hat{h}_1(-\xi) \hat{h}_2^*(\xi)\right] \right\} d\xi = 0.$$

Since $|(b/a)(\xi)| < 1$, we have
$$\left| 2\,\text{Re}\left[\frac{b}{a}(\xi) \hat{h}_1(-\xi) \hat{h}_2^*(\xi)\right] \right| < 2|\hat{h}_1(-\xi)||\hat{h}_2^*(\xi)| \leq |\hat{h}_1(-\xi)|^2 + |\hat{h}_2(\xi)|^2.$$

Hence we must have
$$h \equiv 0,$$
and the solution of the original integral equation (1.3.24) exists and is unique.

It should also be pointed out that complete and rigorous study of the inverse problem associated with (1.2.7a) has not yet been undertaken. Hence the characterization of precisely which class of scattering data leads to "nice" potentials has not yet been resolved. (The formulae and analysis in this section, however, show that if $r = \pm q^*$ and q obeys $\int_{-\infty}^{\infty} x^n |q| \, dx < \infty$ for all n, then inverse formulae can be derived which will yield a potential; moreover, the potential is associated with nice analytic behavior in the appropriate half plane.)

At this point we note that the inverse scattering results ((1.3.24) and following) may be derived using the concepts used to solve standard Riemann–Hilbert problems. Specifically, once we have derived the facts that $\phi\, e^{i\zeta x}$, $\psi\, e^{-i\zeta x}$, $a(\zeta)$ are analytic in the upper half plane, and $\bar{\psi}\, e^{i\zeta x}$, $\bar{\phi}\, e^{-i\zeta x}$, $\bar{a}(\zeta)$ are analytic in the lower half plane, then the statements (1.3.3) may be transformed into what is essentially a Riemann–Hilbert problem. For example (1.3.3a) implies

(1.3.31) $$\frac{\phi e^{i\zeta x}}{a} = \bar{\psi} e^{i\zeta x} + \frac{b}{a} \psi e^{i\zeta x}.$$

Thus if $a(\zeta)$ had no zeros in the upper half plane, (1.3.31) would be a statement about the jump $(b/a)\psi e^{i\zeta x}$ between two functions, each of which is analytic in a half plane i.e., a Riemann–Hilbert problem. With or without zeros of $a(\zeta)$ in the upper half plane, the usual ideas necessary to solve such problems yield a linear integral equation between the eigenfunctions. Specifically, we operate on (1.3.31), with the projection operator $P_+ = \frac{1}{2}(1 + iH)$, H being the Hilbert transform $H(U(\xi)) = (1/\pi) \int_{-\infty}^{\infty} U(\xi')/(\xi' - \xi) \, d\xi'$.
Note that

$$P_+\left(\frac{\phi}{a} e^{i\xi x}\right) = \frac{1}{2}\binom{1}{0} + \sum_K \tilde{C}_k \frac{\psi_k e^{i\zeta_k x}}{\xi - \zeta_k},$$

$$P_+(\bar{\psi} e^{i\xi x}) = \bar{\psi}(\xi) e^{i\xi x} - \frac{1}{2}\binom{1}{0}.$$

The contribution $\binom{1}{0}$ is from the contour at ∞, and the other contribution results from the zeros of $a(\zeta)$. We have

(1.3.32) $$\bar{\psi}(\xi) e^{i\xi x} = -P_+\left(\frac{b}{a}(\xi)\psi(x,\xi) e^{i\xi x}\right) + \binom{1}{0} + \sum \frac{\tilde{C}_k \psi_k e^{i\zeta_k x}}{\xi - \zeta_k}.$$

(1.3.32) is a linear singular integral equation relating the eigenfunctions $\bar{\psi}$, ψ. A second such equation could be derived from (1.3.3b), thus giving a "complete" linear description of ψ, $\bar{\psi}$ from which one could, in principle, obtain the potential via the scattering problem. Since global results are often difficult to obtain from such equations, we usually go to a Gel'fand–Levitan–Marchenko representation. By multiplying (1.3.32) by $e^{i\xi y}$ ($y > x$), and taking the Fourier transform (i.e., operate with $(1/2\pi) \int_{-\infty}^{\infty} d\xi \, e^{i\xi y}$) it is a straightforward calculation to show that (1.3.32) reduces to (1.3.24a), thus providing us with an alternative derivation. However, it is very important to note in all of this that once knowledge of the analytic properties of the eigenfunctions is obtained, the inverse equations may then be readily derived.

We complete this section by giving the results of inverse scattering for the Schrödinger eigenvalue problem,

(1.3.33) $$v_{xx} + (\lambda + q)v = 0.$$

The methods to derive the inverse scattering formulae are similar to those presented earlier, hence we omit the derivation. Details of the rigorous theory may be found in Faddeev (1963) and Deift and Trubowitz (1979). For $\lambda = k^2$

we define the eigenfunctions ϕ, ψ, $\bar\psi$,

(1.3.34a) $$\phi \sim e^{-ikx}, \quad x \to -\infty,$$

(1.3.34b) $$\psi \sim e^{ikx}, \quad \bar\psi \sim e^{-ikx}, \quad x \to +\infty.$$

The Wronskian of ψ and $\bar\psi$ shows that for $k \ne 0$ they are linearly independent; hence we may write

(1.3.35a) $$\phi = a(k)\bar\psi + b(k)\psi.$$

The reflection coefficient is defined by $\rho(k) \equiv b(k)/a(k)$, and the transmission coefficient by $\tau(k) \equiv 1/a(k)$. Equation (1.3.35a) and its derivative yield equations for a, b and hence ρ. The name "reflection coefficient" is used because of the wave analogy to the equation (obtained by dividing (1.3.35a) by a)

(1.3.35b) $$\tau\phi = \bar\psi + \rho\psi.$$

Here $\bar\psi$ denotes the incoming wave $\sim e^{-ikx}$ as $x \to \infty$, etc. The eigenvalues $\lambda_n = -\kappa_n^2$ are those numbers for which $\phi_n = \phi(x, \kappa_n)$ and ψ_n vanish as $|x| \to \infty$ and are related by

(1.3.36a) $$\phi_n = \tilde C_n \psi_n.$$

The normalizing coefficients we will use are defined by

(1.3.36b) $$C_n = \frac{\tilde C_n}{a'(i\kappa_n)}$$

($a(k)$ may be analytically extended into the upper half plane). The eigenvalues can be found by solving $a(i\kappa_n) = 0$.

With this information, the theory of inverse scattering allows us to reconstruct the potential u. The essential results are as follows. First we compute

(1.3.37a) $$F(x) = \frac{1}{2\pi} \int_{-\infty}^{\infty} \rho(k) e^{ikx}\, dk - i \sum_1^N C_n e^{-\kappa_n x}.$$

(in § 1.4 we shall show that $-iC_n$ may be replaced by positive quantities, c_n^2). Then solve the integral equation ($y > x$)

(1.3.37b) $$K(x, y) + F(x + y) + \int_x^{\infty} K(x, z) F(z + y)\, dz = 0$$

for $K(x, y)$. The potential is reconstructed (hence the inversion) by the relation

(1.3.37c) $$q(x) = 2 \frac{d}{dx} K(x, x).$$

Equations (1.3.37) are the Schrödinger analogues of the 2×2 matrix equations (1.3.27) for the eigenvalue problem (1.2.7a).

In the next section we will show that the scattering data $S(k)$ ($\rho(k)$ the reflection coefficient, $\{\lambda_i\}_{i=1}^n$ the discrete eigenvalues, $\{C_i\}_{i=1}^n$ the normalization coefficients) have simple time dependence under the nonlinear evolution equations discussed in § 1.1. Moreover, the bound states give rise to the solitons. Indeed, when $\rho(k) = 0$, the integral equation has a degenerate kernel and $K(x, y)$ has a solution expressible in closed form (see, for example, Kay and Moses (1956), also § 1.4).

Finally, from a pedagogical standpoint it is worth mentioning that for certain potentials the scattering data (reflection coefficient, etc.) can be calculated in closed form. For example, if we have the initial condition

(1.3.38) $$q(x) = Q\delta(x),$$

solving (1.3.33) yields the following results:

(1.3.39a)
$$a(k) = \frac{2ik + Q}{2ik}, \qquad b(k) = -\frac{Q}{2ik},$$
$$\rho(k) = -\frac{Q}{2ik + Q}, \qquad \tau(k) = \frac{2ik}{2ik + Q}.$$

If $Q > 0$, $a = 0$ somewhere in the upper half plane. Hence, there is an eigenvalue $\lambda_1 = -\kappa_1^2$, and a normalization constant given by

(1.3.39b) $$\kappa_1 = \frac{Q}{2}, \qquad C_1 = \frac{iQ}{2} = \operatorname{Res} \rho(i\kappa_1).$$

If $Q < 0$ there are no discrete eigenvalues.

Another particularly easy potential which leads to simple results for the scattering data is a "square well."

In addition to the equations and inverse scattering results presented here, there are numerous alternative questions that have been pursued by researchers. For example, Zakharov and Shabat (1973) investigated the nonlinear Schrödinger equation with nonzero boundary values at infinity (the boundary conditions give rise to so-called envelope hole solitons, i.e., dark pulses; see also Hasegawa and Tappert (1973)). The question of finding solutions to problems with periodic boundary values has been investigated by numerous authors (see § 2.3); semi-infinite problems have been explored by Ablowitz and Segur (1975), and Moses (1976). Similarly, many other second order scattering problems have been investigated in the literature, e.g., Jaulent (1976) and Kaup and Newell (1978a).

1.4. Time dependence and special solutions. In the previous section we developed the inverse scattering equations associated with the generalized Zakharov–Shabat scattering problem and the Schrödinger scattering problem.

THE INVERSE SCATTERING TRANSFORM ON THE INFINITE INTERVAL 29

This means that, given the scattering data

$$S(\zeta) = \left\{\{\zeta_i\}_{i=1}^N, \{C_i\}_{i=1}^N, \frac{b}{a}(\zeta)\right\}$$

(i.e., discrete eigenvalues, normalization constants, and reflection coefficient), we can then, in principle, solve the associated integral equation. The solution of the integral equation then gives the potentials (e.g., (1.3.29b) or (1.3.37c)). This may be done at *any time* t, so that t is a parameter in the process. Since we are interested in solving an evolution equation, we proceed as follows. At $t=0$ we are given initial values of a function satisfying some nonlinear evolution equation such as that discussed in § 1.2. We then use the direct scattering problem (e.g., (1.2.7a)) to map this initial potential into the scattering data $S(\zeta; t=0)$ (i.e., at $t=0$ we solve for the eigenfunctions and from this information we obtain the scattering data). In this section we shall show how one may obtain the scattering data for all times $t > 0$, $S(\zeta, t)$. With this information we may obtain the potential and have the solution of the nonlinear evolution equation for all times t.

We begin by constructing solutions to *both* (1.2.7a) and (1.2.7b). From § 1.2 we have seen that requiring that $q, r \to 0$ as $|x| \to \infty$ gives us a large class of equations with the property that $A \to A_-(\zeta)$, $D \to -A_-(\zeta)$, $B, C \to 0$ as $|x| \to \infty$. The time-dependent eigenfunctions are defined as

(1.4.1.) $$\phi^{(t)} = \phi e^{A_- t}, \qquad \psi^{(t)} = \psi e^{-A_- t},$$
$$\bar{\phi}^{(t)} = \bar{\phi} e^{-A_- t}, \qquad \bar{\psi}^{(t)} = \bar{\psi} e^{A_- t},$$

where $\phi, \bar{\phi}, \psi, \bar{\psi}$ satisfy (1.2.7a) with the fixed boundary conditions (1.3.1). It should be noted that the time evolution equation (1.2.7b) does not allow for fixed boundary conditions. Hence $\phi \sim \binom{1}{0} e^{-i\zeta x}$, etc., cannot satisfy (1.2.7b). For example, the time evolution of $\phi^{(t)}$,

(1.4.2) $$\frac{\partial \phi^{(t)}}{\partial t} = \begin{pmatrix} A & B \\ C & D \end{pmatrix} \phi^{(t)},$$

shows that ϕ satisfies

(1.4.3) $$\frac{\partial \phi}{\partial t} = \begin{pmatrix} A - A_-(\zeta) & B \\ C & D - A_-(\zeta) \end{pmatrix} \phi.$$

If we use the relation

(1.4.4) $$\phi = a\bar{\psi} + b\psi \underset{x \to \infty}{\sim} a\binom{1}{0} e^{-i\zeta x} + b\binom{0}{1} e^{i\zeta x},$$

then (1.4.3), as $x \to \infty$, yields

(1.4.5) $$\begin{pmatrix} a_t e^{-i\zeta x} \\ b_t e^{i\zeta x} \end{pmatrix} = \begin{pmatrix} 0 \\ -2A_-(\zeta) b e^{i\zeta x} \end{pmatrix}.$$

Thus

(1.4.6a) $$b(\zeta, t) = b(\zeta, 0) e^{-2A_-(\zeta)t},$$

(1.4.6b) $$a(\zeta, t) = a(\zeta, 0).$$

From (1.4.6b) the eigenvalues, ζ_k, are fixed in time.

In a similar way, using the definition of the normalizing constant $C_j(t)$: $\phi(\zeta_j) = \tilde{C}_j(t)\psi(\zeta_j)$, $C_j = \tilde{C}_j/a'_j$, we find

(1.4.7) $$C_j(t) = C_{j,0} e^{-2A_-(\zeta_j)t}, \quad j = 1, 2, \cdots, N,$$

where $C_{j,0} \equiv C_j(t=0)$. It should be noted that in those cases where the scattering data can be extended in the upper half plane, the result $C_j = b_j/a'_j$ immediately yields

$$C_j(t) = \frac{b_j(\zeta, t)}{a'_j(\zeta, t)} = \frac{b_j(\zeta, 0)}{a'_j(\zeta, 0)} e^{-2A_-(\zeta_j)t} = C_{j,0} e^{-2A_-(\zeta_j)t}.$$

Similarly, the time dependence of the scattering data $\bar{S}(\zeta; t)$ is obtained using $\bar{\phi} = -\bar{a}\psi + \bar{b}\bar{\psi}$. We find

(1.4.8) $$\bar{b}(\zeta, t) = \bar{b}(\zeta, 0) e^{2A_-(\zeta)t},$$
$$\bar{a}(\zeta, t) = \bar{a}(\zeta, 0),$$
$$\bar{C}_j(t) = \bar{C}_{j,0} e^{2A_-(\bar{\zeta}_j)t}.$$

These results give us the time dependence of $F(x; t)$ and $\bar{F}(x; t)$ in the inverse scattering equations (1.3.24) relative to the evolution equations derived in § 1.2:

(1.4.9a) $$F(x; t) = \frac{1}{2\pi} \int_{-\infty}^{\infty} \frac{b}{a}(\zeta, 0) e^{i\xi x - 2A_-(\xi)t} d\xi - i \sum_{j=1}^{N} C_{j,0} e^{i\zeta_j x - 2A_-(\zeta_j)t},$$

(1.4.9b) $$\bar{F}(x; t) = \frac{1}{2\pi} \int_{-\infty}^{\infty} \frac{\bar{b}}{\bar{a}}(\zeta, 0) e^{-i\xi x + 2A_-(\xi)t} d\xi - i \sum_{j=1}^{N} \bar{C}_{j,0} e^{-i\bar{\zeta}_j x - 2A_-(\bar{\zeta}_j)t}.$$

The formulae (1.4.9) play the role of Fourier integral solutions of a corresponding linear problem (specifically, the linearized version of any nonlinear equation derived in § 1.2). The solution of the nonlinear evolution equation is obtained via solving (1.3.24). Moreover, it should be noted that the integral equations simplify for $x \to \infty$. For example, from (1.3.24), assuming for convenience that $r = \mp q^*$, we have $K_1(x, y) \propto \pm F^*(x, +y)$.

Hence, from (1.3.29b),

$$q(x; t) \propto \frac{1}{2\pi} \int_{-\infty}^{\infty} \left(\frac{b}{a}\right)^* (\zeta, 0) e^{-2i\xi x - 2A_-^*(\xi)t} d\xi.$$

Note that the ζ_j are in the upper half plane and hence the bound state contribution is exponentially small as $x \to \infty$. Thus, as $x \to \infty$, the problem tends to a linear problem whose solutions q can be expressed in the form $(1/2\pi) \int_{-\infty}^{\infty} \alpha(k) e^{i[kx-\omega(k)t]} dk$. Here

$$\omega(\xi) = -2iA_-^*\left(\frac{-\xi}{2}\right).$$

When $r = \mp q^*$, $A_-(\xi)$ is pure imaginary, we have

$$\omega(\xi) = 2iA_-\left(\frac{-\xi}{2}\right).$$

So, for example, in the case of the nonlinear Schrödinger equation (1.2.11) we have from (1.2.12) that $A_-(\zeta) = \lim_{|x| \to \infty} A(\zeta) = 2i\zeta^2$ and hence $\omega(\xi) = -\xi^2$ (which is the dispersion relation of the linearized equation).

We now have worked out the conceptual steps in order to obtain a solution to those evolution equations associated with the scattering problem (1.2.7a) (and indeed these are the same steps one uses for the more general scattering problems associated with IST).

Before turning our attention to special solutions, we shall simply state how the scattering data for the KdV equation (1.2.23) evolves in time. Using an analysis similar to that previously discussed, but for the Schrödinger scattering problem, we find that $a(k, t)$, $b(k, t)$ are defined by (1.3.35),

(1.4.10)
$$a(k, t) = a(k, 0),$$
$$b(k, t) = b(k, 0) e^{8ik^3t},$$
$$C_n(t) = C_{n,0} e^{8\kappa_n^3 t}, \quad n = 1, \cdots, N.$$

The discrete eigenvalues are the zeros of $a(k, t)$ in the upper half plane, and it is clear that the eigenvalues are constant in time and the reflection coefficient $\rho(k, t) = b(k, t)/a(k, t)$ evolves in time in a simple way. Specifically, we note that the time dependence associated with the KdV equation (1.2.23) is such that $d\rho/dt = -8i\omega(k)\rho$. Moreover, $\omega(k) = -k^3$ is the dispersion relation of the linearized problem associated with KdV (i.e., $q_t + q_{xxx} = 0$).

In the Schrödinger problem it can be shown that the discrete eigenvalues, i.e., the zeros of a, occur only on the imaginary axis and must be simple (see, for example, Deift and Trubowitz (1979)). Moreover, it can also be easily shown that the normalizing coefficients appearing in (1.3.37) must be positive. Henceforth we shall write

(1.4.11)
$$-iC_n = c_n^2.$$

The positivity of the normalizing coefficients can be established as follows. We assume all functions can be extended into the upper half plane. Then, at an eigenvalue $k_n = i\kappa_n$, we have as $x \to +\infty$

(1.4.12)
$$\phi(k_n) \equiv \phi_n = \tilde{C}_n \psi_n \sim \tilde{C}_n e^{-\kappa_n x},$$
$$\frac{d}{dk}\phi(\kappa_n) \sim a'(\kappa_n)\bar{\psi}_n \sim a'(\kappa_n) e^{k_n x},$$

where $\tilde{C}_n$ is necessarily real. But for all x

(1.4.13)
$$\frac{d}{dx}(\phi_k \phi_x - \phi \phi_{xk}) = -2\kappa_n \phi^2.$$

We use the asymptotics (1.4.12) in the left-hand side of (1.4.13) at κ_n to find

(1.4.14)
$$a'(\kappa_n) = -i\left(\frac{\int_{-\infty}^{\infty} \phi_n^2 \, dx}{\tilde{C}_n}\right);$$

thus

(1.4.15)
$$-iC_n = \frac{-i\tilde{C}_n}{a'(\kappa_n)} = \frac{\tilde{C}_n^2}{\int_{-\infty}^{\infty} \phi_n^2 \, dx} = c_n^2$$

(usually $\int_{-\infty}^{\infty} \phi_n^2 \, dx$ is normalized to unity). Hence for the discrete portion of the kernel in the Gel'fand–Levitan equation (1.3.37) we have

(1.4.16)
$$F_D(x) = \sum_{i=1}^{n} c_n^2(t) e^{-\kappa_n x},$$

where from (1.4.10) $c_n(t) = c_n(0) e^{4\kappa_n^3 t}$.

Having obtained the time dependence of the scattering data, we are in a position to discuss solutions and properties of the evolution equations. In this section we shall discuss the special soliton solutions, and in § 1.7 we shall investigate the asymptotic solution corresponding to the continuous spectrum (i.e., no discrete eigenvalues).

First we return to the generalized Zakharov–Shabat eigenvalue problem (1.2.7a). When $r = q^*$ there are no discrete eigenvalues. In this case the eigenvalue problem is formally "self-adjoint;" i.e., the scattering problem is such that for $LV = \zeta V$, $L^H = (L^A)^* = L$, where L^H is the Hermitian conjugate of L and L^A is the adjoint of L. If q, r decay rapidly then the self-adjointness implies that the eigenvalues can only be real. Since discrete eigenvalues must be nonreal complex, they cannot appear. Hence we consider only $r = -q^*$. We shall use the equation (1.3.29a). For $F(x)$ we take $(b/a)(t=0) = 0$ (no continuous spectrum) and $N = 1$, one discrete eigenvalue. Hence (for convenience in notation we will suppress the time dependence)

(1.4.17)
$$F(x) = -ic\, e^{i\zeta x}, \quad \zeta = \xi + i\eta, \quad \eta > 0.$$

Substituting (1.4.17) into (1.3.29) yields

(1.4.18)
$$K_1(x, y) = ic^* e^{-i\zeta^*(x+y)}$$
$$- \int_x^\infty \int_x^\infty K_1(x, z)|c|^2 e^{i\zeta z} e^{is(\zeta - \zeta^*)} e^{-i\zeta^* y} \, ds \, dy.$$

Define $\hat{K}_1(x) = \int_x^\infty K_1(x, z) e^{i\zeta z} \, dz$, multiply (1.4.18) by $e^{i\zeta y}$ and operate with $\int_x^\infty e^{i\zeta y} \, dy$. This yields an equation for $\hat{K}_1(x)$ whose solution is given by

(1.4.19)
$$\hat{K}_1(x) = -\frac{c^* e^{i(\zeta - 2\zeta^*)x}}{(\zeta - \zeta^*)[1 - |c|^2 e^{2i(\zeta - \zeta^*)x}/(\zeta - \zeta^*)^2]}.$$

From (1.4.18) we now can find $K_1(x, y)$:

(1.4.20a)
$$K_1(x, y) = ic^* e^{i\zeta^*(x+y)} \left[1 - \frac{|c|^2}{(\zeta - \zeta^*)^2} e^{2i(\zeta - \zeta^*)x} \right].$$

Thus, the potential $q(x)$ is given by

(1.4.20b)
$$q(x) = -2K_1(x, x) = -\frac{2ic^* e^{-2i\xi x}}{e^{2\eta x} + \left(\frac{|c|^2}{4\eta^2}\right) e^{-2\eta x}}.$$

Defining $|c|^2/4\eta^2 = e^{4\phi}$ gives this as

(1.4.20c)
$$q(x) = -i\frac{c^*}{|c|} 2\eta \, e^{-2i\xi x} \operatorname{sech} 2(\eta x - \phi).$$

This in turn is a soliton solution to all the evolution equations with $r = -q^*$, subject to the conditions $A \to A_-(\zeta)$, etc., discussed previously. Noting that $c = c(t)$ obeys

(1.4.21)
$$c = c_0 e^{-2A_-(\zeta)t},$$

we find that $q(x, t)$ is given by

(1.4.22) $q(x, t) = 2\eta \, e^{-2i\xi x} e^{2i\operatorname{Im} A_-(\zeta)t} e^{-i(\psi_0 + \pi/2)} \operatorname{sech}[2\eta x + 2\operatorname{Re} A_-(\zeta)t - x_0],$

where $c_0 \equiv |c_0| e^{i\psi_0}$, $x_0 \equiv \ln|c_0|/2\eta$. Thus, in the case of the nonlinear Schrödinger equation (1.2.11), $A_-(\zeta) = 2i\zeta^2$ and (1.4.21) is given by

(1.4.23) $q(x, t) = 2\eta \, e^{-2i\xi x} e^{4i(\xi^2 - \eta^2)t - i(\psi_0 + \pi/2)} \operatorname{sech}(2\eta x - 8\xi\eta t - x_0)$

(see Fig. 4.16, p. 325). Note that the velocity of the solution is given by 4ξ, and its amplitude by η.

It should be noted that the general case for q, r (when q, r are not related) allows singularities to develop in finite time. Here we have one eigenvalue ζ in the upper half plane, and one $\bar{\zeta}$ in the lower half plane. The analogue to (1.4.17) is

$$F(z, t) = -ic \, e^{i\zeta z}, \qquad \bar{F}(z, t) = i\bar{c} \, e^{-i\bar{\zeta} z};$$

c is given by (1.4.21) and $\bar{c} = \bar{c}_0 e^{2A_-(\bar{\zeta})t}$. Again we solve the degenerate integral equations and find

$$q(x, t) = +\frac{2i\bar{c}_0 e^{2A_-(\bar{\zeta})t - 2i\bar{\zeta}x}}{D(x, t)},$$

(1.4.24) $$r(x, t) = -\frac{2ic_0 e^{-2A_-(\zeta)t + 2i\zeta x}}{D(x, t)},$$

$$D(x, t) = 1 - \frac{c_0 \bar{c}_0}{(\zeta - \bar{\zeta})^2} e^{2(A_-(\bar{\zeta}) - A_-(\zeta))t + 2i(\zeta - \bar{\zeta})x}.$$

It may be verified that even if $D(x, 0) \neq 0$, we may have $D(x, T) = 0$ for some time $T > 0$. We refer to these as exploding solitons. This result motivates our emphasis on $r = \pm q^*$.

The procedure applies, in principle, in the more general case where we have N distinct eigenvalues (a double eigenvalue can be analyzed by taking the limit of two nearby ones). The integral equation is again degenerate and may be solved in closed form (in terms of determinants); see, for example, Zakharov and Shabat (1972) or Wadati (1973). It should also be noted that Hirota has an alternative procedure which also produces N-soliton solutions. However Hirota's method may be applied directly upon the nonlinear evolution equation and does not employ the notation of inverse scattering. We shall discuss Hirota's method later in § 3.3. Moreover, in that section we shall discuss the N-soliton case for the KdV equation.

In some applications two or more eigenvalues are such that the values of Re $A_-(\zeta)/\eta$ are equal (formally, the solitons in (1.4.22) have the same velocity). In this case a new type of soliton is found: a multiple bound state which is periodic in time. In the case of nonlinear Schrödinger we can see that Re $\{A_-(\zeta)\}/\eta \propto \xi$. Hence discrete eigenvalues with the same real part form a multiple bound state. (A further discussion of this appears in the paper of Zakharov and Shabat (1972).) Another example where periodic multiple bound states occur is that of the sine-Gordon equation. From § 1.2 we see that for the sine-Gordon equation $A_-(\zeta) = i/4\zeta$. Hence Re $A_-(\zeta) = \eta/(4(\xi^2 + \eta^2))$, and the condition for two eigenvalues to form a bound state is for them to lie on a circle $\xi^2 + \eta^2 = $ constant. Indeed, in the case of sine-Gordon, when we have only one eigenvalue it must lie on the imaginary axis $\xi = 0$ (when $q = -r$ and q is real the eigenvalues either lie on the imaginary axis or they are in pairs $\zeta, -\zeta^*$; see § 1.3). Hence

$$\text{Re } A_-(\zeta) = \frac{1}{4\eta}, \quad \text{Im } A_-(\zeta) = 0 \quad \left(\text{also } \psi_0 = \frac{\pi}{2}\right).$$

We have, from (1.4.22),

$$\frac{u_x}{2} = -q(x,t) = 2\eta \operatorname{sech}\left(2\eta x + \frac{1}{2\eta}t + x_0\right),$$

and the single kink solution for sine-Gordon is given by

$$u = 4\tan^{-1} \exp\left(2\eta x + \frac{1}{2\eta}t + x_0\right).$$

In terms of "laboratory" coordinates $x = (X+T)/2$, $t = (X-T)/2$ where the sine-Gordon equation becomes

(1.4.25) $$u_{TT} - u_{XX} + \sin u = 0,$$

the single kink solution has the form

(1.4.26) $$u(X,T) = 4\tan^{-1} \exp\left(\left(\eta + \frac{1}{4\eta}\right)(X-X_0) + \left(\eta - \frac{1}{4\eta}\right)T\right).$$

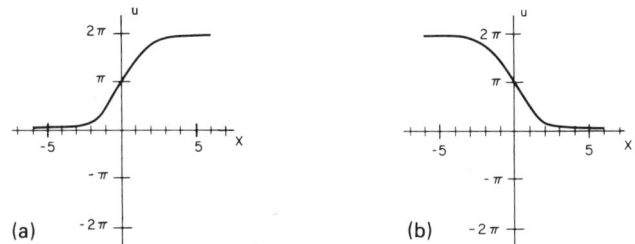

FIG. 1.2. (a) *typical kink* ($\eta = +\frac{1}{2}$), (b) *typical anti-kink* ($\eta = -\frac{1}{2}$).

Similarly, the calculation for two discrete eigenvalues can be worked out in complete detail. We shall only list pertinent results here. Given two paired eigenvalues $\zeta = \xi + i\xi$ and $-\zeta^*$, the solution is given by

(1.4.27) $$u(X,T) = 4\tan^{-1}\left[\frac{\eta}{\xi}\operatorname{sech}\left(\frac{\eta}{2}\nu(X-X_0) - (4-\nu)T\right)\right. \\ \left. \times \cos\frac{\xi}{2}(\nu(T-T_0) - (4-\nu)X)\right],$$

where $\nu = 2 + 1/(2|\zeta|^2)$. If $\xi^2 + \eta^2 = |\zeta|^2 = \frac{1}{4}$, then $\nu = 4$ and we have the "breather" solution

$$(1.4.28) \quad u(X, T) = 4 \tan^{-1}\left(\frac{\sqrt{1-\omega^2}}{\omega^2} \sin(\omega(T - \tilde{T}_0)) \operatorname{sech} \sqrt{1-\omega^2}(X - X_0)\right),$$

where $\omega = 2\xi$, $\xi^2 + \eta^2 = \frac{1}{4}$ (see Fig. 1.3). Similarly a typical kink, anti-kink solution with $\eta_1 = \frac{1}{2}((1-v)/(1+v))^{1/2}$, $\eta_2 = \frac{1}{2}((1+v)/(1-v))^{1/2}$ ($\zeta_j = i\eta_j$, $= 1, 2$) is found to be

$$(1.4.29) \quad u(X, T) = 4 \tan^{-1}\left(\frac{-1}{v} \sinh\left(\frac{vT}{\sqrt{1-v^2}}\right) \operatorname{sech}\left(\frac{X-X_0}{\sqrt{1-v^2}}\right)\right)$$

(we may obtain (1.4.29) from (1.4.28) by setting $v = i\sqrt{\omega^2/1-\omega^2}$ and $T_0 = \pi$). A double pole solution may be obtained by taking $v \to 0$, i.e.,

$$(1.4.30) \quad u(X, T) = 4 \tan^{-1}(-T \operatorname{sech}(X - X_0))$$

(see Fig. 1.4). Spectrally speaking (1.4.29) is a solution which corresponds to two eigenvalues η_1, η_2 located on either side of $\eta = \frac{1}{2}$. They coalesce to a double zero (1.4.30) at $\eta_1 = \eta_2 = \frac{1}{2}$ and then split into negative complex conjugate pairs ζ, $-\zeta^*$ as in (1.4.28). In terms of the energy functional

$$(1.4.31a) \quad E = \int_{-\infty}^{\infty} \left(\frac{1}{2}(u_T^2 + u_X^2) + 1 - \cos u\right) dx,$$

the energy of a kink traveling in laboratory coordinates with velocity v (1.4.29) is given by

$$(1.4.31b) \quad E = \frac{2E_0}{\sqrt{1-v^2}},$$

and that of a breather (1.4.28) oscillating in laboratory coordinates with frequency ω is given by

$$(1.4.31c) \quad E = 2E_0\sqrt{1-\omega^2},$$

where E_0 is the energy of a stationary kink (1.4.26) with $v = 0$.

At this point it probably is wise to reemphasize that the class of problems giving rise to such soliton solutions is indeed quite special. Often a working test of whether an equation is solvable by IST is to examine the interaction of two solitary waves. If they do not interact elastically, then it is generally believed that the governing equation is not solvable by inverse scattering

THE INVERSE SCATTERING TRANSFORM ON THE INFINITE INTERVAL 37

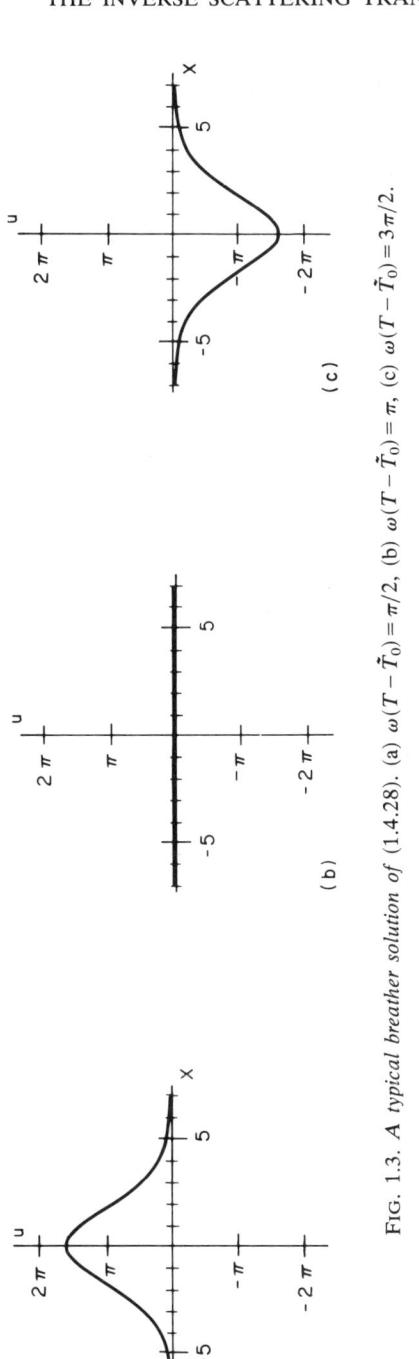

FIG. 1.3. *A typical breather solution of* (1.4.28). (a) $\omega(T-\tilde{T}_0)=\pi/2$, (b) $\omega(T-\tilde{T}_0)=\pi$, (c) $\omega(T-\tilde{T}_0)=3\pi/2$.

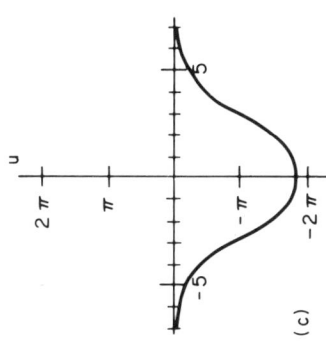
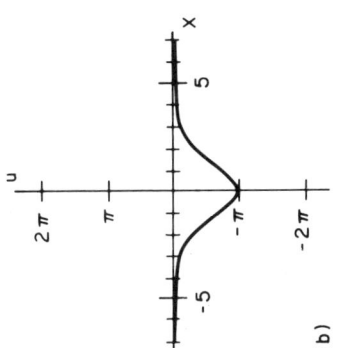
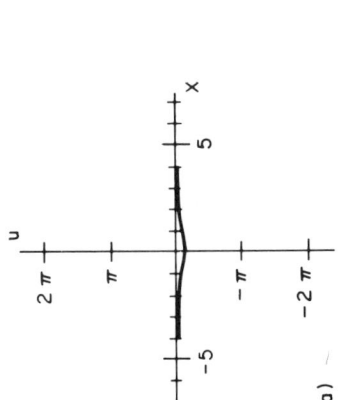

FIG. 1.4. *A typical double pole solution of* (1.4.30). (a) $T=0.1$, (b) $T=1.0$, (c) $T=10.0$.

techniques. As examples of this situation, Ablowitz, Kruskal and Ladik (1979) studied the equations

$$u_{tt} - u_{xx} + F(u) = 0$$

for various functions $F(u)$, namely,

$$F_1(u) = \sin u + \lambda \sin 2u,$$

$$F_2(u) = \begin{cases} \dfrac{\pi}{4}, & 2n\pi < u < (2n+1)\pi, \\ 0, & u = n\pi, \\ -\dfrac{\pi}{4}, & (2n+1)\pi < u < 2(n+1)\pi, \end{cases}$$

$$F_3(u) = -u + \dfrac{u^3}{\pi^2}.$$

Each of these equations has special solitary wave solutions expressible in closed form. Only in the case of $F_1(u)$ with $\lambda = 0$ (i.e., sine-Gordon) did the solitary waves interact elastically. Although at large relative velocities it appeared that the interactions ($\lambda \neq 0$) were elastic, nevertheless for low relative velocities the interactions were very significant. At low enough relative velocities the solitary waves even "destroyed" each other's identity, i.e., formed quasi-breather states.

A number of authors have investigated such questions numerically (early work was done by Zabusky and Kruskal (1965), Hardin and Tappert (1973)). A review of some of this work can be found in Eilbeck (1978). Other work in this direction has been carried out by Kudryavasev (1975) and Makhankov (1978).

The multisoliton solutions associated with the Schrödinger eigenvalue (e.g., KdV) problem are less varied than those associated with the generalized Zakharov–Shabat scattering problem. In this case the possibility of multiple bound states are eliminated due to the self-adjointness of the eigenvalue problem.

In what follows next, we shall derive the N-soliton solution associated with the KdV equation. Our derivation follows that of Gardner, Greene, Kruskal and Miura (1974). For this purpose we consider (1.3.37b) with the reflection coefficient $\rho(k)$ taken to be zero (i.e., reflectionless potentials). Given (1.4.16) we have that $K(x, y)$ satisfies

$$(1.4.32) \quad K(x, y) + \sum_{m=1}^{N} c_m^2 e^{-\kappa_m(x+y)} + \sum_{m=1}^{N} c_m^2 e^{-\kappa_m y} \int_x^\infty e^{-\kappa_m z} K(x, z)\, dz = 0.$$

THE INVERSE SCATTERING TRANSFORM ON THE INFINITE INTERVAL 39

We assume a natural form for $K(x, y)$,

(1.4.33) $$K(x, y) = -\sum_{m=1}^{N} c_m \psi_m(x) e^{-\kappa_m y}.$$

Substituting (1.4.33) into (1.4.32), carrying out an integration, and equating coefficients of $e^{-\kappa_m y}$ yields

(1.4.34a) $$\psi_m(x) + \sum_{m=1}^{N} c_m c_n \psi_n(x) \frac{e^{-(\kappa_n + \kappa_m)x}}{(\kappa_n + \kappa_m)} = c_m e^{-\kappa_m x}.$$

Let E, ψ be column vectors with mth entries $c_m e^{-\kappa_m x}$ and ψ_m respectively, and C an $N \times N$ matrix with its m, n entry

$$C_{mn} = \frac{c_m c_n e^{-(\kappa_m + \kappa_n)x}}{\kappa_n + \kappa_m};$$

then (1.4.34a), in matrix form, is given by

(1.4.34b) $$(I + C)\psi = E.$$

(I is the $N \times N$ identity matrix). We can be sure that (1.4.34b) has a solution ψ since C is positive definite; i.e.,

$$P^T CP = \sum_{n=1}^{N} \sum_{m=1}^{N} P_m c_m c_n \frac{e^{-(\kappa_m + \kappa_n)x}}{(\kappa_m + \kappa_n)} P_n = \int_x^{\infty} \left(\sum_{m=1}^{N} P_m c_m e^{-\kappa_m z} \right)^2 dz \geq 0.$$

The solution of (1.4.34b), by Cramer's rule and standard expansions of determinants, is given by

(1.4.35a) $$\psi_n(x) = \Delta^{-1} \sum_{m=1}^{N} C_m e^{-\kappa_m x} Q_{mn},$$

where $\Delta = \det(I + C)$. We have expanded along the nth column, with Q_{mn} denoting the cofactors of the matrix $(I + C)$. Similarly, Δ is given by the expansion

(1.4.35b) $$\Delta = \sum_{m=1}^{N} \left(\delta_{mn} + c_m c_n \frac{\exp(-(\kappa_m + \kappa_n)x)}{\kappa_m + \kappa_n} \right) Q_{mn}.$$

Hence, from (1.4.33), evaluating $K(x, y)$ at $y = x$ yields

(1.4.36) $$K(x, x) = -\Delta^{-1} \sum_{m,n=1}^{N} c_m c_n e^{-(\kappa_m + \kappa_n)x} Q_{mn}.$$

From the fact that the derivative of a determinant is the sum of N determinants with one column of each determinant differentiated, (once for each column) we have, after factoring and expanding, that

$$\frac{d\Delta}{dx} = -\sum_{m,n=1}^{N} c_m c_n e^{-(\kappa_m + \kappa_n)x} Q_{mn}.$$

Thus we have the final result that

(1.4.37) $$K(x,x) = \Delta^{-1} \frac{d}{dx} \Delta,$$

and that the potential from (1.3.37c) obeys

(1.4.38) $$q(x) = 2 \frac{d^2}{dx^2} \log \Delta.$$

This form of the potential has been found by numerous researchers (e.g., Kay and Moses (1956), Zakharov (1971), Hirota (1971), Wadati and Toda (1972)), and it holds in even wider generality for problems with continuous spectrum contributions included (e.g., Rosales (1978)). Similarly, the form (1.4.38) holds also for certain associated similarity equations (see Ablowitz, Ramani and Segur (1978)). Moreover, under appropriate assumptions an analogous statement for the nonlinear Schrödinger equation can be made (see, for example, Zakharov and Shabat (1972)).

The simplicity of this N-soliton solution is at the root of why there exist other, more direct approaches to the question of finding these special soliton solutions. In Chapter 3 we shall discuss these alternatives in more detail (e.g., see Hirota's method in § 3.3). However it should be noted that these alternative approaches do not relate the solutions to arbitrary initial data.

We now discuss phase shifts of KdV solitons. For the case of $N = 2$ (i.e., two solitons) the situation is quite straightforward. As we have seen, the time dependence of the normalizing constants for the KdV evolution equation is given by

(1.4.39) $$c_m^2(t) = c_{m,0}^2 e^{8\kappa_m^3 t}$$

(recall $\lambda_m = -\kappa_m^2 < 0$ is the discrete eigenvalue). From the above solution we have by direct calculation (note that the definition of η is different than earlier in this section)

(1.4.40) $$\Delta = \det(I + C) = 1 + e^{\eta_1} + e^{\eta_2} + e^{\eta_1 + \eta_2 + A_{12}},$$

where

$$\eta_m = -2\kappa_m(x - 4\kappa_m^2 t) + \eta_{m,0},$$

$$\eta_{m,0} = \log c_{m,0}^2,$$

$$e^{A_{12}} = \left(\frac{\kappa_1 - \kappa_2}{\kappa_1 + \kappa_2}\right)^2.$$

Let us move along the trajectory $\eta_1 = $ const., and assume $\kappa_1 > \kappa_2$. Then as $t \to -\infty$, $\eta_2 \to -\infty$ and we have

(1.4.41) $$\Delta \sim 1 + e^{\eta_1},$$

whereas for $t \to +\infty$, $\eta_2 \to +\infty$

(1.4.42) $$\Delta \sim e^{\eta_2}(1 + e^{\eta_1 + A_{12}}).$$

Thus from (1.4.38) $q(x, t)$ (η_1 fixed) v is given, in either limit, $t \to \pm\infty$, by

(1.4.43a) $$q(x, t) \sim 2\kappa_1^2 \operatorname{sech}^2(\eta_1 + \phi_\pm),$$
$$\phi_+ = A_{12}, \quad \phi_- = 0, \quad \eta_1 \text{ fixed}, \quad t \to \pm\infty.$$

(1.4.43a) is a pure soliton solution, and there is no amplitude (or velocity) change upon interaction. The only effect of the interaction is a phase shift. From these formulae we see that the soliton trajectory is shifted by an amount A_{12} or

(1.4.43b) $$\Delta\phi = \phi_+ - \phi_- = \log\left(\frac{\kappa_1 - \kappa_2}{\kappa_1 + \kappa_2}\right)^2.$$

In the general case for KdV (see, for example, Gardner, Greene, Kruskal and Miura (1974)) each eigenvalue $\lambda_p = -\kappa_p^2$, $\kappa_1 > \kappa_2 > \cdots > \kappa_p > 0$, has associated with it a soliton which approaches a solitary wave of the form

$$q \sim 2\kappa_p^2 \operatorname{sech}^2(\eta_p + \phi_\pm)$$

with velocity $4\kappa_p^2$ and amplitude $2\kappa_p^2$. The phase shift is given by

(1.4.44) $$\Delta\phi = \phi_+ - \phi_- = \sum_{m=p+1}^{N} \log\left(\frac{\kappa_p - \kappa_m}{\kappa_p + \kappa_m}\right) - \sum_{m=1}^{p-1} \log\left(\frac{\kappa_m - \kappa_p}{\kappa_m + \kappa_p}\right).$$

As is evident from (1.4.44), the total phase shift is the sum of the shifts that would be undergone if the pth soliton had pairwise interactions with every other soliton. (See also Zakharov (1971), Wadati and Toda (1972) and Tanaka (1972a)). The general question of interaction of N-solitons and continuous spectra has been examined by Ablowitz and Kodama (1980).)

In the latter reference the general phase shift of the pth soliton due to the other solitons and continuous spectra is given by

(1.4.45) $$\phi_+ - \phi_- = \log\left\{\left(\frac{c_p^+ c_p^-}{2\kappa_p}\right)^2 \prod_{m=1}^{p-1}\left(\frac{\kappa_m - \kappa_p}{\kappa_m + \kappa_p}\right)^4\right\}.$$

In (1.4.45), c_p^+ is the normalizing coefficient when we are using the "right" scattering data (above) and c_p^- is the normalizing coefficient when we are using the "left" scattering data; i.e., the eigenfunctions are defined by

$$\psi_{xx} + (u(x,0) - \kappa_p^2)\psi = 0,$$

(1.4.46)
$$\psi \sim \begin{cases} c_p^+ e^{-\kappa_p x}, & x \to +\infty, \\ c_p^- e^{\kappa_p x}, & x \to -\infty, \end{cases}$$

$$\int_{-\infty}^{\infty} \psi^2\, dx = 1.$$

(1.4.45) reduces to the above N-soliton formula when no continuous spectrum is present, since for pure N-soliton solutions we have

(1.4.47) $$(c_p^+)^2 (c_p^-)^2 = (2\kappa_p)^2 \prod_{\substack{m=1 \\ m \neq p}}^{N} \left(\frac{\kappa_p + \kappa_m}{\kappa_p - \kappa_m}\right)^2.$$

Moreover, when there is only one soliton plus continuous spectra we have

(1.4.48) $$\phi_+ - \phi_- = \log\left(\frac{c_1^+ c_1^-}{2\kappa_1}\right)^2,$$

which was found in Ablowitz and Segur (1977a).

1.5. General evolution operator. In this section we find a general class of nonlinear evolution equations associated with the generalized Zakharov–Shabat scattering problem. It turns out that, subject to certain conditions, a general relation can be found which gives directly a class of solvable nonlinear equations. This relationship depends on the dispersion relation of the linearized form of the nonlinear equation and a certain integrodifferential operator. The derivation here is based on the work of Ablowitz, Kaup, Newell and Segur (1974).

1.5a. Deriving the general evolution equations. We shall work with (1.2.7), (1.2.8) and specific eigenfunctions ϕ, $\bar{\phi}$ having boundary conditions (1.3.1). By multiplying the first and second equations of (1.2.7a) (replacing v by ϕ) by ϕ_1, ϕ_2 respectively we find that the squared eigenfunctions satisfy

(1.5.1)
$$(\phi_1^2)_x + 2i\zeta\phi_1^2 = 2q\phi_1\phi_2,$$
$$(\phi_2^2)_x - 2i\zeta\phi_2^2 = 2r\phi_1\phi_2,$$
$$(\phi_1\phi_2)_x = q\phi_2^2 + r\phi_1^2.$$

By inspection we see that the squared eigenfunctions obey the homogeneous form of (1.2.8); i.e.,

(1.5.2) $$A_x = qC - rB, \quad B_x + 2i\zeta B = -2Aq, \quad C_x - 2i\zeta C = 2Ar.$$

THE INVERSE SCATTERING TRANSFORM ON THE INFINITE INTERVAL 43

Thus one solution of the homogeneous system of (1.2.8) is given by

(1.5.3a)
$$\begin{pmatrix} \phi_1\phi_2 \\ -\phi_1^2 \\ \phi_2^2 \end{pmatrix}.$$

The other two are found similarly, and are given by

(1.5.3b)
$$\begin{pmatrix} \bar{\phi}_1\bar{\phi}_2 \\ -\bar{\phi}_1^2 \\ \bar{\phi}_2^2 \end{pmatrix}, \quad \begin{pmatrix} \frac{1}{2}(\phi_1\bar{\phi}_2+\bar{\phi}_1\phi_2) \\ -\phi_1\bar{\phi}_1 \\ \phi_2\bar{\phi}_2 \end{pmatrix}.$$

With all three homogeneous solutions to (1.5.2) known, we can, (e.g. by variation of parameters) find the general solution of the inhomogeneous A, B, C equations (1.2.8) (where q_t, r_t are the inhomogeneous terms). Motivated by the results of § 1.2, we take for boundary conditions on A, B, C

(1.5.4) $\quad A \to A_-(\zeta), \quad B, C \to 0 \quad \text{as } |x| \to \infty.$

While we could have imposed more general boundary conditions, we have chosen these since *all* of the evolution equations derived in § 1.2 (with the proviso $r, q \to 0$ as $|x| \to \infty$) obey (1.5.4). In any event, the main ideas will be presented here, and this should form the basis of many generalizations.

Since we are requiring (1.5.4) at both $x = \pm\infty$, the solution to (1.2.8) will not exist unless certain "orthogonality" conditions are satisfied. Here we only state the results. Later in this section we shall outline the derivation. With $r, q \to 0$ as $|x| \to \infty$ the orthogonality conditions which must be satisfied are

(1.5.5) $\quad \displaystyle\int_{-\infty}^{\infty} \left[\begin{pmatrix} r \\ -q \end{pmatrix}_t + 2A_-(\zeta)\begin{pmatrix} r \\ q \end{pmatrix} \right] \cdot \Phi_i \, dx = 0, \quad i = 1, 2,$

where

(1.5.6) $\quad \Phi_1 = \begin{pmatrix} \phi_1^2 \\ \phi_2^2 \end{pmatrix}, \quad \Phi_2 = \begin{pmatrix} \bar{\phi}_1^2 \\ \bar{\phi}_2^2 \end{pmatrix}$

and $\mathbf{u} \cdot \Phi_1 \equiv r\phi_1^2 + q\phi_2^2$, with $\mathbf{u} = \binom{r}{q}$ etc. The orthogonality conditions (1.5.5) determine the evolution equation. To show this, let us first derive the equations for Φ_1 (the analysis for Φ_2 is similar). We have, from (1.5.1),

(1.5.7) $\quad \phi_1\phi_2 = \displaystyle\int_{-\infty}^{x} (q\phi_2^2 + r\phi_1^2) \, dx,$

whereby

(1.5.8)
$$(\phi_1^2)_x = -2i\zeta\phi_1^2 + 2qI_-(q\phi_2^2 + r\phi_1^2),$$
$$(\phi_2^2)_x = 2i\zeta\phi_2^2 + 2rI_-(q\phi_2^2 + r\phi_1^2),$$

with the definition

(1.5.9)
$$I_- \equiv \int_{-\infty}^{x} dy.$$

(1.5.8) can now be written as an operator relation

(1.5.10)
$$\zeta \begin{pmatrix} \phi_1^2 \\ \phi_2^2 \end{pmatrix} = \frac{1}{2i} \begin{pmatrix} -\partial_x + 2qI_-r & 2qI_-q \\ -2rI_-r & \partial_x - 2rI_-q \end{pmatrix} \begin{pmatrix} \phi_1^2 \\ \phi_2^2 \end{pmatrix},$$

or more simply,

(1.5.11)
$$\zeta \Phi_i = \mathscr{L} \Phi_i, \qquad i = 1, 2,$$

((1.5.11) is also true for $i = 2$), where

(1.5.12)
$$\mathscr{L} = \frac{1}{2i} \begin{pmatrix} -\partial_x + 2qI_-r & 2qI_-q \\ -2rI_-r & \partial_x - 2rI_-q \end{pmatrix}.$$

Here $\mathscr{L}$ is a differential-integral operator which operates on Φ. If $A_-(\zeta)$ is analytic we have

(1.5.13)
$$A_-(\zeta) \Phi_i = A_-(\mathscr{L}) \Phi_i$$

inside its radius of convergence. In this case the orthogonality conditions (1.5.5) immediately give

(1.5.14)
$$\int_{-\infty}^{\infty} \left[\begin{pmatrix} r \\ -q \end{pmatrix}_t \cdot \Phi_i + 2 \begin{pmatrix} r \\ q \end{pmatrix} A_-(\mathscr{L}) \cdot \Phi_i \right] dx = 0, \qquad i = 1, 2.$$

The objective now is to exchange the operator $A_-(\mathscr{L})$ which operates on Φ_i, for its adjoint operator which acts on the vector $(r, q)^T$ (T is the transpose).

To do this we define the inner product in the usual way: $\langle \mathbf{u}, \mathbf{v} \rangle = \int_{-\infty}^{\infty} \mathbf{u} \cdot \mathbf{v} \, dx$. The (non-Hermitian) adjoint $\mathscr{L}^A$ of an operator $\mathscr{L}$ is defined by the relation $\langle \mathscr{L}^A \mathbf{u}, \mathbf{v} \rangle = \langle \mathbf{u}, \mathscr{L} \mathbf{v} \rangle$. The usual examples are: (i) the adjoint of ∂_x is $-\partial_x$ (with decaying boundary conditions on $\mathbf{u}, \mathbf{v}$); (ii) the adjoint of a square matrix $M = [m_{ij}]$ is $M^A = [m_{ji}^A]$. The only unusual case is finding the adjoint of the scalar operator $\hat{L} = \alpha(x) I_- \beta(x)$. From the above definition and interchanging integrals, we have

$$\langle u, \alpha I_- \beta v \rangle \equiv \int_{-\infty}^{\infty} dx \, u(x) \alpha(x) \int_{-\infty}^{x} dy \, \beta(y) v(y)$$
$$= \int_{-\infty}^{\infty} dy \, \beta(y) v(y) \int_{y}^{\infty} dx \, \alpha(x) u(x).$$

From this we see that

$$\hat{L}^A = \beta I_+ \alpha = \beta(x) \int_{x}^{\infty} dy \cdot \alpha(y).$$

Using these results (1.5.14) gives

(1.5.15a) $$\int_{-\infty}^{\infty}\left[\binom{r}{-q}_t + 2A_-(\mathscr{L}^A)\binom{r}{q}\right]\cdot\Phi_i\,dx = 0, \quad i=1,2,$$

where

(1.5.15b) $$\mathscr{L}^A = \frac{1}{2i}\begin{pmatrix}\partial_x+2rI_+q & -2rI_+r \\ 2qI_+q & -\partial_x-2qI_+r\end{pmatrix},$$

and $A_-(\mathscr{L}^A)$ acts only on $\binom{r}{q}$. Thus, sufficient conditions to solve (1.2.8) $A \to A_-(\zeta), B, C \to 0$ as $|x| \to \infty$ are

(1.5.16a) $$\binom{r}{-q}_t + A_-(\mathscr{L}^A)\binom{r}{q} = 0,$$

or in matrix form

(1.5.16b) $$\sigma_3 \mathbf{u}_t + 2A_-(\mathscr{L}^A)\mathbf{u} = 0,$$

where

$$\sigma_3 = \begin{pmatrix}1 & 0 \\ 0 & -1\end{pmatrix}, \quad \mathbf{u} = \binom{r}{q}.$$

To this point, we have only shown that (1.5.16) are sufficient to satisfy the orthogonality conditions (1.5.5). They are also necessary, so that (1.5.16) is the most general evolution equation solvable by (1.2.7, 8) with: (i) $q \to 0$, $A \to A_-(\zeta)$, $B \to 0$, $C \to 0$ as $|x| \to \infty$; (ii) $r = \pm q^*$; and (iii) $A_-(\zeta)$ entire. This assertion follows from the fact that any evolution equation must hold for $\int |q|\,dx$ arbitrarily small. But if $\int |q|\,dx$ is small enough, there are no bound states by (1.3.16). Kaup (1976a) showed that without bound states, Φ_1, Φ_2 are complete. Therefore (1.5.15) implies (1.5.16), which is the general evolution equation, as asserted. Similar arguments apply if $A_-(\zeta)$ is a ratio of entire functions, but then the evolution equation is subject to additional constraints (cf. Ablowitz, Kaup, Newell and Segur (1974)).

It is significant that we can relate $A_-(\zeta)$ to the dispersion relation of the associated *linearized* problem. In the limit $x \to \infty$, $I_+ \to 0$ (recall that $I_+ = \int_x^\infty dy$) and

$$\mathscr{L}^A \sim \frac{1}{2i}\begin{pmatrix}\partial_x & 0 \\ 0 & -\partial_x\end{pmatrix}.$$

Hence in this limit we have

(1.5.17) $$r_t + 2A_-\left(\frac{1}{2i}\partial_x\right)r = 0,$$

$$-q_t + 2A_-\left(\frac{1}{2i}\partial_x\right)q = 0.$$

(1.5.17) are linear (decoupled) evolution equations and may be solved by Fourier transforms. Moreover the wave solutions $q = \exp(i(kx - \omega_q(k)t))$, $r = \exp(i(kx - \omega_r(k)t))$ results in the conditions

$$(1.5.18) \qquad A_-(\zeta) = \frac{1}{2i}\omega_q(-2\zeta) = -\frac{1}{2i}\omega_r(2\zeta).$$

(a) This means that we have a relationship between $A_-(\zeta)$ and the dispersion relation of the associated *linearized* problem (1.5.17). Hence, we have that the general evolution equation (1.5.16) is expressible in terms of only the linearized dispersion relation and the operator $\mathcal{L}^A$ (see (1.5.15)).

(b) There is a necessary requirement on the linearized forms of the r, q equation in order for us to solve these equations by IST.

With the above results the general evolution equation (1.5.16) takes the form

$$(1.5.19) \qquad \sigma_3 \mathbf{u}_t - i\omega(-2\mathcal{L}^A)\mathbf{u} = 0,$$

where $\omega(k)$ is derived from the linearized problem with $q = \exp(i(kx - \omega t))$ ($\omega(k)$ plays the role of $\omega_q(k)$ above). An example is the nonlinear Schrödinger equation $iq_t = q_{xx} \pm 2q^2 q^*$, which has as its linearized form $iq_t = q_{xx}$. Setting $q = \exp(i(kx - \omega t))$, we find $\omega(k) = -k^2$. From (1.5.19) we have

$$(1.5.20) \qquad \begin{pmatrix} r \\ -q \end{pmatrix}_t = -4i(\mathcal{L}^A)^2 \begin{pmatrix} r \\ q \end{pmatrix} = -2\mathcal{L}^A \begin{pmatrix} r_x \\ -q_x \end{pmatrix} = i\begin{pmatrix} r_{xx} - 2r^2 q \\ q_{xx} - 2q^2 r \end{pmatrix}.$$

When $r = \pm q^*$, both of these equations are compatible with the nonlinear Schrödinger equation (1.2.11).

The derivation of (1.5.19) requires $r, q \to 0$ as $|x| \to \infty$. We cannot simply take $r = -1$ to get the equivalent result for the Schrödinger scattering problem (1.2.20). However, in Ablowitz, Kaup, Newell and Segur (1974, App. 3) the general evolution equation for this case is found to be

$$(1.5.21a) \qquad q_t + \gamma(\mathcal{L}_s)q_x = 0,$$

where

$$(1.5.21b) \qquad \mathcal{L}_s = -\tfrac{1}{4}\partial_x^2 - q + \tfrac{1}{2}q_x I_+,$$

$$(1.5.21c) \qquad \gamma(k^2) = \frac{\omega(2k)}{2k}.$$

$\omega(k)$ is the dispersion relation of the associated linearized problem where $q = \exp(i(kx - \omega(k)t))$. For example, the KdV equation, $q_t + 6qq_x + q_{xxx} = 0$, has as its linearized equation $q_t + q_{xxx} = 0$. Wave-like solutions give the dispersion relation $\omega = -k^3$. Hence $\gamma(k^2) = -4k^2$ and thus $\gamma(\mathcal{L}_s) = -4\mathcal{L}_s$, whereupon (1.5.21) yields

$$q_t - 4\mathscr{L}_s q_x = 0 \Rightarrow q_t - 4(-\tfrac{1}{4}\partial_x^2 - q + \tfrac{1}{2}q_x I_+)q_x = 0$$

$$\Rightarrow q_t + q_{xxx} + 4qq_x + 2qq_x = 0 \quad \text{(KdV)}.$$

From the above analysis it is clear that the squared eigenfunctions play an important role. In this regard we note that these eigenfunctions evolve in a particularly simple way. For example, in the case of KdV, from (1.2.24) and (1.2.20a) (i.e., multiply (1.2.24b) by $2v$ and (1.2.20a) by $2v_x$) we find that any such eigenfunction evolves as (cf. Gardner, Greene, Kruskal and Miura (1974))

(1.5.22) $$(v^2)_t + (v^2)_{xxx} + 6u(v^2)_x = 0.$$

This is an associated linearized form to the KdV equation.

1.5b. Nonlinear Fourier analysis—the inverse scattering transform. What is quite striking is the remarkable analogy to linear Fourier analysis (cf. § A.1). In the linear theory the equations are also characterized by the dispersion relation, i.e.,

(1.5.23a) $$q_t = -i\omega(-i\,\partial_x)q.$$

For example, $iq_t = q_{xx} \Rightarrow \omega(k) = -k^2$. A solution procedure on $(-\infty, \infty)$, assuming $u(x, 0) \to 0$ sufficiently rapidly as $|x| \to \infty$, is by Fourier transforms:

(1.5.23b) $$q(x, t) = \frac{1}{2\pi} \int_{-\infty}^{\infty} b(k, t) e^{ikx}\, dk,$$

with

(1.5.23c) $$b(k, 0) = \int_{-\infty}^{\infty} q(x, 0) e^{-ikx}\, dx$$

and

(1.5.23d) $$b(k, t) = b(k, 0) e^{-i\omega(k)t}.$$

Thus at $t = 0$ $q(x, 0)$ is given. We find $b(k, 0)$ by the direct Fourier transform. $b(k, t)$ satisfies a simple relation in time, and finally $u(x, t)$ is obtained by the inverse Fourier transform. Schematically, we have

$$\begin{array}{ccc}
\text{at } t=0: q(x,0) & \xrightarrow{\text{FT}} & b(k,0) \\
 & & \Big\downarrow \omega(k): \text{ dispersion relation} \\
q(x,t) & \xleftarrow{\text{Inverse FT}} & b(k,t) = b(k,0) e^{i\omega(k)t}
\end{array}$$

The close analogy between the method of inverse scattering and the above Fourier transform procedure led Ablowitz, Kaup, Newell and Segur (1974) to call the solution procedure the *Inverse Scattering Transform* (IST). As in the linear problem, the form of each evolution equation is characterized by a dispersion relation (the dispersion relation of the associated linearized equation, e.g., (1.5.19)). Moreover, the solution procedure is entirely analogous. The IST procedure is as follows. We take, for convenience, $r = \pm q^*$. At $t = 0$, $q(x, 0)$ is given. The direct scattering problem must then be solved for the required scattering data (§ 1.3). The scattering data evolve according to relatively simple equations (§ 1.4) that depend on the linearized dispersion relation. Finally the potential is reconstructed by inverse scattering (§ 1.3). Schematically, we have

$$\text{at } t = 0: q(x, 0) \xrightarrow{\text{direct scattering}} S(\zeta, 0) = \begin{Bmatrix} (b/a)(\zeta, 0) \\ \{\zeta_j; C_{j,0}\}_{j=1}^N \end{Bmatrix}$$

$$\downarrow \omega(k)$$

$$q(x, t) \xleftarrow{\text{inverse scattering}} S(\zeta, t) = \begin{Bmatrix} (b/a)(\zeta, t) = (b/a)(\zeta, 0)\, e^{i\omega(-2\xi)t} \\ \{C_j(t) = C_{j,0}\, e^{i\omega(-2\zeta_j)t} \\ \zeta_j = \text{constant} \end{Bmatrix}_{j=1}^N$$

The inverse scattering follows by the integral equation

(1.5.24a)
$$K(x, y; t) \mp F^*(x + y; t)$$
$$\pm \int_x^\infty \int_x^\infty K(x, z; t) F^*(z + s; t) F(s + y; t)\, ds\, dz = 0,$$

where

(1.5.24b)
$$F(x; t) = \frac{1}{2\pi} \int_{-\infty}^\infty \frac{b}{a}(\xi, t)\, e^{i\xi x}\, d\xi - i \sum_{j=1}^N C_j(t)\, e^{i\zeta_j x}$$

is computed from the initial data and the solution $q(x, t)$ is obtained via

(1.5.24c)
$$q(x, t) = -2K(x, x; t).$$

1.5c. Orthogonality conditions. Finally, let us return to outline the derivation of the integral conditions (1.5.5). We follow closely the derivation presented in Ablowitz, Kaup, Newell and Segur (1974, App. 1).

Consider the eigenvalue problem and time dependence in matrix form

(1.5.25a) $$\mathbf{v}_x = i\zeta D\mathbf{v} + N\mathbf{v},$$

(1.5.25b) $$\mathbf{v}_t = Q\mathbf{v},$$

THE INVERSE SCATTERING TRANSFORM ON THE INFINITE INTERVAL 49

where

(1.5.25c) $\quad \mathbf{v} = \begin{pmatrix} v_1 \\ v_2 \end{pmatrix}, \quad D = \begin{pmatrix} -1 & 0 \\ 0 & 1 \end{pmatrix}, \quad Q = \begin{pmatrix} A & B \\ C & -A \end{pmatrix}, \quad N = \begin{pmatrix} 0 & q \\ r & 0 \end{pmatrix}.$

Cross differentiation and setting $\zeta_t = 0$ yields

(1.5.26) $\quad N_t = Q_x + i[Q, D] + [Q, N],$

where $[A, B] = AB - BA$. Next, we form the fundamental solution matrix

(1.5.27a) $$P = \begin{bmatrix} \phi_1 & \bar{\phi}_1 \\ \phi_2 & \bar{\phi}_2 \end{bmatrix},$$

and the inverse P^{-1} is computed to be

(1.5.27b) $$P^{-1} = \begin{bmatrix} -\bar{\phi}_2 & \bar{\phi}_1 \\ \phi_2 & -\phi_1 \end{bmatrix}.$$

It is convenient to define S such that $Q = PSP^{-1}$ (it is easier to work with S than Q). Hence

$$Q_x = P_x S P^{-1} + P S_x P^{-1} - P S P^{-1} P_x P^{-1}.$$

Substituting these relations into (1.5.26) and using (1.5.25) we find the simple formula $N_t = PS_x P^{-1}$, or

(1.5.28) $$S = S(-\infty) + \int_{-\infty}^{x} P^{-1} N_t P \, dx.$$

The boundary conditions at $x = -\infty$ imply

$$S(-\infty) = A_-(\zeta) \begin{pmatrix} 1 & 0 \\ 0 & -1 \end{pmatrix}.$$

From (1.5.28) we can (if we wish) compute the solutions for A, B, C (i.e., from Q). However, here we shall only derive the integral conditions needed for a solution to exist.

The scattering data $(a, b, \bar{a}, \bar{b})$ are defined by

(1.5.29a) $\quad \phi = a\bar{\psi} + b\psi, \quad \bar{\phi} = -\bar{a}\psi + \bar{b}\bar{\psi},$

where as $x \to \infty$

(1.5.29b) $\quad \phi \sim \begin{pmatrix} a e^{-i\zeta x} \\ b e^{i\zeta x} \end{pmatrix}, \quad \bar{\phi} \sim \begin{pmatrix} \bar{b} e^{-i\zeta x} \\ -\bar{a} e^{i\zeta x} \end{pmatrix}.$

From this and $S = P^{-1}QP$ we easily find

$$(1.5.30) \qquad S(+\infty) = \begin{pmatrix} a\bar{a} - b\bar{b} & 2\bar{a}\bar{b} \\ 2ab & -(a\bar{a} - b\bar{b}) \end{pmatrix} A_-(\zeta).$$

Evaluating (1.5.28) at $x = +\infty$ and substituting (1.5.30) into (1.5.28), we find

$$(1.5.31) \qquad \begin{aligned} A_-(\zeta)(a\bar{a} - b\bar{b} - 1) &= \int_{-\infty}^{\infty} (\phi_1 \bar{\phi}_1 r_t - \phi_2 \bar{\phi}_2 q_t) \, dx, \\ A_-(\zeta) 2\bar{a}\bar{b} &= \int_{-\infty}^{\infty} (\bar{\phi}_1^2 r_t - \bar{\phi}_2^2 q_t) \, dx, \\ A_-(\zeta) 2ab &= \int_{-\infty}^{\infty} (-\phi_1^2 r_t + \phi_2^2 q_t) \, dx. \end{aligned}$$

But from the definitions of the scattering data and the scattering problem the following can be established:

$$(1.5.32) \qquad \begin{aligned} \int_{-\infty}^{\infty} (\phi_1 \phi_2)_x \, dx &= \int_{-\infty}^{\infty} (q\phi_2^2 + r\phi_1^2) \, dx = ab, \\ \int_{-\infty}^{\infty} (\bar{\phi}_1 \bar{\phi}_2)_x \, dx &= \int_{-\infty}^{\infty} (q\bar{\phi}_2^2 + r\bar{\phi}_1^2) \, dx = -\bar{a}\bar{b}, \\ \int_{-\infty}^{\infty} (\phi_1 \bar{\phi}_2 + \phi_2 \bar{\phi}_1)_x \, dx &= 2 \int_{-\infty}^{\infty} (q\phi_2 \bar{\phi}_2 + r\phi_1 \bar{\phi}_1) \, dx = a\bar{a} - b\bar{b} - 1. \end{aligned}$$

Using (1.5.32) in (1.5.31) we have

$$(1.5.33) \qquad \begin{aligned} \int_{-\infty}^{\infty} (\phi_1^2 r_t - \phi_2^2 q_t) \, dx &= -2A_-(\zeta) \int_{-\infty}^{\infty} (q\phi_2^2 + r\phi_1^2) \, dx, \\ \int_{-\infty}^{\infty} (\bar{\phi}_1^2 r_t - \bar{\phi}_2^2 q_t) \, dx &= -2A_-(\zeta) \int_{-\infty}^{\infty} (q\bar{\phi}_2^2 + r\bar{\phi}_1^2) \, dx, \\ \int_{-\infty}^{\infty} (\phi_1 \bar{\phi}_1 r_t - \phi_2 \bar{\phi}_2 q_t) \, dx &= -2A_-(\zeta) \int_{-\infty}^{\infty} (q\phi_2 \bar{\phi}_2 + r\phi_1 \bar{\phi}_1) \, dx. \end{aligned}$$

Thus, defining the squared eigenfunctions

$$(1.5.34) \qquad \Phi_1 = \begin{pmatrix} \phi_1^2 \\ \phi_2^2 \end{pmatrix}, \quad \Phi_2 = \begin{pmatrix} \bar{\phi}_1^2 \\ \bar{\phi}_2^2 \end{pmatrix}, \quad \Phi_3 = \begin{pmatrix} \phi_1 \bar{\phi}_1 \\ \phi_2 \bar{\phi}_2 \end{pmatrix},$$

we have that (1.5.33) reduces to

$$(1.5.35) \qquad \int_{-\infty}^{\infty} \left[\begin{pmatrix} r \\ -q \end{pmatrix}_t + 2A_-(\zeta) \begin{pmatrix} r \\ q \end{pmatrix} \right] \Phi_i \, dx = 0, \qquad i = 1, 2, 3,$$

i.e., (1.5.5). This was essentially the starting point of our analysis in deriving the general evolution operator.

It should also be noted that instead of the squared combinations of the ϕ_i, $\bar{\phi}_i$ in (1.5.35), we also could have found analogous expressions resolving ψ_i, $\bar{\psi}_i$ by integration to $x = +\infty$ in (1.5.28) and appropriately using results at $x = +\infty$ instead of $x = -\infty$. We would have found

$$\text{(1.5.36a)} \qquad \int_{-\infty}^{\infty} (\sigma_3 u_t + 2A_-(\zeta)u)\Psi_i \, dx = 0,$$

where

$$\text{(1.5.36b)} \qquad \sigma_3 = \begin{pmatrix} 1 & 0 \\ 0 & -1 \end{pmatrix}, \quad u = \begin{pmatrix} r \\ q \end{pmatrix},$$

$$\text{(1.5.36c)} \qquad \Psi_1 = \begin{pmatrix} \psi_1^2 \\ \psi_2^2 \end{pmatrix}, \quad \Psi_2 = \begin{pmatrix} \bar{\psi}_1^2 \\ \bar{\psi}_2^2 \end{pmatrix}, \quad \Psi_3 = \begin{pmatrix} \psi_1 \bar{\psi}_1 \\ \psi_2 \bar{\psi}_2 \end{pmatrix}.$$

This is an appropriate point in the analysis to discuss briefly the question of completeness of the squared states. Kaup (1976a) has shown that Ψ_1, Ψ_2 given in (1.5.36c) are not complete. We must add to these two vectors Ψ_3 evaluated at the discrete eigenvalues $\{\zeta_n\}_{k=1}^N$, $\{\bar{\zeta}_k\}_{k=1}^{\bar{N}}$ in order to obtain completeness.

His results allow us to expand certain combinations of the original potential and obtain some simple answers. Specifically, using the Ψ_i, we have

$$\text{(1.5.37)} \qquad \begin{pmatrix} q \\ -r \end{pmatrix} = -\frac{1}{\pi} \int_{-\infty}^{\infty} \left\{ \frac{b}{a}(\xi)\Psi_2(x,\xi) + \frac{\bar{b}}{\bar{a}}(\xi)\Psi_2(x,\xi) \right\} d\xi$$
$$+ 2i \sum_{1}^{N} \frac{b_k}{a_k'} \Psi_1(x, \zeta_k) - 2i \sum_{1}^{\bar{N}} \frac{\bar{b}_k}{\bar{a}_k'} \Psi_2(x, \zeta_k).$$

In a similar manner the adjoint eigenfunctions may be used to show that

$$\text{(1.5.38)} \qquad \begin{pmatrix} r \\ q \end{pmatrix} = \frac{1}{\pi} \int_{-\infty}^{\infty} \left\{ \frac{\bar{b}}{\bar{a}}(\xi) \begin{pmatrix} \phi_2^2 \\ -\phi_1^2 \end{pmatrix}(x,\xi) + \frac{b}{a}(\xi) \begin{pmatrix} \bar{\phi}_2^2 \\ -\bar{\phi}_1^2 \end{pmatrix}(x,\xi) \right\} d\xi$$
$$- 2i \sum_{1}^{N} \frac{1}{b_k a_k'} \begin{pmatrix} \phi_2^2 \\ -\phi_1^2 \end{pmatrix}(x, \zeta_k) + 2i \sum_{1}^{\bar{N}} \frac{1}{\bar{b}_k \bar{a}_k'} \begin{pmatrix} \bar{\phi}_2^2 \\ -\bar{\phi}_1^2 \end{pmatrix}(x, \zeta_k).$$

We also note that an analogous theory may be constructed for the Schrödinger scattering problem (see Kaup and Newell (1978b) and Kodama and Taniuti (1978)). In this case the expansion of the potential in terms of the squared eigenfunctions is given by

$$\text{(1.5.39)} \qquad q(x,t) = \frac{2}{i\pi} \int_{-\infty}^{\infty} k\rho(k)\psi^2(x,k) \, dk - 4 \sum_{i=1}^{N} \gamma_i \kappa_i \psi^2(x, \kappa_i),$$

where $\gamma_i = b_i/a_i'$ and κ_i is the ith discrete eigenvalue. Equations (1.5.37–39) are the starting point of the work of Deift, Lund and Trubowitz (1980), who view the equations of inverse scattering as being those of infinitely many oscillators constrained to lie on an infinite dimensional sphere.

Finally we note that an alternative derivation of the general evolution equations (1.5.16, 21) has been given by Calogero and Degasperis in a series of papers (e.g., Calogero (1978b) and the references cited therein). As we saw in § 1.3 for the generalized Zakharov–Shabat eigenvalue problem, one may rewrite the Schrödinger scattering problem (1.2.20) as an integral equation, and conclude

$$(1.5.40) \qquad 2ik\rho(k) = \int_{-\infty}^{\infty} q(x)\psi(x, k) e^{-ikx} dx,$$

which gives the reflection coefficient as an integral over the potential and its eigenfunction. More generally, if $q_1(x)$ and $q_2(x)$ are two potentials, one may show that

$$(1.5.41) \qquad 2ik[\rho_1(k) - \rho_2(k)] = \int \psi_1(x, k)[q_1(x) - q_2(x)]\psi_2(x, k) \, dx,$$

which reduces to (1.5.50) if $q_2 \equiv 0$. Further generalizations of (1.5.51) also may be given, and are used by Calogero and Degasperis. The important point, however, is that $q_1(x)$ and $q_2(x)$ are independent.

If $q(x, t)$ satisfies some evolution equation, we may let $q_1(x) = q(x, t_0)$, $q_2(x) = q(x, t + \Delta t)$ and let $\Delta \tau \to 0$. Then (1.5.51) relates $\partial_t \rho$ to $\partial_t q$, and relations like (1.5.51) are used to derive (1.5.21), the general evolution equation for (1.2.20). To this point, there is no advantage to their approach over that presented here. The generalizations that they exploit are: (i) to treat (1.2.20) as an $N \times N$ matrix equation; and (ii) to allow $q = q(x, t, y)$ and so to obtain multidimensional evolution equations. Moreover, Chiu and Ladik (1977) used this approach to obtain the general evolution equation for a discrete scattering problem discussed in § 2.2.

1.6. Conservation laws and complete integrability. One of the important events in the early development of IST was the discovery by Miura, Gardner and Kruskal (1968) that the KdV equation has an infinite set of local conservation laws. This discovery, coupled with a similar result for mKdV, led to Miura's transformation relating solutions of the two equations, and finally to the Schrödinger scattering problem (1.2.20). In this section we show that the existence of this infinite set of conserved quantities is a direct consequence of the fact that $a(k)$, the inverse of the transmission coefficient, is time independent. The conserved quantities turn out to be the coefficients in a large

THE INVERSE SCATTERING TRANSFORM ON THE INFINITE INTERVAL 53

k-expansion of $\log a(k)$. Moreover, they can be expressed simply in terms of the scattering data (the trace formulae), which will be useful in § 1.7 and § 4.5. Finally, we show that the nonlinear evolution equations for which $\log a$ is time independent are completely integrable Hamiltonian systems, and that IST is a canonical transformation to action-angle variables, with $\log |a|$ as the action variable. For convenience, we will concentrate on evolution equations of the form (1.5.16), related to (1.2.7). The calculations are similar for those related to (1.2.20), and we will simply state the results.

1.6.a. Conservation laws. We first derive the infinite set of conserved quantities for the nonlinear Schrödinger equation, or any of the other equations solvable by (1.2.7), using a method due to Zakharov and Shabat (1972). Recall that if (ϕ_1, ϕ_2) is the solution of (1.2.7a) that satisfies the boundary conditions (1.3.1) then, for $\operatorname{Im} \zeta \geq 0$, $\phi_1 e^{i\zeta x}$ is analytic and approaches 1 as $|\zeta| \to \infty$. Moreover,

$$(1.6.1) \qquad a(\zeta) = \lim_{x \to \infty} \phi_1 e^{i\zeta x}$$

has these two properties and is also time independent. Eliminate ϕ_2 from (1.2.7a), and then substitute

$$(1.6.2) \qquad \phi_1 = \exp\{-i\zeta x + \hat{\phi}\}$$

into that equation. The result is a Riccati equation for $\mu = \hat{\phi}_x$:

$$(1.6.3) \qquad 2i\zeta\mu = \mu^2 - qr + q\left(\frac{\mu}{q}\right)_x.$$

Because $\hat{\phi}$ vanishes as $|\zeta| \to \infty$ ($\operatorname{Im} \zeta > 0$), we may expand:

$$(1.6.4) \qquad \mu = (2i\zeta)^{-1} \sum_{n=0}^{\infty} \frac{\mu_n(x,t)}{(2i\zeta)^n}.$$

Substituting this into (1.6.3) yields

$$(1.6.5) \qquad \mu_0 = -qr, \qquad \mu_1 = -qr_x,$$

$$\mu_{n+1} = q\left(\frac{\mu_n}{q}\right)_x + \sum_{k=0}^{n-1} \mu_k \mu_{n-k-1}, \qquad n \geq 1.$$

From (1.6.1) and the fact that $\hat{\phi}$ vanishes as $x \to -\infty$ it follows that

$$(1.6.6\text{a}) \qquad \log a(\zeta) = \hat{\phi}(x = +\infty) = \sum_{n=0}^{\infty} \frac{C_n}{(2i\zeta)^{n+1}},$$

where

(1.6.6b) $$C_n = \int_{-\infty}^{\infty} \mu_n \, dx.$$

But $\log a(\zeta)$ is time independent (for *all* ζ with Im $\zeta > 0$) so C_n must be time independent as well. Thus, the first few (global) constants of the motion are

(1.6.7)
$$C_0 = \int \{-qr\} \, dx, \qquad C_1 = \int \{-qr_x\} \, dx,$$
$$C_2 = \int \{-qr_{xx} + (qr)^2\} \, dx, \qquad C_3 = \int \{-qr_{xxx} + 4q^2 rr_x + r^2 qq_x\} \, dx.$$

If r is proportional to q, one shows by induction that $C_{2n+1} \equiv 0$. Note that this derivation does not use (1.2.7b), so these integrals are constants of the motion for *any* of the equations solvable by (1.2.7), i.e., for the equations defined by (1.5.16).

The local conservation laws (both densities and fluxes) can be derived by a related method, due to Konno et al. (1974), that uses both (1.2.7a) and (1.2.7b), but does not directly involve $\log a(\zeta)$. (See also Sanuki and Konno (1974), Wadati et al. (1975) and Haberman (1977).) In this method, we substitute (1.6.2) into both (1.2.7a, b), after eliminating ϕ_2 appropriately. The result is

(1.6.8a) $$2i\zeta \hat{\phi}_x = \hat{\phi}_x^2 - qr + q\left(\frac{\hat{\phi}_x}{q}\right)_x,$$

(1.6.8b) $$\hat{\phi}_t = A + \frac{B}{q}\hat{\phi}_x.$$

With the definition, $\mu = \hat{\phi}_x$, (1.6.8a) is identical with (1.6.3), and we may use (1.6.4) and (1.6.5) again. Substituting (1.6.4) into the x-derivative of (1.6.8b) yields

(1.6.9) $$\partial_t \left\{ \sum_0^\infty \frac{\mu_n}{(2i\zeta)^{n+1}} \right\} = \partial_x \left\{ A + \frac{B}{q} \sum_0^\infty \frac{\mu_n}{(2i\zeta)^{n+1}} \right\}.$$

The particular evolution equation is specified by identifying A and B. Then its conservation laws follow from (1.6.9) by equating coefficients of $(2i\zeta)^{-n}$, and using (1.6.5).

As an example, if

(1.6.10) $$r = \sigma q^*, \quad \sigma = \pm 1,$$
$$A = -2i\zeta^2 - i\sigma|q|^2, \quad B = 2\zeta q + iq_x,$$

the evolution equation is

(1.6.11) $$iq_t + q_{xx} - 2\sigma|q|^2 q = 0$$

and (1.6.9) becomes

(1.6.12) $$\partial_t \left\{ \sum_0^\infty \frac{\mu_n}{(2i\zeta)^{n+1}} \right\} + i\partial_x \left\{ 2\zeta^2 + \sigma|q|^2 + \left(2i\zeta - \frac{q_x}{q} \right) \sum_0^\infty \frac{\mu_n}{(2i\zeta)^{n+1}} \right\} = 0.$$

The coefficients of $(2i\zeta)^{-n}$ are trivial for $n \leq -1$. For $n \geq 0$, we find

(1.6.13) $$n = 0, \quad \partial_t\{|q|^2\} + i\,\partial_x\{qq_x^* - q^*q_x\} = 0,$$
$$n = 1, \quad \partial_t\{-qq_x^*\} + i\,\partial_x\{|q_x|^2 - qq_{xx}^* + \sigma|q|^4\} = 0,$$

etc. In general, the conserved densities are the same for all of the equations defined by (1.5.16), but the corresponding fluxes differ for each equation.

Note that this derivation of an infinite set of conservation laws is appropriate either for the infinite interval (in x) or for the periodic problem (cf. § 2.3). It follows that the *local* conservation laws are valid for either set of boundary conditions. Moreover, there is no need for inverse scattering by this approach.

The conservation laws associated with the Schrödinger scattering problem (1.2.20) can be derived by similar means (Miura, Gardner and Kruskal (1968), Zakharov and Faddeev (1971)). With the substitution $v = \exp\{\phi + ikx\}$, (1.2.20a) becomes a Riccati equation for ϕ_x,

(1.6.14a) $$(\phi_x)_x + (\phi_x)^2 + q + 2ik\phi_x = 0,$$

and (1.2.20b) becomes

(1.6.14b) $$\phi_t = A + (\phi_x + ik)B.$$

After expanding $\mu = \phi_x$ in inverse powers of $(2ik)$,

(1.6.15) $$\mu = \sum_{n=1}^\infty \frac{\mu_n}{(2ik)^n},$$

we find from (1.6.14a) that μ_{2n} is an exact derivative and that

(1.6.16) $$\mu_1 = -q, \quad \mu_3 = -(q^2 + q_{xx}), \cdots,$$
$$\mu_{n+1} = \sum_{p=1}^{n-1} \mu_p \mu_{n-p} + \partial_x(\mu_n), \quad n \geq 2.$$

These are the conserved densities. The local conservation laws for KdV come from substituting

$$A = q_x, \quad B = 4k^2 - 2q$$

into the x-derivative of (1.6.14b), and using (1.6.15, 16):

(1.6.17a) $\quad \partial_t \left\{ \sum_{n=1}^{\infty} \frac{\mu_n}{(2ik)^n} \right\} = \partial_x \left\{ q_x + (4k^2 - 2q) \left(\sum_{n=1}^{\infty} \frac{\mu_n}{(2ik)^n} - ik \right) \right\}.$

The nontrivial conservation laws are the coefficients of

(1.6.17b)
$\quad k^{-1}: \quad \partial_t \{q\} = -\partial_x \{3q^2 + q_{xx}\},$
$\quad k^{-3}: \quad \partial_t \{q^2 + q_{xx}\} = -\partial_x \{4q^3 + 8qq_x - 5q_x^2 - q_{xxxx}\}, \quad$ etc.

We have seen from (1.6.6) that the constants of the motion are known once $\log a(\zeta)$ is known for Im $\zeta > 0$. We next derive the *trace formulae*, first given by Zakharov and Faddeev (1971) for KdV. These give the constants of the motion in terms of $a(\xi)$ with ξ real, and are useful both in this section and in § 1.7. The derivation which follows is based on the original work of Zakharov and Manakov (1974); cf. Flaschka and Newell (1975), Kodama (1975).

Recall that $a(\zeta)$ is analytic for Im $\zeta > 0$ with a finite number of zeros there (at $\zeta = \zeta_m$, $m = 1, \cdots, N$), and $a \to 1$ as $|\zeta| \to \infty$, Im $\zeta > 0$. We also assume that: (i) these zeros are simple; (ii) none occur on the real axis; (iii) for real ξ, $\xi^n \log a(\xi) \to 0$ as $|\xi| \to \infty$ for all $n \geq 0$. Let

(1.6.18a) $\quad \alpha(\zeta) = a(\zeta) \prod_{m=1}^{N} \frac{\zeta - \zeta_m^*}{\zeta - \zeta_m};$

$\alpha(\zeta)$ shares these same properties but with no zeros for Im $\zeta \geq 0$. Similarly, $\bar{a}(\zeta)$ is analytic in the lower half plane, and

(1.6.18b) $\quad \bar{\alpha}(\zeta) = \bar{a}(\zeta) \prod_{l=1}^{\bar{N}} \frac{\zeta - \bar{\zeta}_l^*}{\zeta - \bar{\zeta}_l}$

is analytic with no zeros in Im $\zeta \leq 0$, and $\bar{\alpha} \to 1$ as $|\zeta| \to \infty$ there. By Cauchy's integral theorem, for Im $\zeta > 0$,

$$\log \alpha(\zeta) = \frac{1}{2\pi i} \int_{-\infty}^{\infty} \frac{\log \alpha(\xi)}{\xi - \zeta} d\xi,$$

$$0 = \frac{1}{2\pi i} \int_{-\infty}^{\infty} \frac{\log \bar{\alpha}(\xi)}{\xi - \zeta} d\xi.$$

By adding these, we find that, for Im $\zeta > 0$,

(1.6.19) $\quad \log a(\zeta) = \sum_{m=1}^{N} \log \left\{ \frac{\zeta - \zeta_m}{\zeta - \zeta_m^*} \right\} + \frac{1}{2\pi i} \int_{-\infty}^{\infty} \frac{\log \alpha \bar{\alpha}}{\xi - \zeta} d\xi.$

If we further assume that $r = \pm q^*$, this simplifies to

$$\log a(\zeta) = \sum_{m=1}^{N} \log \left\{ \frac{\zeta - \zeta_m}{\zeta - \zeta_m^*} \right\} + \frac{1}{2\pi i} \int_{-\infty}^{\infty} \frac{\log |a(\xi)|^2}{\xi - \zeta} d\xi.$$

Now we may let $|\zeta| \to \infty$ (keeping Im $\zeta > 0$) and expand the right side of (1.6.15) in inverse powers of ζ. The result is

(1.6.20)
$$\log a(\zeta) = \sum_{n=0}^{\infty} \zeta^{-(n+1)} \left\{ \sum_{m=1}^{N} (n+1)^{-1} [(\zeta_m^*)^{n+1} - (\zeta_m)^{n+1}] \right.$$
$$- (2\pi i)^{-1} \int_{-\infty}^{\infty} \xi^n \left[\log a\bar{a}(\xi) + \sum_{1}^{N} \log \frac{\xi - \zeta_m^*}{\xi - \zeta_m} \right.$$
$$\left. \left. + \sum_{1}^{N} \log \frac{\xi - \bar{\zeta}_l^*}{\xi - \bar{\zeta}_l} \right] d\xi \right\}.$$

This expansion must coincide with that in (1.6.6), so that, for $n = 0, 1, 2, \cdots$,

(1.6.21a)
$$C_n = -\frac{1}{\pi} \int_{-\infty}^{\infty} (2i\xi)^n \left[\log a\bar{a}(\xi) + \sum_{1}^{N} \log \frac{\xi - \zeta_m^*}{\xi - \zeta_m} + \sum_{1}^{N} \log \frac{\xi - \bar{\zeta}_l^*}{\xi - \bar{\zeta}_l} \right] d\xi$$
$$+ \sum_{m=1}^{N} (n+1)^{-1} [(2i\zeta_m^*)^{n+1} - (2i\zeta_m)^{n+1}].$$

If $r = \pm q^*$, these simplify to

(1.6.21b)
$$C_n = -\frac{1}{\pi} \int_{-\infty}^{\infty} (2i\xi)^n \log |a(\xi)|^2 d\xi$$
$$+ \sum_{m=1}^{N} (n+1)^{-1} [(2i\zeta_m^*)^{n+1} - (2i\zeta_m)^{n+1}];$$

moreover, $N = 0$ if $r = +q^*$ with $q \in L_1$. These are the "trace formulae" for (1.2.7), with $r = \pm q^*$. They relate the infinite set of motion constants, C_n, to moments of $\log |a(\xi)|$ and powers of the discrete eigenvalues. We note again that they are identical for every equation in (1.5.16) with $r = \pm q^*$.

For the Schrödinger scattering problems (1.2.20), the scattering data consist of $\{\rho(k), k \text{ real}, \kappa_n, C_n, n = 1, \cdots, N\}$. The corresponding trace formulae are as follows (Zakharov and Faddeev (1971)). Define $C_m = \int_{-\infty}^{\infty} \mu_m \, dx$, where μ_m is defined in (1.6.15). Then

(1.6.22)
$$C_{2m+1} = \frac{2}{2m+1} \sum_{n=1}^{N} (2\kappa_n)^{2m+1}$$
$$+ \frac{(-1)^m}{\pi} \int_{-\infty}^{\infty} (2k)^{2m} \log \{1 - |\rho(k)|^2\} \, dk,$$
$$C_{2m} = 0.$$

1.6.b. Complete integrability. Let us now consider one of the fundamental descriptions of IST: the equations solvable by IST (i.e., (1.5.16)) are completely integrable Hamiltonian systems, and IST amounts to a canonical transformation from physical variables to (an infinite set of) action-angle variables. This description may be considered an alternative to the "nonlinear Fourier analysis" description that was developed in earlier sections of this book. The fundamental question in this section is not "What new problems can be solved with IST?" but rather "Why should IST work at all?"

The description of IST as a canonical transformation to action-angle variables was first developed for KdV by Zakharov and Faddeev (1971), following preliminary work by Gardner (1971). It was applied to the nonlinear Schrödinger equation by Takhtadzhyan (1972) and by Zakharov and Manakov (1974). The results in the last paper were generalized by numerous authors (see, e.g., Flaschka and Newell (1975), Kodama (1975), McLaughlin (1975) and Flaschka and McLaughlin (1976b). The derivation given here draws upon all of these works, despite some minor discrepancies among their results.

We begin with some of the basic concepts and notation of Hamiltonian mechanics that will be required for the formal extension to the infinite dimensional cases under consideration. (Readers with no prior knowledge of Hamiltonian mechanics may wish to consult Goldstein (1950) or Arnold (1978)). Let $p(x, t, \alpha)$, $q(x, t, \beta)$ be analytic functions of x on $-\infty < x < \infty$ which decrease rapidly as $|x| \to \infty$ for all values of (t, α, β) in their appropriate domains. Let

$$H(p, q, t) = \int_{-\infty}^{\infty} h\{p(x, t, \alpha), q(x, t, \beta), t\} \, dx$$

be a complex-valued functional of (p, q) and their x-derivatives. Its functional (or Fréchet, or variational) derivative $\delta H/\delta p$ is defined by

(1.6.23) $$\frac{\partial H}{\partial \alpha} = \int_{-\infty}^{\infty} \frac{\delta H}{\delta p} \frac{\partial p}{\partial \alpha} \, dx;$$

$\delta H/\delta q$ has a similar definition.

Example. From the identity

$$p(y) = \int_{-\infty}^{\infty} \delta(x - y) p(x) \, dx,$$

it follows that

$$\frac{\delta p(y)}{\delta p(x)} = \delta(x - y),$$

where $\delta(x)$ is the Dirac delta function.

For the cases we will consider, h in (1.6.23) is an infinitely differentiable function of (p, q) and all of their x-derivatives. Then by direct computation we have

(1.6.24) $$\frac{\delta H}{\delta p} = \sum_{n=0}^{\infty} (-1)^n \frac{\partial^n}{\partial x^n} \frac{\partial h}{\partial p_n}, \qquad p_n \equiv \frac{\partial^n p}{\partial x^n}.$$

For some applications it is convenient to define $p = \phi_x$. Then the identity

(1.6.25) $$\frac{\delta H}{\delta \phi} = -\frac{\partial}{\partial x}\frac{\delta H}{\delta p}$$

follows from the fact that h depends only on derivatives of ϕ, rather than on ϕ itself.

DEFINITION. A dynamical system is *Hamiltonian* if it is possible to identify generalized coordinates $[q]$ and momenta $[p]$, and a Hamiltonian $[H(p, q, t)]$, such that the equations of motion of the system can be written in the form

(1.6.26a, b) $$\frac{\partial q}{\partial t} = \frac{\delta H}{\delta p}, \qquad \frac{\partial p}{\partial t} = -\frac{\delta H}{\delta q}.$$

These are Hamilton's equations, and the variables (p, q) are called *conjugate*. (There are generalizations of (1.6.27), but this is almost general enough for our purposes; see also (1.6.31).)

Example. The system

(1.6.27) $$\begin{aligned} iq_t + q_{xx} - 2q^2 r &= 0, \\ ir_t - r_{xx} + 2qr^2 &= 0, \end{aligned}$$

is Hamiltonian, as may be seen from the identification:

coordinates (q): $\quad q(x, t)$

momenta (p): $\quad r(x, t)$

Hamiltonian (H): $\quad -i \int_{-\infty}^{\infty} \{q_x r_x + (qr)^2\}\, dx.$

If at $t = 0$, $r(x, 0) = \pm q^*(x, 0)$, then this relation holds for all t according to (1.6.27), which may be replaced by the nonlinear Schrödinger equation

(1.6.28) $$iq_t + q_{xx} \mp 2|q|^2 q = 0.$$

Alternatively, we may assert that (1.6.28) is Hamiltonian, and identify coordinates $[q(x, t)]$, momenta $[q^*(x, t)]$ and a Hamiltonian $[-i \int_{-\infty}^{\infty} \{\pm |q_x|^2 + |q|^4\}\, dx]$. This is correct provided we define independent variations of q and q^*. In the presentation given in this section, we will regard (1.6.28) as a special case of (1.6.27) in which the initial data satisfy an additional constraint $(r = \pm q^*)$.

Many of the equations solvable by IST are first order in time. It is convenient in these cases to use a variation of Hamilton's equations.

LEMMA. *Let $H(p, q, t)$ be the Hamiltonian of a dynamical system such that H does not involve q explicitly, although it may involve x-derivatives of q. Then the relation*

$$p = q_x$$

is consistent with (1.6.26), *both of which reduce to*

(1.6.29)
$$\frac{\partial p}{\partial t} = \frac{\partial}{\partial x} \frac{\delta H}{\delta p}\bigg|_{q_x = p}.$$

Proof. (1.6.29) is the x-derivative of (1.6.26a). Reduction of (1.6.26b) follows from (1.6.25) and the fact that $\partial h/\partial q$ vanishes. □

Example.

$$H = -\int \{pq_x^2 + p^2 q_x - p_x q_{xx}\}\, dx.$$

The dynamical equations are

(1.6.30)
$$q_t = -q_x^2 - 2pq_x - q_{xxx},$$
$$p_t = -(2pq_x)_x - (p^2)_x - p_{xxx}.$$

If at $t = 0$ $p(x, 0) = q_x(x, 0)$, then this relation holds for all t; i.e., the evolution equations for p and q_x are identical. Moreover, both are the KdV equation (1.2.2) for $p(x, t)$.

Thus, a dynamical system is Hamiltonian *either* if its equations of motion can be put in the form (1.6.26) for a Hamiltonian $H(p, q, t)$, *or* if they can be put in the form

(1.6.31)
$$\frac{\partial p}{\partial t} = \frac{\partial}{\partial x} \frac{\delta \hat{H}}{\delta p}$$

for a Hamiltonian, $\hat{H}(p, t)$. The reader should note that $\hat{H}(p, t) \neq H(p, q)|_{q_x = p}$, although they are clearly related.

Example. The KdV equation has the form (1.6.31) with the Hamiltonian

(1.6.32)
$$\hat{H} = -\int \left(p^3 - \frac{1}{2}p_x^2\right) dx.$$

Here $\hat{H}$ differs from $H|_{q_x = p}$ in the previous example by a factor of 2.

Next, in order to change variables from (p, q) to some other set of conjugate variables, (P, Q), we define the *Poisson brackets*:

(1.6.33)
$$\langle A, B \rangle \equiv \int_{-\infty}^{\infty} \left\{ \frac{\delta A}{\delta q} \frac{\delta B}{\delta p} - \frac{\delta A}{\delta p} \frac{\delta B}{\delta q} \right\} dx.$$

If only the derivatives of q, rather than q itself, enter into the Hamiltonian, then we may identify $q_x = p$ and use (1.6.25). Then (1.6.33) should be replaced by

(1.6.34) $$\langle A, B \rangle = \int_{-\infty}^{\infty} \left\{ \frac{\delta A}{\delta p} \frac{\partial}{\partial x} \frac{\delta B}{\delta p} \right\} dx.$$

A transformation from (p, q) to (P, Q) is defined to be *canonical* if

(1.6.35) $$\langle Q(x), Q(y) \rangle = 0, \quad \langle P(x), P(y) \rangle = 0,$$
$$\langle Q(x), P(y) \rangle = \delta(x - y).$$

It follows from (1.6.35) that the volume of the phase space has not changed under the transformation. The question of whether a transformation to a new set of variables is canonical is analogous to whether a new set of basis vectors is complete in a linear vector space.

Example. The identity map

$$P(x, t) = p(x, t), \quad Q(x, t) = q(x, t)$$

is a canonical transformation.

This completes the presentation of the necessary background material. Now we come to the main point, that IST is a canonical transformation of a Hamiltonian system to a set of action-angle variables. In order to keep the presentation as simple as possible, we discuss here only equations of the form (1.5.16), and simply state the results for equations of the form (1.5.21). The main points in this development are the following.

1. Evolution equations of the form (1.5.16) represent (infinite dimensional) Hamiltonian dynamical systems, in which (q, r) play the roles of conjugate variables.
2. There is a subset S of scattering data from which the rest of the scattering data can be reconstructed.
3. The mapping: $(q, r) \to S$ is a canonical transformation.
4. The conjugate variables in $S(=P, Q)$ are of action-angle type, i.e., $H = H(P)$, so that (1.6.26) becomes

(1.6.36) $$\frac{\partial P}{\partial t} = 0, \quad \frac{\partial Q}{\partial t} = \frac{\delta H}{\delta P} = \text{constant}.$$

The infinite set of conservation laws are a direct consequence of (1.6.36a), and are equivalent to it when the initial data for (q, r) are sufficiently restricted.

Recall from § 1.5 that for every dispersion relation that is real for real k and entire, there is a system of nonlinear evolution equations of the form

(1.5.16) $$\begin{pmatrix} r \\ -q \end{pmatrix}_t + 2A_-(\mathscr{L}^A)\begin{pmatrix} r \\ q \end{pmatrix} = 0$$

that can be solved by IST; here $\mathscr{L}^A$ is defined by (1.5.15), A_- is related to the linearized dispersion relation by (1.5.18) and (1.2.7a) is the appropriate scattering problem.

THEOREM. *Let A_- be an entire function of ζ in the form*

(1.6.37) $$A_-(\zeta) = \frac{1}{2i} \sum_{n=0}^{\infty} (-2\zeta)^n a_n,$$

where every a_n is real. Then the system defined by (1.5.16) is Hamiltonian; $q(x, t)$ and $r(x, t)$ are the conjugate variables and the appropriate Hamiltonian is

(1.6.38) $$H(q, r) = i \sum_{n=0}^{\infty} a_n(i)^n C_n(q, r),$$

where C_n is defined by (1.6.6).

Note that it is easy to demonstrate that any particular example, such as (1.6.28), is Hamiltonian, simply by displaying the Hamiltonian and generalized coordinates. The rather lengthy proof that follows is required to show that *every* equation of the form (1.5.16) must be Hamiltonian.

Proof.

(i) From (1.2.7a), for real ζ

$$\phi_1(x, \zeta) e^{i\zeta x} = 1 + \int_{-\infty}^{x} q(y) \phi_2(y, \zeta) e^{i\zeta y} dy,$$

so that

$$\frac{\delta \phi_1(x, \zeta)}{\delta q(y)} = \theta(x-y) \phi_2(y, \zeta) e^{i\zeta(y-x)},$$

where θ is the Heaviside step-function.

(ii) For any function $A(x)$ that is differentiable except at finitely many points, we may define

$$\frac{\delta A(x)}{\delta q(x)} = \lim_{y \uparrow x} \frac{\delta A(x)}{\delta q(y)},$$

so that

$$\frac{\delta \phi_1(x, \zeta)}{\delta q(x)} = \phi_2(x, \zeta).$$

Similarly,

$$\frac{\delta \phi_2(x, \zeta)}{\delta r(x)} = \phi_1(x, \zeta)$$

and

$$0 = \frac{\delta \phi_1}{\delta r} = \frac{\delta \phi_2}{\delta q} = \frac{\delta \psi_{1,2}}{\delta q} = \frac{\delta \psi_{1,2}}{\delta r}.$$

THE INVERSE SCATTERING TRANSFORM ON THE INFINITE INTERVAL 63

Therefore

(1.6.39)
$$\frac{\delta a(\zeta)}{\delta q(x)} = \frac{\delta}{\delta q}\{\phi_1\psi_2 - \phi_2\psi_1\} = \phi_2(x, \zeta)\psi_2(x, \zeta),$$
$$\frac{\delta a(\zeta)}{\delta r(x)} = -\phi_1(x, \zeta)\psi_1(x, \zeta).$$

All of these quantities are defined in the upper half ζ-plane (cf. § 1.3), and these relations may be extended there as well.

(iii) From (1.2.7a), for real ζ,
$$(\phi_1\psi_1)_x + 2i\zeta\phi_1\psi_1 = q(\phi_1\psi_2 + \phi_2\psi_1),$$
$$(\phi_2\psi_2)_x - 2i\zeta\phi_2\psi_2 = r(\phi_1\psi_2 + \phi_2\psi_1),$$
$$(\phi_1\psi_2 + \phi_2\psi_1)_x = 2q\phi_2\psi_2 + 2r\phi_1\psi_1.$$

Using the boundary conditions as $x \to +\infty$, we have
$$\phi_1\psi_2 + \phi_2\psi_1 = a - 2\int_x^\infty \{q\phi_2\psi_2 + r\phi_1\psi_1\}\, dy,$$

so that
$$\zeta\begin{pmatrix}\phi_2\psi_2 \\ -\phi_1\psi_1\end{pmatrix} = \mathscr{L}^A\begin{pmatrix}\phi_2\psi_2 \\ -\phi_1\psi_1\end{pmatrix} - \frac{a}{2i}\begin{pmatrix}r \\ q\end{pmatrix}.$$

These relations also may be extended to the upper half plane.

(iv) Let $|\zeta| \to \infty$, $\operatorname{Im}\zeta > 0$, to obtain

(1.6.40)
$$\begin{pmatrix}\phi_2\psi_2 \\ -\phi_1\psi_1\end{pmatrix} = -\frac{a}{2i\zeta}\left(1 - \frac{\mathscr{L}^A}{\zeta}\right)^{-1}\begin{pmatrix}r \\ q\end{pmatrix}$$
$$= -\frac{a}{i\zeta}\sum_{n=0}^\infty \left(\frac{\mathscr{L}^A}{\zeta}\right)^n \begin{pmatrix}r \\ q\end{pmatrix}.$$

(v) The gradient of $\log a$ follows from (1.6.39) and (1.6.40):
$$\frac{\delta \log a}{\delta q} = -(2i\zeta)^{-1} \sum_0^\infty \left(\frac{\mathscr{L}^A}{\zeta}\right)^n r,$$
$$\frac{\delta \log a}{\delta r} = -(2i\zeta)^{-1} \sum_0^\infty \left(\frac{\mathscr{L}^A}{\zeta}\right)^n q.$$

(vi) Finally, from (1.6.6), it follows that (1.5.19) in expanded form can be written in the form

$$q_t = \frac{\delta H}{\delta r}, \quad r_t = -\frac{\delta H}{\delta q},$$

where

$$H(q, r) = i \sum_{n=0}^{\infty} a_n C_n(q, r)(i)^n.$$

This is the desired result. □

Thus we may consider evolution equations of the form (1.5.19) as Hamiltonian systems whenever the linearized dispersion relation has the form (1.6.37). (Note that (1.6.27) is an example of this theorem, with $\omega = k^2$.) As one might expect, there is a corresponding result for evolution equations of the form (1.5.21), related to (1.2.20), (e.g., see Flaschka and Newell (1975) for details). We may state it as follows. Let

$$\gamma(k^2) = \sum_{n=0}^{\infty} a_n (k^2)^n$$

be an entire function. The nonlinear evolution equation (1.5.21) is Hamiltonian, in the form (1.6.31). The Hamiltonian is

(1.6.41) $$H = -\frac{1}{2} \sum_{n=0}^{\infty} a_n C_{2n+3}(-4)^n,$$

where C_n is defined by (1.6.22). The Hamiltonian for the KdV equation is an example.

Next, we restrict our attention to Hamiltonian systems of the form (1.5.19), and define their Poisson brackets,

(1.6.42) $$\langle A, B \rangle = \int_{-\infty}^{\infty} \left\{ \frac{\delta A}{\delta q} \frac{\delta B}{\delta r} - \frac{\delta A}{\delta r} \frac{\delta B}{\delta q} \right\} dx.$$

DEFINITION. *Given a Hamiltonian system, two functionals A, B of the conjugate variables are in* involution *if*

$$\langle A, B \rangle = 0.$$

For finite dimensional Hamiltonian systems (with N coordinates and N momenta), a theorem of Liouville asserts that if there exist N functionals with linearly independent gradients that are in involution, the equations of motion can be integrated by quadrature (cf. Arnold and Avez (1968)).

LEMMA. *The infinite set of constants of the motion, C_n, defined by (1.6.6), all are in involution for any motion defined by (1.5.19) with an appropriate $\omega(k)$.*

Proof. By computation,

$$0 = \frac{dC_n}{dt} = \int_{-\infty}^{\infty} \left\{ \frac{\delta C_n}{\delta q} \frac{\partial q}{\partial t} + \frac{\delta C_n}{\delta r} \frac{\partial r}{\partial t} \right\} dx$$

$$= \int_{-\infty}^{\infty} \left\{ \frac{\delta C_n}{\delta q} \frac{\partial H}{\partial r} - \frac{\delta C_n}{\delta r} \frac{\delta H}{\delta q} \right\} dx$$

$$= \langle C_n, H \rangle.$$

In particular, this is valid for $H = C_m$. □

Thus, the existence of an infinite set of motion constants for any of the infinite dimensional Hamiltonian systems defined by (1.5.19) *suggests* that these systems may be completely integrable as well. This suggestion turns out to be correct, but its validity relies on somewhat more than just this set of motion constants. In infinite dimensional Hamiltonian systems, the question of "how many" functionals in involution are required to assure complete integrability is not obvious.

Let us now define a subset S of the scattering data, from which the remaining scattering data can be constructed. Here we assume that $a(\zeta)$, $\bar{a}(\zeta)$ have only simple zeros in their respective half planes, and that none occur on the real axis. Further, we assume that $b(\zeta)$, $\bar{b}(\zeta)$ can be extended off the real axis. Then, for real ξ, define

(1.6.43a) $\qquad P(\xi) = \log\{a(\xi)\bar{a}(\xi)\}, \qquad Q(\xi) = -\frac{1}{\pi} \log b(\xi).$

There may be discrete eigenvalues for Im $\zeta > 0$,

$$a(\zeta_m) = 0, \qquad c_m = \left.\frac{b}{a'}\right|_{\zeta_m}, \qquad m = 1, \cdots, N,$$

and for Im $\zeta < 0$,

$$\bar{a}(\bar{\zeta}_l) = 0, \qquad \bar{c}_l = \left.\frac{\bar{b}}{\bar{a}'}\right|_{\bar{\zeta}_l}, \qquad l = 1, \cdots, \bar{N}.$$

Let

(1.6.43b) $\qquad P_m = \zeta_m, \qquad Q_m = -2i \log c_m,$

(1.6.43c) $\qquad \bar{P}_l = \bar{\zeta}_l, \qquad \bar{Q}_l = -2i \log \bar{c}_l.$

Denote by S the variables defined in (1.6.43).

It is easy to see that the remaining scattering data can be constructed from S. For Im $\zeta > 0$, $\log a(\zeta)$ may be found from S via (1.6.20). By similar means we also obtain $\log a$ for Im $\zeta = 0$, and $\log \bar{a}(\zeta)$, Im $\zeta \leq 0$. Then $b(\xi)$ may be obtained from $Q(\xi)$, and $\bar{b}(\xi)$ from the Wronskian relation, $a\bar{a} + b\bar{b} = 1$. If $r = \pm q^*$, so that $\bar{a}(\xi) = a^*(\xi)$ for real ξ, then $P(\xi)$ is real, $|b(\xi)|$ is determined by $P(\xi)$, and $Q(\xi)$ may be replaced by its imaginary part. Moreover, (1.6.43c) is redundant.

Thus the scattering data are determined by S. According to the results in § 1.3, the potentials (q, r) may be reconstructed from S as well. In this sense, S is "complete". In order to show that the mapping $(q, r) \to S$ is a canonical transformation, we must verify (1.6.35). This is a somewhat tedious calculation, which is outlined in Exercise 1. Its consequence, however, is essential:

The mapping $(q, r) \to S$ is a canonical transformation. Therefore, the dynamics of any of the systems defined by (1.5.19) may be described either in terms of (q, r) or in terms of S.

It remains to write the Hamiltonian in terms of S. This is easily done by substituting (1.6.21a) into (1.6.38) and combining terms. The result is that, for a linearized dispersion relation given by (1.6.37),

$$(1.6.44a) \quad H = \frac{2}{\pi} \int_{-\infty}^{\infty} A_-(\xi) \left[\log a\bar{a}(\xi) + \sum_{m=1}^{N} \log \frac{\xi - \zeta_m^*}{\xi - \zeta_m} + \sum_{l=1}^{\bar{N}} \log \frac{\xi - \bar{\zeta}_l^*}{\xi - \bar{\zeta}_l} \right] d\xi$$
$$+ 4i \sum_{m=1}^{N} \int_{\zeta_m^*}^{\zeta_m} A_-(\zeta) \, d\zeta.$$

If $r = \pm q^*$, this simplifies to

$$(1.6.44b) \quad H = \frac{2}{\pi} \int_{-\infty}^{\infty} A_-(\xi) \log |a(\xi)|^2 \, d\xi + 4i \sum_{m=1}^{N} \int_{\zeta_m^*}^{\zeta_m} A_-(\zeta) \, d\zeta.$$

Now it is apparent that H depends on the generalized momenta $(P(\xi), P_m, \bar{P}_l$ in (1.6.43)), but not on the coordinates $(Q(\xi), Q_m, \bar{Q}_l)$. This is the defining property of *action-angle variables*. It is also apparent that Hamilton's equations take the form

$$(1.6.45) \quad \frac{\partial P}{\partial t} = 0, \quad \frac{\partial Q}{\partial t} = \frac{\delta H}{\delta P}.$$

Thus, every P is time independent, while the Q's vary linearly in time, i.e., the motion is uniform in these variables.

Let us make (1.6.45) explicit, using (1.6.43). For real ξ,

$$\frac{\partial}{\partial t}\{\log a\bar{a}(\xi)\} = 0, \quad \frac{\partial}{\partial t}\{\log b(\xi)\} = -2A_-(\xi);$$

for $m = 1, \cdots, N$,

$$(1.6.46) \quad \frac{\partial}{\partial t}\zeta_m = 0, \quad \frac{\partial}{\partial t}\{\log c_m\} = -2A_-(\zeta_m);$$

for $l = 1, \cdots, \bar{N}$,

$$\frac{\partial}{\partial t}\bar{\zeta}_l = 0, \quad \frac{\partial}{\partial t}\{\log \bar{c}_l\} = 2A_-(\bar{\zeta}_l).$$

(Calculation of $\delta H/\delta \zeta_m$ and $\delta H/\delta \bar{\zeta}_l$, which is not obvious from (1.6.44a), is outlined in Exercise 2.) These are precisely the results obtained in § 1.4.

Finally, we may state the corresponding results for evolution equations of the form (1.5.21), such as the KdV equation (Zakharov and Faddeev (1971), Flaschka and Newell (1975)). For real k, define

(1.6.47a)
$$P(k) = \frac{k}{\pi} \log |a(k)|^2 = -\frac{k}{\pi} \log \{1 - |\rho(k)|^2\},$$
$$Q(k) = \arg \rho(k),$$

and for the discrete eigenvalues

(1.6.47b)
$$P_n = -2\kappa_n^2, \quad Q_n = \log c_n$$

(see § 1.3 for notation). Then one can show that the mapping

$$q \to \{P(k), Q(k), P_n, Q_n\}$$

is canonical, i.e., the Poisson brackets satisfy (1.6.35) with (1.6.34). Moreover, it is evident from (1.6.22) and (1.6.41) that the variables in (1.6.47) are of action-angle type. For KdV, Hamilton's equations are equivalent to (1.4.10) in these variables.

In summary, we may view IST as a concrete method to effect a canonical transformation into action-angle variables. The dynamics of the system are very simple when described in terms of these variables. The relatively simple picture that emerges in physical variables (solitons with pairwise interactions, etc.) is a direct consequence of the existence of the action-angle variables. In particular, no stochastic motion is possible in a problem solvable by IST.

This is not to suggest that the entire problem has become trivial. Certainly the problem of inverse scattering (i.e., unscrambling the canonical transformation) is not trivial, and this approach offers no help on this point. Moreover, we are not even guaranteed that the solution is especially well behaved. For example, if $r \neq q^*$ for some constant α, the exact 1-soliton solution of (1.6.27) blows up in a finite time (cf. (1.4.24)). Thus singularities may develop even in completely integrable systems, provided they do not violate any of the conservation laws. From this standpoint, the reason for our interest in the case $r = \pm q^*$ is that $C_0 = \int rq\, dx$ becomes definite, and serves as a (time-independent) norm.

1.7. Long-time behavior of the solutions. The objective of this section is to provide methods to find the dominant behavior of the solutions of completely integrable problems in the limit $t \to \infty$. Among other things, this information is useful if these equations are to be used as models of physical phenomena.

It is not difficult to show that if a solution of an initial value problem on

$-\infty < x < \infty$ contains solitons, then as $t \to \infty$ the solitons remain $O(1)$, while the rest of the solution (i.e., the "radiation" corresponding to the continuous spectrum) slowly disperses away. In this sense, the dominant asymptotic solution consists only of solitons (e.g., Tanaka (1975)). However, this description is not uniformly valid in space because there are large regions where the solitons are negligibly small and the radiation dominates. Moreover, the solitons may contain only a small fraction of such physically important conserved quantities as the momentum of the waves.

N-soliton solutions were discussed in § 1.4. Here we concentrate primarily on problems in which the initial data generate no solitons. These problems are of interest in their own right:

(i) There are no solitons if $r = +q^*$ in (1.5.16) and $|q| \to 0$ rapidly as $|x| \to \infty$.

(ii) Solitons are possible in (1.5.16) if $r = -q^*$, but none arise if the initial data are sufficiently small (cf. (1.3.16e)):

$$(1.7.1) \qquad \int_{-\infty}^{\infty} |q(x,0)| \, dx < 0.904.$$

(iii) In (1.5.21), no solitons arise if $q(x, 0) < 0$. Moreover, most of the literature on the subject is restricted to these problems (Ablowitz and Newell (1973), Shabat (1973), Manakov (1974a), Segur and Ablowitz (1976), Zakharov and Manakov (1976b), Ablowitz and Segur (1977a), Miles (1979), Ablowitz, Kruskal and Segur (1979), Segur and Ablowitz (1981)). After we know how the solitons and radiation behave separately, we will consider how they interact.

Two warnings should be made before we begin the analysis. The first is that almost none of the results to be described in this section are known rigorously. These results are formal, and have great practical value, but proofs of asymptoticity are yet to be given. The second (related) warning is that some of the existing literature on this question contains errors.

The evolution equations in question may be divided into three groups, on the basis of the qualitative behavior of their asymptotic solutions:

(a) equations of the form (1.5.16) whose linearized dispersion relation is even, such as the nonlinear Schrödinger equation;

(b) equations of the form (1.5.16) whose linearized dispersion relation is odd, such as the modified Korteweg–deVries equation;

(c) equations of the form (1.5.21) such as the Korteweg–deVries equation.

1.7.a. The nonlinear Schrödinger equation. We begin with

$$(1.7.2) \qquad iq_t + q_{xx} - 2\sigma |q|^2 q = 0,$$

where $\sigma \pm 1$ and $q \to 0$ as $|x| \to \infty$. Because its (formally) asymptotic solution is comparatively simple, some of the methods can be explained more easily here. To preclude solitons, we require that the initial data satisfy (1.7.1) for $\sigma = -1$. We also assume that the initial data are smooth and vanish rapidly as $|x| \to \infty$.

There is a closed-form similarity solution to (1.7.2),

(1.7.3) $$q(x,t) = t^{-1/2} A \exp\left\{it\left[\frac{1}{4}\left(\frac{x}{t}\right)^2 - 2\sigma A^2 \frac{\log t}{t} + \frac{\phi}{t}\right]\right\}.$$

Guided by this solution, we seek a "slowly-varying" similarity solution where the arbitrary constants A, ϕ, are now slowly varying functions (cf. § A.1). As $t \to \infty$, the expansion is given by

(1.7.4a) $$q(x,t) \sim t^{-1/2} R\, e^{it\theta},$$

(1.7.4b) $$\begin{aligned}R &= f + \sum_{n=1}^{\infty} \sum_{k=0}^{n} \frac{(\log t)^k}{t^n} f_{n,k}\left(\frac{x}{t}\right) \\ &= f\left(\frac{x}{t}\right) + \frac{1}{t}\left(\log t f_{1,1}\left(\frac{x}{t}\right) + f_{1,0}\left(\frac{x}{t}\right)\right) + O\left(\left(\frac{\log t}{t}\right)^2\right),\end{aligned}$$

(1.7.4c) $$\begin{aligned}\theta &= \frac{1}{4}\left(\frac{x}{t}\right)^2 + \sum_{n=1}^{\infty}\sum_{k=0}^{n} \frac{(\log t)^k}{t^n}\theta_{n,k}\left(\frac{x}{t}\right) \\ &= \frac{1}{4}\left(\frac{x}{t}\right)^2 + \frac{1}{t}\left[\log t\, \theta_{1,1}\left(\frac{x}{t}\right) + \theta_{1,0}\left(\frac{x}{t}\right)\right] + O\left(\left(\frac{\log t}{t}\right)^2\right),\end{aligned}$$

where

$f = f\left(\frac{x}{t}\right)$ is real and nonnegative but otherwise arbitrary,

$\theta_{1,1} = -2\sigma f^2,$

$\theta_{1,0} = g = g\left(\frac{x}{t}\right)$ is arbitrary and real,

$f_{1,1} = -4\sigma f(3(f')^2 + f f''),$

$f_{1,0} = fg'' + 2g'f' - 4\sigma f(2(f')^2 + f f''),$ etc.

This expansion can be carried to any desired order in n. All of the subsequent coefficients in the expansion can be found explicitly in terms of two arbitrary functions, $f(x/t)$ and $g(x/t)$, by substituting (1.7.4) into (1.7.2) and collecting terms. The functions f and g are unrestricted by (1.7.2), and we will show that they are determined by the initial data.

We now *assume* that, as $t \to \infty$, the solution of (1.7.2) that evolves from appropriate initial data tends to the form (1.7.4). Then the asymptotic solution of (1.7.2) is known once $f(x/t)$, $g(x/t)$ are specified in terms of the initial data. This can be done by several methods; we will present two.

First, we show that the conservation laws uniquely determine f but place no restriction on g. This is consistent with the formulation given in § 1.6, in the sense that the conserved quantities pin down the action part of the action-angle variables, and provide "half" of the information required to describe the dynamical systems.

Recall from § 1.6 that the motion constants for (1.7.2) take the form

$$C_n = \int_{-\infty}^{\infty} \mu_n(x, t)\, dx,$$

where

$$\mu_0 = -\sigma |q|^2, \qquad \mu_1 = -\sigma q q_x^*$$

and, for $n > 1$,

(1.7.5) $$\mu_{n+1} = q\left(\frac{\mu_n}{q}\right)_x + \sum_{k=0}^{n-1} \mu_k \mu_{n-k-1}.$$

Substituting (1.7.4) into (1.7.5) shows that

$$C_0 = -\sigma \int_{-\infty}^{\infty} f^2(X)\, dX, \qquad \left(X \equiv \frac{x}{t}\right),$$

$$C_1 = -\sigma \int_{-\infty}^{\infty} \left(\frac{X}{2i}\right) f^2(X)\, dX,$$

$$C_2 = -\sigma \int_{-\infty}^{\infty} \left(\frac{X}{2i}\right)^2 f^2(X)\, dX.$$

In fact, one proves (by induction on (1.7.5)) that

$$\mu_n(x, t) = -\sigma \left(\frac{X}{2i}\right)^n f^2(X) + O\left(\frac{\log t}{t}\right),$$

so that as $t \to \infty$

(1.7.6) $$C_n = -\sigma \int_{-\infty}^{\infty} \left(\frac{X}{2i}\right)^n f^2(X)\, dX, \qquad n = 0, 1, 2, \cdots.$$

But the motion constants are given in terms of the scattering data by the trace formulae (1.6.17),

(1.7.7) $$C_n = -\frac{1}{\pi} \int_{-\infty}^{\infty} (2i\xi)^n \log |a(\xi)|^2\, d\xi.$$

If all of these integrals converge, then the infinite set of moment equations obtained by equating (1.7.6) and (1.7.7) has exactly one solution:

(1.7.8)
$$X = \frac{x}{t} = -4\xi,$$

$$f^2(X) = \frac{\sigma}{4\pi} \log |a(\xi)|^2 = \frac{-\sigma}{4\pi} \log \left\{1 - \sigma \left|\frac{b}{a}(\xi)\right|^2\right\}.$$

The dominant behavior of the asymptotic solution of (1.7.2) is given by (1.7.4) along with (1.7.8), even though $g(x/t)$ is still unknown. In many applications the actual phase is much more difficult to measure than the wave "intensity" ($=|f|^2$). For these applications knowledge of g may be unnecessary for practical purposes.

The result in (1.7.8) was obtained by Segur and Ablowitz (1976) using this method, and by Manakov (1974) and by Zakharov and Manakov (1976) using two other methods. As an alternative to the method of conservation laws, we now outline (a slight variation of) the method of Zakharov and Manakov (1976) because it determines both f and g in terms of the scattering data. For real ξ, define

$$w_1(x, t; \xi) = \phi_1(x, t; \xi) \exp(i\xi x), \qquad w_2 = \phi_2(x, t; \xi) \exp(-i\xi x),$$

so that the scattering problem (1.2.7a) becomes

(1.7.9) $$(w_1)_x = q w_2 e^{2i\xi x}, \qquad (w_2)_x = -\sigma q^* w_1 e^{-2i\xi x},$$

with the boundary conditions

$$\begin{pmatrix} w_1 \\ w_2 \end{pmatrix} \to \begin{pmatrix} 1 \\ 0 \end{pmatrix} \text{ as } x \to -\infty, \qquad \begin{pmatrix} w_1 \\ w_2 \end{pmatrix} \to \begin{pmatrix} a(\xi) \\ b(\xi, t) \end{pmatrix} \text{ as } x \to +\infty.$$

If q satisfies (1.7.2), then

(1.7.10) $$\begin{aligned} b(\xi, t) &= b(\xi, 0) \exp(4i\xi^2 t) \\ &= b(\xi) \exp(4i\xi^2 t). \end{aligned}$$

Again we assume that as $t \to \infty$ the solution of (1.7.2) tends to the form (1.7.4), and we substitute this into (1.7.9). Then we have, to leading order,

(1.7.11) $$\begin{aligned} (w_1)_x &\sim t^{-1/2} f\left(\frac{x}{t}\right) w_2 \exp\{it\theta + 2i\xi x\}, \\ (w_2)_x &\sim \sigma t^{-1/2} f\left(\frac{x}{t}\right) w_1 \exp\{-it\theta - 2i\xi x\}, \end{aligned}$$

with the same boundary conditions. Here we assume that $x/t \le O(1)$ as $t \to \infty$, and we have neglected terms that are $O(\log t/t)$ in (1.7.11). For fixed large t, (1.7.11) are coupled ordinary differential equations, with a rapid phase $(t\theta + 2\xi x)$ and a slow phase (x/t); we may use classical WKBJ methods. Thus we look for a solution of (1.7.11) in the form

(1.7.12) $$w_i(x, t; \zeta) = w_{i,0}(\psi, X) + t^{-1/2} w_{i,1}(\psi, X) + O\left(\frac{\log t}{t}, t^{-1}\right),$$

where

$$X = \frac{x}{t}, \quad \psi = t\theta + 2\xi x, \quad i = 1, 2, \quad \theta \text{ given by (1.7.4)}.$$

Thus
$$\partial_x \to \psi_x \partial_\psi + X_x \partial_X \sim (\tfrac{1}{2}X + 2\xi)\partial_\psi + t^{-1}\partial_X.$$
If $(\tfrac{1}{2}X + 2\xi) \neq 0$, then (1.7.11) becomes to leading order,
$$\partial_\psi(w_{1,0}) \sim 0, \qquad \partial_\psi(w_{2,0}) \sim 0,$$
so that for $i = 1, 2$
(1.7.13) $$w_{i,0} = w_{i,0}(X).$$

At order $t^{-1/2}$
$$(\tfrac{1}{2}X + 2\xi)\partial_\psi(w_{1,1}) \sim f(X)w_{2,0}(X)\, e^{i\psi},$$
$$(\tfrac{1}{2}X + 2\xi)\partial_\psi(w_{2,1}) \sim \sigma f(X)w_{1,0}(X)\, e^{-i\psi}.$$

Omitting homogeneous terms, we obtain

(1.7.14)
$$w_{1,1}(\psi, X) \sim \frac{if(X)w_{2,0}(X)}{(\tfrac{1}{2}X + 2\xi)}\, e^{i\psi},$$
$$w_{2,1}(\psi, X) \sim \frac{i\sigma f(X)w_{1,0}(X)}{(\tfrac{1}{2}X + 2\xi)}\, e^{-i\psi}.$$

Secular terms arise at $O(t^{-1})$, where (1.7.9) becomes
$$(\tfrac{1}{2}X + 2\xi)\partial_\psi(w_{1,2}) \sim fw_{2,1}\, e^{i\psi} - \partial_X(w_{1,0}),$$
$$(\tfrac{1}{2}X + 2\xi)\partial_\psi(w_{2,2}) \sim \sigma fw_{1,1}\, e^{-i\psi} - \partial_X(w_{2,0}).$$

Substituting (1.7.14) into these relations shows that suppression of secular terms at this order requires that, for $(\tfrac{1}{2}X + 2\xi) \neq 0$,
$$\partial_X(w_{1,0}) \sim \frac{i\sigma f^2 w_{1,0}}{\tfrac{1}{2}X + 2\xi}, \qquad \partial_X(w_{2,0}) \sim -\frac{i\sigma f^2 w_{2,0}}{\tfrac{1}{2}X + 2\xi}.$$

These may be integrated, with the constants of integration determined from the boundary conditions for w_1, w_2. Thus, for $(\tfrac{1}{2}X + 2\xi) < 0$,

(1.7.15a) $$w_{1,0}(X) = \exp\left\{i\sigma \int_{-\infty}^{X} \frac{f^2(y)}{\tfrac{1}{2}y + 2\xi}\, dy\right\}, \qquad w_{2,0} \sim 0$$

and, for $(\tfrac{1}{2}X + 2\xi) > 0$,

(1.7.15b) $$w_{1,0}(X) = a(\xi) \exp\left\{-i\sigma \int_{X}^{\infty} \frac{f^2(y)}{\tfrac{1}{2}y + 2\xi}\, dy\right\},$$

(1.7.15c) $$w_{2,0}(X) = b(\xi, t) \exp\left\{i\sigma \int_{X}^{\infty} \frac{f^2(y)}{\tfrac{1}{2}y + 2\xi}\, dy\right\}.$$

These results break down near $(\frac{1}{2}X+2\xi)=0$, where an expansion different from (1.7.12) is required. From (1.7.15), as $X\uparrow(-4\xi)$,

(1.7.16)
$$w_{1,0}(X) \sim |X+4\xi|^{2i\sigma f_0^2} \exp\left\{-2i\sigma \int_{-\infty}^{-4\xi} (f^2)_y \log|y+4\xi|\, dy\right\},$$
$$w_{2,0}(X) \sim 0,$$

and as $X\downarrow(-4\xi)$

(1.7.17)
$$w_{1,0}(X) \sim a(\xi)(X+4\xi)^{2i\sigma f_0^2} \exp\left\{-2i\sigma \int_{-4\xi}^{\infty} (f^2)_y \log(y+4\xi)\, dy\right\},$$
$$w_{2,0}(X) \sim b(\xi,t)(X+4\xi)^{-2i\sigma f_0^2} \exp\left\{-2i\sigma \int_{-4\xi}^{\infty} (f^2)_y \log(y+4\xi)\, dy\right\},$$

where $f_0 = f(-4\xi)$.

At $(\frac{1}{2}X+2\xi)=0$, let

(1.7.18)
$$x_0 = 4\xi t,$$
$$\psi_0 = t\theta + 2\xi x|_{x=x_0} \sim \frac{x_0^2}{4t} + 2\xi x_0 - 2\sigma f_0^2 \log t + g(-4\xi)$$

(note that g enters the calculation at this point). Near $(\frac{1}{2}X+2\xi)=0$, using Taylor series, we obtain

$$\psi(x,t) \sim \psi_0 + \frac{(x-x_0)^2}{4t}.$$

Thus in this region we define

$$Z = \frac{x-x_0}{\sqrt{2t}}, \qquad w_i = w_i(Z;t),$$

so that (1.7.9) becomes

(1.7.19)
$$(w_1)_Z \sim \sqrt{2} f_0 w_2 \exp\left\{i\psi_0 + \frac{iZ^2}{2}\right\},$$
$$(w_2)_Z \sim \sqrt{2}\sigma f_0 w_1 \exp\left\{-i\psi_0 - \frac{iZ^2}{2}\right\}.$$

The general solution of (1.7.19) may be written in terms of parabolic cylinder functions. It must match (1.7.16) as $Z \to -\infty$, and (1.7.17) as $Z \to +\infty$. Omitting the intervening details, the final result is that on

$$X = \frac{x}{t} = -4\xi,$$

$$f^2\left(\frac{x}{t}\right) = \frac{\sigma}{4\pi} \log |a(\xi)|^2 = f_0^2,$$

(1.7.20)
$$g\left(\frac{x}{t}\right) = -\arg\{-\sigma b(\xi, 0)\} + \frac{3\pi}{4} - \arg\{\Gamma(1 - 2i\sigma f_0^2)\}$$

$$+ 2\sigma f_0^2 \log 2 - 2\sigma \int_{-\infty}^{-4\xi} \log |y + 4\xi|(f^2)_y \, dy$$

$$+ 2\sigma \int_{-4\xi}^{\infty} \log (y + 4\xi)(f^2)_y \, dy.$$

These results, when coupled with (1.7.4), provide the complete asymptotic expansion of the solution (1.7.2) in the limit $t \to \infty$. They may also be written entirely in terms of $(b/a)(\xi)$: on $x/t = -4\xi$,

$$f^2\left(\frac{x}{t}\right) = -\frac{\sigma}{4\pi} \log \left\{1 - \sigma \left|\frac{b}{a}(\xi)\right|^2\right\},$$

(1.7.21)
$$g\left(\frac{x}{t}\right) = -\arg\left\{-\sigma \frac{b}{a}(\xi)\right\} + \frac{3\pi}{4} - \arg\left\{\Gamma\left(1 - 2i\sigma f^2\left(\frac{x}{t}\right)\right)\right\}$$

$$+ 2\sigma f^2\left(\frac{x}{t}\right) \log 2 + 4\sigma \int_{-4\xi}^{\infty} \log (y + 4\xi)(f^2)_y \, dy.$$

We have seen (in § 1.4) that solitons are distinctly nonlinear objects; they cannot be linearized. On the other hand it is often asserted that the radiating part of the solution, which we are now considering, behaves qualitatively like the solution of the linearized problem. We may now assess this assertion on the basis of the asymptotic solutions of the two problems; i.e., (A.1.39) vs. (1.7.4) with (1.7.20). Note first that in both cases, two of the most important aspects of the solutions are that:

(i) the overall decay rate is $t^{-1/2}$;
(ii) information travels with the group velocity of the linear dispersion relation, $\omega = k^2 = (-2\xi)^2$.

Second, in the limit of small amplitudes ($f \to 0$ uniformly in x in (1.7.20)) one may show that $b^*(\xi) \to \sigma \hat{q}(-2\xi)$, and that the asymptotic solution of the nonlinear problem reduces exactly to that of the linear problem (see Exercise 1). Thus, the two limits ($t \to \infty$, $\int |q(x, 0)| \, dx \to 0$) commute for solutions of (1.7.2). Among the nonlinear features of (1.7.20) is that the phase contains a $\log t$ term and g depends globally on the scattering data through the two integrals.

THE INVERSE SCATTERING TRANSFORM ON THE INFINITE INTERVAL 75

The general solution of (1.7.2) with $\sigma = -1$ consists of both solitons and radiation. For the nonlinear Schrödinger equation, the solitons travel in the midst of the radiation and the interaction of the two is of some importance. Using the conservation laws, Segur (1976) found the asymptotic ($t \to \infty$) solution of (1.7.2 with $\sigma = -1$) in the simplest case, where the scattering data contain one discrete eigenvalue plus continuous spectrum. The dominant behavior is

$$q(x, t) \sim 2\eta \exp(i\phi) \operatorname{sech} \psi$$

(1.7.22)
$$+ t^{-1/2} f\left(\frac{x}{t}\right)\left[\exp(it\theta)\frac{(\xi + x/4t + i\eta \tanh \psi)^2}{(\xi + x/4t)^2 + \eta^2}\right.$$

$$\left. + \exp(2i\phi - it\theta)\frac{\eta^2 \operatorname{sech}^2 \psi}{(\xi + x/4t)^2 + \eta^2}\right] + o(t^{-1/2}),$$

where $(\xi + i\eta)$ is the discrete eigenvalue,

$$\phi = -2[\xi x + 2(\xi^2 - \eta^2)t] + \hat{\phi},$$

$$\psi = 2\eta(x + 4\xi t) + \hat{\psi},$$

$$\theta = \frac{x^2}{4t} + O\left(\frac{\log t}{t}\right),$$

$$f^2\left(\frac{x}{t}\right) = \frac{1}{4\pi}\log\left\{1 + \left|\frac{b}{a}\left(-\frac{x}{4t}\right)\right|^2\right\}.$$

The first term in (1.7.22) is simply the single soliton. The second term represents the radiation, with a correction in the neighborhood of the soliton, and the third term may be thought of as representing the interaction between these two components. Because this result was obtained via the conservation laws, various terms in the phases, including $\hat{\phi}$ and $\hat{\psi}$, remain undetermined. In particular, the total phase shift of the soliton (from $t = -\infty$ to $t = +\infty$) due to its interaction with the radiation is not determined by this method.

1.7.b. The modified Korteweg–de Vries equation. For mKdV,

(1.7.23) $$v_t - 6\sigma v^2 v_x + v_{xxx} = 0, \quad \sigma = \pm 1,$$

the linearized dispersion relation ($\omega = -k^3$) is an odd function, so that the group velocity, $d\omega/dk = -3k^2$, is of one sign for all real k. The consequence is that the decaying oscillations, which covered all of space in the asymptotic solution of (1.7.2) are restricted to $x < 0$ in the asymptotic solution of (1.7.23). This comes as no surprise, since it is also true in the linearized mKdV equation (cf. § A.1), but it means that a separate analysis is required for $x > 0$.

Again we require that the initial data for (1.7.23) satisfy (1.7.1) for $\sigma = -1$ in order to preclude soliton formation. In this problem we also require that the initial data decay rapidly enough in x that $b(\xi, 0)$ can be extended off of the real axis.

The linear integral equation for (1.7.23) may be written as

$$K(x, y; t) + \sigma F(x+y; t)$$

(1.7.24)
$$= \sigma \int_x^\infty \int_x^\infty K(x, z; t) F(z+s; t) F(s+y; t) \, dz \, ds = 0, \qquad y > x,$$

where

$$F(x; t) = \frac{1}{(2\pi)} \int_{-\infty}^\infty \frac{b}{a}(\xi) \exp\{i\xi x + 8i\xi^3 t\} \, d\xi$$

and

$$v(x, t) = -2K(x, x; t).$$

Because $F(2x; t)$ satisfies the linearized mKdV equation, its asymptotic ($t \to \infty$) behavior is known from § A.1. In particular, for $x \gg (3t)^{1/3}$,

(1.7.25)
$$F(\chi; t) \sim \frac{r(ik/2)k^{-1/2}}{4(3\pi t)^{1/2}} \exp(-2tk^3),$$

where $r(\xi) = (b/a)(\xi)$, and $k^2 = (\chi/6t)$. In this region

$$K(x, y; t) \sim -\sigma F(x+y; t)$$

in (1.7.24); the integral term is transcendentally smaller. Thus for $x \gg (3t)^{1/3}$,

(1.7.26)
$$v(x, t) \sim \frac{\sigma r(ik/2)k^{-1/2}}{2(3\pi t)^{1/2}} \exp(-2tk^3).$$

These representations are not uniformly valid as $x/t \to 0$. To obtain such representations, we write

$$F(\chi; t) = \frac{1}{4\pi(3t)^{1/3}} \int_{-\infty}^\infty r\left(\frac{\kappa}{2(3t)^{1/3}}\right) \exp\{i\kappa \tilde{Z}/2 + i\kappa^3/3\} \, d\kappa,$$

where $\tilde{Z} = \chi/(3t)^{1/3}$. Upon expanding $r(\xi)$ in a Taylor series near $\xi = 0$, one finally obtains for $t \to \infty$, $x/t \to 0$, but $Z = x/(3t)^{1/3} \to \infty$,

(1.7.27)
$$\sigma v(x, t) \sim (3t)^{-1/3} r(0) \text{ Ai}(Z) - (3t)^{-2/3} i \frac{r'(0)}{2} \text{ Ai}'(Z) + O((3t)^{-1}),$$

where Ai (Z) is the Airy function. This suggests that where $|x| \leq O((3t)^{1/3})$ we seek an approximate solution of (1.7.23) in the form

(1.7.28)
$$v(x, t) \sim (3t)^{1/3} w(Z) + (3t)^{2/3} w_1(Z) + \cdots.$$

Then $w(Z)$ satisfies the second equation of Painlevé (cf. § 3.7),

(1.7.29)
$$\frac{d^2 w}{dZ^2} = Zw + 2\sigma w^3,$$

THE INVERSE SCATTERING TRANSFORM ON THE INFINITE INTERVAL 77

with the boundary condition that as $Z \to +\infty$

(1.7.30) $$w(Z) \to \sigma r(0) \, \text{Ai}(Z).$$

The fact that (1.7.29) is nonlinear means that the solution in this middle region ($|x| \leq O((3t)^{1/3})$) remains nonlinear as $t \to \infty$, even though its amplitude vanishes in that limit. Such a nonlinear region exists for every mKdV solution for which $r(0) \neq 0$.

At $\zeta = 0$, the scattering problem (1.2.7a) may be solved in closed form; for $\sigma = +1$

$$\phi_1(x, 0) = \cosh\left(\int_{-\infty}^{x} v \, dx\right), \quad \phi_2 = \sinh\left(\int_{-\infty}^{x} v \, dx\right),$$

so that

$$r(0) = \tanh\left(\int_{-\infty}^{\infty} v \, dx\right).$$

Similarly, for $\sigma = -1$,

$$r(0) = -\tan\left(\int_{-\infty}^{\infty} v \, dx\right).$$

But v is required to be absolutely integrable (cf. § 1.3), so we have that

(1.7.31) $\quad |r(0)| < 1 \quad \text{for } \sigma = +1, \quad |r(0)| < \infty \quad \text{for } \sigma = -1.$

In either case, the bounds in (1.7.31) guarantee that the solution of (1.7.29) with (1.7.30) is bounded for all real Z (cf. § 3.7). Thus, in the region $|x| \leq O((3t)^{1/3})$, the solution of (1.7.23) is approximately self-similar; the governing differential equation is (1.7.29), with the matching condition (1.7.30). A typical solution is shown in Fig. 3.2, p. 246.

The solution of (1.7.23) is rapidly oscillatory for $-x \gg (3t)^{1/3}$, as was the solution of the linearized problem. In this region we may use either of the methods we applied to (1.7.2), provided that we make the appropriate modifications. In either case it is convenient to make use of the slowly-varying similarity solution of mKdV. The general solution of (1.7.29) cannot be written in closed form, but for $Z \to -\infty$ (i.e., in the oscillatory region), a formally asymptotic solution is

$$w(Z) \sim (-Z)^{-1/4} d \sin \theta + O(|Z|^{-7/4}),$$
$$\theta \sim \tfrac{2}{3}(-Z)^{3/2} - \tfrac{3}{4}\sigma d^2 \log(-Z) + \bar{\theta} + O(|Z|^{-3/2}),$$

where d and $\bar{\theta}$ are the two constants of integration; $d \geq 0$ by convention. Then, by letting $d, \bar{\theta}$ depend on $X = -x/3t$, we obtain the desired form for $-x \gg (3t)^{1/3}$:

(1.7.32)
$$v(x, t) \sim (3t)^{-1/2} X^{-1/4} d \sin \theta + O((3t)^{-1}),$$
$$\theta \sim 2tX^{3/2} - \frac{\sigma}{2} d^2 \log 3t - \frac{3}{4} \sigma d^2 \log X + \bar{\theta}.$$

This is the analogue of (1.7.4) to leading order.

The conservation laws determine $d(X)$. The only modification required from the previous application of this method is that in this case one must use (1.7.26) and (1.7.28) to show that only the region $-x \gg (3t)^{1/3}$ contributes to the motion constants as $t \to \infty$. The final result is that along

(1.7.33a) $$X = -\frac{x}{3t} = \xi^2,$$

(1.7.33b) $$d^2(X) = \frac{\sigma}{\pi} \log \left| a\left(\frac{\xi}{2}\right) \right|^2 = -\frac{\sigma}{\pi} \log \left\{ 1 - \sigma \left| \frac{b}{a}\left(\frac{\xi}{2}\right) \right|^2 \right\}.$$

Note again that (1.7.33a) identifies the group velocity of the linearized dispersion relation as the velocity with which information travels in the limit $t \to \infty$. Note further that (1.7.33b) is virtually identical with (1.7.8b).

To obtain both $d(X)$ and $\theta(X)$ simultaneously, substitute (1.7.32) into the scattering problem (1.2.7a). The analogue of (1.7.11) is

(1.7.34)
$$(w_1)_x \sim (3t)^{-1/2} X^{-1/4} d(X) \sin \theta \cdot e^{2i\xi x} w_2,$$
$$(w_2)_x \sim (3t)^{-1/2} X^{-1/4} d \sin \theta \cdot e^{-2i\xi x} w_1.$$

The appropriate boundary conditions for (1.7.34) are that

$$\begin{pmatrix} w_1 \\ w_2 \end{pmatrix} \to \begin{pmatrix} 1 \\ 0 \end{pmatrix} \quad \text{as } x \to -\infty$$

and

$$\begin{pmatrix} w_1 \\ w_2 \end{pmatrix} \to \begin{pmatrix} a(\xi) \\ b(\xi, t) \end{pmatrix} \quad \text{as } x \to 0.$$

Here we have used (1.7.26) and (1.7.28) to show that w_1, w_2 are constant (to leading order) for $x > 0$.

The important difference between (1.7.34) and (1.7.11) is that as $t \to \infty$ with $X = O(1)$, (1.7.34) contains explicitly *two* rapid phases ($\theta + 2\xi x, \theta - 2\xi x$) and a slow phase, X. Taking account of these differences, the analysis now proceeds analogously to that leading to (1.7.20). We omit the details, which may be found in Segur and Ablowitz (1981). The result is (1.7.33) again, along with (for $\xi \geq 0$),

$$\bar{\theta}(X) = -\arg\left\{-\sigma b\left(\frac{\xi}{2}\right)\right\} - \frac{3\pi}{4} - \arg\left\{\Gamma\left(1 - \frac{i\sigma d_0^2}{2}\right)\right\}$$

(1.7.35)
$$-\frac{3}{2}\sigma d_0^2 \log 2 - \frac{\sigma}{2}\int_0^\xi \log\left(\frac{\xi-y}{\xi+y}\right)(d^2)_y\, dy$$

$$+\frac{\sigma}{2}\int_\xi^\infty \log\left(\frac{y-\xi}{y+\xi}\right)(d^2)_y\, dy.$$

In the limit $X = -x/3t \to 0$, but $Z = x/(3t)^{1/3} \to -\infty$, the approximate solution of mKdV given in (1.7.32, 33, 35) matches smoothly to that given in (1.7.28–30). This matching amounts to a partial confirmation of our analysis here, and it plays a more important role in § 3.7.

Let us summarize our results for (1.7.23). In the absence of solitons, the asymptotic solution of mKdV that evolves from appropriate initial data is given by (1.7.26) for $x \gg (3t)^{1/3}$, by (1.7.28, 29, 30) for $|x| \leq O((3t)^{1/3})$, and by (1.7.32, 33, 35) for $(-x) \gg (3t)^{1/3}$. If $r(0) \neq 0$, the solution remains nonlinear in the middle region, no matter how small the amplitude becomes. Even so, one may show that this asymptotic solution collapses to that of the linear problem if the initial amplitude is taken to zero (in L_1-norm). Thus, the two limits ($t \to \infty$, initial amplitude $\to 0$) commute for mKdV, as they did for (1.7.2). Note that this commutation of limits does not depend on our having excluded solitons; they are excluded by the small amplitude limit, according to (1.7.1).

1.7.c. The Korteweg–de Vries equation. Every solution of mKdV (1.7.23), $\sigma = +1$, generates a solution of the KdV equation

(1.7.36)
$$u_t + 6uu_x + u_{xxx} = 0,$$

through Miura's (1968) transformation

(1.7.37)
$$u = -v^2 - v_x.$$

In this way, our results for the asymptotic behavior of mKdV solutions determine the asymptotic behavior of an infinite set of KdV solutions as well. However, these KdV solutions turn out to be practically irrelevant in the following sense: given almost any initial data such that KdV can be solved by IST, the solution that evolves *cannot* be obtained from a rapidly decaying (in x) solution of mKdV through (1.7.37).

We may examine this transformation in more detail. Let $v(x, 0)$ be smooth, rapidly decaying (in x) initial data for (1.7.23), $\sigma = +1$. Let $r(k)$ denote the reflection coefficient corresponding to $v(x, 0)$. From (1.7.31), $|r(0)| < 1$. Now let $\rho(k) = r(k)$, where $\rho(k)$ is the reflection coefficient corresponding to some solution of the KdV equation. Then $r(k)$ generates an mKdV solution through (1.7.24), while $\rho(k)$ generates a KdV solution through (1.3.37). By comparing these two integral equations, it is easy to show that the two solutions are related by (1.7.37).

On the other hand, Ablowitz, Kruskal and Segur (1979) showed that for almost any smooth initial data for KdV satisfying $\int_{-\infty}^{\infty} (1+|x|^2)|u|\, dx < \infty$,

$$\rho(0) = -1. \tag{1.7.38}$$

Thus for almost all initial data for KdV, we *cannot* equate $\rho(k) = r(k)$, and we cannot use (1.7.37) effectively.

Hence, a separate analysis of the asymptotic behavior of KdV solutions is required. This analysis follows our previous work on mKdV, but some significant differences appear, due to (1.7.38). As always, we assume the initial data to be rapidly decaying as $|x| \to \infty$ and smooth, and we assume that no solitons exist. Here we only state the main results; more complete details may be found in Ablowitz and Segur (1977a).

For $x \gg (3t)^{1/3}$, the integral term in (1.3.37), as in mKdV, is exponentially smaller than the other two, and

$$u(x, t) \sim \frac{\rho(ik/2)k^{1/2}}{2(3\pi t)^{1/2}} \exp\{-2tk^3\}, \tag{1.7.39}$$

where $k^2 = x/3t$. An alternative representation in this region that remains valid as $x/3t \to 0$ is

$$u(x,t) \sim -(3t)^{-2/3} \left[\rho(0)\, \mathrm{Ai}'(Z) + \sum_{n=1}^{\infty} \frac{(-i)^n \rho^{(n)}(0)}{2^n n!\,(3t)^{n/3}} \left(\frac{d}{dZ}\right)^n \mathrm{Ai}(Z) \right], \tag{1.7.40}$$

where $Z = (x+x_0)/(3t)^{1/3}$ and x_0 is constant; $x_0 = -i\rho'(0)/2\rho(0)$ is an especially convenient choice.

In the region $|x| \le O((3t)^{1/3})$, we seek a solution of KdV in the form

$$u(x, t) \sim (3t)^{-2/3}[f(Z) + (3t)^{-1/3} f_1(Z) + (3t)^{-2/3} f_2(Z) + \cdots]. \tag{1.7.41}$$

(One of the few rigorous results in this section is the proof of Shabat (1973) that (1.7.41) is truly asymptotic to leading order.) These functions satisfy ordinary differential equations:

$$\begin{aligned} f''' + 6ff' - (2f + Zf') &= 0, \\ f_1''' + 6(ff_1)' - (3f_1 + Zf_1') &= 0, \\ f_2''' + 6(ff_2)' - (4f_2 + Zf_2') &= -3(f_1^2)', \end{aligned} \tag{1.7.42}$$

with boundary conditions obtained by matching to (1.7.40) as $Z \to +\infty$. For example

$$f(Z) \sim -\rho(0)\,\mathrm{Ai}'(Z) \quad \text{as } Z \to +\infty. \tag{1.7.43}$$

Now the significance of (1.7.38) appears. The behavior of the solution of (1.7.42a, 43) depends on $\rho(0)$:

(i) If $|\rho(0)|<1$, the solution is bounded for all finite Z, and oscillates as $Z \to -\infty$. These solutions of (1.7.42) are related to the bounded solutions of (1.7.29) through (1.7.37).

(ii) Solutions corresponding to $|\rho(0)|>1$ or to $\rho(0) = +1$ are of no interest.

(iii) If $\rho(0) = -1$, then as $Z \to -\infty$, $f(Z)$ asymptotically approaches

$$(1.7.44) \qquad f(Z) = \frac{Z}{2} - \frac{1}{2}(-2Z)^{-1/2} + \frac{1}{2}(-2Z)^{-2} + O((-2Z)^{-7/2}).$$

Thus the first term in (1.7.42) grows linearly as $Z \to -\infty$ if $\rho(0) = -1$. However, $f_2(Z)$ grows exponentially in the same limit and $f_4(Z)$ grows even faster; i.e., the expansion (1.7.41) becomes disordered. One may show that as $Z \to -\infty (\rho(0) = -1)$

$$(1.7.45) \qquad u(x,t) \sim \frac{-2Z}{(3t)^{2/3}}\left[-\frac{1}{4} + \left\{(3t)^{-2/3}C(-2Z)^{-5/4} \exp\left(\frac{(-2Z)^{3/2}}{3}\right)\right\} \right.$$
$$\left. -\left\{(3t)^{-2/3}C(-2Z)^{-5/4} \exp\left(\frac{(-2Z)^{3/2}}{3}\right)\right\} + \cdots\right],$$

where

$$C = 0.118\{\rho''(0) + [\rho'(0)]^2\},$$

the coefficient being determined by a numerical integration (Ablowitz and Segur (1977a), Miles (1979)). It is clear that (1.7.45) breaks down as $Z \to -\infty$. This breakdown signals the existence of a new region, which may be called a "collisionless shock" layer, or a transition layer. The existence of this new region is a direct consequence of (1.7.38); there is no corresponding region in the typical asymptotic solutions of mKdV.

The location and decay rate of the minimum of the leading wave (see Fig. 4.4) has some importance in applications. The location of this minimum may be found by differentiating the expression in (1.7.45). For very large times, it occurs at

$$(1.7.46a) \qquad (-2Z) \sim (2 \log 3t)^{2/3},$$

and near this (moving) point

$$(1.7.46b) \qquad u \sim -\frac{1}{2}\left(\frac{2 \log 3t}{3t}\right)^{2/3}.$$

This asymptotic decay rate is entirely independent of initial data, except through (1.7.38). However, the time required to reach (1.7.46) is long and this final state may not be attained in many applications.

A separate analysis is required to determine the structures of the collisionless shock layer. Details may be found in Ablowitz and Segur (1977a), Miles (1979). The appropriate scaling in the layer is implicit in the breakdown of (1.7.45). It is convenient to define a new spacelike variable ξ,

(1.7.47a) $$(3t)^{-2/3}(-2Z)^{-5/4} \exp\left(\frac{(-2Z)^{3/2}}{3}\right) = e^{2\xi/3},$$

and a dependent variable $g(\xi, t)$,

(1.7.47b) $$u(x, t) = \frac{-2Z}{(3t)^{2/3}} g(\xi, t).$$

One may show that, to leading order, the front of the collisionless shock is given by

(1.7.48) $$g(\xi, t) \sim -\tfrac{1}{4} + \tfrac{1}{2}\operatorname{sech}^2\{\tfrac{1}{3}(\xi - \xi_0)\}.$$

This matches to (1.7.45) if $\xi_0 = -\tfrac{3}{2}\log(C/2)$. However, it is not uniformly valid for ξ large $[(\log t)^2 \gg \xi \gg \log t$ or $t^{1/3}(\log t)^{4/3} \gg (-x) \gg t^{1/3}(\log t)^{2/3}]$, where it must be replaced by another slowly varying solution:

(1.7.49) $$g \sim a(Y) + b(Y)cn^2(\Phi + \Phi_0; \nu(Y)),$$

where Φ is an (appropriately defined) fast variable and Y is a slow variable. This matches to (1.7.48) as $Y \to 0$. As $Y \to \infty$, the solution of the KdV equation at the downstream end of this layer is given by

(1.7.50) $$u(x, t) \sim (3t)^{-2/3}(-Z)^{1/4}\left(\frac{2\log 3t}{3\pi}\right)^{1/2} \cos\theta,$$
$$\theta \sim \tfrac{2}{3}(-Z)^{3/2},$$

This is the behavior of the solution which must match into the solution in the oscillatory region.

Finally, we consider the purely oscillatory region. There are no major differences between the asymptotic analysis of KdV and mKdV solutions in this region, and we may quote the final results. The asymptotic form of the solution of (1.7.36) in this region is

(1.7.51) $$u(x, t) \sim (3t)^{-1/2} X^{1/4}(2d)\cos\theta - (3t)^{-1} X^{-1/2}(2d^2)(1 - \cos 2\theta),$$

where

$$\theta \sim 2tX^{3/2} - 2d^2 \log 3t - 3d^2 \log X + \bar{\theta},$$

$$X = -\frac{x}{3t}, \quad d = d(X), \quad \bar{\theta} = \bar{\theta}(X).$$

Note that every solution of this form has a nonoscillatory term that is negative and of order $(3t)^{-1}X^{-1/2}$. Thus for KdV, solitons are intrinsically positive, while the radiation has a negative mean. The relation between the $d(X)$, $\bar{\theta}(X)$ and the reflection coefficient, $\rho(k)$, is that on

$$X = -\frac{x}{3t} = 4k^2,$$

(1.7.52)
$$d^2(X) = -\frac{1}{4\pi}\log\{1 - |\rho(k)|^2\},$$

$$\bar{\theta}(X) = \frac{\pi}{4} - \arg\{\rho(k)\} - \arg\{\Gamma(1 - 2id^2)\}$$

$$- 6d^2 \log 2 - 4\int_0^{2k} \log\left(\frac{2k-y}{2k+y}\right)(d^2)_y \, dy.$$

If $|\rho(0)| < 1$, this solution matches as $X \to 0$ to the one in (1.7.41); the entire solution is qualitatively similar to an mKdV solution, and it may be obtained from one via (1.7.37). If $\rho(0) = -1$, the solution in this region matches as $X \to 0$ to the one in (1.7.50). A typical KdV solution is similar to that shown in Fig. 4.3, p. 285.

This completes the analysis of the solutions of KdV without solitons. Its general solution also may contain N solitons, but as $t \to \infty$ they are expected to be confined to N regions where the rest of the solution is exponentially small. Thus as $t \to \infty$, there is no interaction (to leading order) of solitons with the nonsoliton part of the solution; we may simply add the N solitons (all at $x > 0$) to the solution we have already discussed to obtain the asymptotic solution in the general case. This result must then be corrected by the (constant) phase shifts discussed in § 1.4. Ablowitz and Kodama (1980) have given the asymptotic solution of solitons and continuous spectrum. They use the notion of scattering off a given N-soliton solution.

Finally, let us compare the asymptotic solutions of the KdV equation (1.7.36) and the corresponding linear equation (A.1.49). In contrast to our results for both (1.7.2) and (1.7.23)—effectively, in contrast to the results for all solutions of equations of the form (1.5.16) for which $a(k)$ has no zeros for real k—the two limits ($t \to \infty$, initial amplitude $\to 0$) do *not* commute for KdV. Part of the problem is that KdV solitons can be made arbitrarily small (in any L_p-norm, $1 \le p \le \infty$), but the limits do not commute even if no solitons are present. In the linear problem, the asymptotic solution is dominated by the leading wave, whose decay rate is $t^{-1/3}$ if $\int u\,dx \ne 0$; otherwise the decay rate of this wave is $t^{-2/3}$ if $\int xu\,dx \ne 0$, etc. But for almost all KdV solutions, no matter how small, $\rho(0) = -1$, and the asymptotic decay rate of the leading wave is $(\log t/t)^{2/3}$, by (1.7.46). Thus the asymptotic behavior of the two solutions is always different in this region. Moreover, except in special cases where $\int u\,dx = 0$, the slowest decay rate is $t^{-1/3}$ for the solution of the linear problem, but $t^{-1/2}$ for the corresponding solution of KdV.

EXERCISES

Section 1.1

1. (a) Show that (1.1.2) is obtained from (1.1.1) for $h \to 0$.
   (b) What are the terms proportional to h^4; εh^2?
   (c) What happens when $\varepsilon/h^2 \ll 1$; $\varepsilon/h^2 \gg 1$?

2. Given $K(u)$ in (1.1.5) and $M(v)$ in (1.1.7), show that if $u = -(v^2 + v_x)$ we have

$$K(u) = -\left(2v + \frac{\partial}{\partial x}\right)M(v).$$

3. Given (1.1.11–13), verify that $\psi_{txx} = \psi_{xxt}$ yields (1.1.14), and hence $K(u) = 0$ if and only if $\lambda_t = 0$.

Section 1.2

1. Derive (1.2.13) from (1.2.8).

2. (a) Find an exact solution of (1.2.8) in terms of a fifth order expansion of (A, B, C); i.e., $A = \sum_0^5 a_n \zeta^n$, etc. What is the nonlinear evolution equation for $r = q$, with the arbitrary constants chosen so that the linearized form of the equation is $[\partial_t + (\partial_x)^5]v = 0$?
   (b) Find a solution of (1.2.21) in terms of a second order expansion of (A, B); i.e., $A = \sum_0^2 \alpha_n \lambda^n$. What is the nonlinear evolution equation that linearizes to $[\partial_t + (\partial_x)^5]u = 0$?
   (c) Show that these are related by Miura's transformation,

$$u = -v^2 - v_x.$$

Section 1.3

1. Show from (1.3.3, 4) that $\bar{\psi} = \bar{a}\phi + b\bar{\phi}$, $\psi = \bar{b}\phi - a\bar{\phi}$.

2. Prove that if r, q in (1.2.7a) each satisfy $|r(x)| < Ce^{-2K|x|}$ for some $C, K > 0$, then $a(\zeta)$ is analytic for Im $(\zeta) > -K$, $b(\zeta)$, $\bar{b}(\zeta)$ are analytic for $|\text{Im}(\zeta)| < K$, and $\bar{a}$ is analytic for Im $(\zeta) < K$.

3. (a) Why does $r = +q^* \in L_1$ guarantee that (1.2.7a) has no discrete eigenvalues for Im $(\zeta) > 0$?
   (b) Show from (1.3.4, 14) that for $r = +q^*$, $|a|^2 \geq 1$ for Im $(\zeta) = 0$. Conclude that $a(\zeta) \neq 0$ for Im $(\zeta) \geq 0$ if $r = +q^*$.

4. For $r = -q^* \in L_1$ in (1.2.7a), (1.3.16e) was shown to be sufficient to guarantee that $a(\zeta) \neq 0$ for Im $(\zeta) \geq 0$. Let $r = -q$, real. Solve (1.2.7a) explicitly at $\zeta = 0$. Show that $a(0) = \cos(\int_{-\infty}^{\infty} q\, dx)$. Therefore it is necessary that $Q_0(\infty) < \pi/2$ to exclude discrete eigenvalues. (Satsuma and Yajima (1974) showed that for a family of real potentials with $r = -q$, one discrete eigenvalue appears when $Q_0(\infty) = \pi/2$, a second eigenvalue appears when $Q_0(\infty) = 3\pi/2$, etc.)

5. Let $r = -q^*$ in (1.2.7a), where $q(x) = Q$ constant if $0 < x < L$, and $q(x) = 0$ otherwise. Find the scattering data $\{a(\zeta), b(\zeta)\}$ explicitly. Discuss the number and position of the eigenvalues as functions of Q and L. How does $\arg(Q)$ affect the eigenvalues? Show that $\text{Im}(\zeta) < Q$ for each discrete eigenvalue.

6. Prove that $a(k)$, defined in (1.3.35), has an analytic continuation into the upper half plane.

7. Show that if $q(x) \leq 0$ in (1.2.20a), there are no discrete eigenvalues (with $\lambda < 0$). (Hint: use an oscillation theorem.)

8. (a) In (1.2.20a) let $q(x) = Q$ (real) if $0 < x < L$ and $q(x) = 0$ otherwise. Find the scattering data explicitly. Discuss the number and magnitude of the discrete eigenvalues as functions of Q and L. Show that $|\lambda| < |Q|$ for each discrete eigenvalue.

(b) Repeat for $q(x) = Q \, \text{sech}^2 mx$, in (1.2.20a), with (Q, m) real. (Hint: let $t = \tanh mx$, and use associated Legendre functions.)

(c) In (a) and (b), compute $\rho(k = 0)$ explicitly, and show that $\rho(0) = -1$ for almost all potentials in these families. What is the significance of a potential for which $\rho(0) \neq -1$?

(Other bounds on the location and number of the discrete eigenvalues may be found in Segur (1973) for (1.2.20a), and in Ablowitz, Kaup, Newell and Segur (1974), Satsuma and Yajima (1974) and Karney, Sen and Chu (1979) for (1.2.7a).

Section 1.4

1. (a) Find the one-soliton, two-soliton, and breather solutions for the mKdV equation (1.2.2) by solving the linear integral equation (1.3.29) with the appropriate scattering data.

(b) There are actually two forms of mKdV that have solitons,

$$q_t + 6q^2 q_x + q_{xxx} = 0, \quad q \text{ real},$$
$$q_t + 6|q|^2 q_x + q_{xxx} = 0.$$

How do these special solutions compare for these two problems?

2. (a) Pick one of the coupled pairs of evolution equations (for r, q) solved by (1.2.7) and let r, q be unrelated initially; i.e., let (a, b) and $(\bar{a}, \bar{b})$ be unrelated initially. Show that the one-soliton solution blows up for infinitely many real points (x, t), unless $r = \alpha q^*$; α constant. Show that this blowup may evolve from infinitely smooth initial data.

(b) Show that $\int rq \, dx$ is conserved by the evolution equations. If $r = \alpha q^*$, α real, show that the solution remains in L_2 if it started there. Show that no L_2-function can have the kind of singularity given in (a).

3. For (1.2.20a) let $q(x) = N(N+1)m^2 \, \text{sech}^2 mx$ (as in Exercise 8, § 1.3).
(a) For $N = 2$, find explicitly the discrete eigenfunctions ψ_1, ψ_2. Find

c_i, d_i ($i = 1, 2$) from these, and compute the phase shift of two KdV solitons from (1.4.5). Verify (1.4.44) in this special case.

(b) Repeat for $N = 3$.

(c) How do these phase shifts change if the evolution equation were not KdV, but fifth order KdV? What if the evolution equation were first order KdV (i.e., $q_t + c q_x = 0$)?

Section 1.5

1. Show that if $A_-(\zeta) = (1/2i)(2\zeta)^3$, then (1.5.16) reduces to a coupled pair of mKdV equations.

2. Show if $A_-(\zeta) = (1/2i)(2\zeta)^5$, (1.5.16) reduces to a coupled pair of fifth order equations. Show if $\gamma(k^2) = -(2k^4)$, (1.5.21) reduces to another fifth order equation. If $r = q$, real, show that the solutions of these two equations are related by Miura's transformation,

$$u = -v^2 - v_x.$$

3. Is (1.5.16) implied by (1.5.5) or merely consistent with it?

4. What is the general evolution equation, corresponding to (1.5.16), for the scattering problem in (1.2.25)?

Section 1.6

1. Show that the mapping $(q, r) \to S$, defined in (1.6.42), is a canonical transformation for any of the Hamiltonian systems described by (1.5.16). (Most of this analysis follows that of Zakharov and Manakov (1974)).

(a) As done for $a(\xi)$ in (1.6.39), compute the gradients of $\bar{a}, b, \bar{b}$ with respect to (q, r).

(b) Let $u^{(1)} = (u_1^{(1)}, u_2^{(1)})$ and $w^{(1)}$ be two-soliton solutions of (1.2.7a) for $\xi = \xi_1$, and let $u^{(2)}, w^{(2)}$ be two solutions for $\xi = \xi_2$. Establish the identity

$$2i(\xi_2 - \xi_1)[u_1^{(1)} u_2^{(2)} w_1^{(1)} w_2^{(2)} - u_2^{(1)} u_1^{(2)} w_2^{(1)} w_1^{(2)}]$$
$$= \partial_x[(u_1^{(1)} u_2^{(2)} - u_2^{(1)} u_1^{(2)})(w_1^{(1)} w_2^{(2)} - w_2^{(1)} w_1^{(2)})].$$

(c) Because of this identity, the integrals in all of the Poisson brackets involving $(a, \bar{a}, b, \bar{b})$ can be evaluated exactly. Show that

$$\langle a(\xi_1), b(\xi_2) \rangle = -\frac{1}{2i(\xi_1 - \xi_2)} [a(\xi_1) b(\xi_2) - b(\xi_1) a(\xi_2) \lim_{x \to +\infty} \exp\{2i(\xi_1 - \xi_2)x\}],$$

Find the Poisson brackets of all other combinations of $a, \bar{a}, b, \bar{b}$.

(d) The identity

$$\lim_{x \to \infty} \frac{e^{ikx}}{k} = \pi i \delta(k),$$

where the left side should be intepreted in the principal value sense, can be established by evaluating $\lim_{x \to \infty} \int_{-\infty}^{\infty} dk \int_{-x}^{x} dy \, \phi(y) \cos ky$, where $\phi(y)$ is a test function. Use the identity to show that

$$\langle \log a(\xi_1), \log b(\xi_2) \rangle = -\frac{1}{2i(\xi_1 - \xi_2)} + \frac{\pi}{2} \delta(\xi_1 - \xi_2),$$

$$\langle \log \bar{a}(\xi_1), \log b(\xi_2) \rangle = \frac{1}{2i(\xi_1 - \xi_2)} + \frac{\pi}{2} \delta(\xi_1 - \xi_2),$$

$$\langle a(\xi_1), \bar{a}(\xi_2) \rangle = 0.$$

Using (1.6.43a), show that

$$\langle P(\xi_1), P(\xi_2) \rangle = 0 = \langle Q(\xi_1), Q(\xi_2) \rangle,$$

$$\langle P(\xi_1), Q(\xi_2) \rangle = \delta(\xi_1 - \xi_2).$$

(e) Show that $\langle \log b, \log \bar{b} \rangle \neq 0$. (This can be done easily, using the Wronskian relation.)

(f) By extending $b, \bar{b}$ off the real axis, show that several other Poisson brackets vanish.

(g) $\delta \zeta_m / \delta q(x)$ follows almost directly from the observation that $a = a(\zeta, q, r)$, so that

$$\delta a \equiv \frac{\partial a}{\partial \zeta} \delta \zeta + \frac{\partial a}{\partial q} \delta q + \frac{\partial a}{\partial r} \delta r.$$

By definition, $\delta \zeta_n / \delta q$ comes by requiring $\delta a = 0$, $\delta r = 0$ and extending (1.6.39) off the real axis;

$$\left. \frac{\partial a}{\partial \zeta} \right|_{\zeta_m} \frac{\delta \zeta_m}{\delta q(x)} = -\phi_2(x, \zeta_m) \psi_2(x, \zeta_m).$$

Compute $\delta \zeta_m / \delta r$, $\delta \bar{\zeta}_l / \delta q$, $\delta \bar{\zeta}_l / \delta r$. Use (1.6.43) and show that

$$\langle P_m, Q_n \rangle = \delta_{mn}, \qquad \langle \bar{P}_k, \bar{Q}_l \rangle = \delta_{kl},$$

and that all other Poisson brackets of the variables in (1.6.43) vanish.

2. Derivation of (1.6.46) requires $\delta H / \delta \zeta_m$, $\delta H / \delta \bar{\zeta}_l$.

(a) From (1.6.19), show that for Im $\zeta > 0$

$$\frac{\partial (\log a)}{\partial \zeta_m} = -\frac{1}{\zeta - \zeta_m} + \frac{1}{2\pi i} \int_{-\infty}^{\infty} \frac{d\xi}{\xi - \zeta} \frac{1}{\xi - \zeta_m} = -\frac{1}{\zeta - \zeta_m},$$

$$\frac{\partial (\log a)}{\partial \bar{\zeta}_l} = \frac{1}{\zeta - \bar{\zeta}_l}.$$

(b) Expand these as $|\zeta| \to \infty$, Im $\zeta > 0$, and use (1.6.6) to show

$$\frac{\delta C_n}{\delta \zeta_m} = -2i(2i\zeta_m)^n, \qquad \frac{\delta C_n}{\delta \bar{\zeta}_l} = 2i(2i\bar{\zeta}_l)^n.$$

(c) Use (1.6.38) to show that

$$\frac{\delta H}{\delta \zeta_m} = 4iA_-(\zeta_m), \qquad \frac{\delta H}{\delta \bar{\zeta}_l} = -4iA_-(\bar{\zeta}_l).$$

3. (a) What conditions must q, r satisfy in order that $a, b, \bar{a}, \bar{b}$ may be extended off the real axis as we have done in this section?
  (b) What conditions must they satisfy in order that

$$\int_{-\infty}^{\infty} \xi^n \log \alpha \bar{\alpha}(\xi) \, d\xi$$

exist for all $n \geq 0$, so that the integral in (1.6.21) exists?
  (c) If $A_-(\zeta)$ is polynomial, what conditions must q, r satisfy in order that the integral in (1.6.44) exists?
  (d) Under what conditions may we use the polynomial conserved densities (only) as the action variables? What are the corresponding angle variables?

4. The sine-Gordon equation, (1.2.3), has the linearized dispersion relation $\omega(k) = k^{-1}$, and does not satisfy (1.6.37). Even so, the results of this section apply, provided $a\bar{a}(0) = 1$.
  (a) Show that (1.2.3) is Hamiltonian in the form (1.6.31) with

$$H = \int_{-\infty}^{\infty} \left\{ \cos\left(\int_{-\infty}^{x} v(y) \, dy \right) - 1 \right\} dx.$$

  (b) Show that the transformation to the variables defined in (1.6.43) is canonical.
  (c) Write the Hamiltonian in terms of these variables (McLaughlin (1975)).

5. When the dispersions relation for (1.5.16) is an odd function of ζ, the equations admit real solutions with $r = \pm q$. As an alternative to the methods developed here, impose this restriction on (1.5.16) a priori so that r and q do not vary independently, and develop a parallel theory for these restricted equations.
  (a) Show that (1.5.16) now takes the form (1.6.31).
  (b) Show that the transformation to a subset of the scattering data is canonical.
  (c) Show that these are action-angle variables.

6. Here are two interesting problems that fall somewhat outside of the formulation of completely integrable Hamiltonian systems that we have presented here.
  (a) The problem of self-induced transparency (SIT) is discussed in §§ A.2 and 4.4. As noted there, $a(\zeta)$ is *not* time-independent, although the locations of its zeros are. There is only one polynomial conserved density. Nevertheless, the problem is solved by IST. This suggests either that IST is not restricted to

completely integrable Hamiltonian systems, or that the latter needs a broader definition.

(b) The long-wave equations of Benney (1973) are

$$u_t + uu_x - u_x\left(\int_0^y u_x\,dy\right) + h_x = 0,$$

$$h_t + \left(\int_0^h u\,dy\right)_x = 0.$$

Here $u = u(x, y, t)$, $h = h(x, t)$, with $-\infty < x < \infty$, $0 < y < h$, $t \geq 0$. These equations have an infinite set of conservation laws involving polynomials of $\int_0^h u^n\,dy$. Further, there is a sense in which this sytem is Hamiltonian and the conservation laws are in involution. Nevertheless, Kuperschmidt and Manin have shown that:

(i) there are no more local (in x, t) conservation laws involving only the moments of u beyond those found by Benney;

(ii) there are not enough of these polynomial conserved densities to establish complete integrability;

(iii) there seem to be no solitons.

For more details, the reader may consult Benney (1973), Miura (1974b), Kuperschmidt and Manin (1977), (1978), Manin (1981) and Zakharov (1981).

Section 1.7

1. Linearization.

(a) Show that in the limit of $q(x)$, $r(x)$ small,

$$b(\xi) \to \hat{r}(2\xi), \qquad \bar{b}(\xi) \to -\hat{q}(-2\xi),$$

where $\hat{r}(\xi)$ is the Fourier transform of $r(x)$ and $b, \bar{b}$ are defined by (1.3.3). Use the L_1-norm, which is a natural measure of "smallness" in this problem.

(b) Show that the asymptotic solution of the nonlinear Schrödinger equation collapses to that of the linear Schrödinger equation in this limit; i.e., that (1.7.4), with (1.7.20), →(A.1.39). What must be small for this limit to be valid? What is the next term in the expansion? Does the general solution of (1.7.2) collapse to the general solution of (A.1.23) in the limit of small amplitudes (independent of time)?

(c) Treat the nonlinear term in (1.7.2) as a small perturbation and solve (1.7.2) by perturbing about the linear solution. Show that secular terms appear in this expansion. How do you reconcile this result with that in (b)?

2. What is the asymptotic solution of (1.7.2) if a zero of $a(\xi)$ is allowed on the real axis? (Ablowitz and Segur (1977a) conjectured that the decay rate of the solution should be $(\log t/t)^{1/2}$ in this case.) Do the same for (1.7.23).

90   CHAPTER 1

3. Derive (1.7.33), (1.7.35).

4. (a) Let

$$q(x) = \begin{cases} Q, & 0 < x < L, \\ 0, & \text{otherwise}, \end{cases}$$

be initial data for (1.7.2). Find $a(\xi)$, $b(\xi)$ explicitly for $\sigma = +1$ and $\sigma = -1$. What inequality must Q, L satisfy to exclude solitons?

(b) Find the asymptotic solution of (1.7.2), both for $\sigma = +1$ and $\sigma = -1$, that evolves from these initial conditions when no solitons exist. Sketch the envelopes of these solutions based only on the dominant terms in the asymptotic expansions. (You should find well-defined wave-packets, or groups.)

(c) By computing the second terms in the asymptotic expansions and comparing them to the first, estimate the time required for the asymptotic solutions that you sketched in (b) to become valid.

(d) Manakov (1974a) noted the following application of (1.7.2). Let a steady, uniform plane-wave of intense monochromatic light (with intensity $|Q|^2$) shine on a very long transparent slit (of width L) in an otherwise opaque wall. If x measures the distance parallel to the wall, and t the distance normal to it, then the solution of (1.7.2) represents the complex amplitude of the steadily diffracted wave beyond the slit. The nonlinear term in (1.7.2) represents the change in the index of refraction of the medium caused by the intense light beam. The solution that you sketched in (b) corresponds to the diffraction pattern one would observe by placing a screen normal to the beam well beyond the slit. The time estimates in (c) determine how far behind the slit the screen should be placed. In this application, a soliton corresponds to a diffraction fringe whose intensity does not diminish as the screen is moved away from the slit.

5. Repeat problem 4 for (1.7.23) instead of (1.7.2).

6. Find the asymptotic solution of the sine-Gordon equation, $\phi_{xt} = \sin \phi$, in the absence of solitons.

(a) Find the similarity solution. Show that inside the light cone, the equation has a slowly-varying similarity solution of the form

$$\phi \sim -2t^{-1/2} X^{-1/4} d \sin \theta,$$

$$\theta \sim 2tX^{1/2} - \frac{1}{2} d^2 \log t - \frac{d^2}{2} \log X + \bar{\theta},$$

where $X = -x/t$, $d = d(X)$, $\bar{\theta} = \bar{\theta}(X)$.

(b) Show that along $-x/t = \zeta^{-2}$

$$d^2(X) = \frac{1}{\pi}\log\left\{1 + \left|r\left(\frac{\zeta}{2}\right)\right|^2\right\},$$

$$\bar{\theta}(X) = \frac{\pi}{4} - \arg\left\{r\left(\frac{\zeta}{2}\right)\right\} - \arg\left\{\Gamma\left(1 - \frac{id^2}{2}\right)\right\} - \frac{3}{2}d^2\log 2$$

$$- \int_0^{\zeta^{-1}} \log\left(\frac{\zeta^{-1} - y}{\zeta^{-1} + y}\right)(d^2)_y\, dy.$$

(c) Express your final results in terms of laboratory coordinates, $\chi = (x+t)$, $\tau = (x-t)$.

(d) What happens near the light-cone? What happens outside the light-cone?

7. Derive (1.7.52).

8. (a) Show that for large k (i.e., $\lambda \gg |q|$ in (1.3.33))

$$\rho\left(\frac{k}{2}\right) \sim \frac{i\hat{q}(k)}{k},$$

where $\hat{q}$ is the Fourier transform of $q(x)$ and $\rho(k)$ is the reflection coefficient defined by (1.3.33–35).

(b) By comparing the asymptotic solutions of KdV and its linear counterpart (A.1.49), show that the limits ($t \to 0$; initial amplitude $\to 0$) *do* commute for KdV solutions without solitons *except* in the vicinity of the collisionless shock layer. In other words, this layer is intrinsically nonlinear; away from it, KdV solutions without solitons are only weakly nonlinear.

Chapter 2

IST *in Other Settings*

2.1. Higher order eigenvalue problems and multidimensional scattering problems. Up until this point we have only considered those nonlinear evolution equations associated with second order eigenvalue problems. The work of Zakharov and Manakov (1973) effectively showed that there are physically interesting nonlinear evolution equations connected with higher order scattering problems. Specifically, they demonstrated that the well-known three-wave interaction equations were associated with a certain *third* order scattering problem. Somewhat later Kaup (1976b) and Zakharov and Manakov (1976a) examined the related questions of inverse scattering and solutions to the equations of motion.

In this section we shall first show how the ideas of § 1.2 can be simply extended to higher order problems, and then we shall discuss some of the relevant details associated with actually working out the inverse scattering.

2.1.a. Deriving one-dimensional evolution equations. We begin with the matrix formulation (Ablowitz and Haberman (1975b)):

(2.1.1a) $\qquad \mathbf{v}_x = i\zeta D \mathbf{v} + N \mathbf{v},$

(2.1.1b) $\qquad \mathbf{v}_t = Q \mathbf{v},$

where $\mathbf{v}$ is an $n \times 1$ matrix (vector)

$$\mathbf{v} = \begin{Bmatrix} v_1 \\ \vdots \\ v_n \end{Bmatrix},$$

and D, N, Q are $n \times n$ matrices with D diagonal, $D = d_i \delta_{ij}(\text{diag}(d_1, d_2, \cdots, d_n))$, d_i constant and N is such that $N_{ii} = 0$ (this latter assumption is not really necessary, but it simplifies the analysis). Cross differentiation such that $\mathbf{v}_{xt} = \mathbf{v}_{tx}$, and requiring $\zeta_t = 0$ yields

(2.1.2) $\qquad Q_x = N_t + i\zeta[D, Q] + [N, Q]$

$([A, B] \equiv AB - BA)$. As in § 1.2 we wish to find Q given D, N such that (2.1.2) is satisfied and so that (2.1.1a, b) are consistent. In general this requires further restrictions, which are the nonlinear evolution equations. Expanding Q in a power series in ζ is the easiest way to proceed. As an example we shall derive the three-wave interaction equation. It turns out that the three-wave interaction is virtually the easiest system to obtain. We expand Q as follows:

(2.1.3) $$Q = Q^{(1)}\zeta + Q^{(0)}$$

(since the three-wave interaction has first order spatial derivatives). The results of § 1.2, § 1.5 motivate the form (2.1.3). Substitution of (2.1.3) into (2.1.2) yields at order ζ^2:

(2.1.4) $$i[D, Q^{(1)}] = i \sum_{k=1}^{n} (D_{lk}Q^{(1)}_{kj} - Q^{(1)}_{lk}D_{kj}) = 0.$$

Using $D_{ik} = \delta_{ik}d_i$, we have $(d_i - d_j)Q^{(1)}_{ij} = 0$. We shall assume

$$d_1 > d_2 > d_3,$$

and hence

(2.1.5) $$Q^{(1)}_{lj} \equiv q_l\delta_{lj}.$$

We take the q_i each to be constant (it can be shown that when q_i is a function of time the resulting equations are transformable to those when q_i is constant). At order ζ we have

(2.1.6a) $$Q^{(1)}_x = i[D, Q^{(0)}] + [N, Q^{(1)}];$$

or

(2.1.6b) $$Q^{(1)}_{lj,x} = i \sum_k (D_{lk}Q^{(0)}_{kj} - Q^{(0)}_{lk}D_{kj}) + \sum_k (N_{lk}Q^{(1)}_{kj} - Q^{(1)}_{lk}N_{kj}).$$

Substitution of (2.1.5) into (2.1.6b) yields (note that we are not using summation convention on repeated indices)

(2.1.7) $$Q^{(0)}_{lj} = \frac{q_l - q_j}{i(d_l - d_j)}N_{lj}, \quad l \neq j.$$

When $l = j$ we take $Q^{(0)}_{lj} = 0$. Defining

(2.1.8) $$a_{lj} = \frac{1}{i}\frac{q_l - q_j}{d_l - d_j} = a_{jl}$$

reduces (2.1.7) to

(2.1.9) $$Q^{(0)}_{lj} = a_{lj}N_{lj}, \quad l \neq j.$$

At ζ^0 we have $Q^{(0)}_x = N_t + [N, Q^{(0)}]$, from which we obtain

(2.1.10) $$N_{lj,t} - a_{lj}N_{lj,x} = \sum_k (a_{lk} - a_{kj})N_{lk}N_{kj}.$$

These are $N(N-1)$ equations (we note that $N_{ll} = 0$ is consistent with (2.1.10)). The number of equations can be halved by requiring $N_{lj} = \sigma_{lj}N^*_{jl}$. (This is analogous to the choice $q = \pm r^*$ in § 1.2—the second order case.) Then (2.1.10) and its complex conjugate are mutually consistent if

(2.1.11) $$\sigma_{lk}\sigma_{kj} = -\sigma_{lj}, \quad l > k > j,$$

and the a_{lj} are real.

Equation (2.1.10) may be put into a standard set of three-wave interaction equations by a suitable scaling of variables. For example, we find the system

(2.1.12)
$$Q_{1t} + C_1 Q_{1x} = i\gamma_1 Q_2^* Q_3^*,$$
$$Q_{2t} + C_2 Q_{2x} = i\gamma_2 Q_1^* Q_3^*,$$
$$Q_{3t} + C_3 Q_{3x} = i\gamma_3 Q_1^* Q_2^*,$$

where $\gamma_1\gamma_2\gamma_3 = -1$ and $\gamma_i = \pm 1$ if we take

$$N_{12} = -iQ_3/\sqrt{\beta_{13}\beta_{23}}, \quad N_{31} = -iQ_2/\sqrt{\beta_{12}\beta_{23}}, \quad N_{23} = iQ_1/\sqrt{\beta_{12}\beta_{13}},$$
$$N_{13} = -\gamma_1\gamma_3 N_{31}^*, \quad N_{32} = \gamma_3\gamma_2 N_{23}^*, \quad N_{21} = \gamma_1\gamma_2 N_{12}^*,$$

where

$$q_j = -i\frac{C_1 C_2 C_3}{C_j}, \quad \beta_{lj} = d_l - d_j = C_j - C_l,$$
$$C_3 > C_2 > C_1.$$

In (2.1.12) decay instability (positive definite energy) occurs when we choose one of the γ_n's different in sign from the others, and explosive instability when $\gamma_1 = \gamma_2 = \gamma_3 = -1$. Directly from the equations we can derive the conserved quantities

(2.1.13)
$$\gamma_1 M_1 - \gamma_2 M_2 = \text{const.},$$
$$\gamma_2 M_2 - \gamma_3 M_3 = \text{const.},$$
$$\gamma_1 M_1 - \gamma_3 M_3 = \text{const.},$$

where $M_n = \int_{-\infty}^{\infty} Q_n Q_n^* \, dx$, and see that there is no positive definite energy in the case $\gamma_i = -1, i = 1, 2, 3$.

We also remark that the so-called two-wave interaction case is obtained from (2.1.12) by taking $C_3 = C_2$, $Q_3 = Q_2$. From the standpoint of our derivation this is a singular limit and the eigenvalue problem (2.1.1) does not seem to apply. Kaup (1978) discussed the two-wave case in some detail.

We shall return to the three-wave problem in more detail later in this section. However, we first discuss the situation that occurs when we expand Q differently.

Specifically, if we had taken $Q = Q^{(2)}\zeta^2 + Q^{(1)}\zeta + Q^{(0)}$ and followed the same procedure as earlier, we would find

$$\beta_{lj}N_{lj,xx} + \varepsilon_{lj}N_{lj,x} - \sum_{k \neq l,j} \gamma_{ljk}(N_{lk}N_{kj})_x$$

$$= N_{lj,t} + \sum_{k \neq l,j}(\varepsilon_{kj} - \varepsilon_{lk})N_{lk}N_{kj}$$

(2.1.14)
$$+ N_{lj}\{2\beta_{lj}N_{lj}N_{jl} + \sum_{k \neq l,j}(\beta_{kj} + \gamma_{lkj})N_{jk}N_{kj} - (\beta_{kl} + \gamma_{kjl})N_{lk}N_{kl}\}$$

$$+ \sum_{k \neq l,j}(\beta_{kj}N_{lk}N_{kj,x} - \beta_{lk}N_{kj}N_{lk,x})$$

$$+ \sum_{k \neq l,j}\sum_{m \neq l,j}(\gamma_{lkm}N_{kj}N_{lm}N_{mk} - \gamma_{kjm}N_{lk}N_{km}N_{mj}),$$

where

$$a_{lj} = \frac{q_l^{(2)} - q_j^{(2)}}{i(d_l - d_j)} = a_{jl}, \qquad \beta_{lj} = \frac{a_{lj}}{i(d_l - d_j)} = -\beta_{lj},$$

$$\gamma_{ljk} = \frac{a_{kj} - a_{lk}}{i(d_l - d_j)} = \gamma_{jlk} = \gamma_{klj},$$

$$\varepsilon_{lj} = \frac{q_l^{(1)} - q_j^{(1)}}{i(d_l - d_j)} = \varepsilon_{jl},$$

and $q_l^{(2)}$, $q_l^{(1)}$, are arbitrary constants. Schematically these terms can be interpreted as follows:

$$\frac{\partial A}{\partial t} = \underbrace{\frac{\partial A}{\partial x}}_{\text{group velocity}} + \underbrace{i\frac{\partial^2 A}{\partial x^2}}_{\text{dispersion}} + \underbrace{BC + \frac{\partial}{\partial x}(BC)}_{\text{triad resonance}}$$

$$+ \underbrace{A^2 A^*}_{\substack{\text{self-self}\\\text{interaction}}} + \underbrace{ABB^*}_{\substack{\text{self-modal}\\\text{interaction}}} + \underbrace{BDE}_{\substack{\text{quartic}\\\text{resonance}}}.$$

It should be noted that all of these terms arise in systematic perturbation expansions in which each derivative scales like an amplitude (note that BC would have a different asymptotic order from $\partial_x(BC)$). For use in a particular physical problem, one must verify that each coefficient properly reduces to the equation of interest. For example, if the a_{lj}, β_{lj}, etc. are chosen appropriately, then (2.1.14) can be reduced to a number of physically interesting cases:

(i) Manakov (1975). Coupled nonlinear Schrödinger equations:

(2.1.15)
$$N = \begin{pmatrix} 0 & A_1 & A_2 \\ \sigma_{21}A_1^* & 0 & 0 \\ \sigma_{31}A_2^* & 0 & 0 \end{pmatrix},$$

$$iA_{1t} = A_{1xx} + 2A_1(\sigma_{21}|A_1|^2 + \sigma_{31}|A_2|^2),$$
$$iA_{2t} = A_{2xx} + 2A_2(\sigma_{21}|A_1|^2 + \sigma_{31}|A_2|^2).$$

(ii) Yajima and Oikawa (1976). Interaction of Langmuir waves with ion-acoustic waves in a plasma:

(2.1.16)
$$N = \begin{pmatrix} 0 & Ee^{-ix} & in \\ 0 & 0 & Ee^{ix} \\ -i & 0 & 0 \end{pmatrix},$$

$$iE_t + \tfrac{1}{2}E_{xx} + \tfrac{1}{2}(1-n)E = 0,$$
$$n_t + n_x + (|E|^2)_x = 0$$

(see also Yajima and Oikawa (1975)). A similar system describing an interaction of short capillary waves with long gravity waves had been studied by Djordjevic and Redekopp (1977).

(iii) Zakharov (1974), Ablowitz and Haberman (1975b). A Boussinesq equation:

$$N = \begin{pmatrix} 0 & 0 & 1 \\ N_{21} & 0 & (1+\omega_3)N_{31} \\ N_{31} & 1 & 0 \end{pmatrix},$$

where, if

$$\omega_3 = e^{-2\pi i/3}, \quad N_{31} = \mu\phi_x + \nu, \quad N_{21} = \mu\left(\frac{\omega_3}{2\beta}\phi_t + \frac{\omega_3}{2}\phi_{xx}\right),$$

$$\mu = \omega_3 - 1, \quad \nu = \frac{\omega_3 - 1}{12}, \quad w = \phi_x,$$

then

(2.1.17)
$$W_{tt} - \beta^2(W_{xx} + 6(W^2)_{xx} + W_{xxxx}) = 0.$$

We note that (2.1.17) is ill-posed when $\beta^2 = 1$ even though it is frequently derived in physical situations. However, in these applications (2.1.17) arises as a long-wave limit, and hence short waves which give rise to the ill-posedness are prohibited by the asymptotic derivation.

Moreover, in this case the eigenvalue problem (2.1.1a) is reducible to a single third order equation

$$\psi_{xxx} + (\lambda + M_1)\psi + M_2\psi_x = 0, \tag{2.1.18}$$

where

$$M_1 = N_{31,x} + N_{21}, \qquad M_2 = (2 + \omega_3)N_{31}$$

(analogous to setting $r = -1$ in the $n = 2$ case). Zakharov (1974) starts with an equation of the form (2.1.18). The natural time dependence to associate with (2.1.18) is

$$\Psi_t = A\Psi + B\Psi_x + C\Psi_{xx};$$

expand A, B, C in powers of λ.

It should be noted that an interesting generalization of the above ideas follows if we take, formally, $n \to \infty$ in (2.1.1). In this case the eigenvalue problem and associated time dependence are given by

$$\frac{\partial v}{\partial x}(x, y; t) = i\zeta d(y)v(x, y; t) + \int_{-\infty}^{\infty} N(x, y, z; t)v(x, z; t)\, dz, \tag{2.1.19a}$$

$$\frac{\partial v}{\partial t}(x, y; t) = \int_{-\infty}^{\infty} Q(x, y, z; t)v(x, z; t)\, dz. \tag{2.1.19b}$$

The procedure described here then yields the integrodifferential equation

$$N_t(x, y, z; t) = \alpha(y, z)N_x(x, y, z; t)$$

$$+ \int_{-\infty}^{\infty} (\alpha(y, z') - \alpha(z', z))N(x, y, z'; t)N(x, z', z; t)\, dz', \tag{2.1.19c}$$

where

$$\alpha(y, z) = \alpha(z, y).$$

The symmetry condition $N(x, y, z; t) = \sigma(y, z)N^*(x, z, y; t)$ for $y > z$ is consistent if σ satisfies $\sigma(y, z')\sigma(z', z) = -\sigma(y, z)$ for $y > z' > z$.

2.1.b. Scattering theory. We now discuss the analytical inverse scattering associated with the 3×3 matrix eigenvalue problem, (2.1.1). In addition, we shall examine the three-wave interaction both as an application of some of these ideas and because of its physical significance. We will follow closely the work of Kaup (1976b) (some of these results can also be found in Zakharov and Manakov (1976a)). At this time the complete inverse scattering analysis associated with eigenvalue problems higher than third order is still open. However, it has been shown (see for example, Zakharov and Shabat (1974), Cornille (1976a,b)) that if there is an $n \times n$ Gel'fand–Levitan–Marchenko formulation then it is consistent with the $n \times n$ operator; but it is not clear what restrictions on the initial data are required for these formulations. Similarly we mention work by Shabat (1975) who examined some associated questions about the $n \times n$ scattering problem, and recent work by Zakharov and Shabat (1979) and Zakharov and Mikhailov (1978a,b) on a certain Riemann–Hilbert formulation.

Hence, let us consider (2.1.1a) on the interval $|x| < \infty$, where ζ is the eigenvalue, **v** is a 3×1 column matrix (vector), N is a 3×3 "potential" matrix with zero diagonal elements ($N_{ii} = 0$) and $D = \text{diag}(d_1, d_2, d_3)$.

Provided $N_{ij} \to 0$ sufficiently rapidly as $|x| \to \infty$, three linearly independent eigenstates for real ζ can be defined at each end; i.e., we define $\phi_n^{(j)}$ (where $j = 1, 2$, or 3 designates the jth eigenfunction, $n = 1, 2$, or 3 designates the nth component of $\phi^{(j)}$, by the boundary condition

(2.1.20a) $\qquad \phi_n^{(j)} \sim \delta_{n,j} e^{i\zeta d_j x} \quad \text{as } x \to -\infty$

and the eigenfunctions $\psi_n^{(j)}$ by

(2.1.20b) $\qquad \psi_n^{(j)} \sim \delta_{n,j} e^{i\zeta d_j x} \quad \text{as } x \to +\infty.$

As is standard with such matrix differential equations, the Wronskian is given by

(2.1.21a) $\qquad W(u, v, w) = \text{Det} \begin{pmatrix} u_1 & v_1 & w_1 \\ u_2 & v_2 & w_2 \\ u_3 & v_3 & w_3 \end{pmatrix},$

which is nonzero only if u, v, w are linearly independent. Moreover, we have

(2.1.21b) $\qquad W_x = i\zeta (\text{Tr } D) W$

($\text{Tr } D \equiv \text{Trace } D = d_1 + d_2 + d_3$). Thus the triads of vectors $[\phi^{(1)}, \phi^{(2)}, \phi^{(3)}]$, $[\psi^{(1)}, \psi^{(2)}, \psi^{(3)}]$ are each a set of linearly independent eigenfunctions. Thus

(2.1.22a) $\qquad \phi^{(j)} = \sum_{k=1}^{3} [a(\zeta)]_{jk} \psi^{(k)}.$

The so-called scattering matrix is given by

(2.1.22b) $\qquad S = [a_{jk}] = \begin{bmatrix} a_{11} & a_{12} & a_{13} \\ a_{21} & a_{22} & a_{23} \\ a_{31} & a_{32} & a_{33} \end{bmatrix}.$

It relates the solutions at $x \to +\infty$ to those at $x = -\infty$. In the 2×2 problem (§ 1.3) we had

(2.1.22c)
$$S = \begin{pmatrix} a & b \\ -\bar{b} & \bar{a} \end{pmatrix}.$$

Taking the determinant of (2.1.22a) we find the analogue of $a\bar{a} + b\bar{b} = 1$

(2.1.22d)
$$\det[a_{jk}] = 1.$$

In a similar manner we may define the eigenfunction $\psi^{(j)}$ in terms of the $\phi^{(j)}$:

(2.1.23a)
$$\psi^{(j)} = \sum_{k=1}^{3} b_{jk} \phi^{(k)}.$$

Substituting (2.1.23a) into (2.1.22a) we find, upon taking a determinant,

(2.1.23b)
$$\sum_{k=1}^{3} a_{jk} b_{kl} = \delta_{jl}.$$

The question of analyticity is examined by considering the integral equations associated with (2.1.1a). For example, upon specifying the boundary condition (2.1.20a) we find the following integral equation for $\phi^{(j)}$:

(2.1.24a) $\phi_n^{(j)}(x) e^{-i\zeta d_j x} = \delta_{n,j} - i \int_{-\infty}^{x} dy \, e^{i\zeta \beta_{nj}(x-y)} \sum_{m=1}^{3} iN_{nm}(y)(\phi_m^{(j)}(y) e^{-i\zeta d_j x}),$

where again

(2.1.24b) $\beta_{nj} = d_n - d_j.$

We note that: (a) (2.1.24a) is a Volterra integral equation; (b) $\beta_{1m} > 0$ and $\beta_{3m} < 0$. These facts suggest immediately that the eigenfunction $\phi^{(1)} e^{-i\zeta d_1 x}$ is analytic in the lower half ζ-plane for all real x, and $\phi^{(3)} e^{-i\zeta d_3 x}$ is analytic in the upper half plane. Indeed, by assuming that the potentials are in L_1 one can show that (2.1.24a) has a convergent Neumann series. Moreover since $a_{11} = \lim_{x \to \infty} \phi_1^{(1)} e^{-i\zeta d_1 x}$ we have also that $a_{11}(\zeta)$ is analytic in the lower half plane. The same procedure applies to the $\psi^{(j)}$ etc.

It is found that the following functions are
(i) analytic in the lower half plane ($\zeta = \xi + i\eta$, $\eta < 0$):

$$\phi^{(1)} e^{-i\zeta d_1 x}, \quad \psi^{(3)} e^{-i\zeta d_3 x}, \quad a_{11}, \quad b_{33};$$

(ii) analytic in the upper half plane ($\xi = \xi + i\eta$, $\eta > 0$):

$$\phi^{(3)} e^{-i\zeta d_3 x}, \quad \psi^{(1)} e^{-i\zeta d_1 x}, \quad a_{33}, \quad b_{11}.$$

(When the potentials N_{ik} are on compact support or decay faster than any exponential then all the above functions are entire. The functions in (i) and (ii) are bounded as $|\zeta| \to \infty$ respectively.)

These results suggest natural integral representations for $\phi^{(1)}$, $\phi^{(3)}$, $\psi^{(1)}$, $\psi^{(3)}$ with the required analyticity. However, we have *no* information regarding $\phi^{(2)}$ and $\psi^{(2)}$. This is the first difficult question we have so far encountered. All other ideas (once (2.1.1a) is postulated) follow the 2×2 case in an analogous way.

At this point Kaup (1976b), by considering the adjoint eigenfunctions, showed that the functions χ, $\bar{\chi}$ are analytic in the upper (lower) half plane with the relations:

(2.1.25a) $$\chi = e^{-i\zeta d_2 x}(b_{21}\psi^{(1)} - b_{11}\psi^{(2)}),$$

(2.1.25b) $$\bar{\chi} = e^{-i\zeta d_2 x}(b_{33}\psi^{(2)} - b_{23}\psi^{(3)}).$$

Before turning to the inverse scattering we note that by considering (2.1.1a) as $|\zeta| \to \infty$ with the relevant boundary conditions (WKB) we have

(2.1.25c)
$$\text{Im}\,\zeta > 0, \quad \phi^{(3)} e^{-i\zeta d_3 x} = \begin{pmatrix} 0 \\ 0 \\ 1 \end{pmatrix} + O\!\left(\frac{1}{\zeta}\right), \quad \psi^{(1)} e^{-i\zeta d_1 x} = \begin{pmatrix} 1 \\ 0 \\ 0 \end{pmatrix} + O\!\left(\frac{1}{\zeta}\right),$$

$$\chi = \begin{pmatrix} 0 \\ -1 \\ 0 \end{pmatrix} + O\!\left(\frac{1}{\zeta}\right), \quad a_{33} = 1 + O\!\left(\frac{1}{\zeta}\right), \quad b_{11} = 1 + O\!\left(\frac{1}{\zeta}\right).$$

(2.1.25d)
$$\text{Im}\,\zeta < 0, \quad \phi^{(1)} e^{-i\zeta d_1 x} = \begin{pmatrix} 1 \\ 0 \\ 0 \end{pmatrix} + O\!\left(\frac{1}{\zeta}\right), \quad \psi^{(3)} e^{-i\zeta d_3 x} = \begin{pmatrix} 0 \\ 0 \\ 1 \end{pmatrix} + O\!\left(\frac{1}{\zeta}\right),$$

$$\bar{\chi} = \begin{pmatrix} 0 \\ 1 \\ 0 \end{pmatrix} + O\!\left(\frac{1}{\zeta}\right), \quad a_{11} = 1 + O\!\left(\frac{1}{\zeta}\right), \quad b_{33} = 1 + O\!\left(\frac{1}{\zeta}\right).$$

On the basis of the analyticity we assume the following integral representations for $\psi^{(1)}$, $\psi^{(3)}$ (having the appropriate analytical behavior in each of the upper, lower half planes respectively):

(2.1.26a) $$\psi^{(1)} e^{-i\zeta d_1 x} = \begin{pmatrix} 1 \\ 0 \\ 0 \end{pmatrix} + \int_x^\infty K^{(1)}(x, z)\, e^{i\zeta(s-x)\beta_{12}}\, ds,$$

(2.1.26b) $$\psi^{(3)} e^{-i\zeta d_3 x} = \begin{pmatrix} 0 \\ 0 \\ 1 \end{pmatrix} + \int_x^\infty K^{(3)}(x, s)\, e^{-i\zeta(s-x)\beta_{23}}\, ds.$$

Requiring (2.1.26) to satisfy the eigenvalue problem (2.1.1a) gives us PDE's for the $K^{(1)}$, $K^{(3)}$ (analogous to (1.3.19)); e.g., for $K^{(1)}$:

(2.1.27a) $$\left\{(\partial_x + \partial_y) + \frac{D}{\beta_{12}}(I - d_1 D^{-1})\partial_y\right\} K^{(1)}(x, y) = N(x) K^{(1)}(x, y),$$

with

(2.1.27b)
$$\lim_{s\to\infty} K^{(1)}(x, s) = 0,$$
$$K_n^{(1)}(x, x) = -\frac{\beta_{12}}{\beta_{1n}} N_{n1}(x), \qquad n = 2, 3.$$

Hence $K^{(1)}(x, y)$ is independent of the eigenvalue ζ. From the same analysis for $\psi^{(3)}$ we find that the relation between $K^{(3)}$ and the potential is

(2.1.27c)
$$K_n^{(3)}(x, x) = -\frac{\beta_{23}}{\beta_{n3}} N_{n3}(x), \qquad n = 1, 2,$$

and that $K^{(3)}$ is also independent of the eigenvalues ζ.

In order to derive the inverse scattering equations we first derive certain integral representations for the $\psi^{(j)}$, $j = 1, 2, 3$. We assume that the potentials decay faster than any exponential so that all functions are entire and hence contour integrals can be defined. (This restriction can be removed, in which case we would replace the contour integrals by integrals along the real axis plus pole contributions.)

We define the contour $c(\bar{c})$ in the ζ-plane as a path going from $-\infty + i\varepsilon$ $(-\infty - i\varepsilon)$, $\varepsilon > 0$, to $+\infty + i\varepsilon$ $(+\infty - i\varepsilon)$ passing above (below) all zeros of b_{11}, a_{33} (a_{11}, b_{33}).

We now evaluate the integrals

(2.1.28)
$$I_1 = \int_{\bar{c}} \frac{\phi^{(1)} e^{-i\zeta' d_1 x}}{a_{11}(\zeta')(\zeta - \zeta')} d\zeta',$$
$$I_2 = \left(\int_c - \int_{\bar{c}}\right)\left(\frac{d\zeta'}{\zeta' - \zeta} \psi^{(2)} e^{-i\zeta d_2 x}\right),$$
$$I_3 = \int_c \frac{\phi^{(3)} e^{-i\zeta' d_3 x}}{a_{33}(\zeta')(\zeta' - \zeta)} d\zeta',$$

where for $I_1(I_3)$ ζ lies above (below) $\bar{c}$ (c), and for I_2 ζ lies between c and $\bar{c}$. Using (2.1.22a), the asymptotics (2.1.25), (2.1.24) and the analyticity requirements, we find, by contour integration,

(2.1.29a)
$$\psi^{(1)}(\zeta, x) e^{-i\zeta d_1 x} = \begin{pmatrix} 1 \\ 0 \\ 0 \end{pmatrix} - \frac{1}{2\pi i} \int_{\bar{c}} \frac{d\zeta'}{\zeta' - \zeta} e^{-i\zeta' d_1 x} \rho_1(\zeta') \psi^{(1)}(\zeta', x)$$
$$- \frac{1}{2\pi i} \int_{\bar{c}} \frac{d\zeta'}{\zeta' - \zeta} e^{-i\zeta' d_1 x} \rho_3(\zeta') \psi^{(3)}(\zeta', x),$$

$$\psi^{(2)}(\zeta, x) e^{-i\zeta d_2 x} = \begin{pmatrix} 0 \\ 1 \\ 0 \end{pmatrix} + \frac{1}{2\pi i} \int_{\bar{c}} \frac{d\zeta'}{\zeta' - \zeta} \rho_6(\zeta') e^{-i\zeta' d_2 x} \psi^{(3)}(\zeta', x)$$

(2.1.29b)

$$- \frac{1}{2\pi i} \int_c \frac{d\zeta'}{\zeta' - \zeta} \rho_5(\zeta') e^{-i\zeta' d_2 x} \psi^{(1)}(\zeta', x),$$

$$\psi^{(3)}(\zeta, x) e^{-i\zeta d_3 x} = \begin{pmatrix} 0 \\ 0 \\ 1 \end{pmatrix} + \frac{1}{2\pi i} \int_c \frac{d\zeta'}{\zeta' - \zeta} e^{-i\zeta' d_3 x} \rho_4(\zeta') \psi^{(1)}(\zeta', x)$$

(2.1.29c)

$$+ \frac{1}{2\pi i} \int_c \frac{d\zeta'}{\zeta' - \zeta} e^{-i\zeta' d_3 x} \rho_2(\zeta') \psi^{(2)}(\zeta', x),$$

where

(2.1.29d)
$$\rho_4 = \frac{a_{31}}{a_{33}}, \quad \rho_2 = \frac{a_{32}}{a_{33}}, \quad \rho_3 = \frac{a_{13}}{a_{11}}, \quad \rho_1 = \frac{a_{12}}{a_{11}},$$

$$\rho_5 = \frac{b_{21}}{b_{12}}, \quad \rho_0 = \frac{b_{23}}{b_{33}}.$$

In the formulae (2.1.29a, c) we substitute $\psi^{(2)}$ from (2.1.29b). This gives us *only* integral relations between $\psi^{(1)}, \psi^{(3)}$. Next we substitute the integral representations for $\psi^{(1)}, \psi^{(3)}$ from (2.1.26) into these relations. Operating on the equation (2.1.29a) with $(1/2\pi) \int_{-\infty}^{\infty} e^{i\zeta(x-y)\beta_{12}} d\zeta$ and on (2.1.29c) with $(1/2\pi) \int_{-\infty}^{\infty} e^{-i\zeta(x-y)\beta_{23}} d\zeta$ (Fourier transforms), after considerable algebra we obtain for $y > x$,

(2.1.30a)
$$K^{(1)}(x, y) + \begin{pmatrix} 0 \\ 1 \\ 0 \end{pmatrix} F_1(y) + \begin{pmatrix} 1 \\ 0 \\ 0 \end{pmatrix} F_3(x, y) + \begin{pmatrix} 0 \\ 0 \\ 1 \end{pmatrix} F_5(x, y)$$
$$+ \int_x^{\infty} (K^{(1)}(x, s) F_3(s, y) + K^{(3)}(x, s) F_5(s, y)) \, ds = 0,$$

(2.1.30b)
$$K^{(3)}(x, y) + \begin{pmatrix} 0 \\ 1 \\ 0 \end{pmatrix} F_2(y) + \begin{pmatrix} 0 \\ 0 \\ 1 \end{pmatrix} F_4(x, y) + \begin{pmatrix} 1 \\ 0 \\ 0 \end{pmatrix} F_6(x, y)$$
$$+ \int_x^{\infty} (K^{(1)}(x, s) F_6(s, y) + K^{(3)}(x, s) F_4(s, y)) \, ds = 0,$$

where

(2.1.30c)
$$F_1(x) = \frac{\beta_{12}}{2\pi} \int_{\bar{c}} d\zeta\, \rho_1(\zeta)\, e^{-i\zeta\beta_{12}x},$$
$$F_2(x) = \frac{\beta_{23}}{2\pi} \int_{c} d\zeta\, \rho_2(\zeta)\, e^{i\zeta\beta_{23}x},$$

(2.1.30d) $\quad F_3(x, y) = \dfrac{\beta_{12}}{4\pi i} \int_{c} d\zeta\, \rho_5(\zeta)\, e^{i\zeta\beta_{12}x} \int_{\bar{c}} \dfrac{d\zeta'}{\zeta' - \zeta} \rho_1(\zeta')\, e^{-i\zeta'\beta_{12}y},$

(2.1.30e) $\quad F_4(x, y) = -\dfrac{\beta_{23}}{4\pi^2 i} \int_{\bar{c}} d\zeta\, \rho_6(\zeta)\, e^{-i\zeta\beta_{23}x} \int_{c} \dfrac{d\zeta'}{\zeta' - \zeta} \rho_2(\zeta')\, e^{i\zeta'\beta_{23}y},$

(2.1.30f)
$$F_5(x, y) = \frac{\beta_{12}}{2\pi} \int_{\bar{c}} d\zeta\, \rho_3(\zeta)\, e^{-i\zeta(\beta_{12}y + \beta_{23}x)} - \frac{\beta_{12}}{4\pi^2 i} \int_{\bar{c}} d\zeta\, \rho_6(\zeta)\, e^{-i\zeta\beta_{23}x}$$
$$\times \int_{\bar{c}} \frac{d\zeta'}{\zeta' - \zeta + i\varepsilon} \rho_1(\zeta')\, e^{-i\zeta'\beta_{12}y},$$

(2.1.30g)
$$F_6(x, y) = \frac{\beta_{23}}{2\pi} \int_{c} d\zeta\, \rho_4(\zeta)\, e^{i\zeta(\beta_{12}x + \beta_{23}y)} + \frac{\beta_{23}}{4\pi^2 i} \int_{c} d\zeta\, e^{i\zeta\beta_{12}x} \rho_5(\zeta)$$
$$\times \int_{c} \frac{d\zeta'}{\zeta' - \zeta + i\varepsilon} \rho_2(\zeta')\, e^{i\zeta'\beta_{23}x}.$$

In (2.1.30f, g) the limit $\varepsilon \to 0$ is taken (we obtain this when we interchange integrals, keeping track of the relationship between ζ, ζ'). By considering $|\zeta| \to \infty$ for $\psi^{(1)}$, $\psi^{(2)}$, $\psi^{(3)}$ in (2.1.1a) and in the integral relations (2.1.26) we have

(2.1.31a) $\quad N_{21}(x) = K_2^{(1)}(x, x), \qquad N_{31}(x) = -\dfrac{\beta_{13}}{\beta_{12}} K_3^{(1)}(x, x),$

(2.1.31b) $\quad N_{13}(x) = -\dfrac{\beta_{13}}{\beta_{23}} K_1^{(3)}(x, x), \qquad N_{23}(x) = -K_2^{(3)}(x, x),$

(2.1.31c) $\quad N_{12}(x) = -\beta_{12}\left(E_1(x) - \displaystyle\int_x^{\infty} ds\, (K_1^{(3)}(x, s)E_2(s) - K_1^{(1)}(x, s)E_1(s))\right),$

(2.1.31d) $\quad N_{32}(x) = -\beta_{23}\left(E_2(x) + \displaystyle\int_x^{\infty} ds\, (K_3^{(3)}(x, s)E_2(s) - K_3^{(1)}(x, s)E_1(s))\right),$

where

(2.1.32a)
$$E_1(x) = \frac{1}{2\pi} \int_c \rho_5(\zeta)\, e^{i\zeta\beta_{12}x}\, d\zeta,$$

(2.1.32b)
$$E_2(x) = \frac{1}{2\pi} \int_{\bar{c}} \rho_6(\zeta)\, e^{-i\zeta\beta_{12}x}\, d\zeta.$$

IST IN OTHER SETTINGS 105

In deriving (2.1.31) we used (2.1.29b) in the limit $|\zeta| \to \infty$ along the real axis, and substituted the integral relations (2.1.26) into (2.1.29b).

2.1.c. Three-wave interaction. Now let us relate the above inverse scattering to the specific three-wave problem (2.1.12).

Using the scaling below (2.1.12) we have that the time evolution operator (2.1.1b) reduces to

$$(2.1.33) \qquad v_{it} = \left[-\frac{C_1 C_2 C_3}{C_i C_j} N_{ij} - i \frac{C_1 C_2 C_3}{C_j} \zeta \delta_{ij} \right] v_j.$$

We find the time dependence of the scattering data (e.g., a_{mn}) as follows. Define the *time*-dependent eigenfunctions as $(\phi^{(m),(t)}, \psi^{(m),(t)})$:

$$(2.1.34a) \qquad \phi^{(m),(t)} = \phi^{(m)} \exp\left(-i \frac{C_1 C_2 C_3}{C_m} \zeta t\right),$$

$$(2.1.34b) \qquad \psi^{(m),(t)} = \psi^{(m)} \exp\left(-i \frac{C_1 C_2 C_3}{C_m} \zeta t\right).$$

Similarly at any time (t) $\phi^{(m),(t)}, \psi^{(m),(t)}$ satisfy

$$(2.1.35) \qquad \phi^{(m),(t)} = \sum_{n=1}^{3} a_{mn}(t=0) \psi^{(n),(t)}$$

since both $\phi^{(m),(t)}, \psi^{(n),(t)}$ satisfy both (2.1.1a) and (2.1.1b). Then we have, using (2.1.22),

$$(2.1.36) \qquad a_{mn}(t) = a_{mn}(0) \exp\left(i C_1 C_2 C_3 \zeta \left(\frac{1}{C_m} - \frac{1}{C_n}\right) t\right).$$

Thus the diagonal elements of the scattering matrix are time independent. This is analogous to the 2×2 problem. Using trace formulae as in the 2×2 problem an infinity of conserved quantities can be related to these quantities. Alternatively as we discussed in § 1.6 a direct approach on (2.1.1a, b) can lead to the infinity of local conservation laws (Haberman (1977)).

With the relationships below (2.1.12), the following symmetry conditions are necessary:

$$(2.1.37a) \qquad \rho_5(\zeta) = -\gamma_1 \gamma_2 (\rho_1(\zeta^*))^*,$$

$$(2.1.37b) \qquad \rho_6(\zeta) = -\gamma_2 \gamma_3 (\rho_2(\zeta^*))^*,$$

$$(2.1.37c) \qquad \gamma_1 \rho_4(\zeta) + \gamma_3 (\rho_3(\zeta^*))^* - \gamma_2 (\rho_1(\zeta^*))^* \rho_2(\zeta) = 0,$$

where $\rho_i(\zeta)$ are defined in (2.1.29d). These in turn give

$$F_3^*(x, y) = F_3(y, x),$$
$$F_4^*(x, y) = F_4(y, x),$$
$$F_5^*(x, y) = \frac{-\gamma_1\gamma_2\beta_{12}F_5(y, x)}{\beta_{23}},$$
(2.1.38)
$$E_1(x) = \frac{\gamma_1\gamma_2 F_1^*(x)}{\beta_{12}},$$
$$E_2(x) = \frac{\gamma_1\gamma_3 F_2^*(x)}{\beta_{23}}.$$

A special soliton solution is computed by considering the case where we have degenerate kernels. Even though the three-wave interaction is nondispersive and hence the solitons are no more important than the continuous spectrum contribution, it nevertheless is enlightening to have a closed form solution. We consider the case where both a_{11} and a_{33} have zeros in their respective half planes. Let $\bar{\zeta}_1(\bar{\zeta}_3)$ be a simple zero of $a_{11}(a_{33})$ in the lower (upper) half plane, and $C(\bar{C})$ be the residue of $\rho_2(\rho_1)$ at $\zeta = \zeta_3(\zeta = \bar{\zeta})$ (the residues of ρ_4 at ζ_3, ρ_3 at $\bar{\zeta}_1$, ρ_5 at $\bar{\zeta}_1^*$, ρ_6 at ζ_3^*, are determined by (2.1.37c)). Then the kernels reduce to

$$F_1(x) = i\beta_{12}\bar{C}e^{-i\beta_{12}\bar{\zeta}_1 x},$$
$$F_2(x) = -i\beta_{23}C e^{i\beta_{23}\zeta_3 x},$$
$$F_3(x, y) = -\gamma_1\gamma_2\beta_{12}\frac{\bar{C}\bar{C}^*}{2\eta_1}e^{i\beta_{12}(\bar{\zeta}_1^* x - \bar{\zeta}_1 y)},$$
(2.1.39)
$$F_4(x, y) = -\gamma_2\gamma_3\beta_{23}\frac{C^*C}{2\eta_3}e^{-i\beta_{23}(\zeta_3^* x - \zeta_3 y)},$$
$$F_5 = i\gamma_2\gamma_3\beta_{12}\frac{C\bar{C}^*}{\bar{\zeta}_1^* - \zeta_3^*}e^{-i(\beta_{23}\zeta_3^* x + \beta_{12}\bar{\zeta}_1 y)},$$
$$F_6(x, y) = -i\gamma_1\gamma_2\beta_{23}\frac{C\bar{C}^*}{\zeta_3 - \bar{\zeta}_1^*}e^{i(\beta_{12}\bar{\zeta}_1^* x + \beta_{23}\zeta_3 y)},$$

with the others defined from (2.1.38). From (2.1.3b) the time dependence of $C, \bar{C}$ is given by

(2.1.40a)
$$C = C_0 e^{-i\beta_{23}\zeta_3 C_1 t},$$
$$\bar{C} = \bar{C}_0 e^{i\beta_{12}\bar{\zeta}_1 C_3 t},$$

where we shall represent $C_0, \bar{C}_0$ and the eigenvalues $\bar{\zeta}_1, \zeta_3$ by

(2.1.40b)
$$\bar{\zeta}_1 = \frac{\xi_1 - i\eta_1}{\beta_{12}}, \qquad \zeta_3 = \frac{\xi_3 + i\eta_3}{\beta_{23}},$$

(2.1.40c) $$C_0 = 2\eta_3 \frac{e^{\eta_3 x_1} e^{-i\xi_3 \bar{x}_1}}{\beta_{23}},$$

(2.1.40d) $$\bar{C}_0 = 2\eta_1 \frac{e^{\eta_1 x_3} e^{i\xi_1 \bar{x}_3}}{\beta_{12}}.$$

With these choices, the integral equations (2.1.30) are degenerate. Their solution leads to

$$Q_1 = \sqrt{\beta_{12}\beta_{13}} \frac{2\eta_3}{\mathscr{D}} e^{i\xi_3(x-C_1 t-\bar{x}_1)} \left[e^{\eta_1(x-C_3 t-x_3)} - \gamma_1 \gamma_2 \frac{\bar{\zeta}_1 - \zeta_3}{\bar{\zeta}_1^* - \zeta_3} e^{-\eta_1(x-C_3 t-x_3)} \right],$$

$$Q_2 = \frac{-4\eta_1 \eta_3 \beta_{13} \gamma_2 \gamma_3}{\sqrt{\beta_{12}\beta_{23}}(\bar{\zeta}_1 - \zeta_3^*)\mathscr{D}} e^{-i\xi_1(x-C_3 t-\bar{x}_3)} e^{-i\xi_3(x-C_1 t-\bar{x}_1)},$$

$$Q_3 = \sqrt{\beta_{13}\beta_{23}} \gamma_1 \gamma_2 \frac{2\eta_1}{\mathscr{D}} e^{i\xi_1(x-C_3 t-\bar{x}_3)}$$

$$\times \left[e^{\eta_3(x-C_1 t-x_1)} - \gamma_2 \gamma_3 \frac{\bar{\zeta}_1^* - \zeta_3^*}{\bar{\zeta}_1^* - \zeta_3} e^{-\eta_3(x-C_1 t-x_1)} \right],$$

where

(2.1.41)
$$\mathscr{D} = [e^{\eta_1(x-C_3 t-x_3)} - \gamma_1 \gamma_2 e^{-\eta_1(x-C_3 t-x_3)}]$$
$$\cdot [e^{\eta_3(x-C_1 t-x_1)} - \gamma_2 \gamma_3 e^{-\eta_3(x-C_1 t-x_1)}]$$
$$+ \gamma_1 \gamma_3 \frac{(\bar{\zeta}_1 - \bar{\zeta}_1^*)(\zeta_2 - \zeta_3^*)}{(\bar{\zeta}_1 - \zeta_3^*)(\bar{\zeta}_1^* - \zeta_3)} e^{-\eta_1(x-C_3 t-x_3)} e^{-\eta_3(x-C_1 t-x_1)}.$$

For $t \to -\infty$ we have Q_3 and Q_1 moving along their respective characteristics (Q_3 to the left of Q_1) $x - C_3 t$, $x - C_1 t$ respectively. When t increases, the envelopes come together and Q_2 is produced. As $t \to \infty$, Q_2 decays back to zero and Q_1, Q_3 again move along their respective characteristics (Q_3 to the right of Q_1). The amplitudes of Q_1, Q_3 are unchanged. The solution is not singular if any one of the γ_i is different from the other two. It is singular in some region of space-time if $\gamma_1 = \gamma_2 = \gamma_3 = -1$ (explosive instability).

One of the great simplifications inherent in the three-wave problem is due to its nondispersive nature. If initially (which we shall refer to as $t \to -\infty$) the envelopes are well separated and have no significant overlap, it can be argued from the solution procedure (Kaup (1976b)) that this also occurs when $t \to +\infty$. When they are well separated the envelopes propagate with their characteristic velocities $Q_i \sim Q_i(x - C_i t)$, $i = 1, 2, 3$. Moreover the solution of the three-wave problem reduces to solving a sequence of 2×2 scattering problems already discussed in Chapter 1 (as opposed to the complication of the 3×3 problem!). All significant results can then be argued in terms of the analysis of § 1.3. Since $C_3 > C_2 > C_1$, as $t \to -\infty$ we have that the envelopes are spatially ordered Q_3, Q_2, Q_1. To show why the 3×3 scattering problem effectively reduces to 2×2, consider the envelope Q_3. As $t \to -\infty$ in the region of support of Q_3, the envelopes Q_1, Q_2 are zero. Hence the 3×3 problem takes the form (recall that $N_{12} \propto Q_3$)

(2.1.42)
$$v_{1x} = i\zeta d_1 v_1 + N_{12} v_2,$$
$$v_{2x} = i\zeta d_2 v_2 + N_{21} v_1,$$
$$v_{3x} = i\zeta d_3 v_3.$$

We see that v_3 does not change from the value $e^{i\zeta d_3 x}$; hence the solution effectively depends only on the components v_1, v_2 which satisfy the 2×2 problem (1.2.7a) with $q = N_{12}$, $r = N_{21}$. Due to the fact that $v_3 = e^{i\zeta d_3 x}$, the scattering matrix going from the left of Q_3 to the right of Q_3 (where $Q_3 = 0$), we have that $S^{(3)}$, the scattering matrix over this region, satisfies

(2.1.43)
$$S^{(3)} = \begin{bmatrix} a_{11}^{(3)} & a_{12}^{(3)} & 0 \\ a_{21}^{(3)} & a_{22}^{(3)} & 0 \\ 0 & 0 & 1 \end{bmatrix}.$$

In what follows we recall that the 2×2 scattering problem has a scattering matrix

(2.1.44)
$$S_{(2 \times 2)} = \begin{bmatrix} a & b \\ -\bar{b} & \bar{a} \end{bmatrix}.$$

(Here for purposes of analogy we will consider $\phi^{(1)} \to \phi$ of § 1.3, $\psi^{(1)} \to \psi$ of § 1.3, $\phi^{(2)} \to -\bar{\phi}$ of § 1.3, $\psi^{(2)} \to \psi$ of § 1.3, etc. We also acknowledge a minor modification due to d_i multiplying the eigenvalue.) Continuing this argument we see that as $t \to -\infty$

(2.1.45a)
$$S = S_0^{(3)} S_0^{(2)} S_0^{(1)},$$

where 0 indicates values at $-\infty$ and (omitting the subscripts)

$$S^{(3)} = \begin{bmatrix} a^{(3)} & b^{(3)} & 0 \\ -\bar{b}^{(3)} & \bar{a}^{(3)} & 0 \\ 0 & 0 & 1 \end{bmatrix}, \quad S^{(2)} = \begin{bmatrix} a^{(2)} & 0 & b^{(2)} \\ 0 & 1 & 0 \\ -\bar{b}^{(2)} & 0 & \bar{a}^{(2)} \end{bmatrix},$$

$$S^{(1)} = \begin{bmatrix} 1 & 0 & 0 \\ 0 & a^{(1)} & b^{(1)} \\ 0 & -\bar{b}^{(1)} & \bar{a}^{(1)} \end{bmatrix}.$$

Similarly, as $t \to +\infty$ we have

(2.1.45b)
$$S = S_f^{(1)} S_f^{(2)} S_f^{(3)},$$

where the values at $+\infty$ are to be denoted by subscript f. By equating (2.1.45a) and (2.1.45b) and multiplying on the right first by $(S_f^{(3)})^{-1}$ and then by $(S_f^{(2)})^{-1}$, we find the final values of b/a in terms of the initial values, i.e.,

(2.1.46a) $$\frac{\bar{b}_f^{(3)}}{\bar{a}_f^{(3)}} = \frac{\bar{a}_0^{(1)} \bar{b}_0^{(3)} - \bar{a}_0^{(3)} \bar{b}_0^{(2)} b_0^{(1)}}{\bar{a}_0^{(2)} \bar{a}_0^{(3)}},$$

(2.1.46b) $$\frac{\bar{b}_f^{(2)}}{\bar{a}_f^{(2)}} = \frac{\bar{a}_f^{(3)}}{\bar{a}_0^{(2)} \bar{a}_0^{(3)}} [\bar{a}_0^{(3)} \bar{b}_0^{(2)} a_0^{(1)} - \bar{b}_0^{(1)} \bar{b}_0^{(3)}],$$

(2.1.46c) $$\frac{\bar{b}_f^{(1)}}{\bar{a}_f^{(1)}} = \frac{\bar{a}_f^{(2)}}{\bar{a}_0^{(1)} \bar{a}_0^{(2)}} [a_f^{(3)} \bar{a}_0^{(2)} \bar{b}_0^{(1)} - \bar{b}_f^{(3)} \bar{b}_0^{(2)}].$$

All connection information (from $t \to -\infty$ to $t \to \infty$) regarding the solution can be obtained from these formulae. (Note that b is related to $\bar{b}$ and a is related to $\bar{a}$ because of symmetry; see also § 1.3, e.g., (1.3.14).)

An example, and an amusing application of these ideas, is the "transfer of solitons" between envelopes. Here we caution the reader that in what follows we shall refer to the solitons associated with the 2×2 Zakharov–Shabat eigenvalue problem and not "full solitons," i.e., not the modes associated with discrete eigenvalues of the 3×3 problem, which we discussed earlier in this section. Let us recall that zeros of $\bar{a}(\zeta)$ (or $a(\zeta)$) in the complex plane correspond to the analogous solitons in the 2×2 problem (the reader should also note that we are assuming the scattering data $a, \bar{a}, b, \bar{b}$ are entire functions of ζ). From (2.1.46a) we see that whenever $\bar{a}_0^{(2)}$, or $\bar{a}_0^{(3)}$ have zeros, so does $\bar{a}_f^{(3)}$, unless the numerator in (2.1.46a) vanishes. For general initial data this will virtually never occur, so here we shall ignore this possibility. Hence the final number of solitons in Q_3 equals its original number plus the original number contained in Q_2. Q_2 will end up with no solitons since the zeros of $\bar{a}_0^{(3)}$ match those in $\bar{a}_0^{(2)}, \bar{a}_0^{(3)}$. Similarly, the final number of solitons contained in Q_1 is equal to its initial number plus the initial number contained in Q_2. (It should be stressed that these are only representative results. In principle, we could obtain any results we wish since (2.1.46) contains all the information needed to reconstruct the complete solution as $t \to +\infty$.)

Next we make some remarks on the explosive or decay instability cases. When $\gamma_1 = \gamma_2 = \gamma_3 = -1$ (explosive instability) we have

(2.1.47a) $$Q_3 \propto N_{12} = N_{21}^*, \qquad Q_2 \propto N_{31} = -N_{13}^*, \qquad Q_1 \propto N_{23} = +N_{32}^*,$$
$$r^{(1)} = +q^{(1)*}, \qquad r^{(2)} = -q^{(2)*}, \qquad r^{(3)} = +q^{(3)*}.$$

From the results of § 1.3 only Q_2 can have solitons initially (i.e., $r^{(2)} = -q^{(2)*}$). But if Q_2 does have solitons then, from (2.1.46), $\bar{a}_f^{(3)}, \bar{a}_f^{(1)}$ will have zeros in the complex plane. But if $|Q_f^{(3)}|$ and $|Q_f^{(1)}|$ are integrable the results of § 1.3 show this is impossible (since in these cases $r^{(i)} = +q^{(i)*}, i = 1, 3$). The only way out of this paradox is to lose integrability. Indeed, it is well known that the explosive instability case gives singularities (nonlinear instability) in a finite time (see also the previous discussion regarding the solitons associated with the full 3×3 problem).

In the decay instability cases we have two possibilities:
(a) $\gamma_1 = \gamma_3 = -1$, $\gamma_2 = +1$. Here

(2.1.47b) $\quad Q_1 \propto N_{23} = -N_{32}^*, \quad Q_2 \propto N_{31} = -N_{13}^*, \quad Q_2 \propto N_{12} = -N_{21}^*,$
$r^{(1)} = -q^{(1)*}, \quad r^{(2)} = -q^{(2)*}, \quad r^{(3)} = -q^{(3)*}.$

Since $r^{(i)} = -q^{(i)*}$, $i = 1, 2, 3$, the results of § 1.3 show that a unique solution to this problem always exists (see also Ablowitz, Kaup, Newell and Segur (1974)).
(b) $\gamma_1 = \gamma_2 = +1$, $\gamma_3 = -1$. Here

(2.1.47c) $\quad Q_1 \propto N_{23} = -N_{32}^*, \quad Q_2 \propto N_{31} = +N_{13}^*, \quad Q_3 \propto N_{12} = N_{21}^*,$
$r^{(1)} = -q^{(1)*}, \quad r^{(2)} = +q^{(2)*}, \quad r^{(3)} = +q^{(3)*}.$

Since the envelope associated with Q_2 never has any solitons there are no zeros of $\bar{a}_f^{(3)}$, hence no solitons associated with Q_3 (as with the explosive instability case). Moreover the conserved quantities (2.1.13) imply that we necessarily have L_2 integrability of the envelopes if they were initially L_2. (We also encourage the reader to consult Kaup (1976b) and/or Zakharov and Manakov (1976a) for more information and details regarding the three-wave interaction problem.) This completes our discussion of the three-wave problem.

2.1.d. Multidimensional scattering problems. We now describe a procedure to find multidimensional nonlinear evolution equations associated with linear scattering problems. This procedure is a natural generalization of the one-dimensional case described earlier in this chapter. The ideas follow closely those presented in Ablowitz and Haberman (1975b). Another approach, using the linear integral equation as a starting point, was presented by Zakharov and Shabat (1974), and is discussed later in this chapter as well as in § 3.6.

We begin by considering the following spectral problem and associated time dependence:

(2.1.48) $\quad \mathbf{V}_x = i\zeta \mathbf{V} + N\mathbf{V} + B\mathbf{V}_y,$

(2.1.49) $\quad \mathbf{V}_t = Q\mathbf{V} + C_1 \mathbf{V}_y + C_2 \mathbf{V}_{yy} + \cdots + C_m \underbrace{\mathbf{V}_{yy\cdots y}}_{m}.$

Here $\mathbf{V}$ is a vector $(n \times 1)$ and N, B, Q, C_i are $n \times n$ matrices. We claim that corresponding to *each* assumed form of time dependence for $\mathbf{V}_t$ there is one nonlinear evolution equation. This is as opposed to the one-dimensional case, where a class of evolution equations come from the same time-dependent structure (i.e., $\mathbf{V}_t = Q\mathbf{V}$). Special cases of the scattering problem (2.1.48) have been considered by Zakharov and Manakov (1979), Nizhik (1973) and Kaup (1979) with regard to developing the required inverse scattering formulae.

With these inverse scattering results, the nonlinear evolution equations we shall discuss are solvable (namely the three-wave interaction, Kadomtsev–Petviashvili equations, etc.).

Here we shall present the technique via an example: the three-wave interaction in two spatial dimensions (corresponding results can be found in three spatial dimensions (Ablowitz and Haberman (1975a))). As a special case of (2.1.48–49), consider

(2.1.50) $$\mathbf{V}_x = i\zeta D\mathbf{V} + N\mathbf{V} + B\mathbf{V}_y,$$

(2.1.51) $$\mathbf{V}_t = Q\mathbf{V} + C\mathbf{V}_y.$$

Assuming $\zeta_t = 0$ and also that B, D, C are constant, we have

$$\mathbf{V}_{xt} = i\zeta D(Q\mathbf{V} + C\mathbf{V}_y) + N_t\mathbf{V} + N(Q\mathbf{V} + C\mathbf{V}_y)$$
$$+ B(Q_y\mathbf{V} + Q\mathbf{V}_y + C\mathbf{V}_{yy}),$$
$$\mathbf{V}_{tx} = Q_x\mathbf{V} + Q(i\zeta D\mathbf{V} + N\mathbf{V} + B\mathbf{V}_y)$$
$$+ C(i\zeta D\mathbf{V}_y + N_y\mathbf{V} + N\mathbf{V}_y + B\mathbf{V}_{yy}).$$

Setting the coefficients of $\mathbf{V}, \mathbf{V}_y, \mathbf{V}_{yy}$ equal we have

(2.1.52a) $\mathbf{V}_{yy}: [C, B] = 0,$

(2.1.52b) $\mathbf{V}_y: i\zeta[C, D] + [Q, B] + [C, N] = 0,$

(2.1.52c) $\mathbf{V}: i\zeta[Q, D] + [Q, N] + Q_x + CN_y - BQ_y = N_t.$

A simple case is when

$$C = a_i\delta_{ij}, \quad B = b_i\delta_{ij}, \quad D = d_i\delta_{ij}, \quad N_{ii} = 0,$$

with a_i, b_i, d_i constant. Then (2.1.52a) yields

$$\sum_k (C_{ik}B_{kj} - B_{ik}C_{kj}) = 0 = c_ib_i - c_ib_i$$

(trivially). Equation (2.1.52b) yields

$$i\zeta[C, D] + \sum_k (Q_{ik}B_{kj} - B_{ik}Q_{kj}) + \sum_k (C_{ik}N_{kj} - N_{ik}C_{kj}) = 0,$$

which simplifies to

(2.1.53) $$Q_{ij} = \frac{c_i - c_j}{b_i - b_j}N_{ij}, \quad i \neq j,$$
$$Q_{ii} = q_i \quad \text{(take } q_i = \text{constant)}.$$

Define

$$a_{ij} = \frac{c_i - c_j}{b_i - b_j} = a_{ji};$$

then $Q_{ij} = a_{ij}N_{ij}$, $i \neq j$, and (2.1.52c) yields

$$i\zeta \sum_k (Q_{jk}D_{ki} - D_{ik}Q_{kj}) + \sum_k (Q_{ik}N_{kj} - N_{ik}Q_{kj})$$
$$+ Q_{ij,x} + \sum_k (C_{ik}N_{kj,y} - B_{ik}Q_{kj,y}) = N_{ij,t}.$$

For $i \neq j$ we have

(2.1.54)
$$i\zeta(a_{ij}N_{ij}(d_j - d_i)) + (q_i - q_j)N_{ij} + \sum_{k \neq i,j} (a_{ik} - a_{kj})N_{ik}N_{kj}$$
$$+ a_{ij}N_{ij,x} + c_iN_{ij,y} = N_{ij,t}.$$

When $i = j$ the equation is automatically satisfied. But we also note that (2.1.54) contains ζ. To eliminate the ζ-dependence we take $q_i = q_i(\zeta)$ (q_i is free); specifically, $q_i - q_j = i\zeta a_{ij}(d_i - d_j)$ (the equation is satisfied if we take $d_i = b_i$, $q_i = i\zeta c_i$). With these choices, (2.1.54) reduces to

(2.1.55) $$N_{ij,t} = a_{ij}N_{ij,x} + \beta_{ij}N_{ij,y} + \sum_{k \neq i,j} (a_{ik} - a_{kj})N_{ik}N_{kj},$$

where

$$\beta_{ij} = c_i - b_i a_{ij} = \frac{b_i c_j - c_i b_j}{b_i - b_j} \quad (y \text{ group velocity}),$$

$$a_{ij} = \frac{c_i - c_j}{b_i - b_j} \quad (x \text{ group velocity}).$$

Again, $N_{ij} = \sigma_{ij}N_{ji}^*$ is consistent if

$$\sigma_{ij} = -\sigma_{ik}\sigma_{kj}, \quad i > k > j$$

and a_{ij}, β_{ij} are real. The three-wave equation is obtained if one takes $n = 3$ (third order) and $N_{12} = U_1$, $N_{13} = U_2$, $N_{23} = U_3$:

(2.1.56)
$$U_{1t} = a_{12}U_{1x} + \beta_{12}U_{1y} + \sigma_{32}(a_{13} - a_{23})U_2U_3^*,$$
$$U_{2t} = a_{13}U_{2x} + \beta_{13}U_{2y} + (a_{12} - a_{23})U_1U_3,$$
$$U_{3t} = a_{23}U_{3x} + \beta_{23}U_{3y} + \sigma_{21}(a_{12} - a_{23})U_1^*U_2.$$

If $\sigma_{32} > 0$ and $\sigma_{21} > 0$ we then have explosive instability; otherwise a positive definite energy exists and we have decay instability.

There are many other equations that can be deduced. For example, if we choose the time dependence in (2.1.49) as

(2.1.57) $$\mathbf{V}_t = Q\mathbf{V} + C_1\mathbf{V}_y + C_2\mathbf{V}_{yy},$$

with B, C_1, C_2 diagonal matrices, we find, for

(2.1.58)
$$N = \begin{pmatrix} 0 & A \\ \sigma_1 A^* & 0 \end{pmatrix}, \quad \sigma_1 = \pm 1,$$
$$A_t = D_1 A + WA, \quad D_0 W = \sigma_1 D_1(|A|^2),$$

where
$$D_1 = \frac{(e_1 - e_2)\partial_x^2 + 2(b_1 e_2 - e_1 b_2)\partial_x \partial_y + (e_1 b_2^2 - b_1^2 e_2)\partial_y^2}{(b_1 - b_2)^2},$$
$$D_0 = -\partial_x^2 + (b_1 + b_2)\partial_x \partial_y - b_1 b_2 \partial_y^2.$$

Taking
$$b_1 = -b_2, \quad e_1 = -e_2 = -2i, \quad W = \frac{-2(e_1 - e_2)(|A|^2)}{(b_1 - b_2)} - 2iQ,$$

we have

(2.1.59)
$$iA_t = \left(\frac{1}{b_1^2} A_{xx} + A_{yy}\right) + 2QA - \frac{2\sigma_1}{b_1^2}|A|^2 A,$$
$$Q_{xx} - b_1^2 Q_{yy} = -2\sigma_1(|A|^2)_{yy},$$

or, by choosing $Q = (2\sigma_1/b_1^2)|A|^2 + \Phi_x$,

(2.1.60)
$$iA_t = \frac{1}{b_1^2} A_{xx} + A_{yy} + \frac{2\sigma_1}{b_1^2}|A|^2 A + 2\Phi_x A,$$
$$\Phi_{xx} - b_1^2 \Phi_{yy} = -\frac{2\sigma_1}{b_1^2}(|A|^2)_x.$$

For b_1^2 real these equations are the long-wave limit ($kh \to 0$) of (4.3.27) (see § 4.3), which governs the evolution of a nearly monochromatic, nearly one-dimensional packet of water waves of small amplitude. (The arbitrary depth case appears not to be of IST type (Ablowitz and Segur (1979)). Anker and Freeman (1978) considered the case b_1 pure imaginary via the Zakharov–Shabat procedure, and developed N-plane wave soliton (interacting at angles) solutions, and Satsuma and Ablowitz (1979) show how one may obtain "lump" (i.e., a multidimensional soliton decaying to a constant in all directions) type envelope hole solutions to these equations (see § 3.4).

The L, A pair for the Kadomtsev–Petviashvili equation (this equation also arises in water waves when the y variations are sufficiently slow to balance the long x-direction waves and weak nonlinearity, see § 4.1),

(2.1.61)
$$\partial_x(U_t + 6(UU_x) + U_{xxx}) = -3b^2 U_{yy},$$

was originally found by Zakharov and Shabat (1974) and Dryuma (1974). They found that (2.1.61) is related to the scattering problem

(2.1.62) $$v_{xx} + (\lambda + u)v + bv_y = 0.$$

These results may also be obtained from the above procedure by taking

$$B = \begin{bmatrix} 0 & 0 \\ -b & 0 \end{bmatrix}, \quad N = \begin{bmatrix} 0 & 1 \\ -u & 0 \end{bmatrix}, \quad D = \begin{bmatrix} 1 & 0 \\ 0 & -1 \end{bmatrix},$$

and carrying out the indicated analysis. (2.1.62) is the starting point of the scattering analysis of Zakharov and Manakov (1979). It should also be stressed that the spectral problem here is *not* $\nabla^2 v + (\lambda + u)v = 0$, i.e., the usual multidimensional Schrödinger eigenvalue problem where the inverse scattering analysis is very complicated (see, for example, Newton (1979)).

2.2. Discrete problems. Many interesting physical phenomena can be modeled by discrete nonlinear equations. Examples include vibration of particles in a one-dimensional lattice (Toda (1970)), ladder type electric circuits (Hirota and Suzuki (1970), (1973), Hirota and Satsuma (1976b)), collapse of Langmuir waves in plasma physics (Zakharov, Musher and Rubenchik (1974)), growth of conflicting populations in biological science (Hirota and Satsuma (1976b)), difference simulations of differential equations, etc. Hence it is undoubtedly significant that the ideas of the inverse scattering transform apply to certain types of discrete evolution equations.

The story in this chapter begins with the so-called Toda lattice, a system of unit masses connected by nonlinear springs whose restoring force is exponential (this lattice is sometimes referred to as the exponential lattice). The equations of motion,

(2.2.1) $$Q_{n,tt} = e^{-(Q_n - Q_{n-1})} - e^{-(Q_{n+1} - Q_n)},$$

are derivable from the Hamiltonian

(2.2.2) $$H = \sum_{j=-\infty}^{\infty} \left\{ \frac{1}{2} P_j^2 + (e^{-(Q_n - Q_{n-1})} - 1) \right\},$$

where $P_j = Q_{j_t}$ (recall Hamilton's equations $P_{j_t} = -\partial H/\partial Q_j$, $Q_{j_t} = \partial H/\partial P_j$). This lattice was extensively studied by Toda (see, for example, Toda (1967a), (1970)) who discovered a number of explicit solutions for both the periodic and infinite lattice. Flaschka (1974a,b), using the discrete inverse scattering theory (a discrete Schrödinger equation) of Case and Kac (1973) and Case (1973), was able to solve the lattice (2.2.1) by the inverse scattering transform. Similar results were also obtained by Manakov (1975). Shortly thereafter Ablowitz and Ladik (1975), (1976) proposed a new discrete scattering problem. This scattering problem is a discrete version of the 2 × 2 Zakharov–Shabat problem, and serves as a basis for generating solvable discrete equations (special cases are the Toda lattice, a nonlinear self-dual network (Hirota (1973b), etc.). Moreover these ideas can be extended to nonlinear partial difference equations (Ablowitz and Ladik (1976), (1977)). In this section we shall first discuss the discrete Schrödinger scattering problem and its relationship with the Toda lattice. Here we shall follow the work of Flaschka (1974b) and then proceed to discuss the evolution equations associated with the discrete 2×2 Zakharov-Shabat scattering problem. Finally, we shall return to indicate how to solve the inverse problem.

2.2.a. Deriving evolution equations.
Consider the Schrödinger scattering problem

$$\psi_{xx} + (\lambda_c + q)\lambda = 0 \tag{2.2.3}$$

(λ_c refers to the eigenvalue of the continuous problem—see Case and Kac (1973) and also Case (1973)). A discretization of (2.2.3) is

$$\frac{\psi_{n+1} + \psi_{n-1} - 2\psi_n}{h^2} + (\lambda_c + q)\psi_n = 0, \tag{2.2.4}$$

with $\psi_n = \psi(nh)$, etc. Using the substitution $v_n = g_n \psi_n$, where $g_n = \exp(h^2 q_n/2)$, we have

$$\exp\left(-\frac{h^2}{2}(q_{n+1}+q_n)\right) v_{n+1} + \exp\left(-\frac{h^2}{2}(q_n+q_{n-1})\right) v_{n-1} = \lambda v_n, \tag{2.2.5}$$

where we have used $\exp(h^2 q_n) \sim 1 + h^2 q_n$, $\lambda = e^{-h^2 \lambda_c}$. Defining $a_n = \exp((-h^2/2)(q_{n+1}+q_n))$, we have

$$a_n v_{n+1} + a_{n-1} v_{n-1} = \lambda v_n. \tag{2.2.6}$$

Flaschka (1974a) used the generalization

$$a_n v_{n+1} + a_{n-1} v_{n-1} + b_n v_n = \lambda v_n. \tag{2.2.7}$$

To derive the Toda lattice from (2.2.7) we follow Ablowitz and Ladik (1975). Consider the associated time evolution equation (note that $v_{n_t} \equiv (\partial/\partial t) v_n$ in what follows)

$$v_{n_t} = A_n v_{n+1} + B_n v_n. \tag{2.2.8}$$

Taking the time derivative of (2.2.7), and using (2.2.7) to solve for v_{n+2} and v_{n-1} (e.g., $v_{n+2} a_{n+1} = \lambda v_{n+1} - b_{n+1} v_{n+1} - a_n v_n$), we find two equations by setting the coefficients of the terms v_{n+1} and v_n, respectively, to zero (and assuming $\partial \lambda/\partial t = 0$):

$$A_n(b_n - \lambda) + \frac{a_n}{a_{n+1}}(\lambda - b_{n+1})A_{n+1} \tag{2.2.9a}$$

$$+ a_n(B_{n+1} - B_{n-1}) = \frac{a_{n-1_t} a_n}{a_{n-1}} - a_{n_t},$$

(2.2.9b)
$$\frac{-a_n^2 A_{n+1}}{a_{n+1}} + B_n(b_n - \lambda)$$
$$+ (\lambda - b_n)B_{n-1} + a_{n-1}A_{n-1} = \frac{a_{n-1}(b_n - \lambda)}{a_{n-1}} - b_{n_t}.$$

Expanding A_n, B_n as (for example)

(2.2.10)
$$A_n = A_n^{(0)}(t) + A_n^{(1)}(t)\lambda,$$
$$B_n = B_n^{(0)}(t) + B_n^{(1)}(t)\lambda,$$

and requiring the coefficients of λ^2, λ, λ^0 to vanish independently yields A_n, B_n and two coupled evolution equations. The procedure works in a straightforward manner (e.g., at λ^2 (2.2.9a) yields $-A_n^{(1)} + a_n A_{n+1}^{(1)}/a_{n+1} = 0$, $A_\infty^{(1)} = $ const.), so that $A_n^{(1)} = a_n A_\infty^{(1)}$. Similarly, (2.2.9b) yields $-B_n^{(1)} + B_{n-1}^{(1)} = 0 \Rightarrow B_n^{(1)} = B_\infty^{(1)} = $ const., etc. After some algebra we find ($A_\infty^{(i)}$, $B_\infty^{(i)}$, $i = 0, 1$, are const.)

(2.2.11)
$$A_n = A_\infty^{(1)} a_n \lambda + A_\infty^{(0)} a_n + A_\infty^{(1)} a_n b_n,$$
$$B_n = B_\infty^{(1)} \lambda + B_\infty^{(0)} + A_\infty^{(1)}(1 - a_n^2) + \sum_{k=-\infty}^{n} \partial_t \log a_{k-1},$$

and the evolution equations

(2.2.12)
$$a_{n_t} = \tfrac{1}{2} A_\infty^{(0)} a_n (b_{n+1} - b_n) + \tfrac{1}{2} A_\infty^{(1)} a_n (a_{n+1}^2 - a_{n-1}^2 + b_{n+1}^2 - b_n^2),$$
$$b_{n_t} = A_\infty^{(0)}(a_n^2 - a_{n-1}^2) + A_\infty^{(1)}[a_n^2(b_{n+1} + b_n) - a_{n-1}^2(b_n + b_{n-1})].$$

The Toda lattice is arrived at by taking $A_\infty^{(1)} = 0$, $A_\infty^{(0)} = 2$,

(2.2.13)
$$a_{n_t} = a_n(b_{n+1} - b_n),$$
$$b_{n_t} = 2(a_n^2 - a_{n-1}^2),$$

and relating Q_n in the Toda lattice (2.2.1) to a_n, b_n via

(2.2.14)
$$a_n = \tfrac{1}{2} e^{-(Q_n - Q_{n-1})/2},$$
$$b_n = -\tfrac{1}{2} Q_{n-1_t}.$$

Another interesting lattice equation is obtained by taking $A_\infty^{(0)} = 0$, $b_n = 0$, $A_\infty^{(1)} = 1$ in (2.2.12):

(2.2.15) $$a_{n_t} = \tfrac{1}{2} a_n (a_{n+1}^2 - a_{n-1}^2);$$

choosing $a_n = e^{-u_n/2}$, we find

(2.2.16) $$u_{n_t} = e^{-u_{n-1}} - e^{-u_{n+1}}$$

(Manakov (1975), Kac and Van Moerbeke (1975a)).

We now demonstrate that both (2.2.16) and the Toda lattice (2.2.1) can be related to KdV in the continuum limit via appropriate asymptotic limits. For (2.2.16) (the easier of the two cases) assume $h \to 0$ and $u_n = h^2 \bar{u}_n$. Then the continuum limit yields

(2.2.17) $$\bar{u}_t \sim 2h\bar{u}_x + \frac{h^3}{3}\bar{u}_{xxx} - 2h^3 \bar{u}\bar{u}_x.$$

If we define

$$X = x + 2ht, \qquad T = \frac{h^3}{3}t,$$

then $\bar{u} \sim \bar{u}(X, T)$ solves the KdV equation

$$\bar{u}_T \sim \bar{u}_{XXX} - 6\bar{u}\bar{u}_X.$$

In the case of the Toda lattice we have, assuming $Q_n = h\bar{Q}_n$, $\tau = ht$,

(2.2.18) $$\bar{Q}_{\tau\tau} = \bar{Q}_{xx} + h^2(\bar{Q}_{xxxx} - \bar{Q}_x\bar{Q}_{xx}) + \cdots$$

and that (2.2.18) with $w = \bar{Q}_x$ reduces to the Boussinesq equation:

$$w_{\tau\tau} = w_{xx} + h^2(w_{xxxx} - \partial_x(\tfrac{1}{2}w^2))$$

(see § 2.1). Finally, KdV is obtained by looking for unidirectional waves; i.e., we define

$$Q \sim w(X, T),$$

$$X = x - \tau, \qquad T = \frac{h^2 t}{2},$$

$$u = Q_x,$$

and find

(2.2.19) $$-u_T = u_{XXX} - uu_X.$$

Before considering the question of discrete inverse scattering let us consider the discretization associated with the Zakharov–Shabat eigenvalue problem. As motivation we shall naïvely discretize (1.2.7a) by letting

$$(v_i)_x = = \frac{v_{i,n+1} - v_{i,n}}{h}.$$

Thus (1.2.7a) gives

(2.2.20) $$\begin{aligned}v_{1,n+1} &= v_{1,n}(1 - i\zeta h) + q_n h v_{2,n},\\ v_{2,n+1} &= v_{2,n}(1 + i\zeta h) + r_n h v_{1,n},\end{aligned}$$

where $v_{i,n} \equiv v_i(nh)$, $q_n = q(nh)$, etc. If q_n, r_n were zero it would be natural to define $z = e^{-i\zeta h}$, so that the continuous solution v_1 goes to the discrete limit nicely: $v_1 = e^{-i\zeta x} = e^{-i\zeta n h} = z^{-n}$ and similarly for v_2. Hence here we take $z = e^{-i\zeta h} \sim 1 - i\zeta h$, $1/z \equiv e^{i\zeta h} \sim 1 + i\zeta h$ and if we define $Q_n = q_n h$, $R_n = r_n h$ we find

(2.2.21)
$$v_{1,n+1} = z v_{1,n} + Q_n v_{2,n},$$
$$v_{2,n+1} = \frac{1}{z} v_{2,n} + R_n v_{1,n}.$$

There is a generalization of (2.2.21) which is significant:

(2.2.22)
$$v_{1,n+1} = z v_{1,n} + Q_n v_{2,n} + S_n v_{2,n+1},$$
$$v_{2,n+1} = \frac{1}{z} v_{2,n} + R_n v_{1,n} + T_n v_{1,n+1}$$

(note that in the continuum limit, (2.2.22) *also* relaxes to (1.2.7a)).

First we shall discuss nonlinear differential-difference equations. Associated with either (2.2.21) or (2.2.22) we write the time evolution of v_i as

(2.2.23)
$$\frac{\partial}{\partial t} v_{1,n} = A_n v_{1,n} + B_n v_{2,n},$$
$$\frac{\partial}{\partial t} v_{2,n} = C_n v_{1,n} + D_n v_{2,n}.$$

The analogue of (2.2.9) is obtained by assuming $\partial z/\partial t = 0$ and letting

(2.2.24)
$$\frac{\partial}{\partial t}(E v_{i,n}) = E\left(\frac{\partial v_{i,n}}{\partial t}\right), \quad i = 1, 2,$$

where E is the shift operator $E(v_{i,n}) \equiv v_{i,n+1}$. In what follows we take, for simplicity of presentation, $T_n = S_n = 0$. We have, for example,

$$\frac{\partial}{\partial t}(E v_{1,n}) = z \frac{\partial}{\partial t} v_{1,n} + Q_{n,t} v_{2,n} + Q_n \frac{\partial}{\partial t} v_{2,n}$$
$$= z(A_n v_{1,n} + B_n v_{2,n}) + Q_{n,t} v_{2,n} + Q_n(C_n v_{1,n} + D v_{2,n})$$

and

$$E\left(\frac{\partial}{\partial t} v_{1,n}\right) = A_{n+1} v_{1,n+1} + B_{n+1} v_{2,n+1}$$
$$= A_{n+1}(z v_{1,n} + Q_n v_{2,n}) + B_{n+1}\left(\frac{1}{z} v_{2,n} + R_n v_{1,n}\right).$$

Doing the same for $v_{2,n}$, and setting the coefficients of $v_{i,n}$ equal, we find the analogue of (1.2.8) (which is obtained in the continuum limit):

(2.2.25a)
$$z \Delta_n A_n = Q_n C_n - R_n B_{n+1},$$

(2.2.25b) $\quad \dfrac{1}{z}B_{n+1} - zB_n = Q_{n,t} - A_{n+1}Q_n + D_n Q_n,$

(2.2.25c) $\quad zC_{n+1} - \dfrac{1}{z}C_n = R_{n,t} - R_n(A_n - D_{n+1}),$

(2.2.25d) $\quad \dfrac{1}{z}\Delta_n D_n = -(Q_n C_{n+1} - R_n B_{n+1}),$

where $\Delta_n A_n = A_{n+1} - A_n$, etc.

We now wish to solve (2.2.25) in a manner analogous to that for (1.2.8). Some motivation (not necessary) comes from consideration of the dispersion relation of the linearized problem. In the continuous problem the key function was $\omega(2\zeta)$, i.e., the dispersion relation corresponding to wave number 2ζ of the associated linearized problem. The analogue in the differential-difference case is $\omega(z^2)$; i.e., we look for special solutions of the linearized problem in the form $Q_n = z^{2n} e^{-i\omega(z^2)t}$. As an example let us derive a differential-difference nonlinear Schrödinger equation. Its linearized version is $iQ_{n,t} = Q_{n+1} + Q_{n-1} - 2Q_n$ (an obvious discretization of $iq_t = q_{xx}$), for which $\omega(z^2) = z^2 + 1/z^2 - 2$. As in the continuous case, it turns out that $(A - D)_\infty = \lim_{|n|\to\infty}(A_n - D_n)$ is a quantity which arises in the time dependence of the scattering data. This quantity is proportional to $\omega(z^2)$. This (or alternatively, judicious inspection of (2.2.25)) suggests the expansions

(2.2.26)
$$A_n = z^2 A_n^{(2)} + A_n^{(0)},$$
$$B_n = z B_n^{(1)} + \dfrac{1}{z} B_n^{(-1)},$$
$$C_n = z C_n^{(1)} + \dfrac{1}{z} C_n^{(-1)},$$
$$D_n = D_n^{(0)} + \dfrac{1}{z^2} D_n^{(-2)}.$$

Note that we take only even powers for A, D. The symmetry of the equations (2.2.26) allows us to take B, C in odd powers of z.

Substitution of (2.2.26) into (2.2.25) and equating coefficients of z yields equations for $A_n^{(2)}, A_n^{(0)}, \cdots, D_n^{(0)}, D_n^{(-2)}$. The algebra is straightforward and is listed below. The labels (a)–(d) refer to the satisfaction of a specific equation in (2.2.25).

(a) $\quad Z^3: \quad \Delta_n A_n^{(2)} = 0 \Rightarrow A_n^{(2)} = A_-^{(2)} = \text{const.}$

(b) $\quad Z^{-3}: \quad \Delta_n D_n^{(2)} = 0 \Rightarrow D_n^{(2)} = D_-^{(-2)} = \text{const.}$

(b) $\quad Z^2: \quad -B_n^{(1)} = -A_-^{(2)} Q_n \Rightarrow B_n^{(1)} = A_-^{(2)} Q_n.$

(c) Z^2: $\quad C_{n+1}^{(1)} = A_-^{(2)} R_n \Rightarrow C_n^{(1)} = A_-^{(2)} R_{n-1}$.

(b) Z^{-2}: $\quad B_{n+1}^{(-1)} = D_-^{(-2)} Q_n \Rightarrow B_n^{(-1)} = D_-^{(-2)} Q_{n-1}$.

(c) Z^{-2}: $\quad -C_n^{(-1)} = -D_-^{(-2)} R_n \Rightarrow C_n^{(-1)} = D_-^{(-1)} R_n$.

(a) Z: $\quad \Delta_n A_n^{(0)} = A_-^{(2)}(Q_n R_{n-1} - Q_{n+1} R_n)$
$$\Rightarrow A_n^{(0)} = -A_-^{(2)} Q_n R_{n-1} + A_-^{(0)}.$$

(d) Z^{-1}: $\quad \Delta_n D_n^{(0)} = -D_{-1}^{(-1)}(Q_n R_{n+1} - R_n Q_{n-1})$
$$\Rightarrow D_n^{(0)} = -D_-^{(-2)} R_n Q_{n-1} + D_-^{(0)}.$$

(a) Z^{-1}: $\quad Q_n C_n^{(-1)} - R_n B_{n+1}^{(-1)} = 0$
$$\Rightarrow Q_n(D_-^{(-2)} R_n) - R_n(D_-^{(-2)} Q_n) = 0 \quad \text{(consistent)}.$$

(d) Z: $\quad Q_n C_{n+1}^{(1)} - R_n B_n^{(1)} = 0$
$$\Rightarrow Q_n(A_-^{(2)} R_n) - R_n(A_-^{(2)} Q_n) = 0 \quad \text{(consistent)}.$$

(b) Z^0: $\quad Q_{n,t} = A_-^{(2)}(Q_{n+1} - Q_{n+1} Q_n R_n) + A_-^{(0)} Q_n$
$$- D_-^{(2)}(Q_{n-1} - Q_{n-1} Q_n R_n) - D_-^{(0)} Q_n.$$

(d) Z^0: $\quad R_{n,t} = D_-^{(2)}(R_{n+1} - R_{n+1} R_n Q_n) + D_-^{(0)} R_n$
$$- A_-^{(2)}(R_{n-1} - R_{n-1} R_n Q_n) - A_-^{(0)} R_n.$$

The last two expressions yield the evolution equations

(2.2.27a) $\quad Q_{n,t} = (1 - Q_n R_n)(Q_{n+1} A_\infty^{(2)} - Q_{n-1} D_\infty^{(-2)}) + (A_\infty^{(0)} - D_\infty^{(0)}) Q_n$,

(2.2.27b) $\quad R_{n,t} = (1 - Q_n R_n)(R_{n+1} D_\infty^{(2)} - R_{n-1} A_\infty^{(2)}) + (D_\infty^{(0)} - A_\infty^{(0)}) R_n$.

These equations are consistent with $R_n = \pm Q_n^*$ if $D_\infty^{(-2)} = A_\infty^{(2)*}$ and $(A_\infty^{(0)} - D_\infty^{(0)}) = -(A_\infty^{(0)} - D_\infty^{(0)})^*$. If we take $A_\infty^{(0)} = -i/h^2$ and $A_\infty^{(0)} - D_\infty^{(0)} = 2i/h^2$, we find

(2.2.28a) $\quad Q_{n,t} = \left(\dfrac{-i}{h^2}\right)(Q_{n+1} + Q_{n-1} - 2Q_n) \pm Q_n Q_n^*(Q_{n+1} + Q_{n-1})\left(\dfrac{-i}{h^2}\right)$,

or, if $Q_n = h q_n$,

(2.2.28b) $\quad iq_{n,t} = \dfrac{q_{n+1} + q_{n-1} - 2q_n}{h^2} \pm q_n q_n^*(q_{n+1} + q_{n-1})$.

We refer to this as the differential-difference nonlinear Schrödinger equation.

It should be noted that (2.2.25) with the substitution (2.2.2b) gave 12 equations for 8 unknowns. Two equations are identically satisfied; two others are the evolution equations. This is unlike the continuous theory, where we found 10 equations for 8 unknowns. Summarizing, we have

$$A_n = \left(\frac{i}{h^2}\right)(1 - z^2 \mp Q_n Q_{n-1}^*), \qquad B_n = \left(\frac{i}{h^2}\right)\left(-Q_n z + \frac{Q_{n-1}}{z}\right),$$

$$C_n = \left(\frac{\pm i}{h^2}\right)\left(Q_{n-1}^* z - \frac{Q_n^*}{z}\right), \qquad D_n = \left(\frac{-i}{h^2}\right)\left(1 - \frac{1}{z^2} \mp Q_{n-1} Q_n^*\right).$$

We note also that as $|n| \to \infty$

$$\lim_{|n| \to \infty} (A_n - D_n) = \left(\frac{i}{h^2}\right)\left(2 - z^2 - \frac{1}{z^2}\right).$$

The linear form of (2.2.28b) is $iq_{n,t} = (q_{n+1} + q_{n-1} - 2q_n)/h^2$. Its dispersion relation is obtained by letting $q_n \propto z^{2n} \exp(-i\omega t)$; hence $\omega = (z^2 + z^{-2} - 2)/h^2$. Thus A_n, D_n satisfy the relation

(2.2.29) $$\lim_{|n| \to \infty} (A_n - D_n) = -i\omega(z^2).$$

Here are some other interesting nonlinear differential-difference equations associated with (2.2.25).

(1) *Discrete mKdV.* In (2.2.27) take $A_-^{(0)} = D_-^{(0)} = 0$, $A_-^{(2)} = D_-^{(-2)} = 1$, $Q_n = hq_n$, q_n real ($R_n = \pm Q_n$):

(2.2.30) $$q_{n,t} = (1 \pm h^2 q_n^2)(q_{n+1} - q_{n-1}).$$

(2.2.30) relaxes to the mKdV equation in the same way that (2.2.16) relaxes to KdV.

Using (2.2.22) with the associated time dependence yields other interesting nonlinear differential-difference equations (the algebra is somewhat more tedious; see Ablowitz and Ladik (1975)):

(2) *Self-dual network.*

$$R_n = \pm Q_n = I_n, \qquad T_n = \pm S_n = -V_n,$$

(2.2.31) $$I_{n,t} = (1 \pm I_n^2)(V_{n-1} - V_n),$$

$$V_{n,t} = (1 \pm V_n^2)(I_n - I_{n+1}).$$

(3) *Toda lattice* (2.1.1).

$$R_n = 0, \quad T_n = 1, \quad Q_n = u_{n,t}, \quad S_n = 1 - e^{-(u_{n+1} - u_n)},$$

$$u_{n,tt} = e^{-(u_n - u_{n-1})} - e^{-(u_{n+1} - u_n)}.$$

These equations are obtained by the expansions

$$A_n = A_n^{(1)} z + A_n^{(0)}, \quad B_n = B_n^{(0)} + \frac{B_n^{(-1)}}{z}, \quad C_n = C_n^{(1)} z + C_n^{(0)}, \quad D_n = D_n^{(0)} + \frac{D_n^{(-1)}}{z},$$

etc., substituted into the equivalent of (2.2.25) with S_n, T_n included. Moreover, we note that a scattering problem similar to (2.2.7) is obtained from (2.2.22) whenever we take $R_n = 0$, $T_n = 1$, $Q_n = -\beta_n$, $S_n = 1 - \alpha_n$ and we take $\lambda = z + 1/z$.

(4) *"Discrete* KdV" (2.2.16). $R_n = 0$, $T_n = 1$, $Q_n = 0$, $S_n = 1 - e^{-u_n}$. Here we find
$$u_{n_t} = e^{-u_{n-1}} - e^{-u_{n+1}}.$$

These evolution equations have been found by finite expansions, as in § 1.2. Alternatively Chiu and Ladik (1977) have found generalized evolution operators for some of these problems. Hence, analogous formulae to those in § 1.5 exist for discrete problems.

Next we note that the above ideas also apply to nonlinear partial difference equations, by a discretization of the time variable. Consider, for example, (2.2.21) where all quantities are not evaluated at time t but rather at *time step* m (so $v_{in}(t)$ really is $v_{in}^m = v_i(n\Delta x, m\Delta x, m\Delta t)$, $Q_n(t) \to Q_n^m = Q(n\Delta x, m\Delta t)$, etc.). Moreover, we consider the associated time evolution of the eigenfunctions to be governed by

(2.2.32)
$$\Delta^m v_{1,n}^m = A_n^m v_{1,n}^m + B_n^m v_{2,n}^m,$$
$$\Delta^m v_{2,n}^m = C_n^m v_{1,n}^m + D_n^m v_{2,n}^m.$$

Here $\Delta^m v_{i,n}^m = v_{i,n}^{m+1} - v_{i,n}^m$, $i = 1, 2$. Assuming $\Delta^m z = 0$ and setting
$$\Delta^m(E_n v_{i,n}^m) = E_n(\Delta^m v_{i,n}^m)$$

(following (2.2.24, 25)), we find a completely discrete version of (1.2.8):

(2.2.33)
$$z\Delta_n A_n = Q_n^{m+1} C_n - R_n^m B_{n+1},$$
$$\frac{1}{z}B_{n+1} - zB_n = \Delta^m Q_n^m - (A_{n+1} Q_n^m - D_n Q_n^{m+1}),$$
$$zC_{n+1} - \frac{1}{z}C_n = \Delta^m R_n^m + (A_n R_n^{m+1} - D_{n+1} R_n^m),$$
$$\frac{1}{z}\Delta_n D_n = R_n^{m+1} B_n - Q_n^m C_{n+1},$$

where $A_n = A_n^m$ etc.

Expansions are again suggested by the dispersion relation. The inverse scattering method (Ablowitz and Ladik (1977)) gives the result

(2.2.34)
$$\omega(z^2) = \lim_{|n|\to\infty} \frac{1+A_n}{1+D_n},$$

where $\omega(z^2)$ is the "amplification factor" or dispersion relation of the linearized problem (see also § A.1). For example, the most general *six*-point (constant coefficient) linear partial difference equation (difference scheme) is given by

(2.2.35)
$$\Delta^m Q_n^m = (\alpha_2 Q_{n+1}^m + \alpha_0 Q_n^m + \alpha_{-2} Q_{n-1}^m)$$
$$- (\beta_2 Q_{n+1}^{m+1} + \beta_0 Q_n^{m+1} + \beta_{-2} Q_{n-1}^{m+1}).$$

The corresponding dispersion relation is found, using $Q_n^m = z^{2n}\omega^m(z^2)$, to be

(2.2.36)
$$\omega(z^2) = \frac{1+\alpha_2 z^2 + \alpha_0 + \alpha_{-2}z^{-2}}{1+\beta_2 z^2 + \beta_0 + \beta_{-2}z^{-2}}.$$

A scheme is purely "dispersive" or neutrally stable when $|\omega| = 1$. This is a requirement of inverse scattering. A sufficient condition for this to occur is $\beta_2 = \alpha_{-2}^*$, $\beta_0 = \alpha_0^*$, $\beta_{-2} = \alpha_2^*$. (2.2.36) suggests the expansions $A_n = A_n^{(2)}z^2 + A_n^{(0)} + A_n^{(-2)}z^{-2}$, etc. in (2.2.33).

For example, one interesting partial-difference equation is given by

(2.2.37)
$$i\frac{\Delta^m q_n^m}{\Delta t} = \frac{1}{2\Delta x^2}\left[\left(q_{n+1}^m - 2q_n^m + q_{n-1}^m \prod_{-\infty}^{n-1}\Lambda_k^m\right) + \left(q_{n+1}^{m+1}\prod_{-\infty}^{n}\Lambda_k^m - 2q_n^{m+1} + q_{n-1}^{m+1}\right)\right]$$

$$\pm \frac{1}{4}\left[q_n^m(q_n^{m*}q_{n+1}^m + q_n^{m+1*}q_{n+1}^{m+1}) + q_n^{m+1}(q_{n-1}^m q_n^{m*} + q_{n-1}^{m+1}q_n^{m+1*})\right.$$

$$\left. + 2q_{n+1}^{m+1}q_n^{m*}q_n^m \prod_{-\infty}^{n}\Lambda_k^m + 2q_{n-1}^m q_n^{m+1*}q_n^{m+1}\prod_{-\infty}^{n-1}\Lambda_k^m\right]$$

$$- q_n^m \sum_{-\infty}^{n} \Delta^m S_k^m - q_n^{m+1}\sum_{-\infty}^{n-1}\Delta^m S_k^{m*},$$

where

$$\Lambda_k^m = \frac{1\pm \Delta x^2 q_k^{m+1*} q_k^{m+1}}{1\pm \Delta x^2 q_k^m q_k^{m*}},$$

$$S_k^m = \Delta_x^2(q_{k+1}^m q_k^{m*} + q_k^m + q_{k-1}^{m*}).$$

We note that (2.2.37) is a nonlinear version of a Crank–Nicolson scheme. The truncation error is $O(\Delta t^2, \Delta x^2)$ and in the linear limit the scheme does reduce to the standard Crank–Nicolson scheme. It should also be noted that:

(i) On a periodic interval $[-p, p]$, $-\infty$ in the global terms (containing Λ_k^m, S_k^m) would be replaced by $-p$.

(ii) The associated linear scheme is "neutral" and dispersive: $|\omega| = 1$.

(iii) Corresponding to each linear scheme (2.2.35) we may find a nonlinear difference scheme by taking appropriate expansions for $A_n^m, \cdots, D_n^m$ in (2.2.33).

(iv) The full difference scheme for the nonlinear Schrödinger equation preserves "x-t symmetry," i.e.,

continuum: $x \to -x$, $t \to -t$, $i \to -i$,

discrete: $n \to -n$, $m \to -m$, $i \to -i$.

(v) The order of accuracy of the difference equations as approximations of differential equations (i.e., truncation error) is generally the same for the linear and corresponding nonlinear schemes.

(vi) The schemes are *global*; i.e., they depend on all the mesh points. However, local six-point schemes are suggested (take $\prod \Lambda_k^m = 1$, $\sum \Delta^m S_k^m = 0$) which preserve the linear stability, x-t symmetry and accuracy. Even though the equations are global as a single equation (2.2.37), they can be made into a local system of difference equations.

(vii) Solutions can be calculated by inverse scattering (see following) and can be shown to converge to solutions of the differential equation as $\Delta x, \Delta t \to 0$. Soliton solutions are obtained explicitly.

Recent numerical calculations (Taha and Ablowitz (1981)) have shown that such schemes are quite good from a practical point of view. Moreover, the analysis presented here can be easily extended to other "solvable" nonlinear evolution equations.

2.2.b. Scattering theory. We shall now describe the inverse scattering associated with the discrete 2×2 Zakharov–Shabat scattering problem. These ideas follow in a manner similar to the continuous case § 1.3. However there are some important changes. At the end of this section we shall list the results associated with the discrete Schrödinger equation as it applies to the Toda lattice.

Since it is not very much more difficult, we shall discuss how to do the inverse scattering associated with (2.2.22). Then as a special case we shall take $S_n = T_n = 0$ to obtain results for (2.2.21). The details can be found in Ablowitz and Ladik (1975), (1976). The Jost functions are defined as

$$(2.2.38) \quad n \to -\infty: \quad \begin{aligned} \phi_n &\sim \begin{pmatrix} 1 \\ 0 \end{pmatrix} z^n, \\ \bar{\phi}_n &\sim \begin{pmatrix} 0 \\ -1 \end{pmatrix} z^{-n}, \end{aligned} \qquad n \to \infty: \quad \begin{aligned} \psi_n &\sim \begin{pmatrix} 0 \\ 1 \end{pmatrix} z^{-n}, \\ \bar{\psi}_n &\sim \begin{pmatrix} 1 \\ 0 \end{pmatrix} z^n. \end{aligned}$$

Using ideas similar to those in § 1.3, it can be shown that, for potentials Q_n, R_n, S_n, T_n decaying sufficiently rapidly as $|n| \to \infty$,

$$\phi_n z^{-n}, \psi_n z^n \text{ are analytic for } |z| > 1,$$

$$\bar{\phi}_n z^n, \bar{\psi}_n z^{-n} \text{ are analytic for } |z| < 1$$

(outside and inside the unit circle). This is particularly easy to show when the potentials are on compact support. In this case, by induction one can establish that the above functions are polynomials in $1/z$ (analytic $|z| > 1$) and z (analytic $|z| < \infty$) respectively. The Wronskian relation is given by

$$W_n(\psi, \bar{\psi}) = \prod_{i=n}^{\infty} \frac{1 - S_i T_i}{1 - R_i Q_i} = \psi_{1n} \bar{\psi}_{2n} - \psi_{2n} \bar{\psi}_{1n}.$$

When $T_i = -S_i^*$ and $R_i = -Q_i^*$, W_n is positive definite. Otherwise we assume that $S_i T_i$, $R_i Q_i$ are smaller than 1 initially. The linear independence of ψ_n, $\bar{\psi}_n$

implies

(2.2.39a) $$\phi_n = a\bar{\psi}_n + b\psi_n,$$

(2.2.39b) $$\bar{\phi}_n = -\bar{a}\psi_n + \bar{b}\bar{\psi}_n,$$

and the Wronskian relations imply

(2.2.39c) $$a\bar{a} + b\bar{b} = \prod_{-\infty}^{\infty} \left(\frac{1 - R_i Q_i}{1 - S_i T_i}\right).$$

$a, b, \bar{a}, \bar{b}$ depend parametrically on time through the potentials. We will show that $a\bar{a} + b\bar{b}$ is independent of time. Hence, if $\prod_{-\infty}^{\infty}((1 - R_i Q_i)/(1 - S_i T_i))$ is a nonzero finite number initially, then the Wronskian W_n is finite and does not vanish. We will work with (2.2.39a) on the unit circle; (2.2.39b) is similar. We divide (2.2.39a) by a (assuming $a(z) \neq 0$ on the unit circle):

(2.2.39d) $$\frac{\phi_n}{a} = \bar{\psi}_n + \frac{b}{a}\psi_n,$$

and assume the following representations (which have the desired analytic structure):

(2.2.40a) $$\psi_n = \sum_{n'=n}^{\infty} K(n, n') z^{-n'},$$

(2.2.40b) $$\bar{\psi}_n = \sum_{n'=n}^{\infty} \bar{K}(n, n') z^{n'};$$

in analogy to (1.3.17)

$$\psi = \binom{0}{1} e^{i\zeta x} + \int_x^{\infty} K(x, s) e^{i\zeta s} \, ds.$$

Substitute ψ_n, $\bar{\psi}_n$ into (2.2.39c) and operate on (2.2.39c) with $(1/2\pi i)\oint dz \cdot z^{-m-1}$ (where $\oint$ is the contour integral on the unit circle). We find

(2.2.41) $$\begin{aligned} I &= \frac{1}{2\pi i} \oint \frac{\phi_n}{a} z^{-m-1} \, dz \\ &= \sum_{n'=n}^{\infty} \bar{K}_n(n, n') \frac{1}{2\pi i} \oint z^{n'-m-1} \, dz \\ &\quad + \sum_{n'=n}^{\infty} K(n, n') \frac{1}{2\pi i} \oint \frac{b}{a}(z) z^{-(m+n')-1} \, dz. \end{aligned}$$

Noting that

$$\frac{1}{2\pi i} \oint z^{n'-m-1} \, dz = \delta(n', m)$$

($\delta(n, m) = 1$ when $n = m$, 0 otherwise; $\delta(n, m)$ is the Kronecker delta function), and defining

(2.2.42) $$F_c(m + n') \equiv \frac{1}{2\pi i} \oint \frac{b}{a}(z) z^{-(m+n')-1} \, dz,$$

126   CHAPTER 2

we have

(2.2.43) $$I = \bar{K}(n,m) + \sum_{n'=n}^{\infty} K(n,n')F_c(m+n').$$

We now evaluate the left-hand side of this equation:

(2.2.44) $$I = \frac{1}{2\pi i}\oint \frac{\phi_n}{a}z^{-m-1}\,dz = \frac{1}{2\pi i}\oint \frac{\phi_n z^{-n}}{a}z^{n-m-1}\,dz.$$

$\phi_n z^{-n}$ and $a(z)$ are analytic in the region $|z| > 1$; hence the only singularity is from $a(z_j) = 0$ (i.e., zeros of a). Also, as $z \to \infty$, $\phi_n z^{-n}/a \to J_{\infty,n}$. (We may evaluate $J_{\infty,n}$ by consideration of the limit $z \to \infty$ in (2.2.22). But this formula will not be needed in the sequel.) Assuming a has N simple zeros and noting that at z_k, $\phi_k = \tilde{c}_k \psi_k$, we have

(2.2.45)
$$I = -\sum_{j=1}^{N} \frac{\phi_n(z_j)}{a'(z_j)} z_j^{-m-1} + J_{\infty,n}\delta(n,m)$$
$$= -\sum_{j=1}^{N} \frac{\tilde{c}_j}{a'(z_j)} \sum_{n}^{\infty} K(n,n') z_j^{-(n'+m)-1} + J_{\infty,n}\delta(n,m).$$

Defining $F_D(m+n') \equiv \sum^N \tilde{c}_j z_j^{-(n'+m)-1}$, $c_j = \tilde{c}_j/a'_j$, we have, from (2.2.44) and (2.2.45),

(2.2.46a) $$\bar{K}(n,m) + \sum_{n'=n}^{\infty} K(n,n')F(m+n') = J_{\infty,n}\delta(n,m),$$

where

(2.2.46b)
$$F(m+n') = F_c(m+n') + F_D(m+n')$$
$$= \frac{1}{2\pi i}\oint \frac{b}{a} z^{-(n'+m)-1}\,dz + \sum_{j=1}^{N} \tilde{c}_j z_j^{-(n'+m)-1}.$$

If we do the same for (2.2.39b), we find

(2.2.47a) $$K(n,m) - \sum_{n'=n}^{\infty} \bar{K}(n,n')\bar{F}(m+n') = -\bar{J}_{0,n}\delta(n,m),$$

(2.2.47b) $$\bar{F}(m+n') = \frac{1}{2\pi i}\oint \frac{\bar{b}}{\bar{a}} z^{n'+m-1}\,dz - \sum_{j=1}^{N} \bar{c}_j \bar{z}_j^{n'+m-1},$$

where

$$\bar{J}_{0,n} = \lim_{z \to 0} z^n \frac{\bar{\phi}_n}{\bar{a}}$$

(as before, $\bar{J}_{0,n}$ can be calculated, but it is not needed). We need (2.2.46),

(2.2.47) only for $m > n$. Define

$$K(n, m) = \prod_{n}^{\infty} \frac{1}{1 - R_i Q_i} \kappa(n, m),$$

$$\bar{K}(n, m) = \prod_{n}^{\infty} \frac{1}{1 - R_i Q_i} \bar{\kappa}(n, m),$$

$$\kappa(n, n) = \begin{pmatrix} 0 \\ 1 \end{pmatrix}, \qquad \bar{\kappa}(n, n) = \begin{pmatrix} 1 \\ 0 \end{pmatrix}.$$

(2.2.46), (2.2.47) then yield for $m > n$

(2.2.48a) $\quad \bar{\kappa}(n, m) + \begin{pmatrix} 0 \\ 1 \end{pmatrix} F(m+n) + \sum_{n+1}^{\infty} \kappa(m, n') F(n'+m) = 0,$

(2.2.48b) $\quad \kappa(n, m) - \begin{pmatrix} 1 \\ 0 \end{pmatrix} \bar{F}(m+n) - \sum_{n+1}^{\infty} \bar{\kappa}(n, n') \bar{F}(n'+m) = 0.$

These equations are the discrete (see 1.3.24, 25) analogue of the Gel'fand–Levitan–Marchenko integral equation. They are linear summation equations.

We relate $K(n, m)$, $\bar{K}(n, m)$ to the potentials by substituting $\psi_n = \sum_{n}^{\infty} K(n, n') z^{-n'}$, etc., into the scattering problem (2.2.22). We find discrete partial difference equations for $K(n, m)$, $\bar{K}(n, m)$, and

(2.2.49)
$$Q_n = \kappa_1(n, n+1),$$
$$R_n = -\bar{\kappa}_2(n, n+1),$$
$$S_n = -\frac{1}{1 - R_n Q_n}[\kappa_1(n, n+2) + Q_n \kappa_2(n, n+1)],$$
$$T_n = -\frac{1}{1 - R_n Q_n}[\bar{\kappa}_2(n, n+2) + R_n \bar{\kappa}_1(n, n+1)].$$

When $R_n = \mp Q_n^*$ and $T_n = \mp S_n^*$, we may establish the following symmetry properties:

$$\bar{b} = \pm b^*, \quad \bar{a} = a^*, \quad \bar{z}_j = \frac{1}{z_j^*},$$

$$\bar{c}_j = \pm \frac{c_j^*}{z_j^{*2}}, \quad \bar{F}(n) = \pm F^*(n),$$

$$\bar{\kappa}(n, m) = \begin{pmatrix} \kappa_2^*(n, m) \\ \mp \kappa_1^*(n, m) \end{pmatrix}.$$

Equation (2.2.48) then yields

(2.2.50a)
$$\kappa_1(n, m) - \bar{F}(n+m)$$
$$\pm \sum_{n'=n+1}^{\infty} \sum_{n''=n+1}^{\infty} \kappa_1(n, n'') \bar{F}^*(n''+n') - \bar{F}(n'+m) = 0,$$

(2.2.50b)
$$\bar{F}(n) = \frac{1}{2\pi i} \oint \frac{\bar{b}}{\bar{a}}(z) z^{n-1} - \sum_{j=1}^{N} \bar{c}_j \bar{z}_j^{n-1},$$

(2.2.50c)
$$\kappa_2(n, m) \pm \sum_{n'=n+1}^{\infty} \kappa_1^*(n, n') \bar{F}(n'+m) = 0,$$

(2.2.50d)
$$Q_n = \kappa_1(n, n+1),$$

(2.2.50e)
$$S_n = -\frac{1}{1 \pm Q_n Q_n^*} [\kappa_1(n, n+2) + Q_n \kappa_2(n, n+1)].$$

We now recover $S_n = T_n = 0$ as a special case. When $S_n = T_n = 0$ one can show that $a(z)$, $\bar{a}(z)$ are even in z and $b(z)$, $\bar{b}(z)$ are odd in z. The eigenvalues come in $\pm$ pairs, and $c_j(z_+) = c_j(z_-)$. These properties show that

(2.2.51a) $\bar{F}(n+m) = \begin{cases} 2\bar{F}_R(n+m), & m = n+2p-1, \\ 0, & m = n+2p, \end{cases} \quad p \geq 1,$

(2.2.51b) $\bar{F}_R(n) = \frac{1}{2\pi i} \int_{C_R} \frac{\bar{b}}{\bar{a}} z^{n-1} \, dz - \sum_{1}^{N/2} \bar{c}_j \bar{z}_j^{n-1},$

where C_R is the contour along the right half of the unit circle. We take

(2.2.51c) $\kappa_1(n, m) = \begin{cases} \kappa_{1R}(n, m), & m = n+2p-1, \\ 0, & m = n+2p, \end{cases} \quad p \geq 1,$

and for $m = n + 2p - 1$, we have, from (2.2.50)–(2.2.51),

(2.2.52)
$$\kappa_{1R}(n, m) - 2\bar{F}_R(n+m)$$
$$\pm 4 \sum_{n''=n+1}^{\infty} \sum_{n'=n+1}^{\infty} \kappa_{1R}(n, n'') \bar{F}_R^*(n''+n') \bar{F}_R(n'+m) = 0,$$

where
$$n'' = n + 2p'' = 1, \qquad n' = n + 2p',$$
$$p'' = 1, 2, \cdots, \qquad p' = 1, 2, \cdots.$$

$\kappa_2(n, m)$ also has a symmetry property in this case:

(2.2.53) $\kappa_2(n, m) = \begin{cases} \kappa_{2R}(n, m), & m = n+2p, \\ 0, & m = n+2p-1. \end{cases}$

With $m = n + 2p$, $p = 1, 2, \cdots$, (2.2.50)–(2.2.51) yield

$$(2.2.54) \quad \kappa_{2R}(n, m) \pm 2 \sum_{n'=n+1}^{\infty} \kappa_{1R}^*(n, n') \bar{F}_R(n' + m) = 0,$$

where

$$n' = n + 2p' - 1, \quad p' = 1, 2, \cdots.$$

The potentials are related by

$$(2.2.55a) \quad Q_n = -\kappa_{1R}(n, n+1),$$

$$(2.2.55b) \quad S_n = 0.$$

The (spatial) inverse scattering is the same for both differential-difference and partial difference evolution equations. The only difference in the analysis is the *time dependence*, which we list here.

(1) *Differential-difference.* The time evolution equation is given by (2.2.23). We assume that as $n \to \pm\infty$, $A_n \to A_\pm$, $D_n \to D_\pm$, $B_n, C_n \to 0$. The eigenfunctions that satisfy both (2.2.22) and (2.2.23) are given by

$$(2.2.56) \quad \begin{aligned} \phi_n^{(t)} &= \phi_n e^{A_- t}, & \psi_n^{(t)} &= \psi_n e^{D_+ t}, \\ \bar{\phi}_n^{(t)} &= \bar{\phi}_n e^{D_- t}, & \bar{\psi}_n^{(t)} &= \bar{\psi}_n e^{A_+ t}; \end{aligned}$$

Using similar ideas to those in § 1.4 (the continuous problem) we have

$$(2.2.57a) \quad \begin{aligned} a &= a_0 e^{(A_+ - A_-)t}, & b &= b_0 e^{(D_+ - A_-)t}, \\ \bar{a} &= \bar{a}_0 e^{(D_+ - D_-)t}, & \bar{b} &= \bar{b}_0 e^{(A_+ - D_-)t}; \end{aligned}$$

hence

$$(2.2.57b) \quad \begin{aligned} \frac{b}{a}(t) &= \left(\frac{b}{a}\right)_0 e^{(D_+ - A_+)t}, \\ \frac{\bar{b}}{\bar{a}}(t) &= \left(\frac{\bar{b}}{\bar{a}}\right)_0 e^{(A_+ - D_+)t}, \end{aligned}$$

and similarly,

$$(2.2.57c) \quad \begin{aligned} c_j &= c_{j,0} e^{(D_+ - A_+)(z_j)t}, \\ \bar{c}_j &= \bar{c}_{j,0} e^{(A_+ - D_+)(\bar{z}_j)t}. \end{aligned}$$

The linearized dispersion relation from $Q_n = z^n e^{-i\omega(z)t}$ is found to satisfy

$$(2.2.57d) \quad -i\omega(z^2) = (A_+ - D_+)(z).$$

(2) *Partial difference.* Here the associated time evolution equation is given by (2.2.32). We also assume that $A_n^m \to A_\pm$, $D_n^m \to D_\pm$, $B_n^m \to 0$, $C_n^m \to 0$ as

$n \to \pm\infty$. The eigenfunctions that satisfy both (2.2.22) and (2.2.32) are given by

(2.2.58)
$$\phi_n^{m(t)} = \phi_n^m(1+A_-)^m, \qquad \psi_n^{m(t)} = \psi_n^m(1+D_+)^m,$$
$$\bar{\phi}_n^{m(t)} = \bar{\phi}_n^m(1+D_-)^m, \qquad \bar{\psi}_n^{m(t)} = \bar{\psi}_n^m(1+A_+)^m.$$

As above, we then may deduce

(2.2.59a)
$$a = a_0\left(\frac{1+A_+}{1+A_-}\right)^m, \qquad \bar{a} = \bar{a}_0\left(\frac{1+D_+}{1+D_-}\right)^m,$$

$$b = b_0\left(\frac{1+D_+}{1+A_-}\right)^m, \qquad \bar{b} = \bar{b}_0\left(\frac{1+A_+}{1+D_-}\right)^m,$$

(2.2.59b)
$$\frac{b}{a} = \left(\frac{b}{a}\right)_0 \left(\frac{1+D_+}{1+A_+}\right)^m, \qquad \frac{\bar{b}}{\bar{a}} = \left(\frac{\bar{b}}{\bar{a}}\right)_0 \left(\frac{1+A_+}{1+D_+}\right)^m$$

and

(2.2.59c)
$$c_j = c_{j,0}\left(\frac{1+D_+}{1+A_+}\right)^m (z_j),$$

$$\bar{c}_j = \bar{c}_{j,0}\left(\frac{1+A_+}{1+D_+}\right)^m (\bar{z}_j).$$

Note that here $a_0, \cdots, \bar{c}_{j,0}$ stand for the values of the scattering data at $m = 0$. The linearized dispersion relation $(Q_n^m = z^n \omega^m)$ is found to satisfy

(2.2.59d)
$$\omega(z^2) = \frac{1+A_+}{1+D_+}.$$

The solution procedure for the differential-difference and the partial difference formulations is complete. Special soliton solutions can now be calculated. For example, corresponding to one eigenvalue and no continuous spectrum $(\bar{F}(n) = -\bar{c}\bar{z}_j^{n-1}, S_n, T_n \neq 0)$, the method of solution is to define

$$\hat{\kappa}_1(n) = \sum_m \kappa_1(n, m) \bar{z}^{m*}$$

and reduce the summation equation to finding $\hat{\kappa}_1(n)$ (i.e., operate on (2.2.50) with $\sum_m \bar{z}^{m*}$). In the differential-difference problem a single soliton is given by $(S_n, T_n \neq 0)$

(2.2.60a)
$$Q_n = \left(\frac{\bar{c}_0}{\bar{c}_0^*}\right)^{1/2} \exp\left(-\left(\frac{i}{2}\right)(\omega+\omega^*)t + 2in\theta\right)$$
$$\times \sinh W \operatorname{sech}\left(2nW - \frac{i}{2}(\omega^* - \omega)t + \phi_0\right),$$

(2.2.60b)
$$T_n = -\left(\frac{\bar{c}_0^*}{c_0}\right)^{1/2} \exp\left(-\left(\frac{i}{2}\right)(\omega+\omega^*)t - (2n+1)i\theta\right)$$
$$\times \sinh W \operatorname{sech}\left(2nW - \frac{i}{2}(\omega^* - \omega)t + \phi_0 + W\right),$$

where

(2.2.60c) $\quad \omega = \omega(\bar{z}^2), \quad \bar{z} = e^{-W+i\theta}, \quad \phi_0 = -\log\left(\dfrac{|c_0|}{2\sinh W}\right),$

e.g., in the self-dual network, $\omega = \pm i(z - z^{-1})$.

If $S_n = T_n = 0$, we find

(2.2.61a)
$$Q_n = \left(\frac{\bar{c}_0}{\bar{c}_0^*}\right)^{1/2} \sinh 2W \exp\left(2in\theta - \left(\frac{i}{2}\right)(\omega + \omega^*)t\right)$$
$$\times \operatorname{sech}\left(2nW - \frac{i}{2}(\omega^* - \omega)t + \phi_0\right),$$

where

(2.2.61b) $\quad \phi_0 = -\log\left(\dfrac{|c_0|}{\sinh 2W}\right).$

For example, in the discrete nonlinear Schrödinger equation, $\omega(z^2) = z^2 + z^{-2} - 2$.

In the partial difference case when $S_n = T_n = 0$, we write $(Q_n^m = \Delta x q_n^m)$

$$\omega(z^2) = |\omega| e^{i \operatorname{Arg} \omega},$$

(2.2.62) $\quad Q_n^m = \sinh 2W \, e^{i2n\theta + i2m(\operatorname{Arg}\omega) + i\theta_0} \operatorname{sech}(2nw - mn|\omega| - \phi_0).$

In the partial difference nonlinear Schrödinger equation (2.2.37), we use the linearized dispersion relation

$$\omega = \frac{1 + i\sigma(z^2 - 2 + z^{-2})}{1 - i\sigma(z^2 - 2 + z^{-2})},$$

where $\sigma = \Delta t/(\Delta x)^2$.

It should also be mentioned that the conserved quantities can be worked out as in § 1.6. For example, when $S_n = T_n = 0$, it can be shown from the scattering problem that

(2.2.63a) $\quad \log \bar{a}(z) = \sum\limits_{n=-\infty}^{\infty} \log g_n(z^2),$

where g_n satisfies

(2.2.63b) $\quad g_{n+1}(g_{n+2} - 1) - z^2 \dfrac{R_{n+1}}{R_n}(g_{n+1} - 1) = z^2 R_{n+1} Q_n.$

From (2.2.63b), as $z \to 0$, $g_n(z^2)$ has the expansion

$$g_n = \sum_{i=0}^{\infty} g_n^{(i)} z^{2i}.$$

We find, from (2.2.63b),

(2.2.64a) $\quad g_n \sim 1 + z^2 R_{n-1} Q_{n-2} + z^4 R_{n-1} Q_{n-3}(1 - R_{n-2} Q_{n-2} + \cdots).$

Thus, $\bar{a}(z)$, analytic for $|z|<1$, has the expansion as $z \to 0$

(2.2.64b) $\quad \log \bar{a}(z) \sim \sum_0^\infty c_i z^{2i},$

where the C_i are constant, since $\bar{a}(z)$ is constant. Setting (2.2.64a,b) equal, we have found that

(2.2.65)
$$C_1 = \sum_{-\infty}^{\infty} R_k Q_{k-1},$$
$$C_2 = \sum_{-\infty}^{\infty} \left[R_k Q_{k-2}(1 - R_{k-1} Q_{k-1}) - \frac{1}{2} R_k^2 Q_{k-1}^2 \right].$$

The C_i, $i = 1, 2, \cdots$, are the conserved quantities. It should be noted that (2.2.63a,b) is found from the scattering problem by relating $\bar{a}(z)$ to the eigenfunction $\bar{\phi}_{2n}$ by

$$\bar{a}(z) = \lim_{n \to \infty} (-\bar{\phi}_{2n} z^n)$$

and finding an equation for g_k from

$$\phi_{2n} z^n \equiv \sum_{-\infty}^{n} g_k.$$

Once again, we point out that although the solution process looks formidable, the conceptual program of solution is analogous to that of linear Fourier analysis. Naturally, in the nonlinear problem we have the added difficulty of having to solve either a linear integral equation, in the continuous case, or a linear summation equation, in the discrete case.

Next, we shall simply quote (for completeness) the results for the discrete Schrödinger equation (2.2.7) with specific application to the Toda lattice (Flaschka (1974a,b); see also Case and Kac (1973), Case (1973), Manakov (1975) and Moser (1975a)).

We shall assume that $(a_n - \frac{1}{2})$ and b_n decay rapidly as $|n| \to \infty$. Set $\lambda = (z + 1/z)/2$ and define solutions ϕ_n, ψ_n by the asymptotic conditions

(2.2.66)
$$\phi_n \sim z^n \quad \text{as } n \to +\infty,$$
$$\psi_n \sim z^{-n} \quad \text{as } n \to -\infty,$$

for $|z| = 1$ (these are the discrete Jost functions). By the linear independence of $\phi_n(z)$ and $\phi_n(z^{-1})$ we have

(2.2.67) $\quad \psi_n(z) = \beta(z) \phi_n(z) + \alpha(z) \phi_n(z^{-1}),$

where $|\alpha|^2 = 1 + |\beta|^2$. $R(z) = \beta(z)/\alpha(z)$ is called the reflection coefficient. The eigenvalues are discrete in number and correspond to real values of z in the interval $(-1, 1)$. If $\lambda_j = (z_j + z_j^{-1})/2$ is a discrete eigenvalue, we call $\zeta_n(z_j)$ the normalized eigenfunction, defined by the condition

(2.2.68)
$$\sum_{n=-\infty}^{\infty} \zeta_n^2(z_j) = 1,$$

and it has the behavior $\zeta_n(z_j) \sim c_0 z_j^n$ as $n \to \infty$. The inversion is carried out by computing

(2.2.69a)
$$F(n) = \frac{1}{2\pi i} \oint R(z) z^{n-1} \, dz + \sum_{j=0}^{N} c_j^2 z_j^n$$

and solving, for $m > n$,

(2.2.69b)
$$\kappa(n, m) + F(n+m) + \sum_{n'=n+1}^{\infty} \kappa(n, n') F(n' + m) = 0$$

for $\kappa(n, m)$. Then define

(2.2.70a)
$$(\kappa(n, n))^{-2} = 1 + F(2n) + \sum_{n'=n+1}^{\infty} \kappa(n, n') F(n + n'),$$

and find

(2.2.70b)
$$a_n = \frac{1}{2} \frac{\kappa(n+1, n+1)}{\kappa(n, n)}$$

and

(2.2.70c)
$$b_n = -\frac{1}{2} \frac{\kappa(n, n)\kappa(n-1, n) - \kappa(n, n+1)\kappa(n-1, n-1)}{\kappa(n-1, n-1)\kappa(n, n)}.$$

For the Toda lattice (2.2.1), related to (2.2.7) via (2.2.13–2.2.14), the time dependence is given by

(2.2.71)
$$R(z, t) = R(z, 0) e^{(z - z^{-1})t},$$
$$c_j(t) = c_j(0) e^{t(z_j - z_j^{-1})t}.$$

Finally, in the case of a purely discrete spectrum, $R(z, 0) = 0$, the soliton solutions are computable in closed form. A one-soliton solution corresponding to a single eigenvalue (z_1) is given by

(2.2.72a)
$$e^{-(Q_n - Q_{n-1})} - 1 = \frac{z_1^2 + z_1^{-2} - 2}{A z_1^n + (A z_1^n)^{-1}},$$

where $A = c_1 \exp((z_1 - z_1^{-1})t/2)$. Setting $z = \sigma e^{-W}$, $\sigma = \pm 1$ reduces this to

(2.2.72b) $e^{-(Q_n - Q_{n-1})} = 1 + \sinh^2 W \operatorname{sech}^2 (W(n - n_0) + \sigma \sinh Wt)$,

where n_0 is a constant depending only on c_1, z_1. We note that this soliton solution may travel in either the positive or negative n directions.

2.3. Periodic boundary conditions for the Korteweg–deVries equation. One of the research areas in this field which has attracted great interest is the periodic boundary value problem associated with these special nonlinear evolution equations. Some of the early studies on this problem were done by Lax (1975), Novikov (1974), Kac and van Moerbeke (1975b,c), Dubrovin and Novikov (1975), Its and Matveev (1975), McKean and van Moerbeke (1975), McKean and Trubowitz (1976) and Date and Tanaka (1976a,b). In addition, there have been numerous further studies, with significant results. Many of these are discussed in the survey articles of Matveev (1976) and Dubrovin, Matveev and Novikov (1976). In this chapter we shall only consider the integration of the KdV equation for so-called finite band potentials with periodic boundary values. The resulting solutions are conditionally periodic, or quasiperiodic, i.e., a wave with N phase variables, $\theta_i = \kappa_i x - \omega_i t$, periodic in each θ_i but with, generally speaking, noncommensurate frequencies ω_i. The waves we discuss in this section will be periodic in x and almost periodic in time. (Roughly speaking, a function $f(t)$ is almost periodic if there exists a period $T(\varepsilon)$ such that for any ε $|f(t+T)-f(t)|<\varepsilon$ for all t. For a rigorous definition see Nemytskii and Stepanov (1960) or an equivalent text.) We shall reduce the KdV equation to a finite number of nonlinear ODE's which can be integrated. The integration involves knowledge of some algebraic geometry and hyperelliptic functions. We shall not go into the details of the latter here. In this section we follow the work of Dubrovin and Novikov (1975). Some of the other equations with periodic boundary values that have been considered are the nonlinear Schrödinger (Ma and Ablowitz (1981)), sine-Gordon (McKean (1981)), and Kadomtsev–Petviashvili equations (Novikov and Krichever (1981)).

Mathematically speaking, we consider the question of constructing the solution to KdV,

(2.3.1a) $$u_t - 6uu_x + u_{xxx} = 0,$$

with periodic boundary conditions, period T,

(2.3.1b) $$u(x, t) = u(x + T, t)$$

and given initial values $g(x)$,

(2.3.1c) $$u(x, 0) = g(x).$$

Subsequently $g(x)$ will be more restrictively defined (i.e., $g(x)$ will be an N-band potential).

2.3.a. Direct scattering problem. Associated with (2.3.1) is the Schrödinger scattering problem (see also Chapter 1 and note sign change)

(2.3.2) $$v_{xx} + (E - u)v = 0, \quad E = k^2.$$

Define two solutions of (2.3.2), $\phi(x; x_0, k)$ and $\phi^*(x; x_0, k)$ (ϕ^* is the complex conjugate of ϕ), such that when $x = x_0$ (x_0 is an arbitrary point which we fix to be in the interval $0 \leq x_0 \leq T$)

(2.3.3)
$$\phi(x_0; x_0, k) = 1, \quad \phi^*(x_0; x_0, k) = 1,$$
$$\phi_x(x_0; x_0, k) = ik, \quad \phi_x^*(x_0; x_0, k) = -ik.$$

If $\phi(x; x_0, k)$ is a solution of (2.3.2) then, from (2.3.1b), so is $\phi(x + T; x_0, k)$. Since ϕ, ϕ^* are a complete basis we have that the fundamental solution matrices satisfy

(2.3.4a) $$\Phi(x + T; x_0, k) = \hat{T}(x_0, k)\Phi(x; x_0, k),$$

where

(2.3.4b) $$\Phi(x; x_0, k) = \begin{pmatrix} \phi & \phi_x \\ \phi^* & \phi_x^* \end{pmatrix}(x; x_0, k), \quad \hat{T}(x_0, k) = \begin{pmatrix} a & b \\ b^* & a^* \end{pmatrix}(x_0, k).$$

$\hat{T}$ is often referred to as the *monodromy matrix* (cf. § 3.7). It plays the role of the scattering data in the periodic problem. The Wronskian of two solutions of (2.3.2) is constant in x; i.e., $W(u, v) = uv_x - u_x v = $ const. Since $W(\phi, \phi^*) = -2ik$, we have, upon taking determinants of (2.3.4),

(2.3.4c) $$|a|^2 - |b|^2 = 1.$$

Next the so-called Bloch eigenfunctions, $\psi_\pm(x; x_0, k)$, are defined as solutions of (2.3.2) with the proviso

(2.3.5) $\quad \psi_\pm(x_0; x_0, k) = 1, \quad \psi_\pm(x + T; x_0, k) = \lambda \psi_\pm(x; x_0, k).$

Since $\psi_\pm$ satisfy (2.3.2) they also must be linear combinations of ϕ, ϕ^*:

(2.3.6) $$\psi_\pm(x) = C\phi(x) + D\phi^*(x)$$

(the other arguments are understood; C, D are constants). Employing the definitions (2.3.4a), (2.3.5), we have that C, D satisfy

(2.3.7)
$$(a - \lambda)C + Db^* = 0,$$
$$bC + (a^* - \lambda)D = 0.$$

For nontrivial solutions, $(C, D)\lambda$ must satisfy

(2.3.8a) $$\lambda^2 - \lambda(a + a^*) + |a|^2 - |b|^2 = 0$$

or

(2.3.8b) $$\lambda^2 - 2a_R\lambda + 1 = 0,$$

where a_R is the real part of a. For real $E (E = k^2)$ we have the following cases:

(1) If $|a_R| > 1$, one value of $|\lambda|$ is greater than one, and one value is less than one. Hence the Bloch eigenfunctions are unstable.

(2) If $|a_R| < 1$ then $|\lambda| = 1$ and the Bloch eigenfunctions are stable. In this case if we call $a_R(k) = \cos p(k)$ we have $\lambda = \exp(\pm ip(k))$.

(3) If $|a_R| = 1$ then $\lambda = \pm 1$ and the Bloch eigenfunctions are either periodic or antiperiodic.

Next we define two spectra which will enable us to reconstruct the potential u.

The main spectrum. The main spectrum is composed of the eigenvalues $E_i = k_i^2$ for which at least one of the eigenfunctions is periodic or antiperiodic. The E_i are roots of $|a_R| = 1$. A *stable band* is an (open) line segment between any two adjacent E_i such that $|a_R| < 1$. (In this case the Bloch eigenfunctions are stable.) Likewise an *unstable* band is an open line segment between any two adjacent E_i such that $|a_R| > 1$. The E_i are called the band edges. A typical function $a_R = a_R(E)$ is given in Fig. 2.1.

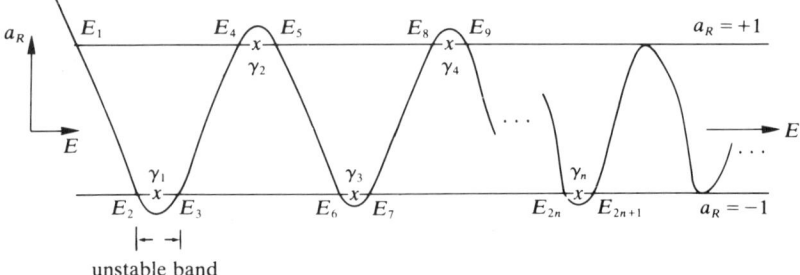

FIG. 2.1. *A typical function* $a_R = a_R(E)$.

Hence in this case (Fig. 2.1) the unstable bands occur between E_{2i} and E_{2i+1}. The plot $a_R(E)$ is sometimes referred to as the Floquet (determinant) diagram. Many of the spectral properties that we shall discuss in this chapter are considered in depth in *Hill's Equation* by Magnus and Winkler (1966), although the point of view is somewhat different.

The auxiliary spectrum γ_i. We shall define these values as those corresponding to locations of E where

(2.3.9) $$a_I + b_I = 0.$$

Since we necessarily have the conditions $|a|^2 - |b|^2 = 1$, (2.3.9) implies $a_R^2 = 1 + b_R^2$. Thus the eigenvalues γ_i lie in the *unstable* bands or possibly at the band edges.

An alternative way to define the auxiliary spectrum is to require an eigenfunction satisfying (2.3.2) (we call it $y(x; x_0, k)$) to satisfy fixed boundary conditions. For example, in this case,

(2.3.10) $$y(x_0) = 0 \quad \text{and} \quad y(x_0 + T) = 0.$$

Then from the fact that $y = A\phi + B\phi^*$ for some nonzero A, B and using (2.3.10) and (2.3.4) we obtain the condition (2.3.9).

Next, we state a number of spectral properties of the E_i and γ_i. The proofs follow essentially standard applications of the theory of ordinary differential equations. (The reader may wish to consult Magnus and Winkler (1966) or McKean and Van Moerbeke (1975) to see the methods of proof.)

Spectral properties.
(1) The main spectrum contains a denumerably infinite number of real eigenvalues. We divide them into nondegenerate band edges E_i, and degenerate band edges $\hat{E}_i$. The property $\partial a_R/\partial E|_{E=E_i} \neq 0$ (i.e., E_i is a simple root of $a_R^2 - 1 = 0$) holds at the nondegenerate band edges, whereas $\partial a_R/\partial E|_{E=\hat{E}_i} = 0$ at the degenerate band edges. Each $\hat{E}_i$ represents a double root of $a_R^2 - 1 = 0$; there are no higher order roots. $a_R'(E) \neq 0$ for $|a_R(E)| < 1$.

(2) The auxiliary spectrum also contains a denumerably infinite number of real eigenvalues. They must be inside the unstable bands or on the band edges. All these eigenvalues are simple roots of $a_I + b_I = 0$. Furthermore, we split the auxiliary spectrum into two portions, γ_i and $\hat{\gamma}_i$. The $\hat{\gamma}_i$ must coincide with the $\hat{E}_i$ and there is one and only one γ_i in each unstable band. (These spectral properties are consequences of the oscillation theorems of ODE's.)

Finite-band potential. An arbitrary periodic potential may have an infinite number of nondegenerate eigenvalues in the main spectrum (the simple roots of $a_R^2 = 1$). Here we consider a finite band potential, i.e., one which has only a finite number of nondegenerate band edges E_i, $i = 1, 2, \cdots, 2n+1$, with all others being degenerate.

The general case has a denumerably infinite number of nondegenerate band edges. This theory was extended to the general case by McKean and Trubowitz (1976), but the extension is far from trivial.

At this point it is convenient to introduce the function $\chi = -i\psi_{\pm x}/\psi_{\pm}$. Since $\psi_{\pm}$ satisfies (2.3.2), we have from this relation that χ satisfies a Riccati equation

(2.3.11a) $$-i\chi' + \chi^2 + u = E.$$

Thus if $\chi = \chi_R + i\chi_I$ we have

(2.3.11b) $$\chi_I = \tfrac{1}{2}(\log \chi_R)_x$$

and the representation

(2.3.11c) $$\psi_{\pm}(x; x_0, E) = \left(\frac{\chi_R(x_0; x_0, E)}{\chi_R(x; x_0, E)}\right)^{1/2} \exp\left(i \int_{x_0}^{x} \chi_R(x; x_0, k)\, dx\right).$$

Later on we shall use the following asymptotic result, as $|E| \to \infty$, $E = k^2$ (we can extend these results into the complex E-plane):

(2.3.12a) $$\psi_{\pm} \sim \exp(ik(x - x_0)),$$

(2.3.12b) $$\chi_{\pm} \sim k\chi_1 + \chi_0 + \frac{1}{k}\chi_{-1} + \frac{1}{k^2}\chi_{-2} + \cdots,$$

with

$$\chi_1 = \pm 1, \quad \chi_0 = 0, \quad \chi_{-1} = \mp u/2,$$

$$\chi_{-2} = \mp \left(\frac{i}{2}\right) u_x, \quad \chi_{-3} = \pm (2u_{xx} - u^2), \cdots;$$

i.e.,

(2.3.13) $$\chi_\pm \sim \pm \left(k - \frac{u}{2k} - \frac{2iu_x}{(2k)^2} + \frac{1}{2k^3}(2u_{xx} - u^2) + \cdots \right).$$

We remark that, generally speaking, the KdV equation and its higher order analogues can be written in the form (cf. Zakharov and Faddeev (1971) or § 1.6)

(2.3.14) $$u_t = \frac{\partial}{\partial x} \frac{\delta \sum_{m=0}^{N} C_m I_{2m+1}}{\delta u(x)},$$

where $I_{2m+1} = \int_{-\infty}^{\infty} \chi_{-(2m+1)}(x) \, dx$, and $\delta I / \delta u$ is the Fréchet derivative of I.

2.3.b. Inverse scattering problem. The function χ has a representation in terms of the scattering data a, b. To see this we use (2.3.6). Applying $\psi(x = x_0)$ and $\psi'(x = x_0) = i\chi(x = x_0)$ (the latter from the definition of χ), we have

(2.3.15a) $$\psi = \frac{1}{2}\left(1 + \frac{\chi_0}{k}\right)\phi + \frac{1}{2}\left(1 - \frac{\chi_0}{k}\right)\phi^*,$$

where $\chi_0 = \chi(x = x_0) = \chi(x_0; x_0, k)$. Then, using the relation (2.3.5) at $x = x_0$, and (2.3.8b), we have

(2.3.15b) $$\chi_{0\pm} = \frac{k(\pm\sqrt{1 - a_R^2} + ib_R)}{a_I + b_I} = \chi_{0R\pm} + i\chi_{0I\pm}.$$

Hence

(2.3.15c) $$\chi_{0R\pm} = \frac{\pm k\sqrt{1 - a_R^2}}{a_I + b_I}.$$

Next, let us introduce another basis for the purpose of discussing *analyticity* of the scattering data. We define eigenfunctions $c(x; x_0, E)$, $s(x; x_0, E)$ ($E = k^2$) such that at $x = x_0$

(2.3.16a) $$c(x_0; x_0, E) = 1, \quad c_x(x_0; x_0, E) = 0$$

and

(2.3.16b) $$s(x_0; x_0, E) = 0, \quad s_x(x_0; x_0, E) = 1.$$

One can write the translation operator as

(2.3.17a) $$c(x + T; x_0, E) = \alpha_{11} c(x; x_0, E) + \alpha_{12} s(x; x_0, E),$$

(2.3.17b) $$s(x + T; x_0, E) = \alpha_{21} c(x; x_0, E) + \alpha_{22} s(x; x_0, E).$$

Then the relationship between the bases (2.3.3-4) and (2.3.16–17) is

(2.3.18)
$$\alpha_{11} = a_R + b_R, \qquad \alpha_{22} = a_R - b_R,$$
$$\alpha_{12} = -k(a_I - b_I), \qquad \alpha_{21} = \frac{a_I + b_I}{k}.$$

By converting the Schrödinger equation (2.3.2) into a Volterra integral equation (from x_0 to x) we can establish that the eigenfunctions c, s are entire functions of E. This implies that the functions α_{ij} in (2.3.18) are entire functions of E. From the theory of complex variables we may write an entire function as the product of its zeros and an entire function with no zeros. Specifically, we write

(2.3.19a) $$1 - a_R^2(E) = g_1(E) \prod_{i=1}^{2N+1} (E - E_i) \prod_{j=1}^{\infty} (E - \hat{E}_j)^2,$$

(2.3.19b) $$\frac{(a_I + b_I)^2}{E} = g_2(E) \prod_{i=1}^{N} (E - \gamma_i)^2 \prod_{j=1}^{\infty} (E - \hat{E}_j)^2.$$

Note that $g_i(E)$, $i = 1, 2$ are entire functions of E with no zeros; the E_i are the simple roots of $1 - a_R^2 = 0$; the $\hat{E}_j$ are the double roots of $1 - a_R^2 = 0$ and simple roots of $a_I + b_I = 0$; the γ_i are the simple roots of $a_I + b_I = 0$ *inside* the unstable bands. Thus χ_{0R}^2 satisfies

(2.3.20) $$\chi_{0R}^2 = \frac{E(1 - a_R^2)}{a_I + b_I} = \frac{\prod_{i=1}^{2N+1} (E - E_i)}{\prod_{i=1}^{N} (E - \gamma_i)^2} g(E),$$

where $g(E) = g_1(E)/g_2(E)$ is entire with no zeros. The asymptotic behavior of $g(E)$ is fixed by considering χ_R as $E \to \infty$ from (2.3.13), namely

(2.3.21) $$\chi_R^2 = E - u + O\!\left(\frac{1}{E}\right).$$

We see that $\lim_{E \to \infty} g(E) = 1$ by comparing (2.3.20–21). Hence, from Liouville's theorem,

(2.3.22) $$g(E) = \frac{g_1(E)}{g_2(E)} = 1.$$

Using (2.3.22), expanding (2.3.20), and comparing with (2.3.21) we have the inverse scattering formula for reconstructing u (at point $x = x_0$):

(2.3.23) $$u = \sum_{i=1}^{2N+1} E_i - 2 \sum_{i=1}^{N} \gamma_i.$$

Next we shall establish that the E_i are independent of the point x_0, whereas the γ_i depend on x_0. Moreover, we shall develop the equations for the $\gamma_i(x_0)$. In this way we shall use (2.3.23) to reconstruct the potential u at any *arbitrary* point x_0.

Consider changing the point x_0 to $x_0 + dx_0$. The fundamental matrix $\Phi(x; x_0 + dx_0, k)$ can then be expanded by Taylor's theorem:

(2.3.24)
$$\Phi(x; x_0 + dx_0, k) \sim \Phi(x; x_0, k) + \Phi_{x_0}(x; x_0, k) \, dx_0$$
$$\equiv (I + Q \, dx_0) \Phi(x; x_0, k).$$

Since both $\Phi(x; x_0 + dx_0, k)$ and $\Phi(x; x_0, k)$ are fundamental solution matrices, and $\Phi(x; x_0, k)$ is already a basis, we know that $I + Q \, dx_0$ must be independent of x. Next we have, from (2.3.4) and (2.3.24),

(2.3.25)
$$\Phi(x + T; x_0 + dx_0) = \hat{T}(x_0 + dx_0) \Phi(x; x_0 + dx_0)$$
$$= (1 + Q \, dx_0) \hat{T}(x_0) \Phi(x; x_0).$$

Then

(2.3.26a) $\quad \hat{T}(x_0 + dx_0)(I + Q \, dx_0) = (I + Q \, dx_0) T(x_0),$

and in the limit $dx_0 \to 0$

(2.3.26b)
$$\frac{d\hat{T}}{dx_0} = [Q, \hat{T}],$$

where $[Q, \hat{T}] = Q\hat{T} - \hat{T}Q$. (The reader may note the analogy to (1.2.4c).)

Next we compute $Q(x_0)$ from

(2.3.27) $\quad Q(x_0) = \Phi_{x_0}(x; x_0) \Phi^{-1}(x; x_0).$

Since the right-hand side must be independent of x, we evaluate it at a convenient location: $x = x_0$. From the boundary conditions we have

(2.3.28a)
$$\Phi(x_0; x_0) = \begin{bmatrix} 1 & ik \\ 1 & -ik \end{bmatrix}.$$

Similarly, the matrix

(2.3.28b)
$$\Phi_{x_0} = \begin{bmatrix} \phi_{x_0} & \phi_{xx_0} \\ \phi_{x_0}^* & \phi_{xx_0}^* \end{bmatrix}$$

is found to be

(2.3.28c)
$$\Phi_{x_0} = \begin{bmatrix} -ik & E - u(x_0) \\ ik & E - u(x_0) \end{bmatrix}.$$

In deriving (2.3.28c) we have used the boundary conditions

(a) $\phi(x_0; x_0, E) = 1$, hence $(d/dx_0)\phi(x_0; x_0, E) = 0$, whereby

$$\phi_{x_0}(x_0; x_0, E) = -\phi_x(x_0; x_0, E) = -ik;$$

(b) $\phi_x(x_0; x_0, E) = ik$, hence $(d/dx_0)\phi_x(x_0; x_0, E) = 0$, whereby

$$\phi_{xx_0}(x_0; x_0, E) = -\phi_{xx}(x_0; x_0, E) = (E - u(x_0))\phi(x_0; x_0, E) = E - u(x_0).$$

Using these results we have, from (2.3.27),

$$(2.3.29) \qquad Q(x_0) = -ik\begin{bmatrix} 1 & 0 \\ 0 & -1 \end{bmatrix} + \frac{iu(x_0)}{2k}\begin{bmatrix} 1 & -1 \\ 1 & -1 \end{bmatrix}.$$

Finally, substituting (2.3.4b) into (2.3.26b) and using the above results, we find

$$(2.3.30a) \qquad \frac{\partial a}{\partial x_0} = -ika + \frac{iu}{2k}(a - b^*) + ika - \frac{iu}{2k}(a + b),$$

$$(2.3.30b) \qquad \frac{\partial b}{\partial x_0} = -ikb + \frac{iu}{2k}(b - a^*) - ikb + \frac{iu}{2k}(a + b)$$

and their complex conjugates (we assume u is real). From (2.3.30a) we find that $a_R = (a + a^*)/2$ satisfies

$$(2.3.31a) \qquad \frac{\partial a_R}{\partial x_0} = 0,$$

which means that the roots (E_i) of $a_R^2 = 1$ are independent of x_0. Moreover, we can now establish the equations for the $\gamma_j(x_0)$. From (2.3.30) we have

$$(2.3.31b) \qquad \frac{\partial}{\partial x_0}(a_I + b_I) = -2kb_R.$$

From $|a|^2 - |b|^2 = 1$ we have at the eigenvalue $E = \gamma_j$, where $(a_I + b_I)(E = \gamma_j) = 0$ (from the definition of γ_j (2.3.9)), that

$$(2.3.31c) \qquad b_R = i\sigma_j\sqrt{1 - a_R^2}, \qquad \sigma_j = \pm 1.$$

Using (2.3.19), (2.3.22) and evaluating (2.3.31) at $k = \gamma_j$ we have

$$(2.3.32) \qquad -\prod_{\substack{k=1 \\ k \neq j}}^{N} \frac{d\gamma_j}{dx_0} = -2i\sigma_j \prod_{i=1}^{2N+1}(\gamma_j - E_i)^{1/2}, \qquad j = 1, \cdots, N,$$

or by defining

$$(2.3.33) \qquad R(E) = \prod_{i=1}^{2N+1}(E - E_i),$$

we have from (2.3.32)

$$(2.3.34) \qquad \frac{d\gamma_j}{dx_0} = \frac{2i\sigma_j R^{1/2}(\gamma_j)}{\prod_{k=1, k\neq j}^{N}(\gamma_j - \gamma_k)}, \qquad \sigma_j = \pm 1, \quad j = 1, \cdots, N.$$

Equations (2.3.34) give the motions of the γ_j with respect to x_0, which in turn determine $u(x_0)$ for all x_0 so long as the γ_j are given at one x_0 and the signs σ_j

are specified at that point. The σ_j will change sign as the γ_j reach the band edge. The E_i, $i = 1, 2, \cdots, 2N+1$ are the branch points. Moreover, for the root $R^{1/2}(E)$ we make branch cuts in the nondegenerate unstable bands between E_{2j} and E_{2j+1} for $j = 1, \cdots, N$, as well as from the point $E = -\infty$ to E_1. From this we can form the Riemann surface of the root $R^{1/2}(E)$.

It is remarkable that there is a transformation by which the N nonlinear ODE's (2.3.34) can be integrated. Before discussing this we shall first show how the auxiliary spectrum evolves in time, due to the KdV equation; again it will turn out that the γ_j satisfy N nonlinear ODE's in time.

2.3.c. Time dependence. We recall from § 1.4 that the time evolution equation of the eigenfunctions associated with (2.3.1), (2.3.2) is given by

$$(2.3.35) \qquad v_t = Mv, \qquad M = (4E + 2u)\frac{\partial}{\partial x} - u_x.$$

At any given time t there are two linearly independent eigenfunctions v_1, v_2 which satisfy (2.3.2) and (2.3.35). In terms of ϕ, ϕ^*, we may express them as follows:

$$(2.3.36) \qquad v_i = f_i(t)\phi + g_i(t)\phi^*, \qquad i = 1, 2.$$

Substituting (2.3.36) into (2.3.35) we find that ϕ satisfies

$$(2.3.37) \qquad \phi_t - M\phi = \lambda\phi + \mu\phi^*,$$

where λ, μ are functions of t only. (Similarly, ϕ^* satisfies the complex conjugate equations.) Next we shall determine λ, μ. At $x = x_0$,

$$(2.3.38) \qquad \phi = 1, \quad \phi_t = 0, \quad \phi_x = ik, \quad \phi_{xt} = 0$$

(similarly for ϕ^*). Thus at $x = x_0$ we have, from (2.3.37),

$$(2.3.39a) \qquad 0 = ik(4E + 2u(x_0)) - u_x(x_0) + \lambda + \mu.$$

Then by taking ∂_x of (2.3.37) and evaluating at $x = x_0$ we have

$$(2.3.39b) \quad 0 = iku_x(x_0) - u_{xx}(x_0) + (4E + 2u(x_0))(u(x_0) - E) + ik(\lambda - \mu).$$

Solving (2.3.39a,b) for λ, μ we find

$$(2.3.40a) \qquad \lambda = \frac{-i}{2k}u_{xx}(x_0) - 4ik^3 + i\frac{u^2}{k},$$

$$(2.3.40b) \qquad \mu = u_x(x_0) - 2iku(x_0) + \frac{i}{2k}u_{xx}(x_0) - \frac{i}{k}u^2.$$

Very generally, these results can be written in the form

$$(2.3.41a) \qquad \Phi_t = Q\Phi + \Lambda\Phi + V,$$

where Φ is given by (2.3.4b), Q by (2.3.27) and Λ, V by

(2.3.41b) $$\Lambda = \begin{bmatrix} \lambda & \mu \\ \mu^* & \lambda^* \end{bmatrix}, \quad V = \begin{bmatrix} 0 & Q\phi_x \\ 0 & Q\phi_x^* \end{bmatrix}.$$

Evaluating (2.3.41) at $x = x_0 + T$ and using (2.3.4), we have

(2.3.42a) $\hat{T}_t \Phi(x_0) + \hat{T}\Phi_t(x_0) = Q\hat{T}\Phi(x_0) + \Lambda \hat{T}\Phi(x_0) + V(x_0 + T),$

which reduces to

(2.3.42b) $$\hat{T}_t = [\Lambda, \hat{T}],$$

where $[\Lambda, \hat{T}] = \Lambda\hat{T} - \hat{T}\Lambda$ (again note the correspondence to 1.2.4c). Evaluating (2.3.42b) yields equations for the scattering data a, b:

(2.3.43)
$$\frac{\partial a}{\partial t} = \mu b^* - \mu^* b,$$
$$\frac{\partial b}{\partial t} = (\lambda - \lambda^*)b + \mu(a^* - a).$$

From this we deduce that a_R, and $a_I + b_I$ satisfy (recall that $a_R = \tfrac{1}{2}(a + a^*)$, etc.)

(2.3.44a) $$\frac{\partial a_R}{\partial t} = 0,$$

(2.3.44b) $$\frac{\partial}{\partial t}(a_I + b_I) = -2\mu_R(a_I + b_I) + 2(\mu_I + \lambda_I)b_R.$$

From (2.3.44a) we immediately have that the eigenvalues E_i of $a_R^2 = 1$ are independent of time. Moreover, from (2.3.44b) we may obtain the motion of the auxiliary spectrum γ_j. We use (2.3.19) and $b_R^2 = a_R^2 - 1 + a_I^2 - b_I^2$. Hence at the eigenvalues $E = \gamma_j$ we have

(2.3.45)
$$-\prod_{k=1}^{\infty}(\gamma_k - \hat{E}_k)\gamma_j^{1/2}g_2^{1/2}(\gamma_j)\prod_{\substack{k=1 \\ j\neq k}}^{N}(\gamma_j - \gamma_k)\frac{d\gamma_j}{dt}$$
$$= -2i(\lambda_I + \mu_I)\sigma_j' g_1^{1/2}(\gamma_j) R^{1/2}(\gamma_j)\prod_{k=1}^{\infty}(\gamma_j - \hat{E}_j),$$

where $\sigma_j' = \pm 1$ and $R(E)$ is given by (2.3.33). Using (2.3.40), and $g_1(E)/g_2(E) = 1$, we have

(2.3.46) $$\frac{d\gamma_j}{dt} = \frac{4i\sigma_j'}{\prod_{k=1, k\neq j}^{N}(\gamma_j - \gamma_k)}(2\gamma_j + u(x_0))R^{1/2}(\gamma_j), \quad j = 1, \cdots, N.$$

Finally, using the inverse formula (2.3.23) in (2.3.46), we have

(2.3.47)
$$\frac{d\gamma_j}{dt} = \frac{8i\sigma_j'}{\prod_{k=1,k\neq j}^{N}(\gamma_j - \gamma_k)} \left(\prod_{k=1}^{N} \gamma_k - \frac{1}{2} \sum_{k=1}^{2N+1} E_k \right) R^{1/2}(\gamma_j),$$

$$\sigma_j' = \pm 1, \quad j = 1, \cdots, N.$$

Equations (2.3.34) and (2.3.47) provide the solution of $\gamma(x_0, t)$ from which we may reconstruct the solution of the KdV equation, $u(x_0, t)$ via (2.3.23). These equations are ordinary differential equations which turn out to be integrable via a suitable transformation (Abel's transformation). Given $\gamma_j, j = 1, \cdots, N$, at some value of x_0, and signs σ_j, we solve (2.3.34) for γ_j. Then we solve (2.3.47) with the appropriate initial values.

Since for $N = 1$ we note that $R(E)$ is a cubic polynomial (2.3.33), it is clear that the solutions of γ_j in x_0 and t (2.3.34, 2.3.47) are simply elliptic functions when $N = 1$. Hence, by integration, the solution $u(x, t)$ from (2.3.23) is the well-known elliptic function solution

$$u = -2(E_3 - E_2) \operatorname{cn}^2 (\sqrt{E_3 - E_1}(x - 2(E_1 + E_2 + E_3)t) + \eta_0/m) + E_1 + E_3,$$

$$m = \frac{E_3 - E_2}{E_3 - E_1}$$

(cn (u/m) is the usual Jacobian elliptic cosine with modulus m). Moreover, this theory now extends known periodic (in x) solutions for KdV to those of the hyperelliptic class.

Geometrically we may think of the γ_j as moving in the nondegenerate unstable bands $l_j = \{E: E_{2j} \leq E \leq E_{2j+1}, j = 1, \cdots, N\}$ in the E-plane. With the branch cuts taken inside the unstable bands, we may form the Riemann surface R of the root $R(E)$, (2.3.33). A path for γ_j has two sections, $[l_j, +1]$ and $[l_j, -1]$. The former is the upper sheet with $\sigma_j = +1$ and the latter is the lower sheet of $R(E)$ with $\sigma_j = -1$. A point transfers sheets when it reaches a band edge (see Fig. 2.2).

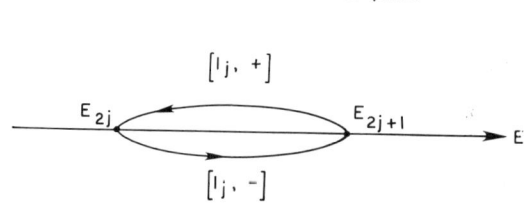

FIG. 2.2. *Branch cut; motion of* γ_j.

Next we describe the integration of (2.3.34) and (2.3.47). Define the coordinates

(2.3.48a) $$P_j = (\gamma_j, \sigma_j),$$

i.e., γ_j with a choice of signs, and the transformation (Abel)

(2.3.48b) $$\Omega_m(E) = \sum_{k=0}^{N-1} C_{km} \frac{E^k \, dE}{R^{1/2}(E)},$$

(2.3.48c) $$\eta_m(P_1, \cdots, P_m) = \sum_{j=1}^{N} \int_{E_{2j}}^{P_j} \Omega_m(E).$$

Typically, the C_{km} are normalized by the condition (α_j is a closed cycle around the unstable band l_j)

(2.3.48d) $$\oint_{\alpha_j} \Omega_m(E) = 2\pi i \, \delta_{jm},$$

which provides N equations in N unknowns. (2.3.48d) indicates that the C_{km} are real ($R^{1/2}(E)$ is purely imaginary if E lies in an unstable band). Moreover, we note that the transformation (2.3.48c) is not uniquely defined (e.g., we can add any multiple of $\oint_{\alpha_j} \Omega_m$ to η_m). In any event, calculating $d\eta_m/dx_0$ using (2.3.48c) and (2.3.34), we find after some manipulation that

(2.3.49) $$\frac{d\eta_m}{dx_0} = \sum_{j=1}^{N} \Omega_m(Q_j) \frac{d\gamma_j}{dx_0} = 2i \sum_{k=0}^{N-1} C_{km} \left(\sum_{j=1}^{N} \frac{\gamma_j^k}{\prod_{n=1, n \neq j}^{N} (\gamma_j - \gamma_n)} \right).$$

But the following relationship is true in general:

(2.3.50) $$\sum_{j=1}^{N} \frac{\gamma_j^k}{\prod_{n=1, n \neq j}^{N} (\gamma_j - \gamma_n)} = \delta_{k, N-1},$$

where $\delta_{k,n}$ is the Kronecker delta function. (We can prove this via contour integration, e.g., by considering the integral

$$\frac{1}{2\pi i} \oint \frac{\gamma^k}{\prod_{n=1, n \neq j}^{N} (\gamma - \gamma_n)} \, d\gamma,$$

with the $\oint$ being a closed contour containing all the γ_n.) Thus with (2.3.50) we immediately have from (2.3.49) that η_m satisfies

(2.3.51) $$\frac{d\eta_m}{dx_0} = 2i C_{N-1, m}.$$

(2.3.51) shows that (2.3.34) is integrable (C_{km} are constant) using the transformation (2.3.48).

In a similar way we may compute the time evolution of the variables η_m (using (2.3.50)):

$$\frac{d\eta_m}{dt} = \sum_{j=1}^{N} \Omega_m(Q_j)\frac{d\gamma_j}{dt}$$

$$(2.3.52) \quad = 8i \sum_{k=0}^{N-1} C_{km} \sum_{j=1}^{N} \frac{\gamma_j^k}{\prod_{n=1,n\neq j}^{N}(\gamma_j - \gamma_n)}\left(\prod_{\substack{s=1\\s\neq j}}^{N}\gamma_s - \frac{1}{2}\sum_{s=1}^{2N+1}E_s\right)$$

$$= 8i\left(\sum_{s=1}^{N}\gamma_s - \frac{1}{2}\sum_{s=1}^{2N=1}E_s\right)C_{N-1,m} - 8i\sum_{k=0}^{N-1}C_{km}\sum_{j=1}^{N}\frac{\gamma_j^{k+1}}{\prod_{s=1,s\neq j}^{N}(\gamma_j - \gamma_s)}.$$

The following identity holds:

$$(2.3.53) \quad \sum_{j=1}^{N}\frac{\gamma_j^{k+1}}{\prod_{i=1,i\neq j}^{N}(\gamma_j - \gamma_i)} = \begin{cases} 0, & k = 0, 1, \cdots, N-3, \\ 1, & k = N-2, \\ \sum_{i=1}^{N}\gamma_1, & k = N-1. \end{cases}$$

The first two results in (2.3.53) ($k \leq N-2$) recapitulate (2.3.50) and the last result can be proven by induction. Hence (2.3.52-3) give

$$(2.3.54) \quad \frac{d\eta_m}{dt} = -8iC_{N-2,m} - 4iC_{N-1,m}\sum_{j=1}^{2N+1}E_j,$$

whereupon (2.3.51, 2.3.54) imply

(2.3.55a) $\quad \eta_m = i(\kappa_m x - \omega_m t + \eta_m^{(0)}),$

(2.3.55b) $\quad \kappa_m = 2C_{N-1,m},$

(2.3.55c) $\quad \omega_m = 8C_{N-2,m} + 4C_{N-1,m}\sum_{j=1}^{2N+1}E_j.$

In (2.3.55) the wavenumbers κ_m, and frequencies ω_m are real since the C_{jk} are real (the C_{jk} are determined from (2.3.48d)).

The transformation (2.3.48) is invertible (see Dubrovin and Novikov (1975) and Dubrovin, Matveev and Novikov (1976)); hence we may write

(2.3.56) $\quad P_j = P_j(\eta_1, \cdots, \eta_N).$

Thus from (2.3.23) and (2.3.48a) we have that u can be written in the general form

(2.3.57) $\quad u(x) = f(\eta_1, \cdots, \eta_N) + \text{const}.$

(i.e., u is a function of $\eta_1, \cdots, \eta_N$). This result indicates that the spatially periodic solution of KdV corresponding to an N-band potential has exactly N phases, and that the solution is, in general, conditionally periodic in time.

Indeed, each γ_j "moves" within its own unstable band and possesses a definite period. Since we have imposed from the beginning that the solution be periodic in x with period T, all the γ_j are periodic in x (here x_0 is replaced by x when (2.3.33) is used).

It can be shown (see, for example, Dubrovin, Matveev and Novikov (1976)) that this solution (2.3.57) of the KdV equation can be expressed in terms of an appropriate algebraic function on a $2N$-dimensional torus, namely

(2.3.58a) $$u = -2\frac{\partial^2}{\partial x^2}\log \Theta_N(\eta_1, \cdots, \eta_N) + \text{const.},$$

where Θ is the so-called Riemann theta function,

(2.3.58b) $$\Theta(\eta_1, \cdots, \eta_N) = \sum_{M=M_1,\cdots,M_N=-\infty}^{\infty} \exp\left(\frac{1}{2}\sum_{j,k=1}^{N} B_{jk}M_jM_k + \sum_{k=1}^{N} M_k\eta_k\right),$$

and the matrix B_{jk} is determined from

(2.3.58c) $$B_{jk} = \oint_{\beta_k} \Omega_j(E).$$

The cycles β_k on the Riemann surface do not intersect the cycles α_j with $j \neq k$, while each β_j intersects cycle α_j at one point E_{2j} (Fig. 2.3).

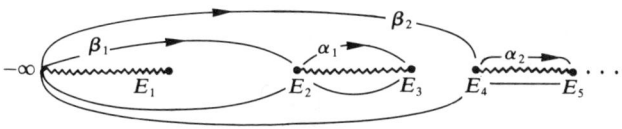

FIG. 2.3

The cycles α_j, β_k represent cycles on the torus of the deformed N-banded Riemann surface. The constant term in (2.3.58a) as well as the phase constant $\eta_m^{(0)}$ in (2.3.55a) can also be given explicit representations on this Riemann surface. The interested reader may refer to Its and Matveev (1975), Matveev (1976) or Dubrovin, Matveev and Novikov (1976). Indeed, Its and Matveev (1975) construct rather general solutions with (2.3.58) as a starting point. These solutions are almost periodic in x as well as in t.

We shall not here discuss in more detail the results pertaining to the periodic problem. However, we do note that many significant results in this direction have been made. In addition to the papers already discussed we suggest that the interested reader consult the following: Krichever (1976), Marchenko (1974), Flaschka and McLaughlin (1976a), Meiman (1977) and Cherednik (1978).

There are not yet many applications of this theory. Recently Flaschka, Forest and McLaughlin (1979) have examined the slow modulation theory of these N-band potentials. Many years earlier Whitham (see Whitham (1974)) worked out the theory for single-phase waves as did Ablowitz and Benney (1970) and Ablowitz (1971), (1972), for multiphase waves. In the latter, numerical integration was used to show the existence of multiply periodic modes (despite the presence of small divisors). The advantage of the present theory is the explicit analytic representations of the solution.

EXERCISES

Section 2.1

1. (a) Prove that $\phi^{(1)} e^{-i\zeta d_1 x}$ and $\psi^{(3)} e^{-i\zeta d_3 x}$ are analytic in the lower half plane and that $\psi^{(1)} e^{-i\zeta d_1 x}$ and $\phi^{(3)} e^{-i\zeta d_3 x}$ are analytic in the upper half plane.

   (b) Take the limit $x \to \infty$ and deduce the corresponding results for a_{11}, b_{33}, a_{33}, b_{11}.

2. In (2.1.1), set

$$N_{12} = N_{21}^* = \begin{cases} iQ = \text{const.} & \text{if } |x| < L_1, \\ 0 & |x| > L_1, \end{cases}$$

$$N_{13} = N_{31}^* = \begin{cases} iR = \text{const.} & \text{if } |x| < L_2, \\ 0 & |x| > L_2, \end{cases}$$

with all other $N_{ij} = 0$. Compute the scattering data explicitly.

3. Show that the time-dependence of a_{11} and a_{33} each yields an infinite set of conserved densities. Are these sets different? Give the first three nontrivial densities in each set. Is there a recursion relation for the nth conserved density? Expansion of a_{22} also gives a set of conservation laws.

4. (a) Under what conditions do zeros of $\bar{a}_0^{(3)}$ in (2.1.46) correspond to zeros of $\bar{a}_f^{(3)}$?

   (b) Show that Q_{2f} has solitons in this case. What happens to Q_{1f}, Q_{3f}?

5. Discuss the differences between the bound states (corresponding to the discrete eigenvalues) in the 2×2 and 3×3 scattering problems.

Section 2.2

1. Aikawa and Toda (1979) have shown that

(*) $\quad \{\partial_t + (a - a^{-1}) \partial_x\}^2 \log(1 + au) = a^{-2}\{u(x + \sqrt{a}) + u(x - \sqrt{a}) - 2u(x)\}$

is a completely integrable equation that contains both the (discrete) Toda lattice and the (continuous) KdV equation as special cases.

   (a) Show that under the transformation

$$a_n - a_{n-1} = -\log(1 + \alpha u_n)$$

the Toda lattice (2.2.1) becomes

(**) $$\partial_t^2 \log(1 + \alpha u_n) = \alpha(u_{n+1} + u_{n-1} - 2u_n).$$

(b) Show that (*) reduces to KdV in the limit $a \to 0$.
(c) Show that under the transformation

$$x = \sqrt{a}\eta + \left(a - \frac{1}{a}\right)t, \quad \tau = a^{3/2}t$$

(*) becomes (**). Because (**) is completely integrable, and the transformation is simply a change of variables with no limits, it follows that (*) is completely integrable. What is the scattering problem?

2. Find the discrete version of the sine-Gordon equation in light-cone coordinates. It is more difficult in lab coordinates; one which is completely integrable in laboratory coordinates has not yet been found

3. Prove solvability of (2.2.48). See § 1.3 (1.3.30)ff.

4. (a) Corresponding to (2.2.63–65), find the recursion relation for the infinite set of motion constants if S_n, $T_n \neq 0$ (see (2.2.22)). Find the first three explicitly. For what choices of Q, R, S, T is there a positive definite motion constant (i.e., an "energy")? How is this related to the solvability of (2.2.48) in Exercise 3?

(b) Find the "trace formulae" for the Toda lattice. How do these compare to Hénon's (1974) integrals?

(c) Find the action-angle variables for the Toda lattice. Differential-difference equations arise as models of one-dimensional crystal lattices. These models are natural candidates for quantization since quantum-mechanical effects often are important in lattice dynamics. The formulation of the problem in terms of action-angle variables may be viewed as a necessary step in the process of quantization; see also § 4.5.

5. Are there any cases in which the scattering data can be worked out explicitly? If so, pick one and find the relevant scattering data.

Section 2.3

Perhaps the exercises in this section should be called "open questions".

1. The word "soliton" has come to mean an exact solution of a completely integrable evolution equation on $-\infty < x < \infty$, represented by one discrete eigenvalue in the IST-spectrum. However, the word originally was coined by Zabusky and Kruskal (1965) to denote an identifiable, localized wave in their numerical experiments on KdV with *periodic* boundary conditions. In terms of the (entirely discrete) spectra of the periodic KdV equation, what distinguishes the "solitons" that Zabusky and Kruskal observed? If one knows both spectra for a particular set of initial data for a periodic KdV problem, can one predict how many "solitons" the numerical experimentalist would observe?

2. (a) What is the recurrence time of an N-band solution of the KdV equation?

(b) Based on his numerical experiments on
$$u_t + uu_x + \delta^2 u_{xxx} = 0$$
on $0 < x < L$ with periodic boundary conditions and initial data of the form
$$u(x, 0) = \sin \frac{2\pi x}{L},$$
Zabusky (1969) found empirically that the recurrence time obeyed
$$T_r = \frac{0.71}{\delta} T_b,$$
where T_b is the time at which the solution with $\delta = 0$ breaks down. Can this formula be derived analytically? For what conditions does it apply? Is there a natural generalization to a wider class of initial conditions?

Chapter 3

Other Perspectives

Overview. We have seen in the previous two chapters that certain nonlinear partial differential equations, when coupled with appropriate boundary conditions and initial data, can be solved exactly by the inverse scattering transform, IST. It is worth noting that one obtains in this way the general solution of the problem posed, which cannot be obtained by any other method known at this time. Even so, IST is not the only possible approach to these problems. In this chapter we consider some other viewpoints and methods for these special equations.

Some order can be imposed on the wide variety of methods available by grouping them in terms of questions they might answer. Here is an attempt at such a grouping, which includes viewpoints that will be discussed in this chapter as well as some that are omitted.

Characterizing IST *problems.* Problems that can be solved by IST possess a great deal of structure, which may include solitons, an infinite set of conservation laws and a complete set of action-angle variables. Other problems do not possess this extra structure, and presumably cannot be solved by IST. Thus the problem arises of *characterizing* the set of partial differential equations that can be solved by IST. The practical question is whether there is a relatively simple test that can be applied directly to a given problem to determine whether it can be solved by some version of IST. This question is relevant for partial differential equations, differential-difference equations, partial difference equations, etc. In this chapter, however, we concentrate almost exclusively on partial differential equations.

It is generally believed at this time that if a problem in $(1+1)$ dimensions has a Bäcklund transformation (§3.1), or a non-Abelian pseudopotential (§3.2), or an N-soliton solution (perhaps $N \geq 3$ is sufficient, but $N = 2$ is not; Hirota (1979a, §3.3, §3.6)), then it should be solvable by some version of IST; i.e., these conditions are thought to be sufficient for IST. On the other hand, the requirement that a partial differential equation have the Painlevé property

(§ 3.7) has been proposed as a necessary condition for a problem to be solvable by IST. Whether any condition is both necessary and sufficient is unknown.

In more than $(1+1)$ dimensions, Zakharov and Shulman (1980) have proposed a method based on whether or not the linearized dispersion relation admits resonant triads. The Painlevé conjecture (§ 3.7) may also be used in higher dimensions. The relation between these two concepts is unknown at this time.

Finding the scattering problem. Given that a problem has the structure required for IST, is there a systematic procedure to find an appropriate scattering problem? Better yet, is there a method to identify scattering problems that is comprehensive enough that its failure implies that the problem in question cannot be solved by IST? That is, can we settle (a) while solving (b)?

Historically, the most successful method for finding scattering problems has been clever guesswork, perhaps with inspiration drawn from a known Bäcklund transformation (§ 3.1). Pseudopotentials (§ 3.2) offer an alternative approach that involves less guessing and is exhaustive in some cases. Chen, Lee and Liu (1979) have proposed another method based on linearization which simultaneously tests the evolution equation and constructs a scattering problem if one exists. Satsuma (1979) has proposed making use of soliton solutions and bilinear forms in construction of Bäcklund transformations and scattering problems. Geometric and group theoretic methods also have been used in special cases.

Special solutions. A viewpoint with some appeal is to abandon the search for general solution, and to concentrate instead on the special solutions that these problems possess (e.g., N solitons on the infinite interval, N-band potentials for the periodic problems). "Direct" methods have been developed (§ 3.3, § 3.6) to find these special solutions. These methods ordinarily are simpler and more direct than IST, and they avoid some of the delicate analytical questions that arise in the study of scattering problems. As an added bonus, the direct method also may generate solutions outside the function-class to which IST applies in its current form. This wider set of solutions includes rational solutions (§ 3.4), higher dimensional solitons and lumps (§ 3.6), and self-similar solutions, including Painlevé transcendents (§ 3.7).

What's going on? Some of the work in this field is aimed not at discovering the next equation solvable by IST, but at learning why this miracle should work at all. Some work suggests that group theory is at the heart of the miracle (e.g., Corones, Markovski and Rizov (1977), Kazhdan, Kostant and Sternberg (1978), Berezin and Perelomov (1980), Hermann (1978)). Related viewpoints focus on differential geometry (Estabrook 1981)) and on algebraic structures of Hamiltonian operators (Gel'fand and Dikii (1977), Adler (1979), Lebedev and Manin (1978), Dorfman and Gel'fand (1979)). Deift and Trubowitz (1980) view problems solvable by IST in terms of infinitely many coupled oscillators constrained to lie on a hypersphere. This description applies both to the periodic problem and to the problem on the infinite interval.

Exploiting the structure. Another viewpoint is simply to accept that these problems possess a great deal of structure, and to use the structure to expand our range of mathematics. The development by McKean and Trubowitz (1976) of hyperelliptic functions with infinitely many branch points is an example, as is the work of Novikov and Krichever (1980) on generalizing hyperelliptic functions.

In summary, there is a wide variety of approaches to problems solvable by IST. Some of these approaches are discussed in this chapter.

3.1. Bäcklund transformations. The focus of this section is on transformations of locally defined solutions of partial differential equations. It may happen that these local solutions can be extended into the global solutions discussed in the previous chapters, but the possibility of such an extension is not germane here. Throughout this section and in § 3.2, a "solution" of a partial differential equation must be defined only on some open connected domain, and does not necessarily satisfy any particular boundary or initial conditions. Moreover, this solution will be understood to be a classical (or "strong") solution. For example, if $u(x, t)$ is to be a solution of KdV, then $(u, u_t, u_x, u_{xx}, u_{xxx})$ all must be defined pointwise in some local domain, and

$$u_t + 6uu_x + u_{xxx} = 0.$$

For simplicity, we will restrict our discussion to (systems of) partial differential equations in two independent variables, (x, t). Because the entire analysis will be local, no distinction between time-like and space-like variables is relevant. It will be convenient to use

$$D(u) = 0 \quad \text{and} \quad E(v) = 0$$

to denote partial differential equations. Depending on the context, these may denote the same equations or different ones.

We begin with some definitions.

DEFINITION. A relation

$$L(u, v, u_x, v_x, u_t, v_t, \cdots ; x, t) = 0$$

(or a set of such relations) is said to *map* $E(v) = 0$ *into* $D(u) = 0$ if every (local) solution of $E(v) = 0$ uniquely defines a (local) solution of $D(u) = 0$.

Example. From Miura's transformation,

(3.1.1) $$u = -v_x - v^2,$$

one computes

$$u_t + 6uu_x + u_{xxx} = -(\frac{\partial}{\partial x} + 2v)(v_t - 6v^2 v_x + v_{xxx}).$$

Hence, every solution of mKdV is mapped under (3.1.1) into a solution of KdV.

Example. The transformation of Cole (1951) and Hopf (1950),

(3.1.2) $$u = -2\nu \frac{\theta_x}{\theta},$$

maps solutions of the heat equation into solutions of Burgers' (1948) equation, because (3.1.2) implies that

(3.1.3) $$u_t + uu_x - \nu u_{xx} = -\frac{2\nu}{\theta}\left(\frac{\partial}{\partial x} - \frac{\theta_x}{\theta}\right)(\theta_t - \nu \theta_{xx}).$$

Two points are worth noting here. The first is that this is the usual definition of a mapping. It is included here only to emphasize that no new terminology is needed to describe (3.1.1) and (3.1.2). Second, mappings do not necessarily identify either $D(u) = 0$ or $E(v) = 0$. Thus,

(3.1.4) $$u_t + cu_x = 0$$

is mapped into itself by either (3.1.1) or (3.1.2). In fact, (3.1.1) maps the infinite sequence of "higher order mKdV's" into the sequence of higher order KdV's, and a similar statement holds for (3.1.2) (cf. Exercises 1,2).

DEFINITION. A set of relations involving $\{x, t, u(x, t)\}$, $\{X, T, V(X, T)\}$ and the derivatives of u and V is a *Bäcklund transformation* (*BT*) between $D(u; x, t) = 0$ and $E(V; X, T) = 0$ if:

(i) BT is integrable for V if and only if $D(u) = 0$;
(ii) BT is integrable for u if and only if $E(V) = 0$;
(iii) given u such that $D(u) = 0$, BT defines V to within a finite set of constants, and $E(V) = 0$;
(iv) given V such that $E(V) = 0$, BT defines u to within a finite set of constants, and $D(u) = 0$.

(Recall that $v_x = f(x, t)$ and $v_t = g(x, t)$ are *integrable* for v iff $v_{xt} = v_{tx}$; i.e., they must be compatible.)

Example. In the theory of complex variables the Cauchy–Riemann conditions,

(3.1.5) $$u_x = v_y, \quad v_x = -u_y,$$

are a BT from the Laplace equation to itself. To see this, eliminate u from (3.1.5) to obtain

(3.1.6) $$v_{xx} + v_{yy} = 0.$$

Then, given v satisfying (3.1.6), (3.1.5) defines u to within one constant $u_0 = u(x_0, y_0)$. Because (3.1.5) is symmetric, this function u must also satisfy (3.1.6).

Example. The transformation actually discussed by Bäcklund (1880) is

(3.1.7) $$\left(\frac{u+v}{2}\right)_x = a \sin\left(\frac{u-v}{2}\right), \quad \left(\frac{u-v}{2}\right)_t = \frac{1}{a} \sin\left(\frac{u+v}{2}\right),$$

which transforms the sine-Gordon equation

(3.1.8) $$\phi_{xt} = \sin \phi$$

into itself. Again, this is verified by cross differentiation of (3.1.7).

Example. The scattering problem for KdV is

(3.1.9a) $$\psi_{xx} + (\zeta^2 + u)\psi = 0,$$
(3.1.9b) $$\psi_t = (\alpha(\zeta) + u_x)\psi + (4\zeta^2 - 2u)\psi_x.$$

These relations also are a BT between the KdV equation and

(3.1.10) $$\psi_t + \psi_{xxx} - \alpha\psi - 6\zeta^2\psi_x - \frac{3\psi_x\psi_{xx}}{\psi} = 0.$$

In this case (3.1.10) is found by solving (3.1.9a) for u and substituting into (3.1.9b), whereas KdV is found by compatibility ($\psi_{xxt} = \psi_{txx}$). Note that u is uniquely determined from ψ (except where ψ vanishes), whereas ψ is only determined by u to within two arbitrary constants (ψ and ψ_x at $\{x_0, t_0\}$).

The distinction between a BT and a mapping is this. Given v, a mapping uniquely defines u but does not specify either $D(u) = 0$ or $E(v) = 0$. A BT need not define u uniquely, even given v, but does specify both $D(u) = 0$ and $E(v) = 0$. Often a BT can be constructed from a mapping by specifying an appropriate evolution equation.

Example.

(3.1.11a) $$v_x = -u - v^2,$$
(3.1.11b) $$v_t = 6v^2 v_x - v_{xxx}$$

is a BT between KdV and mKdV. Note that (3.1.11a) is just (3.1.1). If desired, (3.1.11b) can be rewritten without any x-derivatives of v by using (3.1.11a) repeatedly.

Similarly, (3.1.9a) by itself is a mapping from ψ to u (in a domain in which $\psi \neq 0$).

For the sake of comparison, let us mention some of the other types of transformations possible.

(i) The simplest is a *point transformation*,

(3.1.12) $$u = u(v; x, t).$$

Given a two-dimensional surface defined by $v(x, t)$, (3.1.12) defines a new surface. No differential character is implied. An example is (3.1.1).

(ii) A *contact transformation* (or tangential transformation, or Lie transformation) is characterized by the geometric property that surfaces in one space with a common tangent at a point are transformed into surfaces in another space with a common tangent at the corresponding point. If $v(x, t)$ is transformed into $u(X, T)$, the transformation is a contact transformation if

(3.1.13) $$du - u_X \, dX - u_T \, dT = (dv - v_x \, dx - v_t \, dt)\rho,$$

where ρ is a nonvanishing function of $(v, v_x, v_t; x, t)$. The theory of these transformations was developed by Lie; an ancient reference is Forsyth (1906, Vol. I). An example is the hodograph transformation, used in gas dynamics, in which the role of the dependent and independent variables is reversed. However, these differ from BT's in that neither u nor v is required by (3.1.13) to satisfy any particular differential equation.

(iii) Contact transformations may be generalized to require that higher order contact be preserved under the transformation. Such a transformation has been called a *Lie–Bäcklund transformation* (Anderson and Ibragimov (1979)). The choice of nomenclature is somewhat confusing because it is apparently unrelated to the Bäcklund transformation defined here (but see Fokas (1980), Ibragimov and Shabat (1979)).

(iv) Another definition of a BT, in terms of local jet bundles, was given by Pirani (1979).

Now we come to the main point. What do BT's have to do with solitons and IST? There are a variety of answers to this question, but the most fundamental seems to be this: the scattering problem and (associated time dependence) that constitute an inverse scattering transform also constitute a Bäcklund transformation. We have already seen, in (3.1.9), that the scattering problem for KdV is also a BT. In fact, this identification can be proved rather easily for a large class of problems.

THEOREM. *Let*

(3.1.14) $$v_{1x} + i\zeta v_1 = qv_2, \quad v_{2x} - i\zeta v_2 = rv_1,$$

(3.1.15) $$v_{1t} = Av_1 + Bv_2, \quad v_{2t} = Cv_1 - Av_2,$$

and let $\mathbf{u} = (q, r)$ *satisfy evolution equations* $D(\mathbf{u}) = 0$ *consistent with* (3.1.14) *and* (3.1.15), *and with polynomial linearized dispersion relations. Then for every* ζ, (3.1.14) *and* (3.1.15) *form a Bäcklund transformation between* $D(\mathbf{u}) = 0$ *and some* $E(\mathbf{v}; \zeta) = 0$, *where* $E(\mathbf{v}; \zeta) = 0$ *is a pair of partial differential equations for* $\mathbf{v} = (v_1, v_2)$ *involving* ζ, *but not* $\mathbf{u}$.

Proof. By hypothesis, the integrability condition of (3.1.14) and (3.1.15) for $\mathbf{v}$ is $D(\mathbf{u}) = 0$; this is condition (i) of the definition of a BT. Then, given $\mathbf{u}$ satisfying $D(\mathbf{u}) = 0$, the scattering problem defines $\mathbf{v}$ to within two constants (in Chapter 1, these were fixed by boundary conditions as $x \to +\infty$, say). This is condition (iii) of the definition. Conversely, given any smooth $\mathbf{v}$ in a neighborhood in which neither v_1 nor v_2 vanish, (3.1.14) defines $\mathbf{u}$ uniquely; this is condition (ii). Finally, because $D(\mathbf{u}) = 0$ has a polynomial linearized dispersion relation, (A, B, C) may be found explicitly in terms of $\mathbf{u}$ and its x-derivatives via a finite series expansion (cf. § 1.2). Then from (3.1.14) it follows that (A, B, C) may be given in terms of ζ, v_1, v_2 and their x-derivatives. With this substitution, (3.1.15) becomes $E(\mathbf{v}; \zeta) = 0$, which $\mathbf{v}$ must satisfy. This is condition (iv) and completes the proof. □

The identification of a scattering problem as a BT is not restricted to evolution equations with polynomial dispersion relations, or to this particular scattering problem. We shall not attempt to prove a more general theorem here, but some examples that lie beyond these restrictions are given in the Exercises.

We have now given three different interpretations of IST:

(1) IST is a generalization of the Fourier transform that applies to certain nonlinear problems;

(2) IST is a canonical transformation to action-angle variables of a completely integrable Hamiltonian system.

(3) IST is a Bäcklund transformation.

Each interpretation is valid, and emphasizes some aspect of IST. Whether any of them satisfactorily answers the question "Why should IST work at all?" depends to some extent on the tastes of the reader.

Given that a certain differential equation has a one-parameter family of BT's (i.e., the scattering problem) that relate it to a family of other equations, it may not be surprising that one can construct from these a BT from the equation to itself. The simplest way to do this was developed by Chen (1974), (1976). To illustrate his method, let us derive the BT from KdV to itself, starting with the scattering problem for KdV. If we define $v = \psi_x/\psi$, then (3.1.9a) becomes

$$(3.1.16a) \qquad v_x = -\zeta^2 - u - v^2,$$

and (3.1.10) is equivalent to

$$(3.1.16b) \qquad v_t = 6(v^2 + \zeta^2)v_x - v_{xxx}.$$

This transformation is simply a generalization of (3.1.11), to which it reduces if $\zeta^2 = 0$. The essential point of Chen's method is this: if v is a solution of

(3.1.16b), then so is $(-v)$. Thus, (3.1.16a) produces two different KdV solutions, u and u', from the "same" v:

(3.1.17)
$$v_x = -\zeta^2 - u - v^2,$$
$$(-v)_x = -\zeta^2 - u' - (-v)^2.$$

Adding and subtracting these yields

(3.1.18)
$$-\frac{u+u'}{2} = \zeta^2 + v^2,$$
$$\frac{u-u'}{2} = -v_x.$$

Define a potential function w such that $u = w_x$. Then it follows from (3.1.18) that

$$\frac{w-w'}{2} = -v,$$

(3.1.19)
$$-\left(\frac{w+w'}{2}\right)_x = \zeta^2 + \left(\frac{w-w'}{2}\right)^2.$$

This is the x part of the BT from KdV to itself, originally found by Wahlquist and Estabrook (1973). The other component comes from substituting these into (3.1.16b):

(3.1.20) $\left(\dfrac{w-w'}{2}\right)_t + 6\left(\dfrac{w+w'}{2}\right)_x \left(\dfrac{w-w'}{2}\right)_x + \left(\dfrac{w-w'}{2}\right)_{xxx} = 0.$

Equations (3.1.19, 20) are a BT between

(3.1.21)
$$w_t + 3w_x^2 + w_{xxx} = 0$$

and itself. The solution of KdV is obtained from $u = w_x$.

The main point here is that u' may be constructed from u because $v \to (-v)$ leaves (3.1.16b) invariant but changes (3.1.16a). This same device cannot be applied directly to the scattering problem, because $\psi \to (-\psi)$ leaves both (3.1.10) and (3.1.9a) invariant. The appropriate symmetry to exploit in this case is $\psi \to 1/\psi$, which leaves (3.1.10) invariant but changes (3.1.9a). The equations corresponding to (3.1.17) in this case are

(3.1.22)
$$u = -\zeta^2 - \frac{\psi_{xx}}{\psi},$$
$$u' = -\zeta^2 + \frac{\psi_{xx}}{\psi} - 2\left(\frac{\psi_x}{\psi}\right)^2.$$

Manipulating these equations yields (3.1.19) as before.

It should be emphasized that (3.1.19) is appropriate for all of the higher order KdV equations, which have the same (spatial) scattering problem, (3.1.9a). The t component, (3.1.20), identifies which equation is being transformed.

The derivation of BT's from an equation to itself, starting from the appropriate scattering problem, also was discussed by Ablowitz, Kaup, Newell and Segur (1974) and by Konno and Wadati (1975).

Historically, Bäcklund transformations have been useful both in the discovery of scattering problems and as a means of generating special solutions to these problems (e.g., solitons). We have seen why BT's often have led to scattering problems. Next, let us see how to generate special solutions from BT's. In particular, we consider one-parameter families of BT's between an equation and itself, such as (3.1.7) for the sine-Gordon equation, or (3.1.19) and (3.1.20) for KdV. Then, if one solution of the equation is known, an integration of the BT yields another solution. In each case studied, however, even this integration can be avoided most of the time by showing that the BT in question admits a "theorem of permutability", as discussed by Lamb (1974), (1976). This theorem was first discovered by Bianchi (1902) for the sine-Gordon equation, (3.1.8). From (3.1.7), he showed that four solutions of (3.1.8) are related by

$$(3.1.23) \qquad \tan\left(\frac{\phi_4 - \phi_1}{4}\right) = \frac{a_1 + a_2}{a_1 - a_2} \tan\left(\frac{\phi_2 - \phi_3}{4}\right),$$

where (a_1, a_2) are arbitrary constants. This result can be used to generate an N-soliton solution of (3.1.8), as shown by Lamb (1971). Given a BT from an equation to itself, it is ordinarily not too difficult to find the formula corresponding to (3.1.23), but proving its validity gets rather involved. More details may be found in the papers of Lamb (1974), (1976). (See also § 3.3, where we discuss BT's in terms of Hirota's bilinear equations.)

Roughly, the effect of a BT on a given solution of KdV (say) is to add or subtract one soliton. This may be stated more precisely if the original solution (u_0) satisfies

$$(3.1.24) \qquad \int_{-\infty}^{\infty} |u_0|(1 + |x|)\, dx.$$

We now show that the effect of applying the BT to u_0 is to create u_1, which also satisfies (3.1.24) and whose spectrum (as a potential in (3.1.9a)) differs from that of u_0 by exactly one discrete eigenvalue. The presentation given here is based on work of Deift and Trubowitz (1979); see also Miura (1976b, especially the paper by Wahlquist), Wadati, Sanuki and Konno (1975) and Calogero (1978b). Note that time dependence is irrelevant to this argument, and therefore is suppressed. Because it is irrelevant, the result also applies to any of the higher order KdV's, whose scattering problem is (3.1.9a).

Let $u_0(x)$ be real and satisfy (3.1.24), and have n discrete real eigenvalues $(\lambda = -\kappa_n^2 < \cdots < -\kappa_1^2)$ in

$$\psi_{xx} + u_0\psi = -\lambda\psi, \qquad -\infty < x < \infty.$$

The possibility that $n = 0$ is not excluded, and u_0 may have a continuous spectrum as well. Let $\zeta^2 > \kappa_n^2$, and let $g(x)$ satisfy

(3.1.25) $$g_{xx} + u_0 g = \zeta^2 g,$$

with $g(x) > \varepsilon > 0$ for all x. Deift and Trubowitz (1979), generalizing a theorem of Crum (1955), show that if u_1 is defined by

(3.1.26) $$u_1 = u_0 + 2\frac{d^2}{dx^2}\log g,$$

then u_1 satisfies (3.1.24) and has $n+1$ discrete eigenvalues $(\lambda = -\zeta^2 < -\kappa_n^2 < \cdots < -\kappa_1^2)$ in

(3.1.27) $$\psi_{xx} + u_1\psi = -\lambda\psi.$$

Moreover, the eigenfunction corresponding to $(-\zeta^2)$ is $1/g(x)$, as may be seen by direct substitution in (3.1.27).

To relate (3.1.26) to (3.1.19), set $v = g_x/g$. Because $g > \varepsilon > 0$, v is defined everywhere. Using (3.1.25) we have

$$v_x = \frac{g_{xx}}{g} - \left(\frac{g_x}{g}\right)^2,$$

i.e.,

(3.1.28) $$v_x = -u_0 - (-\zeta^2) - v^2,$$

which is Miura's transformation. With $w_x = u$, (3.1.26) becomes

$$(w_1)_x = (w_0)_x + 2v_x,$$

so that

$$v = \left(\frac{w_1 - w_0}{2}\right),$$

and by (3.1.28)

$$\left(\frac{w_1 + w_0}{2}\right)_x = \zeta^2 - \left(\frac{w_1 - w_0}{2}\right)^2,$$

which is just (3.1.19) with $\zeta^2 \to -\zeta^2$. If u_0 and u_1 satisfy (3.1.19) and each satisfies KdV, then (3.1.20) is necessarily satisfied.

In this case, we have shown that the BT adds one discrete eigenvalue to the spectrum, i.e., it adds one soliton. Alternatively, if we had taken u_1 as given, then u_0 is defined by (3.1.19, 20) and the requirement that $u_0 \to 0$ as $x \to \infty$.

This u_0 satisfies KdV and has the same spectrum as u_1, less one discrete eigenvalue. In general, solitons may be either added or deleted by BT's. More generally, Deift and Trubowitz (1979) derive the N-soliton formula for KdV directly from Crum's theorem.

To simplify the presentation, we have concentrated exclusively on BT's for partial differential equations in this section. However, the close analogy between continuous and discrete problems discussed in § 2.2 suggests that discrete BT's ought to play a role in the theory of partial difference equations analogous to the role of the continuous BT's discussed here. Work on discrete BT's is of much more recent origin, but several examples of discrete BT's between certain partial difference equations and themselves have been found by Chen (1974) and Hirota (1977a,c), (1978), (1979b).

We have not yet addressed the question: How does one determine whether a given partial differential equation has a Bäcklund transformation, relating it to itself or to any other equation? This question was addressed by Clairin (1903), but we defer a discussion of his method to the next section.

3.2. Pseudopotentials and prolongation structures. As in the previous section, our interest here is in local solutions of partial differential equations in two independent variables. Pseudopotentials, originally discussed by Wahlquist and Estabrook (1975), (1976), are closely related to Bäcklund transformations and IST; the terminology is somewhat different. (At first sight, the terminology appears to be very different, because the theory of pseudopotentials usually is developed in terms of exterior differential forms; e.g., see Morris (1979) and the references therein. However, as noted by Corones (1976) and Kaup (1978), that framework is not necessary for the development. The presentation given here does not require familiarity with differential forms.)

3.2.a. The basic concept. The original work of Wahlquist and Estabrook was motivated by the fact that the KdV equation has an infinite set of local conservation laws of the form

$$\frac{\partial T_i}{\partial t} + \frac{\partial F_i}{\partial x} = 0, \quad i = 1, 2, \cdots,$$

where $\{T_i, F_i\}$ are known functions of $u(x, t)$, here a solution of KdV, its x-derivatives, and $\{x, t\}$. Every such conservation law defines a potential function, w_i:

$$\frac{\partial w_i}{\partial t} = F_i, \quad \frac{\partial w_i}{\partial x} = -T_i;$$

i.e.,

(3.2.1) $$dw_i = F_i \, dt - T_i \, dx$$

is an exact differential. Given u and its derivatives, w_i may be obtained by quadrature from (3.2.1). For example, writing the KdV equation as

$$u_t + (3u^2 + u_{xx})_x = 0,$$

we obtain its simplest potential,

(3.2.2) $\qquad w_t = 3u^2 + u_{xx}, \qquad w_x = -u;$

i.e.,

(3.2.3) $\qquad w(x, t) = -\int^x u(\tilde{x}, t) \, d\tilde{x}.$

With appropriate restrictions, $(-w)$ satisfies (3.1.21). Thus, for a given solution of KdV, w is a function of $\{x, t\}$, defined by (3.2.3). Alternatively, considering all possible (local) KdV solutions, one may think of w as a function of five independent variables, $\{x, t, u, u_x, u_{xx}\}$, with w defined to within a constant by (3.2.2).

Once w is fixed (by choosing this constant), one may enlarge the space of independent variables to $\{x, t, u, u_x, u_{xx}, w\}$ and seek new potentials defined on this large space. This is called "prolonging" the original set of variables, and the sequence of potentials obtained by repeated prolongation determines the *prolongation structure* of the original problem.

The next potential, w_1, (if it exists) satisfies equations of the form

(3.2.4)
$$(w_1)_x = A(x, t, u, u_x, u_{xx}; w),$$
$$(w_1)_t = B(x, t, u, u_x, u_{xx}; w),$$

where A and B are to be determined from $(w_1)_{xt} = (w_1)_{tx}$ (integrability), and the fact that u satisfies KdV. Proceeding in this way, one finds a sequence of equations of the form (3.2.4); at each stage, the right-hand sides, A and B, involve the original variables plus all of the new potentials that already have been found. Once A and B are known, the equations may be integrated by quadrature over known functions.

Alternatively, if one allows the unknown potential to enter into the right-hand side as well, then the sequence of equations like (3.2.4) is replaced by a coupled set of equations of the form

(3.2.5) $\qquad \left. \begin{array}{l} \partial_x(w_i) = A_i(x, t, u, u_x, u_{xx}; \mathbf{w}) \\ \partial_t(w_i) = B_i(x, t, u, u_x, u_{xx}; \mathbf{w}) \end{array} \right\}, \quad i = 1, 2, \cdots, N$

where $\mathbf{w} = (w_1, w_2, \cdots, w_N)$. With N fixed and $\mathbf{A}$ and $\mathbf{B}$ known, solving (3.2.5) means solving coupled partial differential equations, in contrast to the previous sequential quadratures. Thus the solutions of (3.2.5) are not necessarily limited to the sequence of potentials discussed above. Wahlquist and Estabrook (1975) called the solutions of (3.2.5) *pseudopotentials*. For reasons to be given later, pseudopotentials that are not (one-to-one) equivalent to a sequence of potentials are called *non-Abelian*.

Examples. (1) The BT between KdV and mKdV, (3.1.11), is also a pseudopotential, because (3.1.1) may be put in the form of (3.2.4), i.e., (3.2.5) with $N = 1$:

$$v_x = -u - v^2,$$
$$v_t = u_{xx} + 2u^2 + 2uv^2 - 2u_x v.$$

(2) The scattering problem for KdV, (3.1.9), may be written as

$$\psi_x = \phi, \qquad \psi_t = (\alpha + u_x)\psi + (4\zeta^2 - 2u)\phi,$$
$$\phi_x = -(\zeta^2 + u)\psi, \qquad \phi_t = [u_{xx} - (4\zeta^2 - 2u)(\zeta^2 + u)]\psi + (\alpha - u_x)\phi.$$

Thus, the eigenfunctions for (3.1.9) are also pseudopotentials, with $N = 2$.

(3) The generalized Zakharov–Shabat scattering problem (1.2.7), with A, B and C determined by a finite series expansion, is in the form (3.2.5), with $N = 2$. All of these examples are non-Abelian. The last two also happen to be linear in the pseudopotential.

It should be clear that pseudopotentials, Bäcklund transformations and IST are all very closely related by the underlying concept of compatibility of ∂_x, ∂_t. In particular, a linear scattering problem (without boundary conditions) for a given partial differential equation is, by definition, also a pseudopotential. Therefore, if it can be determined that a given equation is not consistent with any pseudopotential, linear or not, then there is no scattering problem and the equation cannot be solved by IST.

Thus we come to one of the fundamental open questions in this subject:

> For a given partial differential equation, is there a systematic procedure that always finds a pseudopotential if one exists, and fails conclusively if none exists?

The question is relevant for any number of independent variables, but we restrict our attention here to two: $\{x, t\}$.

3.2.b. Problems with polynomial dispersion relations.

The clearest results apply to equations that are first order in t and of finite order in x; i.e., the linearized dispersion relation, $\omega(k)$, is polynomial. We will show below that, subject to some restrictions on its form, the question of finding a pseudopotential always can be reduced to a certain question in the theory of Lie algebra. As we will show, an important consequence of this reduction is the following fact: if a given equation has no linear pseudopotential, then it has no pseudopotential at all.

The method of Wahlquist and Estabrook (1975), (1976) for finding pseudopotentials also has been used by Corones (1976), Corones and Testa (1976) and Kaup (1980). It overlaps significantly with the much older method of Clairin (1903), which was used by Lamb (1974), (1976) and others. We illustrate the method by considering Burgers' equation,

(3.2.6) $$u_t + uu_x = u_{xx}.$$

Our interest is in all possible (local) solutions of (3.2.6), so $(u, u_x, u_{xx}, \cdots)$ all may be given independently at a particular point, with t-derivatives then found through (3.2.6). If complex solutions were permitted, x-derivatives of u^* also would be independent. We seek a pseudopotential that depends on u and its derivatives, up to but not including the highest derivative in the equation. For (3.2.6), this means

(3.2.7) $$\left. \begin{array}{l} \partial_x(q_i) = A_i(u, u_x, \mathbf{q}) \\ \partial_t(q_i) = B_i(u, u_x, \mathbf{q}) \end{array} \right\}, \quad i = 1, \cdots, N,$$

where $\mathbf{q} = (q_1, \cdots, q_N)$, for some finite N.

(i) If $N = 1$, then q is a scalar and (3.2.7) is in the form of a BT.
(ii) If A_i and B_i are *linear* in $\mathbf{q}$, then

(3.2.8) $$\partial_x(q_i) = A_{ij}q_j,$$
$$\partial_t(q_i) = B_{ij}q_j,$$

where $A_{ij}(i, u_x)$ and $B_{ij}(u, u_x)$ are $N \times N$ matrices (summation over the repeated index in (3.2.8) is implied). If A_{ij} and B_{ij} also contain a free parameter (i.e., an "eigenvalue"), then (3.2.8) is a candidate for a linear scattering problem. We will see below that if any pseudopotential exists, then a linear one (of some finite dimension) exists.

For fixed $\mathbf{A}$, $\mathbf{B}$, (3.2.7) is integrable for $\mathbf{q}$ only if

(3.2.9) $$(\mathbf{q})_{xt} = (\mathbf{q})_{tx}.$$

This is the fundamental requirement for a pseudopotential. It is worth noting that in Chapter 1, the basic equations (1.2.8) were obtained by the same requirement (integrability of (1.2.7)). For Burgers' equation (3.2.6),

(3.2.10) $$\mathbf{q}_{xt} = (u_{xx} - uu_x)\frac{\partial \mathbf{A}}{\partial u} + (u_{xxx} - (uu_x)_x)\frac{\partial \mathbf{A}}{\partial u_x} + (\mathbf{B} \cdot \nabla)\mathbf{A},$$
$$\mathbf{q}_{tx} = u_x \frac{\partial \mathbf{B}}{\partial u} + u_{xx}\frac{\partial \mathbf{B}}{\partial u_x} + (\mathbf{A} \cdot \nabla)\mathbf{B},$$

where $(\mathbf{A} \cdot \nabla) \equiv \sum_i A_i \partial/\partial q_i$. Because u_{xxx} is (locally) independent of u, u_x, u_{xx}, (3.2.9) cannot be satisfied unless

(3.2.11) $$\frac{\partial \mathbf{A}}{\partial u_x} = 0.$$

Similarly, the coefficients of u_{xx} must balance:

$$\frac{\partial \mathbf{B}}{\partial u_x}(u, u_x, \mathbf{q}) = \frac{\partial \mathbf{A}}{\partial u}(u, \mathbf{q}),$$

so that

(3.2.12) $$\mathbf{B}(u, u_x, \mathbf{q}) = u_x \frac{\partial \mathbf{A}}{\partial u}(u, \mathbf{q}) + \mathbf{C}(u, \mathbf{q}).$$

Thus (3.2.9) becomes

(3.2.13) $$u_x^2 \frac{\partial^2 \mathbf{A}}{\partial u^2} + u_x \frac{\partial \mathbf{C}}{\partial u} + u u_x \frac{\partial \mathbf{A}}{\partial u} + u_x \left[\frac{\partial \mathbf{A}}{\partial u}, \mathbf{A}\right] + [\mathbf{C}, \mathbf{A}] = 0,$$

where

(3.2.14) $$[\mathbf{A}, \mathbf{B}] \equiv (\mathbf{B} \cdot \nabla)\mathbf{A} - (\mathbf{A} \cdot \nabla)\mathbf{B}.$$

For $N = 1$,

$$[A, B] = B \frac{\partial A}{\partial q} - A \frac{\partial B}{\partial q} = B^2 \frac{\partial}{\partial q}\left(\frac{A}{B}\right),$$

whereas for linear pseudopotentials $[\mathbf{A}, \mathbf{B}]$ is proportional to the usual commutator of matrices:

(3.2.15) $$[\mathbf{A}, \mathbf{B}] = (A_{ij}B_{jk} - B_{ij}A_{jk})q_k$$

(proof by computation).

In (3.2.13), the dependence on u_x is now explicit. The coefficient of u_x^2 (i.e., $\partial^2 \mathbf{A}/\partial u^2$) must vanish, so that

(3.2.16) $$\mathbf{A} = u\boldsymbol{\alpha}(\mathbf{q}) + \boldsymbol{\beta}(\mathbf{q}).$$

Then the coefficient of u_x in (3.2.13) becomes

$$\frac{\partial}{\partial u} \mathbf{C}(u, \mathbf{q}) + u\boldsymbol{\alpha}(\mathbf{q}) + [\boldsymbol{\alpha}, \boldsymbol{\beta}] = 0,$$

which can be integrated:

(3.2.17) $$\mathbf{C} = -\frac{u^2}{2}\boldsymbol{\alpha} - u[\boldsymbol{\alpha}, \boldsymbol{\beta}] + \boldsymbol{\delta}(\mathbf{q}).$$

Define

(3.2.18) $$\boldsymbol{\gamma}(\mathbf{q}) \equiv [\boldsymbol{\alpha}, \boldsymbol{\beta}],$$

and substitute (3.2.16–18) into (3.2.13). The coefficients of u^2, u and 1, along with (3.2.18), are

(3.2.19)
$$[\alpha, \gamma] + \tfrac{1}{2}\gamma = 0,$$
$$[\alpha, \delta] - [\beta, \gamma] = 0,$$
$$[\delta, \beta] = 0,$$
$$[\alpha, \beta] - \gamma = 0.$$

Thus, (3.2.6) has a pseudopotential of the form (3.2.7) if and only if (3.2.19) has a nontrivial solution. For evolution equations that are first order in time and of finite order in x, the existence of a pseudopotential always reduces to finding a nontrivial solution of a set of relations like (3.2.19). For example, the reader may verify that if we had started with Fisher's (1937) equation (a popular model of population dynamics, cf. Hoppenstaedt (1975)),

(3.2.20) $$u_t = u_{xx} + u - u^2,$$

instead of (3.2.6), a similar calculation would have yielded

(3.2.21a,b,c,d)
$$[\alpha, \gamma] + \alpha = 0,$$
$$[\alpha, \delta] + [\beta, \gamma] - \alpha = 0,$$
$$[\delta, \beta] = 0,$$
$$[\alpha, \beta] - \gamma = 0.$$

instead of (3.2.19). The calculation for KdV also follows these lines (Wahlquist and Estabrook (1975)).

It is not difficult to find solutions for (3.2.19) by taking $N = 1$. One (nearly) trivial solution is

(3.2.22) $$N = 1, \quad \alpha = -\frac{q}{2}, \quad \beta = \gamma = \delta = 0,$$

so that (3.2.7) becomes

(3.2.23a, b) $$q_x = -\frac{u}{2}q, \quad q_t = -\frac{1}{2}\left(u_x - \frac{u^2}{2}\right)q.$$

The reader will recognize (3.2.23a) as the Cole–Hopf transformation (3.1.2), and (3.2.23b) becomes the heat equation after u is eliminated. Note also that ($\log q$) is actually a potential, corresponding to the conservation law of (3.2.6).

A less trivial solution of (3.2.19) is (for $N = 1$)

$$\alpha = -\frac{q}{2},$$

$$\beta = C_1 q^2 - \frac{C_2}{2} q,$$

(3.2.24)

$$\gamma = \frac{C_1}{2} q^2,$$

$$\delta = \frac{C_1 C_2}{2} q^2 - \frac{C_2^2}{4} q,$$

so that (3.2.7) becomes

$$q_x = -\frac{(u + C_2)}{2} q + C_1 q^2,$$

(3.2.25)

$$q_t = -\frac{1}{2}\left(u_x - \frac{u^2}{2} + \frac{C_2^2}{2}\right) q - \frac{C_1}{2}(u - C_2) q^2.$$

In the special case ($C_1 = \frac{1}{2}$, $C_2 = 0$), (3.2.25) is a BT between (3.2.6) and itself. Thus, Burgers' equation has a BT (to itself), and can be solved exactly (via the Cole–Hopf transformation) (cf. Exercise 2). It has traveling wave solutions, but no solitons and (apparently) only one conservation law of polynomial type and independent of x, t.

The same procedure fails when applied to (3.2.21): there are no nontrivial solutions of (3.2.21) with $N = 1$ (Exercise 4). For the general case ($N \neq 1$), it is helpful to introduce some concepts from the theory of Lie algebras. (More information about Lie algebras may be found in the books by Jacobson (1962) or by Samelson (1969), which are reasonably self-contained.)

3.2.c. Lie algebras. Let $(v_1, \cdots, v_n)$ be elements of a linear vector space, V, (dimension $m \leq n$). The vector space can be made into an *algebra* by defining an operation ("multiplication") which associates with every pair of vectors $\{v_1, v_2\}$ in V a product $(v_1 v_2) \in V$. The operation of multiplication must satisfy bilinearity conditions,

(i) $\quad (v_1 + v_2) v_3 = v_1 v_3 + v_2 v_3, \quad v_1(v_2 + v_3) = v_1 v_2 + v_1 v_3,$

(3.2.26)

(ii) $\quad c(v_1 v_2) = (cv_1) v_2 = v_1(cv_2),$

for any scalar c.

Because the space has finite dimension (m), the operation is defined completely by a set of (m) basis vectors and an $m \times m$ multiplication table of these vectors. The algebra is a *Lie algebra* if its multiplication also satisfies

(i) $$v_1 v_1 = 0,$$

(3.2.27)

(ii) $$(v_1 v_2)v_3 + (v_2 v_3)v_1 + (v_3 v_1)v_2 = 0;$$

(ii) is the "Jacobi identity". A Lie algebra is *Abelian* if

(3.2.28) $$v_1 v_2 = 0,$$

for every v_1, v_2 in V, and *non-Abelian* if any two elements of V have a nonzero product.

Examples. (1) The solution (3.2.22) of (3.2.19) is an Abelian Lie algebra on a one-dimensional vector space. A basis vector is α.

(2) The "special unitary Lie algebra," su(2), consists of all complex-valued 2×2 matrices with zero trace, where "multiplication" of two matrices is defined by their commutation:

(3.2.29) $$[A, B] = AB - BA.$$

A basis for this vector space is

$$s_x = \frac{i}{2}\begin{pmatrix} 0 & 1 \\ 1 & 0 \end{pmatrix}, \quad s_y = \frac{1}{2}\begin{pmatrix} 0 & -1 \\ 1 & 0 \end{pmatrix}, \quad s_z = \frac{i}{2}\begin{pmatrix} 1 & 0 \\ 0 & -1 \end{pmatrix},$$

and it is easy to verify that

$$[s_x, s_y] = s_z, \quad [s_y, s_z] = s_x, \quad [s_z, s_x] = s_y,$$

so that su(2) is non-Abelian. Note that the scattering problem in Chapter 1, (1.2.7a), has the form

$$\mathbf{v}_x = X\mathbf{v}, \quad \mathbf{v}_t = T\mathbf{v},$$

where X and T are elements of su(2).

What has all of this to do with pseudopotentials? By hypothesis, the vectors $\{\alpha, \beta, \gamma, \delta\}$ in (3.2.19) or (3.2.21) are elements of an N-dimensional vector space. It is easy to show that the bilinear operation $[\mathbf{A}, \mathbf{B}]$ defined by (3.2.14) satisfies (3.2.26, 27). Thus a set of relations like (3.2.19) or (3.2.21) has a solution if and only if the relations are consistent with some (finite dimensional) Lie algebra. More importantly, the original problem has a pseudopotential (of the form specified) if and only if such a Lie algebra exists.

The general problem of determining whether a partially completed multiplication table can be embedded in any finite dimensional Lie algebra is open at this time. Even so, this connection with Lie algebras can be exploited, using Ado's theorem.

DEFINITIONS. A *representation* of an abstract Lie algebra is an explicit identification of each element of the algebra with an $N \times N$ matrix, such that the multiplication table is satisfied. The representation *is faithful* if the only element identified with the zero-matrix is the zero-element of the original vector space.

ADO'S THEOREM. *Every finite dimensional Lie algebra has a faithful finite dimensional representation.*

The proof may be found in Jacobson (1962, p. 202). The consequence of the theorem is that *it is always sufficient to look for linear pseudopotentials* like (3.2.8), and to look for matrix solutions of systems like (3.2.21). (This result applies to systems that are first order in t, and of finite order in x.)

A system like (3.2.21) always has a trivial solution ($\alpha = \beta = \gamma = \delta = 0$), as well as nearly trivial Abelian solutions

$$\alpha = \gamma = 0, \quad [\beta, \delta] = 0.$$

We now show that every Abelian Lie algebra corresponds to a sequence of potential functions, which in turn corresponds (at best) to a sequence of conservation laws for the original problem. Given an evolution equation for u (first order in t, pth order in x), we seek a (linear) pseudopotential in a form similar to (3.2.8). A calculation similar to that given above further restricts the pseudopotential to

$$(3.2.30) \qquad \mathbf{q}_x = \sum_j^m a_j \underline{\alpha}_j \mathbf{q}, \qquad \mathbf{q}_t = \sum_j^m b_j \underline{\alpha}_j \mathbf{q},$$

where a_j, b_j are known scalar functions of $(u, u_x, \cdots, u_{(p-1)x})$, and $\underline{\alpha}_j$ are $N \times N$ constant matrices (elements of the Lie algebra). The system is Abelian if

$$(3.2.31) \qquad [\underline{\alpha}_i, \underline{\alpha}_j] = 0, \quad \text{for all } i, j.$$

Let λ_1 be a simple eigenvalue of $\underline{\alpha}_1$, and v its eigenvector, i.e.,

$$\underline{\alpha}_1 v = \lambda_1 v.$$

If $[\underline{\alpha}_1, \underline{\alpha}_2] = 0$, then

$$\underline{\alpha}_1(\underline{\alpha}_2 v) = \underline{\alpha}_2 \underline{\alpha}_1 v = \underline{\alpha}_2(\lambda_1 v) = \lambda_1(\underline{\alpha}_2 v);$$

thus $(\underline{\alpha}_2 v)$ must be a multiple of v, i.e.,

$$\underline{\alpha}_2 v = \lambda_2 v,$$

so v is also an eigenvector of $\underline{\alpha}_2$. Thus, two matrices that commute have common eigenvectors. If $\underline{\alpha}_1$ (say) is diagonalizable and has a complete set of eigenvectors, then it follows from (3.2.31) that every $\underline{\alpha}_i$ has the same eigenvectors. Thus, there is a coordinate system for $\mathbf{q}$ in which the right side of (3.2.30) is diagonal, and each component of $\mathbf{q}$ satisfies two scalar equations of the form

$$q_x = \sum_j^m a_j \lambda_j q, \qquad q_t = \sum_j^m b_j \lambda_j q.$$

Thus ($\log q$) is a potential function, corresponding to the conservation law

$$\frac{\partial}{\partial t}\left(\sum a_j \lambda_j\right) + \frac{\partial}{\partial x}\left(-\sum b_j \lambda_j\right) = 0.$$

Different components of **q** that correspond to different eigenvalues represent different conservation laws. Moreover, one may show that an Abelian Lie algebra gives only conservation laws even if none of the matrices $\underline{\alpha}_i$ is diagonalizable (see Exercise 5).

Because Abelian pseudopotentials give only conservation laws, it often is asserted that the pseudopotentials of interest are non-Abelian. This is true from the standpoint of IST, but Abelian pseudopotentials should not necessarily be ignored, as was shown by the fact that the Cole–Hopf transformation follows from one.

Now let us return to Fisher's equation, (3.2.20). It has a nontrivial pseudopotential in the form (3.2.7) if (3.2.21) has a non-Abelian solution in terms of $N \times N$ constant matrices. But (3.2.21) has no such solution (the proof is somewhat involved, and is given in Exercise 7). Therefore (3.2.20) has no pseudopotential, no BT and no linear scattering problem that depends only on u and u_x, i.e., none of the form (3.2.7). However, it is not clear from this approach whether the method failed because of the structure of the equation (3.2.20), or because of the form of the pseudopotential (3.2.7). There is no known way to generalize the method so that its failure guarantees that the other methods discussed in this chapter must also fail.

3.2.d. More general problems. To this point, we have restricted our attention to equations whose linearized dispersion relations are polynomial. The method begins the same way for equations of higher order, but the problem does not necessarily reduce to a purely algebraic one, and it may not be sufficient to look for linear pseudopotentials. To illustrate, consider the second order equation

(3.2.32) $$u_{xt} = f(u),$$

which includes the sine-Gordon equation as a special case. Under various simplifying assumptions, Kruskal (1974), McLaughlin and Scott (1973) and Rund (1976) all showed that (3.2.32) has special structure (extra conservation laws or a BT) if and only if

(3.2.33) $$f'' = kf$$

for some k. However, Mikhailov (1981) has shown that there are other functions for which (3.2.32) is in the IST class (cf. Exercise 8, § 3.7). These are related to a third order eigenvalue problem, whereas those obeying (3.2.33) are related to a second order eigenvalue problem.

Let us seek a pseudopotential for (3.2.32), to find out whether (3.2.33) can be generalized. Following the usual rules, consider

(3.2.34) $\mathbf{q}_x = \mathbf{A}(u, u_x, u_t, \mathbf{q})$, $\mathbf{q}_t = \mathbf{B}(u, u_x, u_t, \mathbf{q})$.

Integrability ($\mathbf{q}_{xt} = \mathbf{q}_{tx}$) requires that

(3.2.35) $$\frac{\partial \mathbf{A}}{\partial u_t} = 0 = \frac{\partial \mathbf{B}}{\partial u_x}$$

and that

(3.2.36) $[\mathbf{A}, \mathbf{B}] + \dfrac{\partial \mathbf{A}}{\partial u} u_t - \dfrac{\partial \mathbf{B}}{\partial u} u_x + \left(\dfrac{\partial \mathbf{A}}{\partial u_x} - \dfrac{\partial \mathbf{B}}{\partial u_t} \right) f = 0$.

At this point, the difference between first order (in t) evolution equations and higher order equations appears: none of the independent variables ($u, u_x, u_t, \mathbf{q}$) is completely explicit in (3.2.36). Thus, whereas (3.2.13) was reduced to the purely algebraic problem (3.2.19), no such reduction is known for (3.2.36). Solutions can be obtained by imposing additional restrictions on the form of $\mathbf{A}, \mathbf{B}$ (e.g., Forsyth (1906, vol. VI), Lamb (1976) or Anderson and Ibragimov (1979)). Subject to these additional restrictions (3.2.32) has a pseudopotential if (3.2.33)) holds. The general question, however, is open.

3.2.e. Summary. Let us summarize what is currently known about pseudopotentials.

(i) The method presented here finds pseudopotentials for some problems. A non-Abelian pseudopotential, once found, typically can be made into a BT. If it is linear and contains an arbitrary parameter, it is a candidate for a scattering problem.

(ii) Failure of this method on a particular problem suggests (only) that none of the other methods discussed in this chapter will work either. (This is also true of the "direct" methods for finding N-soliton formulas presented in §§ 3.5 and 3.6.)

(iii) If the equation in question is first order in t and of finite order in x, it is sufficient to look for a linear pseudopotential. The problem then can be reduced to one of finding a Lie algebra with a specified substructure. For any specific case, it appears that this problem can be conclusively resolved, but no general results seem to be available.

(iv) The situation is murkier if higher order derivatives are allowed to enter into the pseudopotential, or if the equation is higher order. Then the existence of a pseudopotential comes down to a set of differential–algebraic, rather than purely algebraic, relations, and it is not known whether one may look only for linear pseudopotentials without loss of generality.

3.3. Direct methods for finding soliton solutions—Hirota's method. One of the areas of interest in the study of nonlinear wave propagation is the development of techniques for finding special solutions to the underlying equations. For those equations admitting soliton solutions Hirota has made many significant contributions (for a review of some of this work see Hirota (1976), (1979b) and Hirota and Satsuma (1976)). It should be pointed out that these direct methods virtually always work for equations where the IST is known and sometimes work for equations where the IST is still unavailable. In practice these direct methods have often motivated a search for the IST and sometimes have suggested them (e.g., Satsuma and Kaup (1977), Nakamura (1979b), Satsuma, Ablowitz and Kodama (1979), etc.). The basic ideas in this direct method are as follows: (i) Introduce a dependent variable transformation (this may require some ingenuity although there are some standard forms). The transformation should reduce the evolution equation to a so-called bilinear equation, quadratic in the dependent variables. Hirota has developed a novel differential calculus and it is convenient to use it at this stage. (ii) Introduce a formal perturbation expansion into this bilinear equation. In the case of soliton solutions this expansion truncates. (iii) Use mathematical induction to prove that the suggested soliton form is indeed correct.

In this section we shall carefully examine the analysis in the case of the KdV equation. We shall briefly indicate the results for some of the other well-known nonlinear wave equations as well as discuss other problems in which this method gives interesting results.

3.3.a. The KdV equation as an example. Consider the KdV equation

(3.3.1) $$u_t + 6uu_x + u_{xxx} = 0.$$

In § 1.4 we saw that the N-soliton solution takes the form

(3.3.2) $$u = 2 \frac{d^2}{dx^2} \log F,$$

where F is the determinant of a certain matrix. This form suggests transforming (3.3.1) to an equation involving F. Substitution of (3.3.2) into (3.3.1), integrating once and setting the integration constant equal to zero yields

(3.3.3) $$F_{xt}F - F_x F_t + F_{xxxx}F - 4F_{xxx}F_x + 3F_{xx}^2 = 0.$$

Equation (3.3.3) is a quadratic form (Hirota usually refers to this as a bilinear equation); such forms are typical once we have isolated the correct dependent variable transformation. The introduction of the following special operator is helpful in the analysis:

(3.3.4) $$D_x^m D_t^n a \cdot b = (\partial_x - \partial_{x'})^m (\partial_t - \partial_{t'})^n a(x,t) b(x',t')|_{\substack{x'=x \\ t'=t}}.$$

Using this definition (3.3.3) can be written in the form

(3.3.5) $$(D_x D_t + D_x^4) F \cdot F = 0.$$

The following properties, which are easily verified, are useful in working with this new operator:

(3.3.6a) $$D_x^m a \cdot 1 = \partial_x^m a,$$

(3.3.6b) $$D_x^m a \cdot b = (-1)^m D_x^m b \cdot a,$$

(3.3.6c) $$D_x^m a \cdot a = 0, \quad m \text{ odd}$$

(3.3.6d)
$$D_x^m D_t^n e^{(k_1 x - \omega_1 t)} \cdot e^{(k_2 x - \omega_2 t)} = (k_1 - k_2)^m (-\omega_1 + \omega_2)^n e^{(k_1 + k_2)x - (\omega_1 + \omega_2)t}.$$

There are many other relations involving this D operator. The reader may wish to consult some of the above-mentioned review articles. The method proceeds by assuming a formal expansion in powers of ε,

(3.3.7a) $$F = 1 + \varepsilon f^{(1)} + \varepsilon^2 f^{(2)} + \cdots,$$

where

(3.3.7b) $$f^{(1)} = \sum_{i=1}^{N} e^{\eta_i}, \quad \eta_i = k_i x - \omega_i t + \eta_i^{(0)},$$

and k_i, ω_i, $\eta_i^{(0)}$ are constants. In the case of KdV, and indeed all those problems which admit N-soliton solutions, this formal perturbation procedure truncates.

Hence, substituting (3.3.7b) into (3.3.5),

$$(D_x D_t + D_x^4)(1 + \varepsilon f^{(1)} + \varepsilon^2 f^{(2)} + \cdots) \cdot (1 + \varepsilon f^{(1)} + \varepsilon^2 f^{(2)} + \cdots) = 0,$$

equating to zero powers of ε, and using the properties (3.3.6a–d), we have

(3.3.8a) $\quad O(1): \quad 0 = 0,$

(3.3.8b) $\quad O(\varepsilon): \quad 2(\partial_x \partial_t + \partial_x^4) f^{(1)} = 0,$

(3.3.8c) $\quad O(\varepsilon^2): \quad 2(\partial_x \partial_t + \partial_x^4) f^{(2)} = -(D_x D_t + D_x^4) f^{(1)} \cdot f^{(1)},$

(3.3.8d) $\quad O(\varepsilon^3): \quad 2(\partial_x \partial_t + \partial_x^4) f^{(3)} = -2(D_x D_t + D_x^4) f^{(1)} \cdot f^{(2)},$

$\vdots$

The first nontrivial equation above, (3.3.8b), is homogeneous. For the solution we take (3.3.7b). Unfortunately, if we try to continue to higher orders with the solution (3.3.7b) for arbitrary N, the analysis becomes unwieldy. Most frequently one obtains solutions for $N = 1, 2$ (and sometimes $N = 3$) and then hypothesizes a solution for arbitrary N and proves this by induction. For $N = 1$ we take $f^{(1)} = e^{\eta_1}$. Then (3.3.8b) requires that $\omega_1 = -k_1^3$. The equation for $f^{(2)}$ is then given by (3.3.8c). But it is clear that using (3.3.6d) reduces (3.3.8c) to

(3.3.9a) $$(\partial_x \partial_t + \partial_x^4) f^{(2)} = 0,$$

so that

(3.3.9b) $$f^{(2)} = 0,$$

and the expansion then terminates. Hence for $N = 1$ we have
$$F_1 = 1 + e^{\eta_1}, \qquad \omega_1 = -k_1^3$$
and

(3.3.10) $$u = \frac{k_1^2}{2}\operatorname{sech}^2 \frac{1}{2}(k_1 x - k_1^3 t + \eta_1^{(0)}).$$

For the case of $N = 2$, we take for one solution of equation (3.3.8b)

(3.3.11) $$f^{(1)} = e^{\eta_1} + e^{\eta_2}, \qquad \eta_i = k_i x - k_i^3 t + \eta_i^{(0)}.$$

Then (3.3.8c) reduces to

(3.3.12a) $$2(\partial_x \partial_t + \partial_x^4) f^{(2)} = -2((k_1 - k_2)(-\omega_1 + \omega_2) + (k_1 - k_2)^4) e^{\eta_1 + \eta_2},$$

with the solution

(3.3.12b) $$f^{(2)} = e^{\eta_1 + \eta_2 + A_{12}},$$

(3.3.12c) $$e^{A_{ij}} = \left(\frac{k_i - k_j}{k_i + k_j}\right)^2$$

(note that $k_1 \neq k_2$). Substituting $f^{(1)}$, $f^{(2)}$ into (3.3.8d) shows that the right-hand side of (3.3.8d) vanishes, whereupon we take $f^{(3)} = 0$. Hence for $N = 2$

(3.3.13) $$F_2 = 1 + e^{\eta_1} + e^{\eta_2} + e^{\eta_1 + \eta_2 + A_{12}}.$$

The two-soliton solution for KdV is obtained from $u = 2(d^2/dx^2) \log F_2$. Similarly, performing the analogous calculation for $N = 3$ yields

(3.3.14) $$\begin{aligned}F_3 = 1 &+ e^{\eta_1} + e^{\eta_2} + e^{\eta_3} + e^{\eta_1 + \eta_2 + A_{12}} + e^{\eta_1 + \eta_3 + A_{13}}\\ &+ e^{\eta_2 + \eta_3 + A_{23}} + e^{\eta_1 + \eta_2 + \eta_3 + A_{12} + A_{13} + A_{23}},\end{aligned}$$

where A_{ij} is given by (3.3.12c).

On the basis of these solutions we hypothesize the general N-soliton solution

(3.3.15) $$F_N = \sum_{\mu = 0,1} \exp\left(\sum_{i=1}^{N} \mu_i \eta_i + \sum_{1 \leq i < j}^{N} \mu_i \mu_j A_{ij}\right),$$

where the sum over $\mu = 0, 1$ refers to each of the μ_i, $i = 1, \cdots, N$. Before proving by induction that (3.3.15) indeed satisfies (3.3.5), we mention how the A_{ij} can be easily seen to be the phase shifts of the solitons from the above formulae.

Consider the two-soliton case. Let us assume $0 < k_1 < k_2$ and define
$$\xi_i = x - k_i^2 t, \qquad i = 1, 2.$$
Thus
$$\xi_2 = -(k_2^2 - k_1^2)t + \xi_1.$$

In the frame of the first soliton, ξ_1 fixed, we take the limits as $t \to \pm\infty$. First, as $t \to +\infty$, $\xi_2 \to -\infty$, hence $e^{\eta_2} \to 0$ and

$$F_2 \sim 1 + e^{\eta_1},$$

with

(3.3.16) $$u \sim \frac{k_1^2}{2} \operatorname{sech}^2 \frac{\eta_1}{2}.$$

Similarly as $t \to -\infty$, $\xi_2 \to +\infty$, $e^{\eta_2} \to +\infty$ and

$$F_2 \sim e^{\eta_2}(1 + e^{\eta_1 + A_{12}}).$$

From (3.3.2) two solutions are equivalent up to a multiple of e^{ax+bt}. Hence the solution is given by

(3.3.17) $$u \sim \frac{k_1^2}{2} \operatorname{sech}^2 \frac{\eta_1 + A_{12}}{2}.$$

Thus a two-soliton solution is such that a phase shift A_{12} is produced upon interaction (note that this is the same result as that of § 1.4; see (1.4.43)). Similar ideas apply in the N-soliton case.

We now return to verifying the validity of the N-soliton solution (3.3.15) (Hirota (1971)). We caution the reader that this is somewhat tedious. Without loss of continuity one can pick up with the example of mKdV (equation (3.3.32)).

THEOREM. *The function F_N given by* (3.3.15) *satisfies* (3.3.5).

Proof. Substituting (3.3.15) into (3.3.5) and using the relevant properties of the D operators we have

(3.3.18) $$\sum_{\mu=0,1} \sum_{\nu=0,1} \left\{ \left(\sum_i (\mu_i - \nu_i)k_i \right) \left((-) \sum_i (\mu_i - \nu_i)k_i^3 \right) + \left(\sum_{i=1}^N (\mu_i - \nu_i)k_i \right)^4 \right\}$$
$$\times \exp\left(\sum_i (\mu_i + \nu_i)\eta_i + \sum_{1 \le i<j} (\mu_i\mu_j + \nu_i\nu_j)A_{ij} \right) = 0.$$

Since $\mu_i, \nu_i = 0, 1$, it is clear that we only have exponential terms of the form

$$\exp\left(\sum_{i=1}^n \eta_i + \sum_{i=n+1}^m 2\eta_i \right), \quad 0 \le n \le N, \quad n \le m \le N$$

(where we might have to rearrange the indices). Next we show that the coefficient of this general exponential term is zero, the coefficient being given by

(3.3.19a) $$\Delta = \sum_\mu \sum_\nu \left\{ -\left(\sum_{i=1}^N (\mu_i - \nu_i)k_i \right) \left(\sum_{i=1}^N (\mu_i - \nu_i)k_i^3 \right) + \left(\sum_{i=1}^N (\mu_i - \nu_i)k_i \right)^4 \right\}$$
$$\times \exp\left(\sum_{i<j} (\mu_i\mu_j + \nu_i\nu_j)A_{ij} \right) \operatorname{cond}(\mu, \nu).$$

Here cond (μ, ν) is given by:

(3.3.19b)
$$\text{cond}(\mu, \nu) = \begin{cases} \text{for } i = 1, \cdots, n \text{ we take only those } \mu_i, \nu_i \text{ such that} \\ \quad \text{①} \ : \ \mu_i + \nu_i = 1, \quad 0 \leq n \leq N; \\ \text{for } i = n+1, \cdots, m \text{ we take only those } \mu_i, \nu_i \text{ such that} \\ \quad \text{②} \ : \ \mu_i = \nu_i = 1, \quad 0 \leq m \leq N; \\ \text{for } i = m+1, \cdots, N \text{ we take only those } \mu_i, \nu_i \text{ such that} \\ \quad \text{③} \ : \ \mu_i = \nu_i = 0. \end{cases}$$

Let us call

(3.3.20) $$\sigma_i = \mu_i - \nu_i, \quad i = 1, 2, \cdots, n.$$

For $i > n$, $\sigma_i = 0$ (either $\mu_i = \nu_i = 1$, or $\mu_i = \nu_i = 0$). Since for $i = 1, \cdots, n$ we have $\mu_i + \nu_i = 1$ (from (3.3.19b)), with (3.3.20) we have

(3.3.21) $$\mu_i = \frac{1+\sigma_i}{2}, \quad \nu_i = \frac{1-\sigma_i}{2}.$$

With (3.3.21) all of the terms in (3.3.19a) with the exception of the exponential term are readily transformed.

Using the numbering scheme in (3.3.19b) we evaluate the term

$$T = \exp\left(\sum_{i<j}^{N} (\mu_i\mu_j + \nu_i\nu_j)A_{ij}\right)$$

$$= \exp\left(\sum_{①<①} + \sum_{①<②} + \sum_{①<③} + \sum_{②<②} + \sum_{②<③} + \sum_{③<③}\right)(\mu_i\mu_j + \nu_i\nu_j)A_{ij}.$$

The only contribution from this exponential which depends on $\sigma_i\sigma_j$ is the "①<①" term. Hence T is given by

(3.3.22a) $$T = \text{const.} \sum_{i<j}^{n} \left(\frac{k_i - k_j}{k_i + k_j}\right)(1 + \sigma_i\sigma_j)$$

(3.3.22b) $$= \text{const.} \sum_{i<j}^{n} \left(\frac{\sigma_i k_i - \sigma_j k_j}{k_i + k_j}\right)^2$$

the latter relation being true since $\sigma_i = \pm 1$ only. This then indicates that the coefficient Δ is given by $\Delta = \text{const.} \times \hat{\Delta}$, where

(3.3.23) $$\hat{\Delta} = \sum_{\sigma=1} \left\{-\left(\sum_{i=1}^{n} \sigma_i k_i\right)\left(\sum_{i=1}^{n} \sigma_i k_i^3\right) + \left(\sum_{1}^{n} \sigma_i k_i\right)^4\right\} \prod_{i<j}^{n} (\sigma_i k_i - \sigma_j k_j)^2$$

(note that we have absorbed the term $\sum_{i<j}^{n}(1/(k_i + k_j))^2$, into the multiplicative constant).

Now we show, by induction, that $\hat{\Delta} = 0$. $\hat{\Delta}$ is a polynomial in the k_i's; specifically, $\hat{\Delta} = \hat{\Delta}(k_1, \cdots, k_n)$. We shall write this as $\hat{\Delta} = \hat{\Delta}_n$. Also note that the order of the polynomial product $\sum_{i<j}^n (\sigma_i k_i - \sigma_j k_j)$ is $\frac{1}{2}n(n-1)$; hence we have that the order of $\hat{\Delta}$ is such that

(3.3.24) $$\text{order }(\hat{\Delta}_n) \leq n^2 - n + 4.$$

To carry out the induction we note that

(3.3.25a) $$\hat{\Delta}(k_1) = \hat{\Delta}_1 = \sum_{\sigma_1 = 1} \{-(\sigma_1 k_1)(\sigma_1 k_1^3) + \sigma_1^4 k_1^4\} = 0,$$

(3.3.25b)
$$\begin{aligned}\hat{\Delta}(k_1, k_2) = \hat{\Delta}_2 &= \sum_{\sigma_1, \sigma_2 = 1} \{-(\sigma_1 k_1 + \sigma_2 k_2)(\sigma_1 k_1^3 + \sigma_2 k_2^3) + (\sigma_1 k_1 + \sigma_2 k_2)^4\} \\ &\quad \times (\sigma_1 k_1 - \sigma_2 k_2)^2 \\ &= 3k_1 k_2 (k_1^2 - k_2^2)(k_1^2 - k_2^2) \sum_{\sigma_1, \sigma_2 = \pm 1} \sigma_1 \sigma_2 \\ &= 0.\end{aligned}$$

$\hat{\Delta}(k_1, \cdots, k_n)$ has the following properties:

(i) $\hat{\Delta}_n$ is even in the k_j, i.e.,

(3.3.26a) $$\hat{\Delta}(k_1, \cdots, k_j, \cdots, k_n) = \hat{\Delta}(k_1, \cdots, -k_j, \cdots, k_n).$$

(ii) $\hat{\Delta}_n$ is symmetric upon interchanging k_j and k_i; i.e.,

(3.3.26b) $$\hat{\Delta}(k_1, \cdots, k_i, \cdots, k_j, \cdots, k_n) = \hat{\Delta}(k_1, \cdots, k_j, \cdots, k_i, \cdots, k_n).$$

Equation (3.3.26a) is easily seen by replacing k_i by $-k_i$ and σ_i by $-\sigma_i$ (dummy index) in (3.3.23). The only question as to the symmetry (3.3.26b) is with regard to the product term. However, since

(3.3.27) $$\prod_{1 \leq i < j}^n (\sigma_i k_i - \sigma_j k_j)^2 = (-1)^{(n^2-n)/2} \prod_{\substack{i,j=1 \\ i \neq j}}^n (\sigma_i k_i - \sigma_j k_j),$$

it is clear that the product term is symmetric on interchanging k_p, k_q.

Evaluate $\hat{\Delta}_n$ at $k_1 = 0$:

$$\hat{\Delta}_n|_{k_1=0} \equiv \hat{\Delta}(k_1 = 0, k_2, \cdots, k_n)$$

(3.3.28)
$$\begin{aligned}&= \sum_{\sigma_1 = \pm 1} \sum_{\substack{\text{other} \\ \sigma = \pm 1}} \left\{ -\left(\sum_{i=2}^n \sigma_i k_i\right)\left(\sum_{i=2}^n \sigma_i k_i^3\right) + \left(\sum_{i=2}^n \sigma_i k_i\right)^4 \right\} \\ &\quad \times \prod_{i=2}^n k_i^2 \sum_{2 \leq i < j}^n (\sigma_i k_i - \sigma_j k_j)^2 \\ &= 2\left(\prod_{i=2}^n k_i^2\right) \hat{\Delta}_{n-1}\end{aligned}$$

(note that here $\hat{\Delta}_{n-1} = \hat{\Delta}(k_2, k_3, \cdots, k_n)$). Similarly, evaluating $\hat{\Delta}_n$ when $k_1 = k_2$ yields

$$\hat{\Delta}_n|_{k_1=k_2} = \sum_{\sigma_1,\sigma_2=\pm 1} \sum_{\substack{\text{other} \\ \sigma = \pm 1}} \left\{ -\left((\sigma_1+\sigma_2)k_1 + \sum_{i=3}^n \sigma_i k_i\right)\left((\sigma_1+\sigma_2)k_1^3 + \sum_{i=3}^n \sigma_i k_i\right) \right.$$

(3.3.29a)
$$\left. + \left((\sigma_1+\sigma_2)k_1 + \sum_{i=3}^n \sigma_i k_i\right)^4 \right\}$$

$$\times (\sigma_1-\sigma_2)^2 k_1^2 \prod_{i=3}^n (\sigma_1 k_1 - \sigma_i k_i)^2 (\sigma_2 k_2 - \sigma_i k_i)^2 \prod_{3 \leq i < j} (\sigma_i k_i - \sigma_j k_j)^2.$$

We see that the only contribution occurs when $\sigma_1 = -\sigma_2$; hence

$$\hat{\Delta}_n|_{k_1=k_2} = 8k_1^2 \sum_{(\sigma_3,\cdots,\sigma_n)=\pm 1} \left\{ -\left(\sum_{i=3}^n \sigma_i k_i\right) + \left(\sum_{i=3}^n \sigma_i k_i^3\right) + \left(\sum_{i=3}^n \sigma_i k_i\right)^4 \right\}$$

(3.3.29b)
$$\times \prod_{i=3}^n (k_1^2 - k_i^2)^2 \prod_{3 \leq i < j} (\sigma_i k_i - \sigma_j k_j)^2$$

$$= 8k_1^2 \prod_{3}^n (k_1^2 - k_i^2) \hat{\Delta}_{n-2},$$

where $\hat{\Delta}_{n-2} = \hat{\Delta}(k_3, k_4, \cdots, k_n)$. By induction it is clear that both $\hat{\Delta}_n|_{k_1=0} = 0$, $\hat{\Delta}_n|_{k_1=k_2} = 0$ for all n. Hence $\hat{\Delta}_n$ is certainly *factorized* by $k_1(k_1 - k_2)$. But interchanging k_1 with any other k_i, and similarly treating k_2 makes it clear that for any i, j

$$\hat{\Delta}_n|_{k_i=0} = 0, \qquad \hat{\Delta}_n|_{k_i=k_j} = 0,$$

and therefore $\hat{\Delta}_n$ must be factorized by

$$\prod_{i=1}^n k_i \prod_{\substack{i,j=1 \\ i \neq j}}^n (k_i - k_j) \quad \text{or} \quad \prod_{i=1}^n k_i \prod_{1 \leq i < j} (k_i - k_j)^2.$$

But since $\hat{\Delta}_n$ is even in k_i for all i, we have that $\hat{\Delta}_n$ must be factorized by

(3.3.30)
$$\prod_{i=1}^n k_i^2 \prod_{1 \leq i < j} (k_i^2 - k_j^2)^2.$$

This implies that the order of the polynomial for $\hat{\Delta}_n$ must be at least of order $2n^2$; i.e.,

$$\text{order}(\hat{\Delta}_n) \geq 2n^2.$$

Since $2n^2 > n^2 - n + 4$ for all $n \geq 2$, we have a contradiction; i.e., $\hat{\Delta}_n$ cannot satisfy both the restrictions order $(\hat{\Delta}_n) \leq n^2 - n + 4$, order $(\hat{\Delta}_n) \geq 2n^2$. The only way out of this is to have, for $n \geq 2$,

(3.3.31)
$$\hat{\Delta}_n = 0.$$

From (3.3.25a), (3.3.31) is also true for $n = 1$. In this way we have proven that $\hat{\Delta}$ in (3.3.23) and hence Δ in (3.3.19a) vanishes, whereupon we have verified that the N-soliton solution $u = 2d^2(\log F_N)/dx^2$ satisfies the KdV equation. $\square$

3.3.b. Some other nonlinear PDE's. We now turn to another aspect of this direct method. Hirota (1976) has pointed out that very often a useful dependent variable transformation can be deduced. We shall illustrate these ideas for the mKdV equation

(3.3.32) $$v_t + 6v^2 v_x + v_{xxx} = 0.$$

Substituting the expression $v = G/F$ into (3.3.32) and using the definition of the D_x, D_t operators, we find

(3.3.33) $$(D_t + D_x^3)G \cdot F + \frac{6}{F^2}(D_x G \cdot F)\left(\frac{1}{2}D_x^2 F \cdot F - G^2\right) = 0.$$

Since F, G are both arbitrary we decouple the equations as follows:

(3.3.34a) $$(D_t + D_x^3)G \cdot F = 0,$$

(3.3.34b) $$D_x^2 F \cdot F = 2G^2.$$

The motivation for this choice of decoupling is to build in the linearized dispersion relation appropriately by (3.3.34a). The following expansion gives soliton solutions (such that $v \to 0$ as $|x| \to \infty$)

(3.3.35)
$$F = 1 + \varepsilon^2 F_2 + \varepsilon^4 F_4 + \cdots,$$
$$G = \varepsilon G_1 + \varepsilon^3 G_3 + \cdots,$$

where each expansion truncates. For example a one-soliton solution has

(3.3.36)
$$(\partial_t + \partial_x^3)G_1 = 0,$$
$$G_1 = e^{\eta_1}, \quad \eta_1 = k_1 x - k_1^3 t + \eta_1^{(0)},$$
$$\partial_x^2 F_2 = G_1^2 = e^{2\eta_1},$$
$$F_2 = \frac{1}{4k_1^2} e^{2\eta_1},$$
$$F_j = 0, \quad j \geq 4,$$
$$G_j = 0, \quad j \geq 3.$$

With this result, with $\varepsilon = 1$, a one-soliton solution of mKdV is given by

(3.3.37) $$v = \frac{e^{\eta_1}}{1 + e^{2\eta_1}/4k_1^2} = k_1 \text{ sech } \eta_1.$$

Similarly a two-soliton solution has the formulae for G, F $(v = G/F)$

$$G = e^{\eta_1} + e^{\eta_2} + \frac{1}{4k_2^2}\left(\frac{k_1-k_2}{k_1+k_2}\right)^2 e^{\eta_1+2\eta_2} + \frac{1}{4k_1^2}\left(\frac{k_1-k_2}{k_1+k_2}\right)^2 e^{2\eta_1+\eta_2},$$

(3.3.38a)

$$F = 1 + \frac{1}{4k_1^2}e^{2\eta_1} + \frac{1}{4k_2^2}e^{2\eta_2} + \frac{2}{(k_1+k_2)^2}e^{\eta_1+\eta_2} + \frac{(k_1-k_2)^4}{16k_1^2k_2^2(k_1+k_2)^4}e^{2\eta_1+2\eta_2}.$$

Next note that F may be combined as

(3.3.38b)
$$F = \left(\frac{1}{2k_1}e^{\eta_1} + \frac{1}{2k_2}e^{\eta_2}\right)^2 + \left(1 - \frac{(k_1-k_2)^2}{4k_1k_2(k_1+k_2)^2}e^{\eta_1+\eta_2}\right)^2$$
$$= \hat{g}^2 + \hat{f}^2.$$

With this combination G may be written as

(3.3.38c)
$$G = 2D_x\hat{g}\cdot\hat{f},$$

hence v has the form

(3.3.39a)
$$v = 2\frac{D_x\hat{g}\cdot\hat{f}}{\hat{f}^2 + \hat{g}^2} = 2\frac{\hat{g}_x\hat{f} - \hat{f}_x\hat{g}}{\hat{f}^2(1+(\hat{g}/\hat{f})^2)}$$

or

(3.3.39b)
$$v = 2\left(\tan^{-1}\frac{\hat{g}}{\hat{f}}\right)_x \equiv i\left(\log\left(\frac{\hat{f}-i\hat{g}}{\hat{f}+i\hat{g}}\right)\right)_x.$$

If we define $f = \hat{f} + i\hat{g}$, we have the dependent variable transformation

(3.3.40)
$$v = i\left(\log\frac{f^*}{f}\right)_x.$$

With this new dependent variable transformation we find the bilinear equations

(3.3.41a) $\quad\quad\quad\quad\quad (D_t + D_x^3)f^* \cdot f = 0,$

(3.3.41b) $\quad\quad\quad\quad\quad D_x^2 f^* \cdot f = 0.$

After substitution of (3.3.40) into (3.3.32), N-soliton solutions are derived by substituting

$$f_N = 1 + \varepsilon f_N^{(1)} + \varepsilon^2 f_N^{(2)} + \cdots$$

into (3.3.41), equating powers of ε and then taking $\varepsilon = 1$. One- and two-soliton solutions are given by

$$f_1 = 1 + e^{\eta_1 + i\pi/2},$$

(3.3.42) $\quad f_2 = 1 + e^{\eta_1+i\pi/2} + e^{\eta_2+i\pi/2} + e^{\eta_1+\eta_2+i\pi+A_{12}},$

$$\eta_i = k_i x - k_i^3 t + \eta_i^{(0)}, \quad e^{A_{ij}} = \left(\frac{k_i-k_j}{k_i+k_j}\right)^2.$$

The N-soliton solution has the structure

$$(3.3.43) \qquad f_N = \sum_{\mu=0,1} \exp\left(\sum_{i=1}^{N} \mu_i\left(\eta_i + i\frac{\pi}{2}\right) + \sum_{1 \leq i < j} \mu_i \mu_j A_{ij}\right).$$

For completeness we next give the results for the sine-Gordon equation (Hirota (1972); see also Caudrey, Gibbon, Eilbeck and Bullough (1973a,b)) and the nonlinear Schrödinger equation (Hirota (1973a)). First, consider the dependent variable transformation

$$(3.3.44) \qquad u = 2i \log\left(\frac{f^*}{f}\right)$$

for the sine-Gordon equation

$$(3.3.45) \qquad u_{xt} = \sin u.$$

(The reader might wish to recall the results of Chapter 1 which shows the deep relationship between sine-Gordon and mKdV). Then, noting

$$(3.3.46) \qquad \sin u = \frac{1}{2i}\left(\left(\frac{f}{f^*}\right)^2 - \left(\frac{f^*}{f}\right)^2\right),$$

we find

$$(3.3.47) \qquad D_x D_t f \cdot f = -\tfrac{1}{2}(f^{*2} - f^2)$$

(and its complex conjugate) after substituting (3.3.44) into (3.3.45). A one-soliton solution to (3.4.47) is given by

$$(3.3.48a) \qquad f_1 = 1 + e^{\eta_1 + i\pi/2},$$

whereas the N-soliton solution satisfies (3.3.43) with

$$(3.3.48b) \qquad f_N = \sum_{\mu=0,1} \exp\left(\sum_{j=1}^{N} \mu_j\left(\eta_j + i\frac{\pi}{2}\right) + \sum_{1 \leq i < j}^{N} \mu_i \mu_j A_{ij}\right),$$

where

$$\eta_i = k_i x - \omega_i t + \eta_i^{(0)},$$

$$(3.3.48c) \qquad \omega_i = \frac{1}{k_i},$$

$$e^{A_{ij}} = -\frac{(k_1 - k_2)(\omega_1 - \omega_2)}{(k_1 + k_2)(\omega_1 + \omega_2)} = \frac{(k_1 - k_2)^2}{(k_1 + k_2)^2}.$$

Alternatively, we note that if

$$f = F + iG, \qquad F, G \text{ real},$$

then (3.3.47) can be written as

(3.3.49a) $\quad D_x D_t (F \cdot F - G \cdot G) = 0, \quad D_x D_t F \cdot G = FG$

and

(3.3.49b) $\quad u = 2i \log \dfrac{F - iG}{F + iG} = 4 \tan^{-1} \dfrac{G}{F}.$

In the case of the nonlinear Schrödinger equation (Hirota (1973a))

(3.3.50) $\quad iu_t + u_{xx} + |u|^2 u = 0,$

one has a significantly more complicated structure for the N-soliton solution. Substituting $u = G/F$, F real, into (3.3.50) gives

(3.3.51) $\quad \dfrac{1}{F^2}(iD_t + D_x^2)G \cdot F - \dfrac{G}{F^3}(D_x^2 F \cdot F - GG^*) = 0,$

whereupon we decouple the equation by choosing

(3.3.52) $\quad (iD_t + D_x^2)G \cdot F = 0, \quad D_x^2 F \cdot F = GG^*.$

Hence

(3.3.53) $\quad |u|^2 = \dfrac{GG^*}{F^2} = \dfrac{D_x^2 F \cdot F}{F^2} = 2(\log F)_{xx}.$

A one-soliton solution takes the form

(3.3.54)
$$\begin{aligned}
G &= e^{\eta_1}, \\
F &= 1 + e^{\eta_1 + \eta_1^* + \phi_{1,1}^*}, \\
\eta_1 &= p_1 x - \Omega_1 t + \eta_1^{(0)}, \quad \Omega_1 = -ip_1^2, \\
e^{\phi_{1,1}^*} &= \tfrac{1}{2}(p_1 + p_1^*)^{-2}.
\end{aligned}$$

An N-soliton solution is given by

$$F = \sum_{\mu = 0,1} D_1(\mu) \exp\left(\sum_{i=1}^{2N} \mu_i \eta_i + \sum_{1 \le i < j} \phi_{ij} \mu_i \mu_j\right),$$

$$G = \sum_{\mu = 0,1} D_2(\mu) \exp\left(\sum_{i=1}^{2N} \mu_i \eta_i + \sum_{1 \le i < j}^{2N} \phi_{ij} \mu_i \mu_j\right),$$

$$\eta_i = p_i x - \Omega_i t + \eta_i^{(0)}, \quad p_{i+N} = p_i^*, \quad \Omega_{i+N} = \Omega_i^*, \quad i = 1, \cdots, N,$$

$$\eta_{i+N} = \eta_i^*, \quad \Omega_j = -ip_j^2,$$

$$e^{\phi_{ij}} = \begin{cases} \tfrac{1}{2}(p_i + p_j)^{-2} & \text{for } i = 1, 2, \cdots, N \text{ and } j = N+1, \cdots, 2N, \\ \tfrac{1}{2}(p_i - p_j)^{-2} & \text{for } i = N+1, \cdots, 2N \text{ and } j = N+1, \cdots, 2N, \end{cases}$$

and where

(3.3.55)
$$D_1(\underline{\mu}) = \begin{cases} 1 & \text{when } \sum_{i=1}^{N} \mu_i = \sum_{i=1}^{N} \mu_{i+N}, \\ 0 & \text{otherwise}; \end{cases}$$

$$D_2(\underline{\mu}) = \begin{cases} 1 & \text{when } 1 + \sum_{1}^{N} \mu_{i+N} = \sum_{1}^{N} \mu_i, \\ 0 & \text{otherwise}. \end{cases}$$

3.3.c. Discrete evolution equations. It is significant that many of the previous ideas can be extended to handle discrete problems. (See, for example, Hirota and Satsuma (1976a).) Here we shall indicate how the analysis proceeds in the case of the Toda lattice, i.e.,

(3.3.56)
$$\frac{d^2 y_n}{dt^2} = e^{-(y_n - y_{n-1})} - e^{-(y_{n+1} - y_n)}.$$

If we define $r_n = y_n - y_{n-1}$ and subtract two equations involving y_n and y_{n-1}, we have

(3.3.57)
$$\frac{d^2 r_n}{dt^2} = 2e^{-r_n} - e^{-r_{n+1}} - e^{-r_{n-1}}.$$

If we define

(3.3.58)
$$r_n = -\log(1 + V_n),$$

then (3.3.57) obeys

(3.3.59)
$$\frac{d^2 \log(1 + V_n)}{dt^2} = V_{n+1} - 2V_n + V_{n-1}.$$

Physically speaking, (3.3.59) describes a certain nonlinear ladder network where V_n is the voltage across the nth node (Hirota and Suzuki (1970)).

The following dependent variable transformation is useful:

(3.3.60)
$$V_n = \frac{d^2 \log F_n}{dt^2}$$

(note that here the subscript refers to the coordinate location, not the number of solitons). Substituting (3.3.60) into (3.3.59) yields a bilinear (quadratic) equation

(3.3.61)
$$\tfrac{1}{2} D_t^2 F_n \cdot F_n = F_{ntt} F_n - F_{nt}^2 = F_{n+1} F_{n-1} - F_n^2.$$

The analogue of the operator D_x is D_n, which satisfies

(3.3.62)
$$e^{D_n} a_n \cdot b_n = e^{\partial_n - \partial_{n'}} a_n b_{n'}\big|_{n'=n} = a_{n+1} b_{n-1},$$

where $e^{\partial_n}a_n \equiv a_{n+1}$. Thus

(3.3.63) $\quad \cosh D_n a_n \cdot b_n = \frac{1}{2}(a_{n+1}b_{n-1} + a_{n-1}b_{n+1}).$

Using (3.3.62), we may rewrite (3.3.61) in the form

(3.3.64) $\quad \frac{1}{2}D_t^2 F_n \cdot F_n = (\cosh D_n - 1)F_n \cdot F_n$

or

(3.3.65) $\quad \frac{1}{2}D_t^2 F_n \cdot F_n = 2(\sinh \frac{1}{2}D_n)^2 F_n \cdot F_n.$

Again the perturbation approach works; i.e.,

$$F_n = 1 + \varepsilon F_n^{(1)} + \varepsilon^2 F_n^{(2)} + \cdots.$$

A one-soliton solution has $F_n^{(j)} = 0$, $j \geq 2$ and $F_n^{(1)} = e^{\eta_1}$. The result is

(3.3.66)
$$F_1 = 1 + e^{\eta_1}, \qquad \eta_1 = k_1 n - \omega_1 t,$$
$$\omega_1^2 = \left(2 \sinh \frac{k}{2}\right)^2;$$

hence the voltage is given by

$$V_n = \frac{\omega_1^2}{2} \operatorname{sech}^2 \frac{1}{2} \eta_1.$$

(A one-soliton solution may propagate to the left or right.) The N-soliton solution has the form

(3.3.67a) $\quad F_n = \sum_{\mu=0,1} \exp\left(\sum_{i=1}^N \mu_i \eta_i + \sum_{1 \leq i < j} A_{ij} \mu_i \mu_j\right),$

where

$$\eta_i = p_i n - \Omega_i t + \eta_i^{(0)},$$
$$\Omega_i = 2\varepsilon_i \sinh \frac{1}{2} p_i, \qquad \varepsilon_i = \pm 1,$$
$$e^{A_{ij}} = -\frac{(\Omega_i - \Omega_j)^2 - (2 \sinh \frac{1}{2}(p_i - p_j))^2}{(\Omega_i + \Omega_j)^2 - (2 \sinh \frac{1}{2}(p_i + p_j))^2}$$

or

(3.3.67b) $\quad e^{A_{ij}} = \begin{cases} \left(\dfrac{\sinh \frac{1}{4}(p_i - p_j)}{\sinh \frac{1}{4}(p_i - p_j)}\right)^2 & \text{if } \varepsilon_i \varepsilon_j > 0, \\ \left(\dfrac{\cosh \frac{1}{4}(p_i - p_j)}{\cosh \frac{1}{4}(p_i + p_j)}\right)^2 & \text{if } \varepsilon_i \varepsilon_j < 0. \end{cases}$

3.3.d. Bäcklund transformations in bilinear form. It is interesting, and will be useful to us in later sections, to derive the Bäcklund transformations and permutation relations via direct methods for the KdV equation. We begin with the $N(f_N)$- and $(N+1)(f_{N+1})$-soliton solutions (here again the subscript N refers to the number of solitons) satisfying KdV (in bilinear form):

(3.3.68a) $$D_x(D_t + D_x^3)f_N \cdot f_N = 0,$$

(3.3.68b) $$D_x(D_t + D_x^3)f_{N+1} \cdot f_{N+1} = 0.$$

We shall show formally that given f_N we can find f_{N+1} by solving a linear equation (Hirota (1974)).

Multiplying (3.3.68a) by f_{N+1}^2 and subtracting (3.3.68b) multiplied by f_N^2 yields

(3.3.69)
$$P = [D_x(D_t + D_x^3)f_N \cdot f_N]f_{N+1}f_{N+1} - [D_x(D_t + D_x^3)f_{N+1} \cdot f_{N+1}]f_N f_N = 0.$$

Using the identities (which may be verified)

(3.3.70a)
$$(D_x D_t f_N \cdot f_N)f_{N+1}f_{N+1} - f_N f_N D_x D_t f_{N+1} \cdot f_{N+1} = 2D_x[(D_t f_N \cdot f_{N+1}) \cdot f_N f_{N+1}]$$

and

(3.3.70b)
$$\begin{aligned}f_{N+1}f_{N+1}(D_x^4 f_N \cdot f_N) &- f_N f_N D_x^4 f_{N+1} \cdot f_{N+1} \\ &= 2D_x[(D_x^3 f_N \cdot f_{N+1}) \cdot f_N f_{N+1} - 3(D_x^2 f_N \cdot f_{N+1}) \cdot (D_x f_N \cdot f_{N+1})],\end{aligned}$$

we have that P reduces to

(3.3.71) $$P = 2D_x[(D_t + D_x^3)f_N \cdot f_{N+1} - 3(D_x^2 f_N \cdot f_{N+1}) \cdot (D_x f_N \cdot f_{N+1})] = 0.$$

The Bäcklund transformation in bilinear form is obtained by choosing

(3.3.72a) $$D_x^2 f_N \cdot f_{N+1} = \lambda f_N f_{N+1},$$

(3.3.72b) $$(D_t + 3\lambda D_x + D_x^3)f_N \cdot f_{N+1} = 0,$$

since (3.3.72) satisfies $P = 0$ (the new soliton solution contains the parameter λ). The equations (3.3.72) are a bilinear version of the Bäcklund transformations given in § 3.1.

As an example, we show how to compute a one-soliton solution from the vacuum state. A "zero" soliton solution corresponds to $f_0 = 1$ which yields $u = 0$. f_1 then satisfies, by the Bäcklund transformation (3.3.72a),

(3.3.73a) $$\partial_x^2 f_1 = \lambda f_1,$$

(3.3.73b) $$(\partial_t + 3\lambda \partial_x + \partial_x^3)f_1 = 0.$$

The solution satisfying (3.3.73) is

(3.3.74) $$f_1 = e^{\eta_1/2} + e^{-\eta_1/2},$$
$$\eta_1 = k_1 x - k_1^3 t + \eta_1^{(0)}, \qquad \lambda \equiv \frac{k_1^2}{4}.$$

Since $u = 2\partial_x^2 \log f$, it is clear that

(3.3.75) $$f_1 = e^{\eta_1/2}(1+e^{\eta_1}) \asymp 1+e^{\eta_1}$$

(here $f \asymp g$ if and only if $f = e^{\alpha x + \beta} g$, where α, β are independent of x) in the sense that the right and left sides of (3.3.75) yield equivalent solutions for u. Moreover, if we had taken $f_1 = e^{-(1/2)\eta_1}$ then we would see that this is equivalent to $f_0 = 0$. For this reason we take the linear combination (3.3.74).

It is easy to verify that, if we take f_1 given by (3.3.74), then the solution generated by the Bäcklund transformation is given by

$$f_2 = (k_1-k_2)(e^{(\eta_1+\eta_2)/2} + e^{-(\eta_1+\eta_2)/2}) + (k_1+k_2)(e^{(\eta_1-\eta_2)/2} + e^{-(\eta_1-\eta_2)/2}),$$
(3.3.76)

where $\eta_i = k_i x - k_i^3 t + \eta_i^{(0)}$. Note that

$$f_2 = (k_1-k_2) e^{-(\eta_1+\eta_2)/2} \left(1 - \left(\frac{k_1+k_2}{k_1-k_2}\right) e^{\eta_1} - \left(\frac{k_1+k_2}{k_1-k_2}\right) e^{\eta_2} + e^{\eta_1+\eta_2}\right)$$

$$\asymp 1 + e^{\eta_1} + e^{\eta_2} + e^{\eta_1+\eta_2+A_{12}}$$

(the usual two-soliton solution), where $e^{A_{12}} = (k_1-k_2)^2/(k_1+k_2)^2$, so long as the phase constants are chosen appropriately, i.e.,

$$e^{\eta_i^{(0)}} \to -\left[\frac{k_1-k_2}{k_1+k_2}\right] e^{\eta_i^{(0)}}.$$

In general, the N-soliton solution that satisfies the Bäcklund transformation is given by

(3.3.77a) $$f_N = \sum_{\varepsilon=\pm 1} \frac{\prod_{i<j}^{N}(\varepsilon_i k_i - \varepsilon_j k_j)}{\prod_{i=1}^{N} \varepsilon_i k_i} \exp\left(\sum_1^N \frac{1}{2} \varepsilon_i \eta_i\right).$$

(3.3.77a) is equivalent to the usual N-soliton formulas (3.3.15). This may be shown as follows. In (3.3.15) take $\varepsilon_i = 2\mu_i - 1$, $i = 1, 2, \cdots, N$ and take a convenient choice of phase factors $\eta_i^{(0)} = \eta_i^{(1)} + \eta_i^{(2)}$, where $\exp \eta_i^{(1)} = -\prod_{j=1, j\neq i}^{N} (k_i + k_j)/(k_j - k_i)$. Calling $\xi_i = k_i x - k_i^3 t + \eta_i^{(2)}$ we have that (3.3.15) becomes

(3.3.77b) $$F_N = \sum_{\varepsilon=\pm 1} (-1)^{N(N+1)/2} \prod_{i=1}^{N} \varepsilon_i \prod_{i<j}^{N} \left(\frac{\varepsilon_i k_i - \varepsilon_j k_j}{k_i - k_j}\right) \exp\left(\sum_{i=1}^{N} \frac{1}{2}(\varepsilon_i+1)\xi_i\right),$$

where we have used

(3.3.77c) $$(-1)^{\sum_1^N (\varepsilon_i+1)/2} = (-1)^N \prod_{i=1}^{N} \varepsilon_i,$$

$$\prod_{1\leq i<j}^{N} \left(\frac{k_i-k_j}{k_i+k_j}\right)^{(\varepsilon_i+1)(\varepsilon_j+1)/2} \prod_{i,j=1}^{N} \left(\frac{k_i+k_j}{k_i-k_j}\right)^{(\varepsilon_i+1)/2}$$
(3.3.77d)

$$= (-1)^{N(N-1)/2} \prod_{1\leq i<j}^{N} \left(\frac{\varepsilon_i k_i - \varepsilon_j k_j}{k_i - k_j}\right),$$

$$(3.3.77\text{e}) \quad (-1)^{(\varepsilon_i+1)/2}\left(\frac{k_i-k_j}{k_i+k_j}\right)^{[(\varepsilon_i+1)(\varepsilon_j+1)/2-(\varepsilon_i+1)/2-(\varepsilon_j+1)/2]} = -\frac{(\varepsilon_i k_i - \varepsilon_j k_j)}{(k_i-k_j)}.$$

Using $f \asymp g$ if and only if $f = e^{\alpha x + \beta} g$, we have

$$(3.3.78) \quad F_N \asymp \sum_{\varepsilon=\pm 1} \prod_{i<j}^{N} \left(\frac{\varepsilon_i k_i - \varepsilon_j k_j}{(k_i^2 - k_j^2)}\right) \frac{1}{\prod_1^N \varepsilon_i k_i} \exp\left(\sum_{i=1}^{N} \frac{1}{2}\varepsilon_i \eta_i\right).$$

It is clear that (3.3.78) is equivalent to (3.3.77a) since the factor $\prod_{i<j}^{N} 1/(k_i^2 - k_j^2)$ is simply a constant.

Next we use the Bäcklund transformation (3.3.72) in order to derive a soliton superposition (i.e., permutation) formula (Hirota and Satsuma (1978)). First consider four solutions f_{N-1}, f_N, $\tilde{f}_N$ and f_{N+1} with parameters defined by:

$$f_{N-1} = f_{N-1}(k_1, \cdots, k_{N-1}), \qquad \tilde{f}_N = \tilde{f}_N(k_1, \cdots, k_{N-1}, k_{N+1}),$$
(3.3.79)
$$f_N = f_N(k_1, \cdots, k_{N-1}, k_N), \qquad f_{N+1} = f_{N+1}(k_1, \cdots, k_{N-1}, k_N, k_{N+1}),$$

and satisfying the Bäcklund transformations

(3.3.80a) $\qquad (D_x^2 - \tfrac{1}{4}k_N^2)f_{N-1} \cdot f_N = 0,$

(3.3.80b) $\qquad (D_x^2 - \tfrac{1}{4}k_{N+1}^2)f_{N-1} \cdot \tilde{f}_N = 0,$

(3.3.80c) $\qquad (D_x^2 - \tfrac{1}{4}k_{N+1}^2)f_N \cdot f_{N+1} = 0,$

(3.3.80d) $\qquad (D_x^2 - \tfrac{1}{4}k_N^2)\tilde{f}_N \cdot f_{N+1} = 0.$

Multiplying (3.3.80a) by $\tilde{f}_N f_{N+1}$ and (3.3.80d) by $f_{N-1} f_N$ and subtracting one from the other gives

$$(3.3.81) \qquad \tilde{f}_N f_{N+1}(D_x^2 f_{N-1} \cdot f_N) - f_{N-1} f_N (D_x^2 \tilde{f}_N \cdot f_{N+1}) = 0.$$

Given any four smooth functions of $x(a, b, c, d)$, the following identity exists:

$$(3.3.82) \quad (D_x^2 a \cdot b)cd - ab D_x^2 (c \cdot d) = D_x((D_x a \cdot d) \cdot bc + (ad) \cdot (D_x c \cdot b)).$$

With this identity (3.3.81) is reduced to

$$(3.3.83) \quad D_x[(D_x f_{N-1} \cdot f_{N+1}) \cdot \tilde{f}_N f_N + (f_{N-1} f_{N+1}) \cdot (D_x \tilde{f}_N \cdot f_N)] = 0.$$

Similarly we get from (3.3.80b) and (3.3.80c)

$$(3.3.84) \quad D_x[(D_x f_N \cdot \tilde{f}_N) \cdot (f_{N-1} f_{N+1}) + (f_N \tilde{f}_N) \cdot (D_x f_{N-1} \cdot f_{N+1})] = 0.$$

Subtracting (3.3.84) from (3.3.83) and noticing that $D_x a \cdot b = -D_x b \cdot a$, we have

$$(3.3.85) \qquad D_x(D_x f_{N-1} \cdot f_{N+1}) \cdot (f_N \tilde{f}_N) = 0,$$

which means $D_x f_{N-1} \cdot f_{N+1}$ is proportional to $f_N \tilde{f}_N$; i.e.,

(3.3.86) $$D_x f_{N-1} \cdot f_{N+1} = C f_N \tilde{f}_N.$$

The constant C is determined by any three-soliton solutions (at any location). (3.3.86) is referred to as the soliton superposition formula. Hence, given $f_0 = 1$, we use the Bäcklund transformations to develop f_1 and thereafter use (3.3.86) to develop f_N ($N \geqq 2$).

It is instructive to see how the scattering problem and time dependence for KdV can be derived from the Bäcklund transformation in bilinear form. We define

(3.3.87a) $$u = 2(\log f_N)_{xx} = 2\frac{f_{Nxx}f_N - f_{Nx}^2}{f_N^2},$$

(3.3.87b) $$\psi = \frac{f_{N+1}}{f_N},$$

and observe that

(3.3.88a) $$\frac{D_x f_{N+1} \cdot f_N}{f_N^2} = \psi_x,$$

(3.3.88b) $$\frac{D_x^2 f_{N+1} \cdot f_N}{f_N^2} = \psi_{xx} + u\psi,$$

(3.3.88c) $$\frac{D_x^3 f_{N+1} \cdot f_N}{f_N^2} = \psi_{xxx} + u\psi_x.$$

Using (3.3.72) we have the IST pair

(3.3.89a) $$\psi_{xx} + u\psi = \lambda\psi,$$

(3.3.89b) $$\psi_t + 3\lambda\psi_x + \psi_{xxx} + u\psi_x = 0,$$

from which $\psi_{xxt} = \psi_{txx}$. (Note that (3.3.89b) may be reduced to the form $\psi_t = A\psi + B\psi_x$ by making use of (3.3.89a).) Thus we see that it is possible to derive from the evolution equation itself, and its N-soliton solution, both the Bäcklund transformation and the IST pair.

3.3.e. Comments on some multidimensional problems. Finally we note that this direct method of finding soliton solutions has been applied to certain multidimensional nonlinear problems. We shall briefly discuss two such equations.

First we consider a two-dimensional version of the KdV equation, specifically the so-called Kadomtsev–Petviashvili (K–P) equation:

(3.3.90a) $$(u_t + 6uu_x + u_{xxx})_x + \alpha u_{yy} = 0, \quad \alpha = \pm 1.$$

Substituting $u = 2(\log f)_{xx}$ into (3.3.90a), we find

(3.3.90b) $$(D_x D_t + D_x^4 + \alpha D_y^2) f \cdot f = 0.$$

Satsuma (1976) has shown using the method described earlier that the following N-soliton solution holds:

(3.3.91a) $$f = \sum_{\mu=0,1} \exp\left[\sum_{1 \leq i < j}^{N} \mu_i \mu_j A_{ij} + \sum_{i=1}^{N} \mu_i \eta_i\right],$$

where

(3.3.91b)
$$\eta_i = k_i(x + p_i y - C_i t), \quad C_i = k_i^2 + \alpha p_i^2,$$
$$e^{A_{ij}} = \frac{3(k_i - k_j)^2 - \alpha(p_i - p_j)^2}{3(k_i + k_j)^2 - \alpha(p_i - p_j)^2}.$$

This N-soliton solution expresses how N plane wave solitons interact with each other.

Miles (1977a,b) has investigated certain circumstances in which the interaction of two such suitable plane wave solitons resonantly produces a third one. Specifically, he noted that the phase shift caused by the interaction of two solitons could be made arbitrarily large. Miles' idea can be seen from the two-soliton solution of (3.3.90) when $\alpha = +1$. We note that $e^{A_{ij}} = 0$ if we choose the wavenumbers to satisfy

(3.3.92a) $$\sqrt{3}(k_1 - k_2) \pm (p_1 - p_2) = 0,$$

whereupon the two-soliton solution is

(3.3.92b) $$f = 1 + e^{\eta_1} + e^{\eta_2}.$$

If we assume $k_i > 0$, $i = 1, 2$, then as $x \to -\infty$ the only nontrivial portions of the solution occur when $\eta_1 = $ const. or $\eta_2 = $ const., whereupon $u \sim (k_i^2/2) \text{sech}^2 \frac{1}{2}\eta_i$, i.e., two plane wave solitons. However, as $x \to +\infty$ the only nontrivial portion of the solution occurs when $\eta_1 - \eta_2 = $ const. (Note that $f \sim e^{\eta_2}(1 + e^{\eta_1 - \eta_2}) \approx (1 + e^{\eta_1 - \eta_2})$.) Thus we have only *one* plane wave soliton emerging from this interaction: $u = (k_3^2/2) \text{sech}^2 \frac{1}{2}\eta_3$, where

$$\eta_3 = \eta_1 - \eta_2,$$
$$k_3 = k_1 - k_2, \quad k_3 p_3 = k_1 p_1 - k_2 p_2, \quad C_3 = \frac{k_1 C_1 - k_2 C_2}{k_1 - k_2}.$$

If

$$\omega_i = k_i C_i$$

(the dispersion relation), then we may verify that from (3.3.92a) we have

$$\omega_3 = \omega_1 - \omega_2.$$

Thus $\underline{k}_3 = \underline{k}_1 - \underline{k}_2$ and $\omega_3 = \omega_1 - \omega_2$. Hence we have a triad resonance situation and two solitons as $x \to -\infty$ produce a third when $x \to +\infty$! Note that this type of resonant interaction also occurs in ion acoustic waves in a collisionless plasma (Yajima, Oikawa and Satsuma (1978)).

It is interesting to note that Hirota's method works, in a restricted sense, on the sine-Gordon equation in $(2+1)$ dimensions,

(3.3.93) $$u_{xx} + u_{yy} - u_{tt} = \sin u$$

(Hirota (1973a); see also Liebbrandt (1978), Burt (1978), Gibbon et al. (1978) and Whitham (1979)). Applying the dependent variable transformation

(3.3.94a) $$u = i \log\left(\frac{f^*}{f}\right)$$

in (3.3.93), we find

(3.3.94b) $$(D_x^2 + D_y^2 - D_t^2) f \cdot f = \tfrac{1}{2}(f^2 - f^{*2}).$$

Interestingly enough, it is found that the formula

(3.3.95a) $$f = \sum_{\mu=0,1} \exp\left(\sum_{j=1}^{N} \left(\mu_j \eta_j + i\frac{\pi}{2}\right) + \sum_{1 \leq i < j} \mu_i \mu_j A_{ij}\right)$$

with

(3.3.95b) $$\eta_i = k_i x + p_i y - \omega_i t + \eta_i^{(0)}, \qquad k_i^2 + p_i^2 - \omega_i^2 = 1,$$

(3.3.95c) $$e^{A_{ij}} = -\frac{(k_i - k_j)^2 + (p_i - p_j)^2 - (\omega_i - \omega_j)^2}{(k_i + k_j)^2 + (p_i + p_j)^2 - (\omega_i + \omega_j)^2},$$

satisfies (3.3.94b) for arbitrary k_i, p_i when $N = 1, 2$. However, for $N = 3$ the solution necessarily must satisfy the additional constraint

(3.3.95d) $$\det \begin{vmatrix} k_1 & p_1 & \omega_1 \\ k_2 & p_2 & \omega_2 \\ k_3 & p_3 & \omega_3 \end{vmatrix} = 0.$$

The situation for higher orders than $N = 3$ has been considered by Kobayashi and Izutsu (1976).

Finally we note that: (a) Hirota and Wadati (1979) have shown how the linear Gel'fand–Levitan integral equation may be deduced by the direct method; (b) Hirota (1979a) has given examples of equations for which two, but no more than two, "soliton" solutions may be constructed; (c) Nakamura (1979a,b) has used the direct method to find periodic and multiply (two) periodic wave solutions; (d) Oishi (1979) has shown how Fredholm determinants and "continuous spectrum" solutions may be viewed via this direct approach.

3.4. Rational solutions of nonlinear evolution equations.

It turns out that the special nonlinear evolution equations discussed in this monograph admit as solutions a class of rational functions (in x, the spatial variable). These rational solutions were first discovered by Airault, McKean and Moser (1977) (with further results and variants being obtained by, for example, Adler and Moser (1978) and Ablowitz and Satsuma (1978)). In this section we shall follow the methods of Ablowitz and Satsuma (1978). (i) For the KdV equation one can recover these rational solutions by taking (long wave) limits of the one-dimensional soliton solutions obtained via direct methods (i.e., using Hirota's method). In particular we demonstrate the results for the first few soliton cases, and then we show how, by performing the above limiting procedure on the Bäcklund transformation of KdV (in bilinear form), we obtain a recursion formula capable of generating the full class of rational solutions for the KdV equation. The corresponding rational solutions have poles on the real axis. (ii) These ideas also can be applied to the first few soliton (rational) solutions of the modified KdV, Boussinesq and Kadomtsev–Petviashvili (K–P) equations (as examples). For the modified KdV equation there exist real nonsingular rational solutions. This result for mKdV is in agreement with that of Ono (1976).

In the latter (K–P equation) case, one special solution is real, nonsingular and algebraically decaying in all directions. It has solitonic properties. We refer to this multidimensional soliton as a *lump* solution. This result was first noted by Manakov et al. (1977). It should be remarked that these methods also apply to other multidimensional problems of physical significance (see Satsuma and Ablowitz (1979)).

We begin with the KdV equation in the form

(3.4.1) $$u_t + 6uu_x + u_{xxx} = 0.$$

As we have seen in § 3.3 (3.4.1) has an N-soliton solution of the form

(3.4.2) $$u = 2(\log F_N)_{xx},$$

where F_N satisfies

(3.4.3) $$D_x(D_t + D_x^3)F_N \cdot F_N = 0.$$

Recall the definition of the D operator,

(3.4.4) $$D_x^n D_t^m a \cdot b = (\partial_x - \partial_{x'})^n (\partial_t - \partial_{t'})^m a(x, t) b(x', t')|_{x'=x, t'=t}$$

(see also § 3.3), and that F_N can be obtained simply by expanding F_N in a (formal) perturbation series

(3.4.5) $$F_N = 1 + \varepsilon F_N^{(1)} + \varepsilon^2 F_N^{(2)} + \cdots.$$

Substituting (3.4.5) into (3.4.3) and equating powers of ε yields a system of equations to be solved. This expansion *truncates* when $F_N^{(1)}$ is chosen to be of the form

$$(3.4.6) \qquad F_N^{(1)} = \sum_{i=1}^{N} \exp(\eta_i), \qquad \eta_i = k_i x - k_i^3 t + \eta_i^{(0)},$$

where $k_i, \eta_i^{(0)}$ are constant (a posteriori, we can take $\varepsilon = 1$). From § 3.3 we have the first few multisoliton solutions given by

$$(3.4.7a) \qquad F_1 = 1 + e^{\eta_1},$$

$$(3.4.7b) \qquad F_2 = 1 + e^{\eta_1} + e^{\eta_2} + e^{\eta_1 + \eta_2 + A_{12}},$$

$$(3.4.7c) \qquad F_3 = 1 + e^{\eta_1} + e^{\eta_2} + e^{\eta_3} + e^{\eta_1 + \eta_2 + A_{12}} + e^{\eta_1 + \eta_3 + A_{13}}$$
$$+ e^{\eta_2 + \eta_3 + A_{23}} + e^{\eta_1 + \eta_2 + \eta_3 + A_{12} + A_{13} + A_{23}},$$
$$\vdots$$

with the general N-soliton formula

$$(3.4.8a) \qquad F_N = \sum_{\mu=0,1} \exp\left(\sum_{i<j}^{N} A_{ij}\mu_i\mu_j + \sum_{i=1}^{N} \mu_i \eta_i\right),$$

where A_{ij} in (3.4.7) and (3.4.8a) satisfies

$$(3.4.8b) \qquad \exp A_{ij} = \left(\frac{k_i - k_j}{k_i + k_j}\right)^2.$$

It is easy to see that (3.4.7a–c) are special cases of (3.4.8). In addition we note that from (3.4.7a) and (3.4.2) the usual simple soliton is given by

$$(3.4.9) \qquad u = \left(\frac{k_1^2}{2}\right) \operatorname{sech}^2 \frac{1}{2}(k_1 x - k_1^3 t + \eta_1^{(0)}).$$

The fact that one can recover rational solutions relies on our freedom to choose arbitrary phase constants $\eta_i^{(0)}$. For example, in (3.4.9), if we choose $e^{\eta_i(0)} = -1$, then we have the singular soliton

$$(3.4.10) \qquad u = -\left(\frac{k_1^2}{2}\right) \operatorname{cosech}^2 \frac{1}{2}(k_1 x - k_1^3 t).$$

Passing to the limit $k_1 \to 0$ (i.e., the "long wave" limit) we find

$$(3.4.11) \qquad u = \frac{-2}{x^2}.$$

(3.4.11) is the first in this class of rational solutions. Moreover, it turns out if the phase constants are chosen appropriately, all the F_N described above have nontrivial distinguished limits.

OTHER PERSPECTIVES

We now discuss this technique as it pertains to the computation of F_N. One can easily recover u from F_N by use of (3.4.2). For this purpose let us again consider (3.4.7a). Calling $\alpha_i = \exp(\eta_i^{(0)})$, we write (3.4.7a) as

(3.4.12) $$F_1 = 1 + \alpha_1 e^{\xi_1}, \qquad \xi_1 = k_1(x - k_1^2 t).$$

As $k_1 \to 0$ we have

$$F_1 = 1 + \alpha_1(1 + \xi_1) + O(k_1^2).$$

Choosing $\alpha_1 = -1$, we have

$$F_1 = -k_1(x + O(k_1)).$$

Since u is given by (3.4.2), we take

$$F_1 \backsimeq x + O(k_1)$$

(here $f \backsimeq g$ if and only if $f = e^{ax+b} g$, where a, b are independent of x). Then in the limit $k_1 \to 0$, $f_1 \backsimeq \Theta_1$, where

(3.4.13) $$\Theta_1 = x.$$

Hence in the "long wave" limit $k_1 \to 0$, we have recovered the rational solution (3.4.11) (via 3.4.2, of course). The same idea applies to F_2 (as well as the higher F_N).

For $N > 1$, we consider all the k_i to be of the same asymptotic order as $k_i \to 0$ (i.e., $k_i = \varepsilon \bar{k}_i$, $\bar{k}_i = O(1)$). For F_2 we have, from (3.4.7b),

(3.4.14) $$F_2 = 1 + \alpha_1 e^{\xi_1} + \alpha_2 e^{\xi_2} + \alpha_1 \alpha_2 e^{\xi_1 + \xi_2 + A_{12}}.$$

As $k_1, k_2 \to 0$, we require that the coefficients of the $O(1)$, $O(k)$ terms in F_2 vanish:

$$O(1): \quad 1 + \alpha_1 + \alpha_2 + \alpha_1 \alpha_2 e^{A_{12}} = 0,$$

$$O(k): \quad k_1 \alpha_1 + k_2 \alpha_2 + (k_1 + k_2) \alpha_1 \alpha_2 e^{A_{12}} = 0.$$

The solution to these equations is

$$\alpha_1 = -\alpha_2 = \frac{k_1 + k_2}{k_1 - k_2}.$$

As it turns out, this solution also satisfies the equation at $O(k^2)$, and we find

(3.4.15a) $$F_2 = -\tfrac{1}{6} k_1 k_2 (k_1 + k_2)[(x^3 + 12t) + O(k)].$$

Thus as $k \to 0$, F_2 is equivalent to Θ_2:

(3.4.15b) $$\Theta_2 = x^3 + 12t, \qquad u = 2(\log \Theta_2)_{xx}.$$

Θ_2 has three zeros, so u has three poles. In the three-soliton case, if we choose

$$\alpha_1 = \frac{k_1+k_2}{k_1-k_2} \cdot \frac{k_3+k_1}{k_3-k_1},$$

$$\alpha_2 = \frac{k_2+k_3}{k_2-k_3} \cdot \frac{k_1+k_2}{k_1-k_2},$$

$$\alpha_3 = \frac{k_3+k_1}{k_3-k_1} \cdot \frac{k_2+k_3}{k_2-k_3},$$

we find a *six*-zero solution associated with F_3:

(3.4.16a)
$$F_3 = -\tfrac{1}{360} k_1 k_2 k_3 (k_1+k_2)(k_2+k_3)(k_3+k_1)[(x^6+60x^3t-720t^2)+O(k)],$$

or simply, in the limit $k \to 0$, $F_3 \simeq \Theta_3$,

(3.4.16b) $$\Theta_3 = x^6 - 60x^3 t - 720 t^2.$$

In principle this technique applies to any number of solitons. However, the calculations are tedious, and here we shall instead use the Bäcklund transformation (in bilinear form) for KdV to generate a recursion relation for the polynomials.

Before doing this, for convenience, we shall employ a slightly different formula for the N-soliton solution (3.4.8a). From § 3.3 (3.3.78) we have

(3.4.17) $$F_N \simeq \hat{F}_N = \sum_{\varepsilon=\pm 1} \prod_{i<j}^N \frac{(\varepsilon_i k_i - \varepsilon_j k_j)}{(k_i^2-k_j^2)} \cdot \frac{\exp\left(\sum_{i=1}^N \tfrac{1}{2}\varepsilon_i \eta_i\right)}{\prod_{i=1}^N \varepsilon_i k_i},$$

so, for example,

$$\hat{F}_1 = \frac{1}{k_1}(e^{\eta_1/2} - e^{-\eta_1/2}) = -\frac{1}{k_1} e^{-\eta_1/2}(1-e^{\eta_1}),$$

which is seen to be equivalent to (3.4.7a). One of the advantages of (3.4.17) is that the limit as $k_1 \to 0$ directly yields the polynomial in x. We may rewrite $\hat{F}_N$ in the following way (hereafter we shall drop the $\hat{}$ on $\hat{F}_N$):

(3.4.18) $$F_N = \frac{g_N}{\prod_{i<j}^N (k_i^2-k_j^2)\prod_{I=1}^N k_i},$$

where

(3.4.19) $$g_N = \sum_{\varepsilon=\pm 1} \prod_{i<j}^N (\varepsilon_i k_i - \varepsilon_j k_j) \prod_{i=1}^N \varepsilon_i \exp\left(\sum_{i=1}^N \frac{1}{2} \varepsilon_i \eta_i\right).$$

It is important to note that g has the following properties:

(i) $g_N(k_1, k_2, \cdots, k_i, \cdots, k_j, \cdots, k_N) = -g_N(k_1, k_2, \cdots, k_j, \cdots, k_i, \cdots, k_N)$, for $i<j$ (g_N is antisymmetric in the k_i's),

(ii) $g_N(k_1 = 0, k_2, \cdots, k_N) = 0$,
(iii) $g_N(k_1 = k_2, k_3, \cdots, k_N) = g_N(k_1 = -k_2, k_3, \cdots, k_N) = 0$.

(i)–(iii) and (3.4.19) imply that g_N is factorized by

$$\prod_{i<j}^{N} (k_i^2 - k_j^2) \prod_{i=1}^{N} k_i.$$

Thus, the first term as $k_i \to 0$ in F_N is at least order one, and we write

(3.4.20) $\qquad F_N = a_N \Theta_N(x) + O(k).$

We shall later show that $a_N \neq 0$. Moreover, since every k_i in the phase factor of (3.4.17) has an x multiplying it, in order for F_N to be at least order one, the polynomial $\Theta_N(x) = x^P + \cdots$ must have its highest power satisfying

$$P = N + N\left(\frac{N-1}{2}\right) = \frac{N(N+1)}{2}.$$

Next we derive a recursion relation for Θ_N. From § 3.3 we take the following soliton permutation formula derived from the Bäcklund transformations:

(3.4.21) $\qquad D_x F_{N-1} \cdot F_{N+1} = C F_N \tilde{F}_N,$

where the parameters of the N-soliton solution F_N, etc., are defined by

$$F_{N-1} = F_{N-1}(k_1, \cdots, k_{N-1}),$$
$$F_N = F_N(k_1, \cdots, k_{N-1}, k_N),$$
$$\tilde{F}_N = \tilde{F}_N(k_1, \cdots, k_{N-1}, k_{N+1}),$$
$$F_{N+1} = F_{N+1}(k_1, \cdots, k_{N-1}, k_N, k_{N+1}).$$

The constant C is determined by any three soliton solutions given by (3.4.17). For example, using

$$F_0 = 1,$$
$$F_1 = \frac{1}{k_1}(e^{\eta_1/2} - e^{-\eta_1/2}),$$
$$F_2 = \frac{1}{k_1 k_2 (k_1^2 - k_2^2)}[(k_1 - k_2)(e^{(\eta_1+\eta_2)/2} - e^{-(\eta_1+\eta_2)/2})$$
$$+ (k_1 + k_2)(e^{-(\eta_1-\eta_2)/2} - e^{(\eta_1-\eta_2)/2})],$$

we find $C = -\frac{1}{2}$, whereupon the superposition formula is given by

(3.4.22) $\qquad D_x F_{N+1} \cdot F_{N-1} = \frac{1}{2} F_N \tilde{F}_N.$

Using (3.4.20) and (3.4.22) we can obtain a recursion formula for a_N and Θ_N.

Inserting (3.4.20) into (3.4.22) yields

(3.4.23) $\quad a_{N+1}a_{N-1}D_x\Theta_{N+1}\cdot\Theta_{N-1} = \frac{1}{2}a_N^2\Theta_N^2.$

Since $\Theta_N(x)$ is a polynomial in x, (3.4.23) must hold separately for each power, and in particular for the highest power, $x^{N(N+1)/2}$. Hence a_N satisfies the recursion formula

(3.4.24) $\quad a_{N+1}a_{N-1}(4N+2) = a_N^2,$

and Θ_N satisfies

(3.4.25) $\quad D_x\Theta_{N+1}\Theta_{N-1} = (2N+1)\Theta_N^2$

(see also Adler and Moser (1978)). Using our earlier calculations involving the first few solitons, we have already deduced

$$a_0 = 1, \quad a_1 = 1, \quad a_2 = \tfrac{1}{6}, \quad a_3 = \tfrac{1}{360},$$

$$\Theta_0 = 1, \quad \Theta_1 = x, \quad \Theta_2 = x^3 + 12t, \quad \Theta_3 = x^6 + 60x^3 t - 720 t^2.$$

From the recursion relations using a_0, a_1, we see that a_2, a_3 above are recovered, and that $a_N \neq 0$, for all $N \geq 0$. Similarly, we can use (3.4.25) with Θ_0, Θ_1 to deduce Θ_2, Θ_3 and the higher Θ_N, if we supplement (3.4.25) with a time evolution equation. For this purpose we can use either the original PDE (KdV in this case) or the time-dependent equation obtained from the Bäcklund transformation (3.3.72b) (using $\lambda = \frac{1}{4}k_{N+1}^2$ and taking the limit $\underline{k} \to 0$):

(3.4.26) $\quad (D_t + D_x^3)\Theta_N \cdot \Theta_{N+1} = 0.$

This still allows an arbitrary multiple of Θ_{N-1} to be added onto a particular solution Θ_{N+1}. In this problem we take this arbitrary multiple to be zero. In the limiting procedure described above, one can see that to obtain a given asymptotic power of k, each x^3 term corresponds with each power of t. As an example, we know the leading order term in Θ_3 is x^6. Thus the general form of this polynomial must be $x^6 + \alpha x^3 t + \beta t^2$. We determine $\alpha = 60$, $\beta = -720$ from (3.4.25)–(3.4.26). Although we can add the term Cx to this result, and still satisfy (3.4.25)–(3.4.26), we take $C = 0$ since it cannot arise via the limiting procedure on F_3.

Since each x^3 term has a corresponding power of t, all of the above solutions satisfy the similarity equation for $w(z)$, where

$$u = \frac{1}{(3t)^{2/3}} w(z), \qquad z = \frac{x}{(3t)^{1/3}},$$

$$w''' + 6ww' - (2w + zw') = 0.$$

Hence the similarity solution $w(z)$ also has a class of rational solutions. Since we have shown that the rational solutions arise from limits of the soliton solutions and that they may be computed from Bäcklund transformations, we have a direct connection between the solitons and the similarity solutions. Special elementary solutions may be obtained for other equations including the classical Painlevé transcendents (Erugin (1976) (review article), Airault (1979), Boiti and Pempinelli (1979)). These authors also derive Bäcklund transformations between solutions of such nonlinear ODE's. (See also Fokas and Yortsos (1981).)

As mentioned earlier the methods we have employed can be readily adapted to other nonlinear evolution equations possessing soliton solutions. Here we shall only discuss the results of the limiting procedures on the first few soliton formulas of (i) the K–P (two-dimensional KdV) equation; (ii) the Boussinesq equation; and (iii) the modified KdV equation with a nonzero background state.

The K–P equation is given by

(3.4.27a) $$\partial_x(u_t + 6uu_x + u_{xxx}) + \alpha u_{yy} = 0,$$

where α is a constant depending on the dispersive property of the system. We look for a solution of (3.4.27a) with $u \to 0$ as $|x| \to \infty$, of the form (Satsuma (1976))

(3.4.27b) $$u = 2(\log F_N)_{xx},$$

Inserting (3.4.27b) into (3.4.27a) yields

(3.4.28) $$(D_x D_t + D_x^4 + \alpha D_y^2) F_N \cdot F_N = 0.$$

The N-soliton solution can be ascertained by direct methods (cf. § 3.3). Here we will only discuss the $N = 1, 2$ cases. The one- and two-soliton solutions are given by

(3.4.29a) $$F_1 = 1 + e^{\eta_1},$$

(3.4.29b) $$F_2 = 1 + e^{\eta_1} + e^{\eta_2} + e^{\eta_1 + \eta_2 + A_{12}},$$

where

(3.4.29c) $$\eta_i = k_i(x + P_i y - (k_i^2 + \alpha P_i^2)t) + \eta_i^{(0)},$$

(3.4.29d) $$\exp A_{ij} = \frac{3(k_i - k_j)^2 - \alpha(P_i - P_j)^2}{3(k_i + k_j)^2 - \alpha(P_i - P_j)^2}.$$

Taking $e^{\eta_i^{(0)}} = -1$, $k_i \to 0$ (with $P_i = O(1)$, $k_1/k_2 = O(1)$), we find

(3.4.30a) $$F_1 = -k_1 \theta_1 + O(k_1^2),$$

(3.4.30b) $$F_2 = k_1 k_2 \left(\theta_1 \theta_2 + \frac{12}{\alpha(P_1 - P_2)^2}\right) + O(k^3),$$

where

(3.4.30c) $$\theta_i = x + P_i y - \alpha P_i^2 t,$$

and we have used

(3.4.30d) $$\exp A_{12} \sim 1 + \frac{12 k_1 k_2}{\alpha (P_1 - P_2)^2}.$$

Since u is given by (3.4.27b) we have therefore deduced the following rational solutions:

(3.4.31a) $\hat{F}_1 = \theta_1,$

(3.4.31b) $\hat{F}_2 = \theta_1 \theta_2 + B_{12}, \quad B_{12} = \dfrac{12}{\alpha (P_1 - P_2)^2}.$

Although $\hat{F}_1, \hat{F}_2$ are generally singular at some position, a real nonsingular solution is obtained for $\hat{F}_2$ if $\alpha = -1$ and $P_2 = P_1^*$. In this case we have

(3.4.31c) $\hat{F}_2 = \theta_1 \theta_1^* - \dfrac{12}{(P_1 - P_1^*)^2}.$

Letting $P_1 = P_R + i P_I$, we have

(3.4.32a) $$u = 2 \partial_x^2 \log\left[(x' + P_R y')^2 + P_I^2 (y')^2 + \frac{3}{P_I^2} \right],$$

where

$$x' = x - (P_R^2 + P_I^2) t,$$
$$y' = y + 2 P_R t.$$

Alternatively (3.4.32a) may be written explicitly as

(3.4.32b) $$u = \frac{4(-(x' + P_R y')^2 + P_I^2 (y')^2 + 3/P_I^2)}{((x' + P_R y')^2 + P_I^2 (y')^2 + 3/P_I^2)^2}.$$

Hence we have a permanent lump solution decaying as $O(1/x^2, 1/y^2)$ for $|x|, |y| \to \infty$, and moving with the velocity $v_x = P_R^2 + P_I^2$, $v_y = -2 P_R$ (see Fig. 3.1). When $N = 4$ one may obtain in this way a two-lump solution. Here we simply present the results of the calculation for $N = 3, 4$. We find

(3.4.33a) $F_3 = \theta_1 \theta_2 \theta_3 + B_{12} \theta_3 + B_{23} \theta_1 + B_{31} \theta_2,$

(3.4.33b) $F_4 = \theta_1 \theta_2 \theta_3 \theta_4 + B_{12} \theta_3 \theta_4 + B_{13} \theta_2 \theta_4 + B_{14} \theta_2 \theta_3 + B_{23} \theta_1 \theta_4$
$\qquad + B_{24} \theta_1 \theta_3 + B_{34} \theta_1 \theta_2 + B_{12} B_{34} + B_{13} B_{24} + B_{14} B_{23},$

where θ_i is given by (3.4.30c), $B_{ij} = 12/(\alpha (P_i - P_j)^2)$. Taking $\alpha = -1$, $P_3 = P_1^*$, $P_4 = P_2^*$ in (3.4.33b) yields a two-lump solution. We note that with this

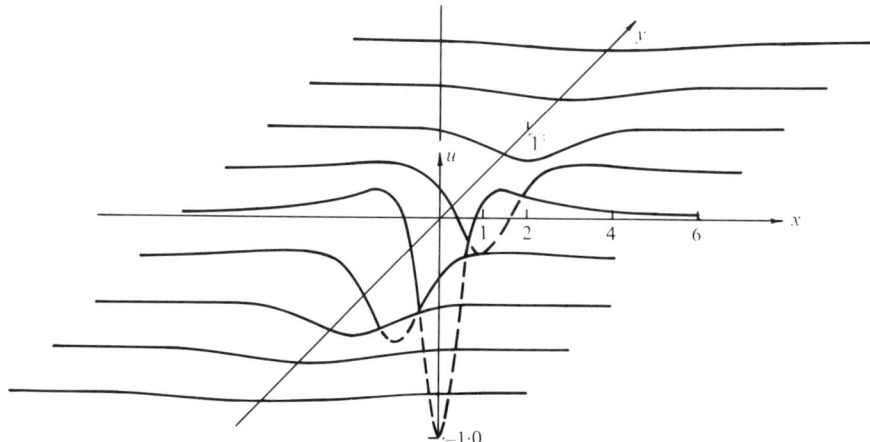

FIG. 3.1. *Lump solution of* (3.4.32) *as seen in two dimensions at a fixed time.* $P_R = 0$, $P_I = \frac{1}{8}$, $\alpha = -1$.

proviso F_4 is positive and yields a solution u via (3.4.27b) which decays as $O(1/x^2, 1/y^2)$ for $|x|, |y| \to \infty$. This two-lump solution yields zero asymptotic phase shift. All of these results agree with those of Manakov et al. (1977). In general when $N = 2M$ this method yields formulae for an M-lump solution (see Satsuma and Ablowitz (1979)). Of course, the methods used here unfortunately do not give strong evidence about the role of these solutions in the general initial value problem, e.g., genericity, stability, etc. Recent work by Zakharov and Manakov (1979) shows, however, that for rapidly decaying initial values (faster than $O(1/x^2, 1/y^2)$) the K–P equation (3.4.27a) is solvable by IST. (See also the long time asymptotic results of Manakov, Santini and Takhtadzhyan (1980).) No permanent soliton solutions are found. This seems to indicate that the above solutions, and in general permanent wave solutions, may not play an important role in (3.4.27a) (as opposed to the one-dimensional version, i.e., KdV).

Our second example is the Boussinesq equation,

(3.4.34) $$u_{tt} - u_{xx} - (3u^2)_{xx} - u_{xxxx} = 0.$$

Note that (3.4.34) is a special case of the K–P equation discussed above. However, the rational solutions we will now obtain are in a different class.

We note that the methods we employ work equally well if the sign of the last term in (3.4.34) is changed. With a positive sign in the last term the equation will be well posed on the infinite interval (even with this change the equation remains an isospectral flow). It should also be noted that (3.4.34) arises in a physical problem (e.g., water waves) as a long wave equation. Thus, in the context of the physical problem, it is, in fact, well posed.

Following Hirota (1973a), we let

(3.4.35) $$u = 2(\log F_N)_{xx}$$

and find the bilinear equation

(3.4.36) $$(D_t^2 - D_x^2 - D_x^4)F_N \cdot F_N = 0.$$

The first two soliton solutions are (as usual) given by

(3.4.37a) $$F_1 = 1 + e^{\eta_1},$$
(3.4.37b) $$F_2 = 1 + e^{\eta_1} + e^{\eta_2} + e^{\eta_1 + \eta_2 + A_{12}},$$

where

(3.4.37c) $$\eta_i = k_i x + \varepsilon_i k_i \sqrt{1 + k_i^2}\, t + \eta_i^{(0)}, \qquad \varepsilon_i = +1 \text{ or } -1$$

and

(3.4.37d) $$e^{A_{12}} = \frac{3(k_1 - k_2)^2 + (\varepsilon_1 \sqrt{1 + k_1^2} - \varepsilon_2 \sqrt{1 + k_2^2})^2}{3(k_1 + k_2)^2 + (\varepsilon_1 \sqrt{1 + k_1^2} - \varepsilon_2 \sqrt{1 + k_2^2})^2}.$$

For $N = 1$, taking $e^{\eta_1^{(0)}} = -1$ and $k_1 \to 0$, we have

(3.4.38) $$F_1 \sim -k_1(x \pm t).$$

For the two-soliton solution, we note

(3.4.39a) $$e^{A_{12}} \sim \begin{cases} \left(\dfrac{k_1 - k_2}{k_1 + k_2}\right)^2 \left(1 + \dfrac{1}{3} k_1 k_2\right) & \text{for } \varepsilon_1 \varepsilon_2 = 1, \\ 1 - 3 k_1 k_2 & \text{for } \varepsilon_1 \varepsilon_2 = -1. \end{cases}$$
(3.4.39b)

In the case of $\varepsilon_1 \varepsilon_2 = 1$, we take

(3.4.40a) $$e^{\eta_1^{(0)}} = \frac{k_1 + k_2}{k_1 - k_2} + \frac{1}{6} k_1 k_2,$$

(3.4.40b) $$e^{\eta_2^{(0)}} = -\frac{k_1 + k_2}{k_1 - k_2} + \frac{1}{6} k_1 k_2,$$

and find

(3.4.41) $$F_2 \sim -\tfrac{1}{6}k_1 k_2(k_1+k_2)\{(x\pm t)^3+(x\pm t)\mp 6t\}.$$

It is interesting to notice that F_2 gives another rational solution for the case, $\varepsilon_1\varepsilon_2 = -1$. In this case, taking $e^{\eta_1^{(0)}} = e^{\eta_2^{(0)}} = -1$ and using (3.4.39a), we obtain

(3.4.42) $$F_2 \sim k_1 k_2(x^2-t^2-3).$$

Thus, the first few polynomial solutions of the Boussinesq equation are given by (3.4.35) with the following F_N:

$$x\pm t, \quad x^2-t^2-3, \quad (x\pm t)^3+(x\pm t)\mp 6t.$$

Higher order polynomials can be obtained in this manner, and presumably the Bäcklund transformation will yield a recursion relation between rational solutions, although this has not yet been carried out.

Finally, let us consider the modified KdV equation

(3.4.43) $$v_t + 6v^2 v_x + v_{xxx} = 0,$$

with a nonzero asymptotic condition, $v \to v_0$ as $|x| \to \infty$. Following Hirota (1974) and Hirota and Satsuma (1976a), we have

(3.4.44) $$v = v_0 + i\left(\log \frac{G_N}{F_N}\right)_x.$$

Substituting (3.4.44) into (3.4.43) and decoupling the resulting equation, we get

(3.4.45a) $$(D_t + 6v_0^2 D_x + D_x^3)F_N \cdot F_N = 0,$$

(3.4.45b) $$(D_x^2 - 2iv_0 D_x)G_N \cdot F_N = 0.$$

In order to obtain soliton solutions, we expand

(3.4.46a) $$F_N = 1 + \varepsilon F_{N,1} + \varepsilon^2 F_{N,2} + \cdots,$$

(3.4.46b) $$G_N = 1 + \varepsilon G_{N,1} + \varepsilon^2 G_{N,2} + \cdots,$$

substitute (3.4.46) into (3.4.45), and equate coefficients of ε. Starting with $F_{N,1} = e^{\eta_1+\phi_1}$ and $G_{N,1} = e^{\eta_1+\psi_1}$, we obtain a one-soliton solution

(3.4.47a) $$F_1 = 1 + e^{\eta_1+\phi_1},$$

(3.4.47b) $$G_1 = 1 + e^{\eta_1+\psi_1},$$

where

(3.4.47c) $$\eta_i = k_i x - (6v_0^2 k_i + k_i^3)t + \eta_i^{(0)},$$

(3.4.47d) $$e^{\phi_i} = 1 + ik_j/2v_0,$$

(3.4.47e) $$e^{\psi_i} = 1 - ik_j/2v_0.$$

Inserting (3.4.47) into (3.4.44), we have an explicit form of the one-soliton solution,

$$\tag{3.4.48} v = v_0 + \frac{k_1^2}{(\sqrt{4v_0^2 + k_1^2}\cosh\eta_1 + 2v_0)},$$

which was also found by Ono (1976). To get a two-soliton solution, we start with

$$F_{N,1} = \sum_{i=1}^{2} e^{\eta_i + \phi_i}, \qquad G_{N,1} = \sum_{i=1}^{2} e^{\eta_i + \psi_i},$$

and find

$$\tag{3.4.49a} F_2 = 1 + e^{\eta_1 + \phi_1} + e^{\eta_2 + \phi_2} + e^{\eta_1 + \eta_2 + \phi_1 + \phi_2 + A_{12}},$$

$$\tag{3.4.49b} G_2 = 1 + e^{\eta_1 + \psi_1} + e^{\eta_2 + \psi_2} + e^{\eta_1 + \eta_2 + \psi_1 + \psi_2 + A_{12}},$$

where

$$\tag{3.4.49c} e^{A_{12}} = \left(\frac{k_1 - k_2}{k_1 + k_2}\right)^2.$$

As before, rational solutions are deduced by taking the limit $k_i \to 0$ and choosing the phase constant adequately. For $N = 1$, choosing $e^{\eta_1^{(0)}} = -1$, we get

$$\tag{3.4.50a} F_1 \sim -k_1\left(x - 6v_0^2 t + \frac{i}{2v_0}\right),$$

$$\tag{3.4.50b} G_1 \sim -k_1\left(x - 6v_0^2 t - \frac{i}{2v_0}\right),$$

which gives the rational solution

$$\tag{3.4.51} v = v_0 - \frac{4v_0}{4v_0^2(x - 6v_0^2 t)^2 + 1}.$$

This solution was also found by Ono (1976) and is a nonsingular one-dimensional algebraic soliton. For $N = 2$, taking

$$e^{\eta_1^{(0)}} = -e^{\eta_2^{(0)}} = \frac{k_1 + k_2}{k_1 - k_2}\left(1 + \frac{k_1 k_2}{8v_0^2}\right),$$

and $k_i \to 0$, we find that

$$\tag{3.4.52a} F_2, G_2 \sim -\frac{1}{6} k_1 k_2 (k_1 + k_2) \cdot \left[\xi^3 + 12t - \frac{3}{4v_0^2}\xi \pm \frac{3i}{2v_0}\left(\xi^2 + \frac{1}{4v_0^2}\right)\right],$$

where

$$\tag{3.4.52b} \xi = x - 6v_0^2 t$$

and the upper (lower) sign stands for $F_2(G_2)$. Substituting (3.4.32a) into (3.4.44), we see that the solution is also a nonsingular algebraic soliton, given by

$$(3.4.53) \qquad v = v_0 - \frac{12v_0(\xi^4 + (3/2v_0^2)\xi^2 - 3/16v_0^4 - 24\xi t)}{4v_0^2(\xi^3 + 12t - (3/4v_0^2)\xi)^2 + 3(\xi^2 + 1/4v_0^2)^2}.$$

3.5. N-body problems and nonlinear evolution equations. The philosophy which we frequently expound in this book is that reductions of "integrable" nonlinear evolution equations are also (in some sense) "integrable." For example, similarity reductions lead to equations of Painlevé type (cf. § 3.7) such as the classical transcendents of Painlevé, or special hyperelliptic function solutions (cf. § 2.3). Here we shall discuss another example of this principle. By looking for algebraic "pole expansion" solutions of various nonlinear evolution equations we obtain a finite dynamical system of ordinary differential equations, i.e., N-body problems. These dynamical systems have an interest in their own right.

The original idea of investigating the motion of the poles of solutions to nonlinear evolution equations is quite old; it was used, for example, in studying the motion of point vortices in hydrodynamics (see, e.g., Onsager (1949)). With regard to problems of IST type, Kruskal (1974) first examined this question in the context of the KdV equation. He reasoned that a soliton is expressible as an infinite array of poles, and hence the interaction of solitons could be studied by the interaction of poles. Thickstun (1976) carried these ideas further and Airault, McKean and Moser (1977) showed how rational and elliptic solutions could be considered from the point of view of finite pole expansions (see also the work of Chudnovsky and Chudnovsky (1977)).

In this section we shall concentrate on pole expansions (i.e., the rational solutions) of three nonlinear evolution equations: the so-called Benjamin–Ono (B–O) equation, KdV and an intermediate equation which has as limiting forms both of the above equations. All of these equations arise in the propagation of long internal waves in a stratified fluid. It should be noted that Moser (1975a,b) has considered another interesting finite dimensional system. This is obtained from the Toda lattice (§ 2.2) with free ends. We shall not go into this problem here.

In each of these cases we shall develop the associated dynamical system. In only the first example shall we carry out an explicit integration. We shall obtain solutions to the dynamical system studied by Calogero (1971), (1976), Sutherland (1972) and Moser (1975a,b). While the solutions to the associated nonlinear evolution equations are rational functions of the spatial variable (x), we are splitting this section off from the previous one since the point of view here is to obtain and study certain interesting dynamical systems.

We begin with the B–O equation (first proposed by Benjamin (1967) and later derived via a formal asymptotic expansion by Ono (1975)): a nonlinear singular integral-differential equation,

(3.5.1) $$u_t + 2uu_x + H(u_{xx}) = 0,$$

where $H(u)$ is the Hilbert transform

(3.5.2) $$H(u) = \frac{1}{\pi} \fint_{-\infty}^{\infty} \frac{u(x')}{x' - x} dx'.$$

$\fint f(x) \, dx$ is the usual principal value integral. As suggested by Case (1978b) and Chen, Lee and Pereira (1979) we look for the motion of the poles (i.e., a pole expansion) of a solution of this equation, namely,

(3.5.3) $$u = \sum_{j=1}^{N} \frac{i}{x - x_j(t)} + \sum_{j=1}^{N} \frac{-i}{x - x_j^*(t)}.$$

We may view as partial motivation the fact that the solitary-wave solution to (3.5.1) is a rational function of x found by Benjamin (1967):

(3.5.4) $$u = \frac{2C}{1 + [C(x - Ct - x_0)]^2} = iC \left[\frac{1}{C(x - Ct - x_0) + i} - \frac{1}{C(x - Ct - x_0) - i} \right].$$

We note that the Hilbert transform of a pole is a pole; i.e.,

$$H\left(\frac{1}{x - x_i}\right) = \frac{i}{x - x_i}, \quad \operatorname{Im} x_i > 0.$$

Substituting (3.5.3) into (3.5.1) yields

(3.5.5) $$\sum_j \frac{1}{(x - x_j)^2} \left\{ i\dot{x}_j + 2 \left(\sum_{k \neq j} \frac{1}{(x - x_k)} - \sum_k \frac{1}{x - x_k^*} \right) \right\}$$
$$+ \sum_j \frac{1}{(x - x_j^*)^2} \left\{ -i\dot{x}_j^* + 2 \left(\sum_{k \neq j} \frac{1}{(x - x_k^*)} - \sum_k \frac{1}{(x - x_k)} \right) \right\} = 0.$$

where $\dot{x} \equiv dx/dt$.

There are various ways to obtain the motion of the poles. Rational fraction decomposition shows that

(3.5.6a) $$\frac{1}{(x-a)^2(x-b)} = \frac{A}{(x-a)^2} + \frac{B}{(x-a)} + \frac{C}{(x-b)},$$

with

(3.5.6b) $$A = \frac{1}{a-b}, \quad B = \frac{-A}{a-b}, \quad C = \frac{1}{(a-b)^2}.$$

Then, using (3.5.6) and the facts that

(3.5.7a) $$\sum_j \sum_{k\neq j} \frac{1}{(x_k - x_j)(x - x_k)(x - x_j)} = 0$$

and

(3.5.7b) $$\sum_j \sum_k \frac{1}{(x_k^* - x_j)(x - x_j)(x - x_k)} + \sum_j \sum_k \frac{1}{(x_k - x_j^*)(x - x_j^*)(x - x_k^*)} = 0,$$

we have

(3.5.8) $$\sum_j \frac{1}{(x - x_j)^2} \left\{ i\overset{\circ\circ}{x}_j + 2\left(\sum_{k\neq j} \frac{1}{(x_j - x_k)} - \sum_k \frac{1}{x_j - x_k^*} \right) \right\}$$
$$+ \sum_j \frac{1}{(x - x_j^*)^2} \left\{ -i\overset{\circ\circ}{x}_j^* + 2\left(\sum_{k\neq j} \frac{1}{(x_j^* - x_k^*)} - \sum_k \frac{1}{x_j^* - x_k} \right) \right\} = 0,$$

whereupon we have the dynamical system (N-body problem)

(3.5.9) $$i\overset{\circ\circ}{x}_j + 2\left(\sum_{\substack{k=1 \\ k\neq j}}^N \frac{1}{x_j - x_k} - \sum_{k=1}^N \frac{1}{x_j - x_k^*} \right) = 0,$$

as well as the complex conjugate system. We also note that (3.5.9) may be derived from (3.5.8) by substituting $x = x_j + \varepsilon$ and expanding for $\varepsilon \to 0$.

It is remarkable that (3.5.9) can be embedded into a Hamiltonian system. By taking another time derivative of (3.5.9) and rearranging terms we have

(3.5.10) $$-\frac{1}{4}\overset{\circ\circ\circ}{x}_j = -2\left(\sum_{k\neq j} \frac{1}{(x_j - x_k)^3} - \sum_{k\neq j} \sum_{l\neq k, j} \frac{1}{(x_k - x_j)(x_k - x_l)(x_j - x_l)} \right.$$
$$- \sum_k \sum_{l\neq k} \frac{1}{(x_k^* - x_j)(x_j - x_l^*)(x_k^* - x_l^*)}$$
$$+ \sum_{k\neq j} \sum_l \frac{1}{(x_k - x_j)(x_k - x_l^*)(x_j - x_l^*)}$$
$$\left. + \sum_k \sum_{l\neq k} \frac{1}{(x_k^* - x_j)(x_j - x_l)(x_k^* - x_l)} - \sum_k \frac{1}{(x_k^* - x_j)^3} + \sum_k \frac{1}{(x_k^* - x_j)^3} \right).$$

We may verify that the second and third terms on the right-hand side of (3.5.10) are reducible to *zero*, and the fourth through seventh thus cancel each other. Hence, we are left with an N-body Hamiltonian system with a pairwise inverse square potential:

(3.5.11) $$\overset{\circ\circ}{x}_j = 8 \sum_{k\neq j} \frac{1}{(x_j - x_k)^3},$$

with Hamiltonian

$$H = \frac{1}{2}\sum_j \dot{x}_j^2 + 2\sum_j \sum_{k\neq j} \frac{1}{(x_k - x_j)^2}.$$

The system (3.5.11) has been studied by numerous authors, including Calogero (1971), (1976), Sutherland (1972), Moser (1975a,b), Olshanetsky and Perelomov (1976a,b,c) and Kazhdan, Kostant and Sternberg (1978). In connection with the B–O equation, see Case (1978b), (1979) and Chen, Lee and Pereira (1979).

At this point it is convenient to rescale (3.5.11) in the form

(3.5.12) $$\ddot{x}_j = 2\sum_{k\neq j} \frac{1}{(x_j - x_k)^3},$$

by simply rescaling time: $t \to 2t$.

The integration of this N-body problem has been given in a more general version by Moser (1975a, b). Consider the L, A pair

(3.5.13a) $$L\psi = \lambda\psi,$$

(3.5.13b) $$\psi_t = A\psi,$$

where L, A are the matrices

(3.5.14a) $$L_{kj} = \delta_{kj}\dot{x}_j + \frac{i(1-\delta_{kj})}{x_k - x_j},$$

(3.5.14b) $$A_{kj} = -i\delta_{kj}\sum_{l\neq k}\frac{1}{(x_k - x_l)^2} + \frac{i(1-\delta_{kj})}{(x_k - x_j)^2}.$$

Then, assuming $\lambda_t = 0$, we have the evolution equation $L_t = [A, L]$ or, after some algebra,

(3.5.15) $$\delta_{kj}\ddot{x}_j = \frac{(1-\delta_{kj})}{(x_k - x_j)}\left(\sum_{l\neq i}(x_k - x_l)^{-2} - \sum_{l\neq j}(x_j - x_l)\right)^{-2}$$
$$+ \sum_l \frac{(1-\delta_{kl})(1-\delta_{lj})}{(x_k - x_l)(x_l - x_j)}\left(-\frac{1}{(x_k - x_l)} + \frac{1}{(x_l - x_j)}\right).$$

When $k = j$ we have (3.5.11), whereas when $k \neq j$ both sides of (3.5.15) vanish.

Hence the dynamical system (3.5.12) is *isospectral*. This immediately implies that we have N invariants of the motion (action variables). To see this, let

(3.5.16a) $$I_n = \text{tr}(L^n), \quad n = 1, \cdots, N;$$

then

(3.5.16b) $$\frac{dI_n}{dt} = 0,$$

since the traces (tr) of L^n are expressible in terms of the eigenvalues of L. Another set of N quantities (angle variables) may also be obtained (Olshanetsky and Perelomov (1976a,b,c)). By direct computation we have that the equation of motion may be written in the form

(3.5.17) $$X_t = [A, X] + L,$$

where $X_{kj} = \delta_{kj} x_j$, and L, A are given in (3.5.14) and we note that $[A, X] = i(1 - \delta_{kj})/(x_j - x_k)$.

Next, by induction we may verify that

(3.5.18) $$\frac{d}{dt}(XL^{n-1}) = [A, XL^{n-1}] + L^n.$$

Calling

(3.5.19) $$J_n = \text{tr}(XL^{n-1}),$$

we see that

(3.5.20) $$\frac{dJ_n}{dt} = I_n,$$

where we have used the fact that tr $[A, B] = 0$ for any A, B. Hence

(3.5.21) $$J_n(t) = I_n t + J_n(0).$$

Thus we have two sets of N quantities given explicitly for all time. These completely determine the motion of the poles.

For example, consider, the case $N = 2$:

(3.5.22a) $$\text{tr } L = I_1 = \mathring{x}_1 + \mathring{x}_2,$$

(3.5.22b) $$\text{tr } L^2 = I_2 = \mathring{x}_1^2 + \mathring{x}_2^2 + \frac{2}{(x_1 - x_2)^2},$$

(3.5.22c) $$\text{tr } X = J_1 = x_1 + x_2 = I_1 t + J_1(0),$$

(3.5.22d) $$\text{tr } XL = J_2 = x_1 \mathring{x}_1 + x_2 \mathring{x}_2 = I_2 t + J_2(0).$$

From (3.5.22d),

(3.5.23a) $$\frac{d}{dt}(x_1^2 + x_2^2) = 2(x_1 \mathring{x}_1 + x_2 \mathring{x}_2) = 2I_2 t + 2J_2(0).$$

Hence

(3.5.23b) $$x_1^2 + x_2^2 = I_2 t^2 + 2J_2(0)t + x_1^2(0) + x_2^2(0).$$

Using (3.5.22c) we have an algebraic equation to solve for $x_1(t)$ or $x_2(t)$ (Z is either x_1 or x_2):

(3.5.23c) $\quad 2Z^2 - 2J_1(t)Z + J_1^2(t) = I_2 t^2 + 2 J_2(0) t + x_1^2(0) + x_2^2(0).$

A detailed analysis of the solution can be given, but we shall not do so here. Instead, we shall show that the eigenvalues of the operator

(3.5.24) $\quad\quad\quad\quad M(t, t_0) = X(t_0) + (t - t_0) L(t_0)$

coincide with the poles $x_i(t)$, $i = 1, \cdots, N$. Then the motion of the poles follows from writing (3.5.24) in diagonal form. Consider

(3.5.25) $\quad\quad\quad\quad K(t) = U^{-1}(t) X(t) U(t),$

where $U(t_0) = I$. Since A is anti-Hermitian, i.e., $A^\dagger \equiv (A^*)^T = -A$, we have that $U(t)$ evolving according to

(3.5.26a) $\quad\quad\quad\quad \dfrac{dU}{dt} = AU$

is unitary, i.e.,

$$U^\dagger U = I.$$

Thus

(3.5.26b) $\quad\quad\quad\quad \dfrac{d}{dt}(U^{-1}) = \dfrac{d}{dt}(U^\dagger) = -U^{-1} A,$

and by direct computation

(3.5.27) $\quad\quad\quad\quad \dfrac{dK}{dt} = U^{-1}\{\mathring{X} + [X, A]\} U.$

From (3.5.17)

(3.5.28) $\quad\quad\quad\quad \dfrac{dK}{dt} = U^{-1} L U.$

Differentiating again and using (3.5.26) and $L_t = [A, L]$ gives

(3.5.29a) $\quad\quad\quad\quad \dfrac{d^2 K}{dt^2} = 0;$

hence

(3.5.29b) $\quad\quad\quad\quad K(t) = C_1 + (t - t_0) C_2,$

where C_1, C_2 are constants. Since $U(t_0) = I$ we have, from (3.5.25), (3.5.28)

(3.5.29c) $\quad\quad\quad K(t_0) = U^{-1}(t_0) X(t_0) U(t_0) = X(t_0) = C_1,$

(3.5.29d) $\quad\quad\quad \dfrac{dK}{dt}(t_0) = U^{-1}(t_0) L(t_0) U(t_0) = L(t_0) = C_2.$

Thus

(3.5.29e) $$K(t) = X(t_0) + (t - t_0)L(t_0) \equiv M(t, t_0).$$

Using (3.5.25) we may solve for $X(t)$ in terms of $K(t)$, or rather $M(t, t_0)$,

(3.5.29f) $$X(t) = UM(t, t_0)U^{-1}.$$

So if we consider the eigenvalues of $X(t)$ we have

(3.5.30a) $$0 = \det(\lambda I - X(t)) = \det(U(\lambda I - M(t, t_0))U^{-1}).$$

But since $X_{kj}(t) = \delta_{kj} x_k(t)$ we have

(3.5.30b) $$\det(\lambda I - X(t)) = \prod_{k=1}^{N} (\lambda - x_k(t)).$$

This then proves that the eigenvalues of $M(t, t_0)$ coincide with the location of the poles $x_k(t)$. Moreover, the pole solution of the Benjamin–Ono equation is obtained by noting that

$$\frac{\partial}{\partial \lambda} \log \det(\lambda I - X(t))\Big|_{\lambda = x} = \sum_{i=1}^{N} \frac{1}{x - x_i(t)},$$

whereupon we have

(3.5.31) $$u = i\frac{\partial}{\partial \lambda} \log \det(\lambda I - M(2t, 2t_0))\Big|_{\lambda = x}$$
$$-i\frac{\partial}{\partial \lambda} \log \det(\lambda I - M^*(2t, 2t_0))\Big|_{\lambda = x}.$$

(Note that we have rescaled the time so as to obtain solutions of B–O in the form (3.5.1).) It is in this sense that the pole dynamics of the Benjamin–Ono equation are solved. The above solutions represent soliton interactions. Indeed, this fact suggests that the B–O equation is, in fact, solvable by inverse scattering. Later in this section we shall briefly discuss some of the known results, and give a linear scattering-like problem which is related to B–O.

For the KdV equation (Airault, McKean and Moser (1977))

(3.5.32) $$u_t + 6uu_x + u_{xxx} = 0,$$

the finite pole expansion is

(3.5.33) $$u = -2 \sum_{i=1}^{N} \frac{1}{(x - x_i(t))^2}.$$

Substitution of (3.3.33) into (3.5.32) yields

(3.5.34) $$\sum_{i=1}^{N} \frac{-2}{(x-x_i)^3}\left[-\overset{\circ}{x}_i - 12 \sum_{\substack{j=1 \\ j \neq i}}^{N} \frac{1}{(x-x_j)^2}\right] = 0.$$

Carrying out a rational fraction decomposition, or more simply just letting $x = x_I + \varepsilon$, $\varepsilon \to 0$ and setting the $O(1/\varepsilon^3)$, $O(1/\varepsilon^2)$ terms to zero, we have

(3.5.35a) $$\overset{\circ}{x}_I = -12 \sum_{\substack{j=1 \\ j \neq I}}^{N} \frac{1}{(x_I - x_j)^2},$$

and a subsidiary *condition* on the solution manifold,

(3.5.35b) $$\sum_{\substack{j=1 \\ j \neq I}}^{N} \frac{1}{(x_I - x_j)^3} = 0.$$

It should be noted that these solutions of KdV correspond to the rational solutions found in § 3.4.

By differentiating (3.5.35a) we may reduce it to the Hamiltonian N-body system

(3.5.36) $$\overset{\circ\circ}{x}_I = -(12)^2 \sum_{\substack{j=1 \\ j \neq I}}^{N} \frac{1}{(x_I - x_j)^5}$$

(see, for example, Chudnovsky and Chudnovsky (1977)). Indeed, the pole expansions can be generalized (ibid) to elliptic function interactions:

(3.5.37a) $$\overset{\circ}{x}_I = -12 \sum_{\substack{j=1 \\ j \neq I}}^{N} \mathcal{P}(x_I - x_j),$$

(3.5.37b) $$\sum_{\substack{j=1 \\ j \neq I}}^{N} \mathcal{P}'(x_I - x_j) = 0,$$

where $\mathcal{P}(x)$ is the usual Weierstrass elliptic function. Taking a derivative we have

(3.5.38a) $$\overset{\circ\circ}{x}_I = -(12)^2 \sum_{\substack{j=1 \\ j \neq I}}^{N} \mathcal{P}'(x_I - x_j)\mathcal{P}(x_I - x_j),$$

with the Hamiltonian

(3.5.38b) $$H = \frac{1}{2} \sum_{i=1}^{N} \overset{\circ}{x}_i^2 + \frac{(12)^2}{2} \sum_i \sum_{j \neq i} \mathcal{P}^2(x_i - x_j).$$

The earlier (3.5.36) pole expansion results may be obtained from the elliptic function solution by taking $\mathcal{P}(x) = x^{-2}$.

Next we briefly mention the pole expansion results (Satsuma and Ablowitz (1980)) for an intermediate equation which describes long internal gravity waves in a stratified fluid with finite depth (Joseph (1977), Kubota, Ko and Dobbs (1978)). Here δ is a parameter representing the distance between the boundary and the internal wave layer. We shall write the equation in the form

(3.5.39a) $$u_t + 2uu_x + \left(1 + \frac{1}{\delta}\right)T(u_{xx}) = 0,$$

where

(3.5.39b) $$T(u_x) = \int_{-\infty}^{\infty} \left[-\frac{1}{2\delta} \coth \frac{\pi(x-\xi)}{2\delta} + \frac{1}{2\delta} \operatorname{sgn}(x-\xi)\right] u_\xi \, d\xi.$$

Alternatively, the equation may be written as

(3.5.40a) $$u_t + 2uu_x + \left(1 + \frac{1}{\delta}\right)\frac{\partial}{\partial x} \int_{-\infty}^{\infty} K(x-\xi) u(\xi) \, d\xi = 0,$$

where

(3.5.40b)
$$K(x) = \frac{1}{2\pi} \int_{-\infty}^{\infty} C(k) e^{ikx} \, dk,$$

$$C(k) = -k \coth k\delta + \frac{1}{\delta}.$$

In the shallow water limit, $\delta \to 0$, (3.5.39) or (3.5.40) reduces to KdV:

(3.5.41) $$u_t + 2uu_x + \tfrac{1}{3} u_{xxx} = 0,$$

and in the deep water limit, $\delta \to \infty$, to the B–O equation:

(3.5.42) $$u_t + 2uu_x + H(u_{xx}) = 0$$

(again $H(u)$ is the Hilbert transform of u). The question of finding N-soliton solutions to (3.5.39) has been considered by Joseph and Egri (1978) and Chen and Lee (1979). Before going into the pole expansion analysis, it is convenient to transform equations (3.5.39–40) into bilinear form (see § 3.3). In (3.5.40), removing the integration and formally replacing k by $-i\,\partial/\partial x$ (Fourier transform) we have the following differential-difference equation:

(3.5.43) $$u_t + 2uu_x + \frac{1}{\delta}\left(1 + \frac{1}{\delta}\right) u_x - i\left(1 + \frac{1}{\delta}\right) \coth\left(i\delta \frac{\partial}{\partial x}\right) u_{xx} = 0.$$

Using the dependent variable transformation (with $e^{\pm i\delta \partial/\partial x} f(x) = f(x \pm i\delta)$)

(3.5.44)
$$u = -2i\left(1 + \frac{1}{\delta}\right) \sinh\left(i\delta \frac{\partial}{\partial x}\right) \frac{\partial}{\partial x} \log f(x)$$

$$= -i\left(1 + \frac{1}{\delta}\right) \frac{\partial}{\partial x} \log \frac{f(x + i\delta)}{f(x - i\delta)},$$

we reduce (3.5.43) to the bilinear equation

(3.5.45) $$\left(\frac{\delta}{1+\delta}iD_t+\frac{1}{\delta}iD_x+D_x^2\right)f^+\cdot f^-=0,$$

where we have used

(3.5.46) $$f^\pm \equiv f(x\pm i\delta),$$

with D_t, D_x defined in § 3.3 (3.3.4).

Although this is certainly suggestive, one must be careful. Specifically, by substitution of (3.5.44) into (3.5.39) or (3.5.40) we find that (3.5.45) is obtained only when the following condition holds.

Condition A. $f(x+i\delta)$ has no zeros in the strip $-2\delta \leq \text{Im } x \leq 0$.

If condition A is satisfied then

$$\int \frac{i}{2\delta}\coth\frac{\pi(x-\xi)}{\delta}\frac{\partial}{\partial \xi}\log\frac{f(\xi+i\delta)}{f(\xi-i\delta)}\,d\xi = -\frac{\partial}{\partial x}\log f(x+i\delta)f(x-i\delta)+\text{const.}$$

follows (if we suppose f is well enough behaved at infinity).

The simplest nontrivial soliton solution may be given by

$$f(x)=1+\exp(k_1x-\omega_1t+\eta_1^{(0)}),$$

(3.5.47) $$\omega_1=\left(1+\frac{1}{\delta}\right)\left(\frac{1}{\delta}k_1-k_1^2\cot\delta k_1\right);$$

k_1, $\eta_1^{(0)}$ are arbitrary parameters. A restriction, $0<k_1\delta<\pi$, is required in order to satisfy condition A. Substituting (3.5.47) into (3.5.44) gives

(3.5.48a) $$u=\left(1+\frac{1}{\delta}\right)\frac{k_1\sin\delta k_1}{(\cos\delta k_1+\cosh\{k_1x-\omega_1t+\eta_1^{(0)}\})}.$$

It is easily seen that (3.5.48a) is reduced to the KdV soliton

(3.5.48b) $$u=\frac{k_1^2}{2}\text{sech}^2\frac{1}{2}\left(k_1x-\frac{1}{3}k_1^3t+\eta_1^{(0)}\right)$$

as $\delta\to 0$. In the limit $\delta\to\infty$, on the other hand, there is no proper limit unless $k_1\to 0$. Taking $\delta k_1=\pi-k_1/C_1$ with a positive real constant $C_1(\sim 0(1))$ yields the B–O rational soliton

(3.5.48c) $$u=\frac{2C_1}{(1+C_1^2(x-C_1t)^2)}.$$

N-soliton solutions may be obtained from (3.5.45), but we shall not go into this here (see Satsuma and Ablowitz (1979)).

We now discuss the dynamical system of the motion of the poles of the intermediate equation. It is particularly convenient to use the equation in the bilinear form (3.5.45). We assume

$$f(x) = \prod_{j=1}^{N} (x - x_j(t)), \qquad |\operatorname{Im} x_j| > \delta;$$

i.e.,

$$u = -i\left(1 + \frac{1}{\delta}\right) \sum_{j=1}^{N} \left[\frac{1}{(x - x_j(t) + i\delta)} - \frac{1}{(x - x_j(t) - i\delta)}\right]$$

($|\operatorname{Im} x_j| > \delta$ is required in order to satisfy condition A). Substituting $f(x)$ into (3.5.45), we find

$$\sum_{j=1}^{N} \frac{1}{(x - x_j)^2 + \delta^2} \left\{ \overset{\circ}{x}_j + 2(1 + \delta) \sum_{k \neq j} \frac{1}{((x - x_k)^2 + \delta^2)} \right\} = 0.$$

This may be rewritten, using partial fraction decomposition,

$$\sum_{j=1}^{N} \left(\frac{1}{x - x_j + i\delta} - \frac{1}{(x - x_j - i\delta)} \right) \overset{\circ}{x}_j$$

$$+ 4(1 + \delta) \sum_{j=1}^{N} \sum_{k \neq j} \left\{ \left(\frac{1}{(x - x_j + i\delta)(x_k - x_j)(x_k - x_j + 2i\delta)} \right) \right.$$

$$\left. - \frac{1}{(x - x_j - i\delta)(x_k - x_j)(x_k - x_j - 2i\delta)} \right\} = 0,$$

whereupon we immediately have

(3.5.49a) $\qquad \overset{\circ}{x}_j + 4(1 + \delta) \sum_{k \neq j} \dfrac{1}{(x_k - x_j)(x_k - x_j + 2i\delta)} = 0,$

(3.5.49b) $\qquad \overset{\circ}{x}_j + 4(1 + \delta) \sum_{k \neq j} \dfrac{1}{(x_k - x_j)(x_k - x_j - 2i\delta)} = 0,$

for $j = 1, \cdots, N$. Adding (3.5.49a, b) yields

(3.5.50a) $\qquad \overset{\circ}{x}_j + 4(1 + \delta) \sum_{k \neq j} \dfrac{1}{(x_k - x_j)^2 + 4\delta^2} = 0,$

and subtracting yields

(3.5.50b) $\qquad \sum_{k=1, k \neq j}^{N} \dfrac{1}{(x_k - x_j)((x_k - x_j)^2 + 4\delta^2)} = 0.$

(3.5.50) is a dynamical system with a constraint. As $\delta \to 0$ we have the dynamical system associated with KdV (3.5.35), suitably rescaled since

(3.5.32) is different from (3.5.41) (t in (3.5.41) $\to 3t$),

(3.5.51a) $$\overset{\circ\circ}{x}_j + 4 \sum_{k \neq j} (x_k - x_j)^{-2} = 0,$$

(3.5.51b) $$\sum_{k \neq j} (x_k - x_j)^{-3} = 0.$$

For $\delta \to \infty$ we introduce $\hat{x}_j = x_j - i\delta$ (here Im $x_j > \delta$ so $\hat{x}_j$ is in the upper half plane) for $j = 1, 2, \cdots, M$ and $\hat{x}_j = x_j + i\delta$ for $j = M+1, M+2, \cdots, N$ (here $x_j < -i\delta$ so Im $\hat{x}_j$ is in the lower half plane). Then as $\delta \to \infty$ we have

(3.5.52a) $$\frac{1}{2i}\overset{\circ}{\hat{x}}_j = \sum_{\substack{k=1 \\ k \neq j}}^{M} \frac{1}{\hat{x}_k - \hat{x}_j} - \sum_{k=M+1}^{N} \frac{1}{\hat{x}_k - \hat{x}_j}, \quad \text{for } j = 1, 2, \cdots, M$$

and

(3.5.52b) $$\frac{1}{2i}\overset{\circ}{\hat{x}}_j = \sum_{k=1}^{M} \frac{1}{\hat{x}_k - \hat{x}_j} - \sum_{\substack{k=M+1 \\ k \neq j}}^{N} \frac{1}{\hat{x}_k - \hat{x}_j}, \quad \text{for } j = M+1, \cdots, N,$$

which is a dynamical system without any constraint. If we take $N = 2M$ and $\hat{x}_j = \hat{x}^*_{M+j}$ for $j = 1, 2, \cdots, M$, we have simply

(3.5.53) $$\frac{1}{2i}\overset{\circ}{\hat{x}}_j = \sum_{\substack{k=1 \\ k \neq j}}^{M} \frac{1}{\hat{x}_j - \hat{x}_k} - \sum_{k=1}^{M} \frac{1}{\hat{x}_j - \hat{x}^*_k},$$

which is equivalent to the complex conjugate system of (3.5.9) (note that here $\hat{x}_j$, $1, 2, \cdots, M$ lies in the upper half plane, whereas the x_j in (3.5.9) lies in the lower half plane).

Moreover, by taking the time derivative of (3.5.52) we have (after some algebra and using identities similar to those used in deriving (3.5.11))

(3.5.54a) $$\overset{\circ\circ}{\hat{x}}_j = 8 \sum_{\substack{k=1 \\ k \neq j}}^{M} (\hat{x}_j - \hat{x}_k)^{-3}, \quad \text{for } j = 1, 2, \cdots, M$$

and

(3.5.54b) $$\overset{\circ\circ}{\hat{x}}_j = 8 \sum_{\substack{k=M+1 \\ k \neq j}}^{N} (\hat{x}_j - \hat{x}_k)^{-3}, \quad \text{for } j = M+1, \cdots, N.$$

It should be noted that the pole representations are expressed as $f^+ = \prod_{j=1}^{M} (x - \hat{x}_j)$ and $f^- = \prod_{j=M+1}^{N} (x - \hat{x}_j)$ in this limit ($\delta \to \infty$). Thus the result (3.5.54) shows that the motion of the poles belonging to f^+ is not affected by those belonging to f^- and vice-versa; i.e., the poles in the lower half plane move independently of those in the upper half plane.

Finally we mention that Bäcklund and generalized Miura transformations may be given for (3.5.39) (Satsuma, Ablowitz and Kodama (1979), Satsuma, Ablowitz (1980)). Calling $W_x = u$, we may write (3.5.59) in the form

(3.5.55)
$$W_t + W_x^2 + \left(1 + \frac{1}{\delta}\right) T W_{xx} = 0,$$

and the Bäcklund transformation is given by

(3.5.56a)
$$(W + W')_x = \lambda + iT(W' - W)_x - i\delta^{-1}(W' - W) + \mu\, e^{i\delta(W' - W)/(1+\delta)},$$

(3.5.56b)
$$(W' - W)_t = -\left\{\frac{1}{\delta}\left(1 + \frac{1}{\delta}\right) + \lambda\right\}(W' - W)_x + i\left(1 + \frac{1}{\delta}\right)(W' + W)_{xx'}$$
$$-i(W' - W)_x T(W' - W)_x + i\delta^{-1}(W' - W)(W' - W)_{x'},$$

where λ, μ are abitrary parameters. If W satisfies (3.5.55) then W' defined by (3.5.56) also satisfies (3.5.55). Introducing $V = W' - W$ (and using $W_x = u$) we may rewrite (3.5.56) to give a generalized Miura transformation:

(3.5.57a)
$$V_x + 2u = \lambda + iT(V_x) - i\delta^{-1}V + \mu\, e^{iV(\delta/1+\delta)},$$

(3.5.57b)
$$V_t = -\left(\left(1 + \frac{1}{\delta}\right)\delta^{-1} + \lambda\right)V_x$$
$$+ i\left(1 + \frac{1}{\delta}\right)(V_{xx} + 2u_x) - iV_x T(V_x) + i\delta^{-1}VV_x.$$

Substituting $V_x + 2u$ from (3.5.57a) into the right-hand side of (3.5.57b) gives a modified internal wave equation,

(3.5.58) $V_t + \lambda V_x + \left(1 + \frac{1}{\delta}\right)T(V_{xx}) + \left\{\mu\, e^{i(\delta/1+\delta)V} + iT(V_x) - i\frac{1}{\delta}V\right\}V_x = 0,$

which has the same dispersion relation as the intermediate equation. Alternatively, solving for u in (3.5.57a), and using the identity

(3.5.59) $T(V_x)T(V_{xx}) - V_x V_{xx} - T(V_x T(V_x))_x = \delta^{-1}(VT(V_{xx}) - T(VV_x)),$

we have (directly from (3.5.39))

(3.5.60)
$$u_t + 2uu_x + T(u_{xx}) = \left[\frac{1}{2}i\frac{\partial}{\partial x}T \cdot -i\delta^{-1} + \mu i\frac{\delta}{1+\delta}e^{i(\delta/1+\delta)V} - \frac{\partial}{\partial x}\right]$$
$$\times \left[V_t + \lambda V_x + \left(1 + \frac{1}{\delta}\right)T(V_{xx})\right.$$
$$\left. + \left\{\mu\, e^{i(\delta/1+\delta)V} + iT(V_x) - \frac{i}{\delta}V\right\}V_x\right].$$

Thus we can see that (3.5.58) plays the same role in this intermediate equation as does the modified KdV equation with respect to the KdV equation.

Imposing $V(\pm\infty) = 0$ and using $\int_{-\infty}^{\infty} fT(f)\,dx = 0$, we note that $(\partial/\partial t)$ $(\int_{-\infty}^{\infty} V\,dx) = 0$. This implies that (3.5.39) has an infinite number of conserved quantities. Substituting $V = -i(1 + 1/\delta)(\chi + \log(-\lambda/\mu))$ into (3.5.57a) yields

$$(3.5.61)\quad e^\chi - 1 = \lambda^{-1}\left[i\left(1 + \frac{1}{\delta}\right)\chi_x + \left(1 + \frac{1}{\delta}\right)T(\chi_x) - \delta^{-1}\left(1 + \frac{1}{\delta}\right)\chi - 2u\right].$$

Expanding $\chi = \sum_1^\infty \lambda^{-n}\chi_n$ for $\lambda \to \infty$ and equating powers of λ gives χ_n recursively. Arguments such as the above show that each χ_n is a conserved density of (3.5.39). The first four χ_n are given by

$$\chi_1 = u, \quad \chi_2 = u^2, \quad \chi_3 = u^3 + \left(1 + \frac{1}{\delta}\right)\frac{3}{2}uTu_x,$$

$$\chi_4 = u^4 + 3\left(1 + \frac{1}{\delta}\right)u^2 Tu_x + \frac{1}{2}\left(1 + \frac{1}{\delta}\right)^2 u_x^2 + \frac{3}{2}\left(1 + \frac{1}{\delta}\right)^2 (Tu_x)^2 + \frac{3}{2}\delta^{-1}\left(1 + \frac{1}{\delta}\right)uTu_x.$$

The conserved densities reduce correctly to those of KdV as $\delta \to 0$ and Benjamin–Ono as $\delta \to \infty$.

A formal linear problem can be derived by defining

$$(3.5.62\text{a})\quad \log\frac{\psi^+}{\psi^-} = i\frac{\delta}{1+\delta}V,$$

$$(3.5.62\text{b})\quad (\log\psi^+\psi^-)_x = \frac{\delta}{1+\delta}[-T(V_x) + \delta^{-1}V].$$

In the limit $\delta \to \infty$, for appropriate V, this amounts to splitting the function V into functions analytically extendable into the upper $(-)$, lower $(+)$ half plane, i.e., $\log\psi^\pm = \pm(i \mp H)V$. Substitution of (3.5.62) into (3.5.57a) gives

$$(3.5.63)\quad \left(1 + \frac{1}{\delta}\right)\psi_x^- - i\left(u - \frac{\lambda}{2}\right)\psi^- = -\frac{i}{2}\mu\psi^+.$$

(There is also an appropriate linear time evolution operator; see Satsuma, Ablowitz and Kodama (1979).)

As $\delta \to 0$, letting $V = 2(1+\delta)(\log\phi)_x$, $\lambda = -(1 + 1/\delta)k\cos k\delta$, $\mu = (1 + 1/\delta)k\,\mathrm{cosec}\,k\delta$, we find that (3.5.63) goes to the Schrödinger scattering problem

$$(3.5.64)\quad \phi_{xx} - \left(\frac{k^2}{4} - u\right)\phi = 0,$$

whereas when $\delta \to \infty$ we have

$$(3.5.65)\quad \psi_x^- - i\left(u - \frac{\lambda}{2}\right)\psi^- = -\frac{1}{2}i\mu\psi^+.$$

Equations (3.5.63, 65) are differential Riemann–Hilbert problems. For finite δ, $\psi\pm$ are the boundary values of functions analytic in strips (+: $-2\delta <$ Im $x < 0$), ($-$; $\delta <$ Im $x < 2\delta$) and periodically extended. Recently, use of these linear problems has led to solutions via IST; see Kodama, Satsuma and Ablowitz (1981). Work in this direction has also been done by Nakamura (1979b) and Bock and Kruskal (1979).

3.6. Direct approaches with the linear integral equation.

In previous sections we have established that related to certain nonlinear evolution equations is a linear integral equation (of Gel'fand–Levitan–Marchenko form). In this section we shall discuss a procedure for deriving the evolution equation directly from the linear integral equation. The derivation applies to partial as well as to ordinary differential equations (see also § 3.7). We need only require that the solutions decay rapidly enough as $x \to +\infty$ (say) that the integral operators are defined. It should be noted, however, that in general a solution which decays rapidly enough as $x \to +\infty$ may diverge at some finite value of x, diverge as $x \to -\infty$ or have weak decay as $x \to -\infty$. In any of these cases, the classical analysis of inverse scattering using the analytic properties of the Jost functions is not applicable, since this requires "nice" properties of the potential on the whole line (see, for example, Faddeev (1963), Deift and Trubowitz (1979)). Because of this freedom, the range of solutions obtained by the present approach is larger than that obtained by IST. In this way the self-similar solutions satisfying ODE's related to these evolution equations can be obtained.

This method was first used in the field of nonlinear evolution equations by Zakharov and Shabat (1974) (see also Shabat (1973)). Blazek (1966) and Cornille (1967) have used similar ideas for various problems associated with inverse scattering; Cornille (1976b), (1979) has also made a number of further contributions using these ideas for nonlinear evolution equations. In this section we shall first follow the presentation of Ablowitz, Ramani and Segur (1980a); subsequently we discuss the work of Zakharov and Shabat (1974), which has a somewhat different point of view.

Consider the linear integral equation

$$(3.6.1) \quad K(x, y) = F(x, y) + \int_x^\infty K(x, z) N(x; z, y)\, dz, \quad y \geq x.$$

Besides the arguments (x, y, z) which appear explicitly in (3.6.1), F, N and K may depend on other parameters $(t, \lambda, \cdots)$. Derivatives with respect to these extra variables may appear in the differential equations that F and K satisfy, but (3.6.1) is understood to be solved at fixed given values of these parameters.

In each specific case N is explicitly given in terms of F. For example,

(A) $\quad N(x; z, y) = F(z, y) \qquad$ (KdV, higher order KdV, Boussinesq, Kadomtsev–Petviashvili, $\cdots$);

(B) $\quad N(x; z, y) = \pm \int_x^\infty F(z, s) F(s, y)\, ds \qquad$ (mKdV and higher order mKdV, sine-Gordon, $\cdots$);

(C) $\quad N(x; z, y) = \pm \int_x^\infty F^*(z, s) F(s, y)\, ds \qquad$ (NLS, higher order NLS, $\cdots$).

In the usual approach, F is constructed from the scattering data of the "direct problem" and the scattering potential $u(x)$ is reconstructed from K (e.g., $u(x) = K(x, x)$ or $u(x) = (d/dx)K(x, x)$). Here we do not give to F any such interpretation, but only demand that it satisfy some linear (partial or ordinary) differential equation.

Define the operator A_x by

(3.6.2) $$A_x f(y) = \begin{cases} \int\int_x^\infty f(z) N(x; z, y) \, dz, & y \geq x, \\ 0, & y < x. \end{cases}$$

We assume that for each specific choice of N, one can prove that $(I - A_x)$ is invertible. More precisely, there is an x large enough and a function space on which $(I - A_x)$ is invertible, and $(I - A_x)^{-1}$ is continuous. Moreover, we assume that the operators obtained by differentiating (3.6.2) with respect to x or y also are defined on this function space. It can be shown that these assumptions are valid in a large variety of problems (see, for example, Ablowitz, Ramani and Segur (1980a)).

Subject to these assumptions and the fact that F satisfies some linear differential equation, we show in this section that $u(x)$ (defined above) satisfies a nonlinear differential equation. We shall say that this equation is solvable by an inverse scattering transform even though no reference is made to the direct scattering problem.

The outline of this approach can be stated rather simply.

(i) F satisfies two linear (partial or ordinary) differential equations

(3.6.3) $$L_i F = 0, \quad i = 1, 2.$$

(ii) K is related to F through (3.6.1) which we may write in the form

(3.6.1′) $$(I - A_x) K = F.$$

(iii) Applying L_i, $i = 1, 2$ to this equation yields

(3.6.4) $$L_i (I - A_x) K = 0, \quad i = 1, 2.$$

This can be rewritten as

(3.6.5) $$(I - A_x)(L_i K) = R_i, \quad i = 1, 2,$$

where R_i, $i = 1, 2$ contains all the remaining terms in (3.6.5). However, (3.6.1) and (3.6.3) are chosen such that we can write

(3.6.6) $$R_i = (I - A_x) M_i(K), \quad i = 1, 2,$$

where $M_i(K)$ is a nonlinear functional of K.

(iv) Therefore

$$(I - A_x)[L_i K - M_i(K)] = 0, \quad i = 1, 2.$$

But $(I - A_x)$ is invertible, so K must satisfy the nonlinear differential equation

(3.6.7) $$L_i K - M_i(K) = 0, \qquad i = 1, 2.$$

Therefore every solution of the linear integral equation (3.6.1) is also a solution of the nonlinear differential equation (3.6.7).

The basic ingredients to this approach are the linear integral equation (3.6.1) and two linear differential operators L_i, $i = 1, 2$. The two linear operators correspond to the linear scattering problem ($i = 1$, say) and the associated linear time evolution of the eigenfunctions ($i = 2$, say). In order to make the method effective we must identify a class of suitable operators L_i, $i = 1, 2$. The operator (L_1), related to the scattering problem, is crucial. In addition we find it convenient to establish a "dictionary" of terms that may appear in the right side of (3.6.5): R_i. These results and those generated from the L_1 operator allow us, in every specific case, to reduce (3.6.7) for $i = 2$ to a nonlinear differential equation along the line $y = x$. Finally we note that: (i) it is straightforward to show that K, defined by (3.6.1) is differentiable enough that $L_i K$ exists; (ii) the equations $L_2 F = 0$ may be either PDE's that depend on time or suitable similarity forms (see Ablowitz, Ramani and Segur (1980a)).

In what follows we shall discuss two prototype examples, namely the KdV and the mKdV equations. The integral equation we shall begin with is

(3.6.8) $$K(x, y) = F(x, y) + \int_x^\infty K(x, z) F(z, y) \, dz$$

(i.e., (3.6.1) with (A)). First we begin by establishing a "dictionary" of results which we shall call on again later.

(3.6.9a) $$\partial_x^n \int_x^\infty K(x, z) F(z, y) \, dz = \int_x^\infty dz \, F(z, y)(\partial_x^n K(x, z)) + A_n,$$

(3.6.9b) $$\partial_x \partial_x^{n-1} \int_x^\infty K(x, z) F(z, y) \, dz = \partial_x \left[\int_x^\infty F(z, y) \partial_x^{n-1} K(x, z) \, dz + A_{n-1} \right].$$

An integration by parts and setting (3.6.9a, b) equal yields

(3.6.9c) $$A_n = A_{n-1 x} - F(x, y) [\partial_x^{n-1} K(x, z)]_{z=x},$$

with

(3.6.9d) $$A_1 = -K(x, x) F(x, y),$$

(3.6.9e) $$A_2 = -\frac{d}{dx}(K(x, x) F(x, y)) - F(x, y) [\partial_x K(x, z)]_{z=x},$$

(3.6.9f) $$A_3 = -\left(\frac{d}{dx}\right)^2 (K(x, x) F(x, y)) - \frac{d}{dx}(F(x, y) [\partial_x K(x, z)]_{z=x})$$
$$- F(x, y) [\partial_x^2 K(x, z)]_{z=x},$$

where $(d/dx)K(x, x) = (\partial_x K(x, z) + \partial_z K(x, z))_{z=x}$. Similarly, by integration by parts, we have

(3.6.10a) $\quad \int_x^\infty K(x, z) \partial_z^n F(z, y)\, dz = (-1)^n \int_x^\infty F(z, y) \partial_z^n K(x, z)\, dz + B_n,$

with

(3.6.10b) $\quad B_1 = -K(x, x) F(x, y),$

(3.6.10c) $\quad B_2 = -K(x, x)(\partial_x F(x, y)) + (\partial_z K(x, z))_{z=x} F(x, y),$

(3.6.10d) $\quad B_3 = -K(x, x) \partial_x^2 F(x, y) + \left.\dfrac{\partial K}{\partial z}\right|_{z=x} \partial_x F(x, y)$
$\qquad\qquad - (\partial_z^2 K(x, z))_{z=x} F(x, y),$

$\qquad \vdots$

Hence

(3.6.11a) $\quad A_1 - B_1 = 0,$

(3.6.11b) $\quad A_2 - B_2 = -2 F(x, y) \partial_x K(x, x),$

(3.6.11c) $\quad A_3 - B_3 = -3 \partial_x F(x, y) \dfrac{d}{dx} K(x, x)$
$\qquad\qquad - 3 F(x, y) [(\partial_x^2 + \partial_z \partial_x) K(x, z)]_{z=x}$

$\qquad \vdots$

Next we introduce the operator L_1 and require that F satisfy

(3.6.12) $\qquad\qquad L_1 F \equiv (\partial_x^2 - \partial_y^2) F(x, y) = 0.$

Then we operate on (3.6.8) with L_1. We find

$$(\partial_x^2 - \partial_y^2) K = (\partial_x^2 - \partial_y^2) \int_x^\infty K(x, z) F(z, y)\, dz$$
$$= \int_x^\infty F(z, y) \partial_x^2 K(x, z)\, dz + A_2 - \int_x^\infty K(x, z) F_{yy}(z, y)\, dz.$$

Using $F_{zz} = F_{yy}$ and (3.6.11b) gives

(3.6.13)
$$(\partial_x^2 - \partial_y^2) K(x, y) = \int_x^\infty F(x, z)(\partial_x^2 - \partial_z^2) K(x, z)\, dz$$
$$- 2 F(x, y) \dfrac{d}{dx} K(x, x).$$

Using $F = (I - A_x)K$ in (3.6.13) and rearranging, we have

(3.6.14) $\quad (I - A_x)\left\{(\partial_x^2 - \partial_y^2)K(x, y) + 2\left[\dfrac{d}{dx}K(x, x)\right]K(x, y)\right\} = 0.$

The invertibility of $(I - A_x)$ yields

(3.6.15a) $\quad (\partial_x^2 - \partial_y^2)K(x, y) + u(x)K(x, y) = 0,$

where $u(x)$ is defined to be

(3.6.15b) $\quad u(x) = 2\dfrac{d}{dx}K(x, x).$

Thus we have shown that if F satisfies a linear differential equation (3.6.12) and if F generates K through (3.6.8), then K satisfies the nonlinear equation (3.6.15). This is an example of (3.6.7). If we take $K(x, y) = \psi(x, k)e^{iky}$, (3.6.15) reduces to

(3.6.15c) $\quad \psi_{xx} + (k^2 + u)\psi = 0,$

i.e., the Schrödinger scattering problem.

We now introduce a second linear operator on F, and require that F satisfy

(3.6.16) $\quad L_2 F = (\partial_t + (\partial_x + \partial_y)^3)F = 0.$

Operating on (3.6.8) with L_2 gives

(3.6.17) $(\partial_t + (\partial_x + \partial_y)^3)K(x, y) = (\partial_t + (\partial_x + \partial_y)^3)\int_x^\infty K(x, z)F(z, y)\, dz.$

On the right-hand side of (3.6.17) we have the term

(3.6.18a) $\quad I = \int_x^\infty K(x, z)F_t(z, y)\, dz + (\partial_x + \partial_y)^3 \int_x^\infty K(x, z)F(z, y)\, dz$

(3.6.18b) $\quad = -\int_x^\infty K(x, z)(\partial_z + \partial_y)^3 F(z, y)\, dz$

$\quad\quad + (\partial_x + \partial_y)^3 \int_x^\infty K(x, z)F(z, y)\, dz.$

We write

(3.6.18c) $\quad I = I_1 + I_2 + I_3.$

First, I_1 is given by (arguments understood)

(3.6.18d) $\quad I_1 = \partial_x^3 \int_x^\infty KF\, dz - \int_x^\infty K\, \partial_z^3 F\, dz;$

from (3.6.9, 10),

(3.6.18e) $$I_1 = \int_x^\infty (\partial_x^3 K + \partial_z^3 K) F\,dz + A_3 - B_3.$$

Then I_2 is given by

$$I_2 = 3\partial_x^2 \partial_y \int_x^\infty KF\,dz - 3\int_x^\infty K\,\partial_z^2 \partial_y F\,dz$$

(3.6.18f) $$= 3\int_x^\infty (\partial_x^2 K - \partial_z^2 K) F_y\,dz + 3\partial_y (A_2 - B_2)$$

(3.6.18g) $$= -3u\partial_y (K(x,y) - F(x,y)) + 3\partial_y (A_2 - B_2),$$

where we have used (3.6.15) and (3.6.8) in the form $\int_x^\infty KF\,dz = K - F$. Finally, for I_3 we have

(3.6.18h) $$I_3 = 3\partial_x \partial_y^2 \int_x^\infty KF\,dz - 3\int_x^\infty K\partial_z \partial_y^2 F\,dz$$

$$= 3\int_x^\infty (\partial_x K) F_{yy}\,dz + 3\int_x^\infty K_z F_{yy}\,dz$$

$$= 3\int_x^\infty (\partial_x K + \partial_z K) F_{yy}\,dz$$

$$= 3\int_x^\infty (\partial_x K + \partial_z K) F_{zz}\,dz$$

$$= [3(K_{xz} + K_{zz})F - 3(K_x + K_z) F_x]_{z=x}$$

(3.6.18i) $$+ 3\int_x^\infty ((\partial_z^2 \partial_x + \partial_z^3) K) F\,dz.$$

Summing the results, and using $K_{zz} = K_{xx} + uK$ in the last term in (3.6.18i), we have

$$I = I_1 + I_2 + I_3 = \int_x^\infty ((\partial_x + \partial_z)^3 K) F\,dz + A_3 - B_3 - 3u\partial_y (K - F)$$

$$+ 3\partial_y (A_2 - B_2) + [3(K_{xz} + K_{zz})F - 3(K_x + K_z)F_x]_{z=x}$$

$$+ 3u\int_x^\infty K_z F\,dz.$$

Inserting this result in (3.6.17), we have

(3.6.19a) $$(\partial_t + (\partial_x + \partial_y)^3 + 3u\partial_y) K = \int_x^\infty (K_t + (\partial_x + \partial_z)^3 K + 3uK_z) F\,dz + T,$$

where

(3.6.19b)
$$T = (A_3 - B_3) + 3\partial_y(A_2 - B_2) + 3uF_y$$
$$+ 3[(K_{xz} + K_{zz})F - (K_x + K_z)F_x]_{z=x}.$$

Substituting for $A_3 - B_3$, $A_2 - B_2$, and using (3.6.15), (3.6.8) we find

$$T = 3u(x)K(x,x)F(x,y) - 3u(x)F_x(x,y)$$
$$= -3u(x)K_x(x,y) + 3u(x)\int_x^\infty K_x(x,z)F(z,x)\,dz.$$

Using T we find that (3.6.19a) reduces to

(3.6.20a) $\quad (I - A_x)\{(\partial_t + (\partial_x + \partial_y)^3 + 3u(\partial_x + \partial_y))K(x,y)\} = 0,$

whereupon

(3.6.20b) $\quad K_t + (\partial_x + \partial_y)^3 K + 3u(\partial_x + \partial_y)K = 0.$

On $y = x$, using $u = 2(d/dx)K(x,x)$ we have KdV (after taking a derivative of (3.6.20b)):

$$u_t + 6uu_x + u_{xxx} = 0.$$

Thus every F that satisfies (3.6.12, 16) and vanishes rapidly as $x \to +\infty$ generates a solution of KdV via (3.6.8). In particular, F need not have originated in a direct scattering problem.

As a second example, we begin with the linear integral equation ($\sigma = \pm 1$; the factor $\frac{1}{4}$ is for convenience)

(3.6.21) $\quad K(x,y) = F(x,y) + \dfrac{\sigma}{4}\int_x^\infty \int_x^\infty K(x,z)F(z,u)F(u,y)\,dz\,du.$

First we introduce the operator L_1 such that

(3.6.22a) $\quad L_1 F = (\partial_x - \partial_y)F = 0;$

hence

(3.6.22b) $\quad F(x,y) = F\left(\dfrac{x+y}{2}\right)$

(again with a factor $\frac{1}{2}$ for convenience). Using (3.6.22b) and shifting to the origin, we rewrite (3.6.21) as

(3.6.23a)
$$K(x,y) = F\left(\frac{x+y}{2}\right) + \frac{\sigma}{4}\int_0^\infty \int_0^\infty K(x, x+\zeta)F\left(\frac{2x+\zeta+\eta}{2}\right)F\left(\frac{x+\eta+y}{2}\right)d\zeta\,d\eta$$

or

(3.6.23b) $$[(I - \sigma A_x)K](x, y) = F\left(\frac{x+y}{2}\right),$$

where A_x is defined as

(3.6.24) $$A_x f(y) = \frac{1}{4}\int_0^\infty \int_0^\infty f(\zeta) F\left(\frac{2x+\zeta+\eta}{2}\right) F\left(\frac{x+\eta+y}{2}\right) d\zeta\, d\eta.$$

It is also useful to define

(3.6.25) $$K_2(x, z) = \int_0^\infty K(x, x+\zeta) F\left(\frac{x+\zeta+z}{2}\right) d\zeta.$$

With (3.6.25) one easily shows that

(3.6.26) $$(I - \sigma A_x)K_2(x, z) = \int_0^\infty F\left(\frac{2x+\zeta}{2}\right) F\left(\frac{x+\zeta+z}{2}\right) d\zeta,$$

and we may write the integral equation as

(3.6.27) $$K(x, y) = F\left(\frac{x+y}{2}\right) + \frac{\sigma}{4}\int_0^\infty K_2(x, x+\eta) F\left(\frac{x+\eta+y}{2}\right) d\eta.$$

Applying the operator $L_1 = (\partial_x - \partial_y)$ to (3.6.27) yields

(3.6.28) $$(\partial_x - \partial_y)K(x, y) = \frac{\sigma}{4}\int_0^\infty [(\partial_1 + \partial_2)K_2(x, x+\eta)] F\left(\frac{x+\eta+y}{2}\right) d\eta,$$

where ∂_1 and ∂_2 are derivatives with respect to the first and second arguments of K. Similarly, applying $(\partial_x + \partial_z)$ to (3.6.23a) yields

(3.6.29)
$$\begin{aligned}(\partial_x + \partial_z)K_2(x, z) &= \int_0^\infty \left\{(\partial_1 + \partial_2)K(x, x+\zeta) F\left(\frac{x+\zeta+z}{2}\right)\right.\\ &\quad \left. + K(x, x+\zeta) F\left(\frac{x+\zeta+z}{2}\right)\right\} d\zeta \\ &= \int_0^\infty [(\partial_1 - \partial_2)K(x, x+\zeta)] F\left(\frac{x+\zeta+z}{2}\right) d\zeta \\ &\quad - 2K(x, x) F\left(\frac{x+y}{2}\right).\end{aligned}$$

Substituting (3.6.28) into (3.6.29) we see that

$$\begin{aligned}(I - \sigma A_x)(\partial_x + \partial_z)K_2(x, z) &= -2K(x, x) F\left(\frac{x+z}{2}\right) \\ &= -2K(x, x)(I - \sigma A_x)K(x, z).\end{aligned}$$

Similarly, substituting (3.6.29) into (3.6.28) leads to

$$(I - \sigma A_x)(\partial_x - \partial_y)K(x, y) = -\frac{\sigma}{2}K(x, x)\int_0^\infty F\left(\frac{2x + \eta}{2}\right)F\left(\frac{x + \eta + y}{2}\right) d\eta$$

$$= -\frac{\sigma}{2}K(x, x)(I - \sigma A_x)K_2(x, y).$$

Note that A_x commutes with multiplication by a function of x. Thus, if $(I - \sigma A_x)$ is invertible, we have proven that

(3.6.30)
$$(\partial_x + \partial_y)K_2(x, y) = -2K(x, x)K(x, y),$$

$$(\partial_x - \partial_y)K(x, y) = -\frac{\sigma}{2}K(x, x)K_2(x, y).$$

These are the results expected from the inverse scattering analysis (cf. (1.3.19)). However, here we obtained them using only invertibility of $(I - \sigma A_x)$, which is a much weaker condition than what is required to apply the usual analytic approach. Moreover, taking $K(x, y) = v_1(x) e^{i\zeta y}$, and $K_2(x, y) = v_2(x) e^{-i\zeta y}$ yields (1.2.7a) for v_1, v_2.

Next apply $(\partial_x + \partial_y)$ to (3.6.23a):

$$(\partial_x + \partial_y)K(x + y)$$

(3.6.31)
$$= F' + \frac{\sigma}{4}\int_0^\infty \int_0^\infty [(\partial_1 + \partial_2)K(x, x + z)]F\left(\frac{2x + \zeta + \eta}{2}\right)F\left(\frac{x + \eta + y}{2}\right) d\zeta\, d\eta$$

$$+ \frac{\sigma}{4}\int_0^\infty \int_0^\infty K(x, x + \zeta)(\partial_x + \partial_y)\left[F\left(\frac{2x + \zeta + \eta}{2}\right)F\left(\frac{x + \eta + y}{2}\right)\right] d\zeta\, d\eta.$$

But

$$(\partial_x + \partial_y)F\left(\frac{2x + \zeta + \eta}{2}\right)F\left(\frac{x + \eta + y}{2}\right)$$

(3.6.32)
$$= F'\left(\frac{2x + \zeta + \eta}{2}\right)F\left(\frac{x + \eta + y}{2}\right) + F\left(\frac{2x + \zeta + \eta}{2}\right)F'\left(\frac{x + \eta + y}{2}\right)$$

$$= 2\partial_\eta\left\{F\left(\frac{2x + \zeta + \eta}{2}\right)F\left(\frac{x + \eta + y}{2}\right)\right\}.$$

Performing the η integration in (3.6.31) leads to

$$(I - \sigma A_x)(\partial_x + \partial_y)K(x, y)$$

$$= F'\left(\frac{x+y}{2}\right) - \left(\frac{\sigma}{2}\int_0^\infty K(x, x+\zeta)F(2x+\zeta)\,d\zeta\right)F(x+y)$$

$$= F'\left(\frac{x+y}{2}\right) - \frac{\sigma}{2}K_2(x, x)(I - \sigma A_x)K(x, y);$$

i.e.,

(3.6.33) $\quad F'\left(\dfrac{x+y}{2}\right) = (I - \sigma A_x)\left\{(\partial_x + \partial_y)K(x, y) + \dfrac{\sigma}{2}K_2(x, x)K(x, y)\right\}.$

This is the "dictionary" required for this problem.

The final step makes use of the fact that F satisfies another linear equation

(3.6.34) $\qquad\qquad L_2 F = (\partial_t + (\partial_x + \partial_y)^3)F = 0.$

Applying L_2 to (3.6.23a) yields

(3.6.35)

$$\{\partial_t + (\partial_x + \partial_y)^3\}K(x, y)$$
$$= 0 + \frac{\sigma}{4}\{\partial_t + (\partial_x + \partial_y)^3\}\int_0^\infty K(x, x+\zeta)F\left(\frac{2x+\zeta+\eta}{2}\right)F\left(\frac{x+\eta+y}{2}\right)d\eta\,d\zeta.$$

The terms on the right side of (3.6.35) proliferate when the differentiation is performed under the integral, but several simplifying cancellations occur. For example, using (3.6.34) leads to

$$\{\partial_t + (\partial_x + \partial_y)^3\}F\left(\frac{2x+\zeta+\eta}{2}\right)F\left(\frac{x+\eta+y}{2}\right) = 6\partial_\eta\left[F'\left(\frac{2x+\zeta+\eta}{2}\right)F'\left(\frac{x+\eta+y}{2}\right)\right].$$

It follows that (3.6.35) is equivalent to

$$(I - \sigma A_x)\{\partial_t + (\partial_x + \partial_y)^3\}K(x, y)$$

(3.6.36) $\quad = -\dfrac{3\sigma}{2}\left[\partial_x \int_0^\infty d\zeta\left\{\dfrac{\partial_1 + \partial_2}{2}K(x, x+\zeta)\right\}F\left(\dfrac{2x+\zeta}{2}\right)\right]F\left(\dfrac{x+y}{2}\right)$

$$-\frac{3\sigma}{2}[\partial_x K_2(x, x)]F'\left(\frac{x+y}{2}\right).$$

But, from (3.6.30),

$$\partial_x K_2(x, x) = -2K(x, x)$$

and

$$\partial_x \int_0^\infty d\zeta F\left(\frac{2x+\zeta}{2}\right)\frac{\partial_1+\partial_2}{2}K(x,x+\zeta)$$

$$= \partial_x[(\partial_x-\partial_y)K_2(x,y)]_{y=x}$$
$$= [(\partial_x+\partial_y)(\partial_x-\partial_y)K_2(x,y)]_{y=x}$$
$$= (\partial_x-\partial_y)\{-2K(x,x)K(x,y)\}_{y=x}$$
$$= -2[\partial_xK(x,x)]K(x,x)+\sigma K^2(x,x)K_2(x,x),$$

where we have used

$$(\partial_x-\partial_y)K_2(x,y) = \int_0^\infty (\partial_1+\partial_2)K(x,x+\zeta)F\left(\frac{x+\zeta+y}{2}\right)d\zeta.$$

Then, via (3.6.23), (3.6.33) and the invertability of $(I-\sigma A_x)$, (3.6.36) becomes (for $y \geq x$)

(3.6.37)
$$\{\partial_t+(\partial_x+\partial_y)^3\}K(x,y) = 3\sigma K(x,x)K(x,y)\partial_x K(x,x)$$
$$+3\sigma K^2(x,x)(\partial_x+\partial_y)K(x,y).$$

If we define

(3.6.38) $$q(x,t) = K(x,x;t)$$

and evaluate (3.6.37) along $y=x$, then

(3.6.39) $$\partial_t q + \partial_x^3 q = 6\sigma q^2 q_x;$$

i.e., q satisfies the modified Korteweg–de Vries equation.

Thus, every solution of these $L_i F = 0$, $i=1,2$, that decays fast enough as $x \to \infty$ defines a solution of (3.6.39), via the linear integral equation (3.6.21). No global properties (on $-\infty < x < \infty$) are required. A special case of interest is obtained if F and K are self-similar:

(3.6.40) $$K(x,y;t) = (3t)^{-1/3}\hat{K}(\xi,\eta), \quad F\left(\frac{x+y}{2};t\right) = (3t)^{-1/3}\hat{F}\left(\frac{\xi+\eta}{2}\right),$$

where

$$\xi = \frac{x}{(3t)^{1/3}}, \quad \eta = \frac{y}{(3t)^{1/3}}.$$

Substituting these into (3.6.23) shows that $\hat{K}$ satisfies an equation of the same form:

(3.6.41) $$\hat{K}(\xi,\eta) = \hat{F}\left(\frac{\xi+\eta}{2}\right) + \frac{\sigma}{4}\int_\xi^\infty \hat{K}(\xi,\zeta)\hat{F}\left(\frac{\zeta+\psi}{2}\right)\hat{F}\left(\frac{\psi+\eta}{2}\right)d\zeta\,d\psi, \quad \eta \geq \xi.$$

Substituting (3.6.40) into (3.6.34) yields

$$\hat{F}'''(\xi) - [\hat{F}(\xi) + \xi\hat{F}'(\xi)] = 0,$$

which can be integrated once:

(3.6.42) $$\hat{F}''(\xi) - \xi\hat{F}(\xi) = C_1.$$

If $C_1 = 0$, the solutions of (3.6.42) that vanish as $\xi \to \infty$ are multiples of the Airy function:

(3.6.43) $$\hat{F}\left(\frac{\xi+\eta}{2}\right) = r \operatorname{Ai}\left(\frac{\xi+\eta}{2}\right).$$

Meanwhile, $Q(\xi) = \hat{K}(\xi, \xi)$ must solve the similarity form of (3.6.38),

$$Q''' - [Q + \xi Q'] = 6\sigma Q^2 Q',$$

which can also be integrated once:

(3.6.44) $$Q'' = \xi Q + 2\sigma Q^3 + C_2.$$

This nonlinear ODE is the second equation of Painlevé (P_{II}).

What we have shown here is that every solution of the linear integral equation (3.6.41) in which $\hat{F}$ is defined by (3.6.42) also gives a solution of (3.6.44). In particular, if $C_1 = 0$ it then follows from (3.6.41) that $Q(\xi)$ is exponentially small as $\xi \to \infty$, so that C_2 vanishes in (3.6.44). Thus (3.6.44) becomes

(3.6.45) $$Q'' = \xi Q + 2\sigma Q^3.$$

Then, if (3.6.43) is used in (3.6.41), a one-parameter family of solutions of (3.6.45) is obtained from the solutions of

(3.6.46) $$[I - \sigma r^2 \tilde{A}_\xi]\hat{K}(\xi, \eta; r) = r \operatorname{Ai}\left(\frac{\xi+\eta}{2}\right),$$

where

(3.6.47) $$\tilde{A}_\xi f(\eta) = \frac{1}{4} \int_\xi^\infty \int_\xi^\infty f(\zeta) \operatorname{Ai}\left(\frac{\zeta+\psi}{2}\right) \operatorname{Ai}\left(\frac{\psi+\eta}{2}\right) d\zeta\, d\psi,$$

since $Q(\xi; r) = \hat{K}(\xi, \xi; r)$. This result was first obtained by Ablowitz and Segur (1977b). In § 3.7 we discuss in more detail some of the properties of P_{II} that can be derived from (3.6.46).

Next we shall discuss the original Zakharov–Shabat (1974) procedure with specific application to the Kadomtsev–Petviashvili equation. It will be convenient to rewrite (3.6.8) in the form

(3.6.48) $$K(x, z; y, t) + F(x, z; y, t) + \int_x^\infty K(x, z; y, t) F(x, z; y, t)\, dz = 0.$$

We shall first state the results for the Kadomtsev–Petviashvili (K–P) equation, and then discuss the actual procedure of Zakharov and Shabat (1974) (which is somewhat different from that described earlier).

If we require F to satisfy the linear equations (L_i, $i = 1, 2$)

(3.6.49a) $$L_1 F = \beta F_y + F_{xx} - F_{zz} = 0,$$

(3.6.49b) $$L_2 F = \alpha F_t + F_{xxx} + F_{zzz} = 0,$$

and follow either the recipe described above or the one that follows later, we will find that $u = 2(d/dx)K(x, x)$ satisfies

(3.6.50a) $$\partial_x(\alpha u_t + \tfrac{1}{4}(u_{xxx} + 6uu_x)) = -\tfrac{3}{4}\beta^2 u_{yy},$$

or, if $\alpha = \tfrac{1}{4}$,

(3.6.50b) $$\partial_x(u_t + 6uu_x + u_{xxx}) = -3\beta^2 u_{yy}.$$

This is the K–P equation (see also § 2.1, § 3.3, § 3.4).

We now derive (3.6.50) using the operator formalism of Zakharov and Shabat (1974). All operators are denoted by a symbol with a caret above them (e.g., $\hat{K}$ is an operator, K is a function, perhaps a matrix function). The operator form of the linear integral equation is written in a factorized form

(3.6.51) $$(1 + \hat{K}_+)(1 + \hat{F}) = (1 + \hat{K}_-),$$

where $\hat{K}_\pm$, $\hat{F}$ are $N \times N$ (matrix) operators acting on a (vector-valued) function $\psi = \{\psi_1, \cdots, \psi_N\}^T$ and

(3.6.52) $$\hat{F}\psi \equiv \int_{-\infty}^{\infty} F(x, z)\psi(z)\,dz$$

(F is an $N \times N$ matrix function). Here $\hat{K}_\pm$ are Volterra operators where $K_+(x, z) = 0$ for $z < x$, $K_-(x, z) = 0$ for $z > x$. It is assumed that the operators $(1 + K_\pm)$ are invertible. Applying (3.6.51) to ψ yields

(3.6.53a) $$K_+(x, z) + F(x, z) + \int_x^\infty K_+(x, z) F(s, z)\,ds = 0,$$

(3.6.53b) $$K_-(x, z) = F(x, z) + \int_{-\infty}^x K_+(x, s) F(s, z)\,ds.$$

Note that (3.6.53a) is equivalent to (3.6.48) (here it is understood that F also depends on auxiliary variables y, t).

Next, certain "unperturbed" operators $\hat{M}_{0,i}$, $i = 1, 2$, are defined. They are required to satisfy

(3.6.54) $$[\hat{M}_{0,i}, \hat{F}] = \hat{M}_{0,i}\hat{F} - \hat{F}\hat{M}_{0,i} = 0.$$

Hence if $\hat{M}_0 = \partial_x^2$ then (3.6.54) requires $\partial_x^2 \int_{-\infty}^\infty F(x, z)\psi(z)\,dz - \int_{-\infty}^\infty F(x, z)\partial_z^2 \psi(z) = 0$, which in turn yields the equation $(\partial_x^2 - \partial_z^2)F = 0$. These "unperturbed" operators induce "perturbed" operators $\hat{M}_i$, $i = 1, 2$ via the condition

(3.6.55) $$\hat{M}_i(1+\hat{K}_+) - (1+\hat{K}_+)\hat{M}_{0,i} = 0, \qquad i = 1, 2.$$

The $\hat{M}_0$ have the form

$$\hat{M}_0 = \alpha \partial_t + \beta \partial_y + \hat{L}_0, \qquad \hat{L}_0 = \sum l_n \partial_x^n$$

(l_n are constant matrices). Equation (3.6.55) induces an $\hat{M}$ of the form

$$\hat{M} = \alpha \partial_t + \beta \partial_y + \hat{L}, \qquad \hat{L} = \sum_n \left\{ l_n \partial_x^n + \sum_{k=1}^n V_k(x) \partial_x^{n-k} \right\},$$

where the $V_k(x)$ are determined recursively, and an equation for K_+ of the type

(3.6.56a) $$\alpha \partial_t K_+ + \beta \partial_y K_+ + \hat{L} K_+ + \sum (-1)^{n-1} \partial_z^2 K_+ l_n = 0$$

(this is analogous to (3.6.7)). For example, if $\alpha = \beta = 0$, $L_0 = \partial_x^2$ (a scalar operator) then $V_1 = 0$, $V_2 = 2(d/dx)K(x,x) = u(x)$, i.e., $\hat{L} = \partial_x^2 + u$, and K_+ satisfies $(\partial_x^2 - \partial_z^2 + u)K_+ = 0$.

Now, condition (3.6.54) with the integral equation (3.6.51) implies

(3.6.57) $$\hat{M}(1+\hat{K}_-) - (1+\hat{K}_-)\hat{M}_0 = \hat{M}(1+\hat{K}_+)(1+\hat{F}) - (1+\hat{K}_+)(1+\hat{F})\hat{M}_0$$
$$= (\hat{M}(1+\hat{K}_+) - (1+\hat{K}_+)\hat{M}_0)(1+\hat{F}).$$

For $z > x$ the left-hand side of (3.6.57) vanishes, hence we must have (3.6.55). Suppose that

$$\hat{M}_{0,1} = \alpha \partial_t + \hat{L}_{0,1},$$
$$\hat{M}_{0,2} = \beta \partial_y + \hat{L}_{0,2},$$

such that $[M_{0,i}, \hat{F}] = 0$, $i = 1, 2$. From (3.6.57) we have

$$\hat{M}_i(1+\hat{K}_+) = (1+\hat{K}_+)\hat{M}_{0,i}, \qquad i = 1, 2.$$

Operating on the $i = 1$ equation with $\hat{M}_2$, the $i = 2$ equation with $\hat{M}_1$ and subtracting yields

(3.6.58a) $$[\hat{M}_1, \hat{M}_2] = 0,$$

or

(3.6.58b) $$\alpha \partial_t \hat{L}_2 - \beta \partial_y \hat{L}_1 + [\hat{L}_1, \hat{L}_2] = 0,$$

or

(3.6.58c) $$\hat{L} \equiv \beta \partial_y + \hat{L}_2, \qquad \frac{\hat{L}^{(1)}}{\alpha} \equiv \hat{A}, \qquad \hat{L}_t = [\hat{L}, \hat{A}].$$

(3.6.58) is the nonlinear evolution equation solvable by the linear integral equation. (3.6.58c) is the standard Lax form, whereas (3.6.58b) shows explicitly the additional auxiliary variable y.

For the case of the K–P equation, two linear operators are defined on $\hat{F}$ via (3.6.54):

(3.6.59a) $$\hat{M}_{0,1} = \alpha \partial_t + \partial_x^3,$$

(3.6.59b) $$\hat{M}_{0,2} = \beta \partial_y + \partial_x^2.$$

Hence F satisfies

(3.6.60a) $$\{\alpha \partial_t + (\partial_x^3 + \partial_z^3)\} F = 0,$$

(3.6.60b) $$\{\beta \partial_y + (\partial_x^2 - \partial_z^2)\} F = 0.$$

Moreover the induced "perturbed" operators are

(3.6.61a) $$\hat{M}_1 = \alpha \partial_t + \hat{L}_1,$$

(3.6.61b) $$\hat{M}_2 = \beta \partial_y + \hat{L}_2,$$

where

(3.6.61c) $$\hat{L}_1 = \partial_x^3 + \tfrac{3}{4}(u\partial_x + \partial_x u) + w,$$

(3.6.61d) $$\hat{L}_2 = \partial_x^2 + u$$

(note that $\partial_x u \equiv u\partial_x + u_x$), and $u = 2(d/dx)K(x, x)$,

$$w \equiv \frac{3}{2}\frac{d}{dx}\left((\partial_x - \partial_z)K(x, z)\Big|_{z=x} + (K(x, x))^2\right).$$

We may verify that

(3.6.62) $$[\hat{L}_2, \hat{L}_1] = \partial_x w + w \partial_x - \tfrac{1}{4}(u_{xxx} + 6uu_x).$$

Hence (3.6.58b) yields (after operating on ψ and setting coefficients of ψ, ψ_x equal to zero, respectively)

(3.6.63) $$\alpha u_t + \tfrac{1}{4}(u_{xxx} + 6uu_x) = \beta w_y, \qquad w_x = -\tfrac{3}{4}\beta u_y.$$

This reduces to the K–P equation (3.6.50). If we set $\alpha = \tfrac{1}{4}$, the $\hat{L}, \hat{A}$ operators (3.6.58c) are given by

(3.6.64a) $$\hat{L} = \partial_x^2 + u + \beta \partial_y,$$

(3.6.64b) $$\hat{A} = 4\partial_x^3 + 3(u\partial_x + \partial_x u) - 3\beta \int_{-\infty}^{x} u_y \, dx'.$$

We also encourage the reader to compare this method with the alternative approach, discussed in § 2.1 (see also Dryuma (1974)) of Ablowitz and Haberman (1975a).

It is clear that the above method is very powerful. It yields both solutions and L, A pairs. The pair in (3.6.64) are the starting point in the work of Zakharov and Manakov (1979) on the inverse scattering solution of the K–P equation (see also Zakharov (1975)).

It should also be noted that by taking the time-independent limit, $\partial_t = 0$, of (3.6.63) we obtain the so-called Boussinesq equation

$$u_{xxxx} + 6(uu_x)_x + 3\beta^2 u_{yy} = 0. \tag{3.6.65}$$

On the other hand, if $\partial_y = 0$ then we have the derivative of the usual KdV equation. Special solutions may be constructed from the linear integral equation. Assuming

$$F = \sum_n M_n(t, y)\, e^{-\kappa_n x - \eta_n z}, \tag{3.6.66}$$

then we have

$$M_n(t, y) = M_n(0) \exp\{(\eta_n^2 - \kappa_n^2)y + (\kappa_n^3 + \eta_n^3)t\} \tag{3.6.67}$$

from (3.6.60). Then putting $K(x, z) = \sum_n K_n(x)\, e^{-\eta_n z}$ into the integral equation (3.6.48) yields the system

$$K_n(x) + M_n e^{-\kappa_n x} + M_n \sum_n K_m(x) \frac{\exp(-(\kappa_n + \eta_m)x)}{(\kappa_n + \eta_m)} = 0. \tag{3.6.68}$$

From this we can verify that the potential $u(x) = 2(d/dx)K(x, x)$ has the form

$$u = 2 \frac{d^2}{dx^2} \log \Delta,$$

$$\Delta = \det\left(M_n \delta_{nm} + M_n \frac{\exp(-(\kappa_n + \eta_m)x)}{\kappa_n + \eta_m}\right). \tag{3.6.69}$$

This is equivalent to the result in § 3.3 by Satsuma (1976). It corresponds to solitons interacting at arbitrary angles to the x-axis (save for the resonant case examined by Miles (1977a,b); see also § 3.3 and Newell and Redekopp (1977)). When $\alpha = \frac{1}{4}$, $\beta^2 = -1$, Manakov et al. (1977) have shown that limits of (3.6.69) yield lump-type solitons decaying like $1/R^2$ ($R^2 = x^2 + y^2$) as $R \to \infty$ (see also § 3.4):

$$u = 2\, \partial_x^2 \log \det B, \tag{3.6.70a}$$

where B is a $2N \times 2N$ matrix

$$B = \delta_{nm}(x - i\nu_n y - \xi_n - 3\nu_n^2 t) + (1 - \delta_{nm})\left(\frac{2}{\nu_n - \nu_m}\right). \tag{3.6.70b}$$

Asymptotically this solution breaks up into lumps having velocity $V_x = 3|\nu_n|^2$, $V_y = -6\text{Im}\, \nu_n$; there is no phase shift upon interaction.

Apart from all of these results, Zakharov and Shabat (1974) discuss certain other solutions, as well as showing how these methods can be used to study the three-wave problem in three spatial dimensions,

$$\frac{\partial u_1}{\partial t} + V_1 \cdot \nabla u_1 = i\gamma_1 u_2^* u_3^*,$$

(3.6.71)
$$\frac{\partial u_2}{\partial t} + V_2 \cdot \nabla u_2 = i\gamma_2 u_1^* u_3^*,$$

$$\frac{\partial u_3}{\partial t} + V_3 \cdot \nabla u_3 = i\gamma_3 u_2^* u_1^*.$$

This idea was generalized by Zakharov (1975) and furthered by Cornille (1979). The inverse scattering solution for appropriately decaying u_i as $|x|$, $|y| \to \infty$ is considered by Kaup (1979).

Finally, it should be noted that finite perturbations from special solutions can be constructed using this method. Namely, if $u_0(x, t)$ is a special solution to one of these equations (say, KdV to be specific) then the perturbation $v(x, t)$ ($u(x, t) = u_0(x, t) + v(x, t)$) can be expressed in terms of a Gel'fand–Levitan–Marchenko integral equation. For example, in this way rational-exponential soliton solutions can be constructed; such perturbation ideas were examined by Shabat (1973), Kuznetsov and Mikhailov (1975), Ablowitz and Cornille (1979) and Ablowitz and Airault (1981).

3.7. Painlevé transcendents. Among the possible solutions of a partial differential equation (PDE), certain special solutions may depend only on a single combination of variables, so that they effectively satisfy an ordinary differential equation (ODE; in this section we use ODE loosely for either a single or a system of ordinary differential equations). For example, the KdV equation

(3.7.1a) $$u_t + 6uu_x + u_{xxx} = 0$$

admits both traveling wave solutions

$$u(x, t) = U(x - ct),$$

where $U(z)$ satisfies the ODE

(3.7.1b) $$U'' + 3U^2 - cU = K,$$

and self-similar (or similarity) solutions

$$u(x, t) = (3t)^{-2/3} f\left(\frac{x}{(3t)^{1/3}}\right),$$

where $f(z)$ satisfies

(3.7.1c) $$f''' + 6ff' = zf' + 2f.$$

Each of these ODE's is an *exact reduction* of the PDE, obtained by suitably restricting the set of solutions.

The methods discussed in this chapter and elsewhere in this book succeed because of the remarkably rich structure of the PDE's in question. The ODE's obtained by exact reductions of these PDE's exhibit rich structure as well, and some of the methods we have used to analyze nonlinear PDE's also can be applied successfully to nonlinear ODE's. Recognition of this fact has provided solutions to some problems in nonlinear ODE's that have been outstanding for almost a century.

The ODE's obtained by these exact reductions have a rather simple characterization: they all have the Painlevé property (which will be defined below). Exploiting this fact provides a direct method to test whether a given PDE can be solved by IST. As we will see, this connection between ODE's with the Painlevé property and PDE's solvable by IST may be used to obtain information about *both* the PDE's and the ODE's.

3.7.a. The Painlevé property. We begin by reviewing some facts about linear ODE's (cf. Ince (1956, Chapt. 15)). Consider the nth order ODE

$$\frac{d^n w}{dz^n} + P_1(z)\frac{d^{n-1} w}{dz^{n-1}} + \cdots + P_{n-1}(z)\frac{dw}{dz} + P_n(z)w = 0.$$

If the n coefficients are all analytic at point z_0 in the complex plane, then z_0 is a regular point of the ODE, which has n linearly independent, analytic solutions in some neighborhood of z_0. Any singularities of the solutions of the ODE must be located at the singularities of the coefficients of the equation. These singularities are called *fixed*, because their locations are independent of the (n) constants of integration. It is a general property of linear ODE's in the complex plane that their solutions have only fixed singularities.

Nonlinear ODE's lose this property. A very simple example of a nonlinear ODE is

(3.7.2a) $$\frac{dw}{dz} + w^2 = 0;$$

its general solution is

(3.7.2b) $$w(z; z_0) = \frac{1}{z - z_0}.$$

Here z_0 is the constant of integration; it also defines the location of the singularity. This singularity is called *movable* because its location depends on the constant of integration. Nonlinear ODE's may exhibit both movable and fixed singularities.

Any singularity of a solution of an ODE that is not a pole (of any order) is called a *critical point*. These include branch points (algebraic and logarithmic) and essential singularities. A problem of interest in the late 19th century was to classify ODE's on the basis of the singularities they admitted. (A review of much of this work may be found in Ince (1956, Chapt. 12–14), or in Hille (1976).) In 1884, Fuchs showed that out of all first order equations in the form

$$\frac{dw}{dz} = F(w, z),$$

where F is rational in w and locally analytic in z, the only equations without movable critical points are generalized Ricatti equations,

(3.7.3) $$\frac{dw}{dz} = P_0(z) + P_1(z)w + P_2(z)w^2.$$

Undoubtedly familar with this result, Kovalevskaya made the next significant advance. She was awarded the Bordin Prize in 1888 for her major contribution to the theory of the motion about a fixed point of a rigid body under the influence of gravity. Her main idea was to carry out the apparently nonphysical calculation of determining the choices of parameters for which the equations of motion admitted no movable critical points. In all such cases she then solved the equations explicitly. In all other cases the solution is still unknown. (For a discussion of this work the reader is recommended to consult Golubov (1953).)

Shortly thereafter, Painlevé and his coworkers examined second order equations of the form

(3.7.4) $$w'' = F(w', w, z),$$

where F is rational in w', w and locally analytic in z. They showed that out of all possible equations of this form, there are only 50 canonical equations with the property of having no movable critical points. We will call this property the *Painlevé property*, and will refer to any equation possessing this property as being *P-type* (P for Painlevé). All 50 of these equations may be reduced either to an equation already solved, or to one of six nonlinear, nonautonomous ODE's. These six equations are

P_I $$\frac{d^2w}{dz^2} = 6w^2 + z,$$

P_{II} $$\frac{d^2w}{dz^2} = zw + 2w^3 + \alpha,$$

P_{III}
$$\frac{d^2w}{dz^2} = \frac{1}{w}\left(\frac{dw}{dz}\right)^2 - \frac{1}{z}\frac{dw}{dz} + \frac{1}{z}(\alpha w^2 + \beta) + \gamma w^3 + \frac{\delta}{w},$$

P_{IV}
$$\frac{d^2w}{dz^2} = \frac{1}{2w}\left(\frac{dw}{dz}\right)^2 + \frac{3w^3}{2} + 4zw^2 + 2(z^2 - \alpha) + \frac{\beta}{w},$$

P_V
$$\frac{d^2w}{dz^2} = \left\{\frac{1}{2w} + \frac{1}{w-1}\right\}\left(\frac{dw}{dz}\right)^2 - \frac{1}{z}\frac{dw}{dz} + \frac{(w-1)^2}{z^2}\left\{\alpha w + \frac{\beta}{w}\right\} + \frac{\gamma w}{z}$$
$$+ \frac{\delta w(w+1)}{w-1},$$

P_{VI}
$$\frac{d^2w}{dz^2} = \frac{1}{2}\left\{\frac{1}{w} + \frac{1}{w-1} + \frac{1}{w-z}\right\}\left(\frac{dw}{dz}\right)^2 - \left\{\frac{1}{z} + \frac{1}{z-1} + \frac{1}{w-z}\right\}\frac{dw}{dz}$$
$$+ \frac{w(w-1)(w-z)}{z^2(z-1)^2}\left\{\alpha + \frac{\beta z}{w^2} + \frac{\gamma(z-1)}{(w-1)^2} + \frac{\delta z(z-1)}{(w-z)^2}\right\}.$$

Painlevé and Gambier proved that these equations cannot be reduced to any simpler ODE's. Therefore they define new functions, the *Painlevé transcendents*.

The question of which ODE's have the Painlevé property is appropriate at any order, but comprehensive results are available only at first and second order. (Bureau (1964), (1972) has given a partial classification of third order equations.)

3.7.b. Relation to IST. How are these ODEs's of P-type related to the nonlinear PDE's that we have been studying? Recall from § 3.6 that a nonlinear PDE is said to be solvable by an inverse scattering transform if $K(x, x)$ [or $(d/dx)K(x, x)$] solves the PDE, where $K(x, y)$ is defined by a linear integral equation of the Gel'fand–Levitan–Marchenko type,

$$(3.7.5) \quad K(x, y) = F(x, y) + \int_x^\infty K(x, y) N(x; z, y)\, dz, \quad y \geq x,$$

where N is known in terms of F. Here we have suppressed the dependence of K on other variables (necessarily t, perhaps also $\tilde{y}$, etc.) to emphasize the distinguished role played by (x) in (3.7.5). Ablowitz, Ramani and Segur (1978) made the following conjecture:

PAINLEVÉ CONJECTURE. *A nonlinear PDE is solvable by an inverse scattering transform only if every nonlinear ODE obtained by exact reduction is of P-type, perhaps after a transformation of variables.* (See also McLeod and Olver (1981) and Exercise 13.)

For the moment, let us assume the validity of the conjecture and spell out how to use it to test PDE's.

(i) Given a nonlinear PDE, find an exact reduction to an ODE. The easiest reductions occur if the PDE admits traveling wave or similarity solutions, but these are not the only possibilities. Often a number of such simple reductions are apparent from the equation.

(ii) Using the singular point analysis that will be described below, determine whether this ODE is of P-type. If the ODE is not of P-type, then the PDE in its present form cannot be solved by IST.

(iii) However, a transformation may be available to make the ODE of P-type; these transformations often are suggested by the details of the singular point analysis. If such a transformation exists, then the transformed PDE is the candidate for IST. The sine-Gordon equation is an example for which such a transformation is necessary. Recall from § 1.2 that the equations actually solved by IST are (1.2.17), which are then transformed into the sine-Gordon equation, (1.2.18). See also Exercise 6.

(iv) If the ODE is of P-type, one may try a different reduction, and test it. However, because there is no systematic way to find every exact reduction of a PDE, the test is definitive only in ruling out PDE's that are candidates for IST. If one or two nontrivial reductions of a given PDE yield ODE's of P-type then one may start looking for a Bäcklund transformation or a scattering problem with some confidence.

(v) At the other extreme, if one knows that the PDE can be solved by IST, then any particular reduction in a variable relating it to the inherent linear integral equation must yield an ODE of P-type.

Here are some examples. Zakharov (1974) showed that the Boussinesq equation,

$$(3.7.6) \qquad u_{tt} = u_{xx} + \left(\frac{u^2}{2}\right)_{xx} + \frac{1}{4} u_{xxxx},$$

is solvable by IST. An exact reduction to an ODE may be obtained by looking for a traveling wave solution

$$u(x, t) = w(x - ct) = w(z),$$

where $w(z)$ satisfies

$$(3.7.7) \qquad (1-c^2)w'' + \left(\frac{w^2}{2}\right)'' + \frac{1}{4} w'''' = 0,$$

which can be integrated twice. Depending on the constants of integration, the result after rescaling is either

$$(3.7.8) \qquad w'' + 2w^2 + \alpha = 0 \quad \text{or} \quad w'' + 2w^2 + z = 0.$$

The first possibility defines an elliptic function, whose only singularities are poles. The second possibility is the equation for P_I. In either case, the ODE has the Painlevé property.

Another example is mKdV,

(3.7.9) $$u_t - 6u^2 u_x + u_{xxx} = 0,$$

which can be solved by IST. An exact reduction to an ODE may be obtained by looking for a self-similar solution:

$$u(x, t) = \frac{w(z)}{(3t)^{2/3}}, \quad z = \frac{x}{(3t)^{1/3}},$$

$$\Rightarrow w''' - 6w^2 w' - (zw)' = 0.$$

This can be integrated once:

P_{II} $$w'' = 2w^3 + zw + \alpha.$$

Again, the ODE is of P-type.

The sine-Gordon equation

(3.7.10) $$u_{xt} = \sin u$$

can be solved by IST (after a transformation). It has a self-similar solution

(3.7.11) $$u(x, t) = f(z), \quad z = xt.$$

If we set $w(z) = \exp(if)$, then

P_{III} $$w'' = \frac{1}{w}(w')^2 - \frac{1}{z}(w') + \frac{1}{2z}(w^2 - 1).$$

Again, the ODE is of P-type.

The derivative nonlinear Schrödinger equation

(3.7.12) $$iq_t = q_{xx} - 4iq^2(q^*)_x + 8|q|^4 q$$

can be solved by IST (Kaup and Newell (1978a)). Its similarity solution eventually reduces to P_{IV} (Ablowitz, Ramani and Segur (1980b)). Other examples have been given by Jimbo (1979), Salihoglu (1980), and McLeod and Olver (1981); also see the exercises at the end of this chapter. In every example known to us, PDE's solvable by IST reduce to ODE's of P-type, and PDE's for which there is general agreement that they are not solvable by IST (e.g., based on numerical experiments in which two solitary waves do not interact like solitons) reduce to ODE's that are not of P-type. To avoid any confusion, we should emphasize that the critical question is not whether the ODE turns out to be one of the six Painlevé transcendents, but only whether it is of P-type (i.e., has no movable critical points).

These examples demonstrate that the conjecture seems to work. Following Ablowitz, Ramani and Segur (1980a), we now sketch a partial proof of the conjecture, which shows why it should work. Consider a special case of the linear integral equation (in which $F(x, y) = F(x+y)$):

(3.7.5a) $\quad K(x, y) = F(x+y) + \int_x^\infty K(x, z) N(x; z, y) \, dz, \qquad y \geq x.$

We require that F satisfies a linear ODE, that F vanishes for large positive values of its argument, and that N depends on F in a known way. (Several examples of how N might depend on F were given in § 3.6; generally, the reasoning here is closely tied to that in § 3.6.) We want to show that every solution of a linear integral equation like (3.7.5a) must have the Painlevé property. Then if K also satisfies an ODE, the family of solutions of this ODE obtained via (3.7.5a) has the Painlevé property as well. Thus, the relation between ODE's of P-type and PDE's solvable by IST is a direct consequence of the role of the linear integral equation (3.7.5a) in the IST formulation.

A rough outline of the proof is as follows (for details see Ablowitz, Ramani and Segur (1980a)):

(i) F satisfies a linear ODE, and therefore has no movable singularities at all.

(ii) If F vanishes rapidly enough as its argument becomes large, then the Fredholm theory of linear integral equations applies. It follows that (3.7.5a) has a unique solution in the form

(3.7.13) $\quad K(x, y) = F(x+y) + \int_x^\infty F(x+z) \frac{D_1(x; z, y)}{D_2(x)} \, dz,$

where D_1 and D_2 are entire functions of their arguments. Therefore the singularities of K (if any) can come only from the fixed singularities of F, or from the movable zeros of D_2. But D_2 is analytic, so these movable singularities must be poles.

(iii) Therefore, K, the solution of the linear integral equation, has the Painlevé property.

This proof relates the Painlevé property to the linear integral equation. McLeod and Olver (1981) give a similar proof. However, the connection to IST may be exploited from other points of view. Flaschka (1980) and Flaschka and Newell (1980) have made use of the scattering problem and associated time dependences. Flaschka's results may be stated quite simply.

(i) As discussed in § 1.2, compatibility of a given scattering problem,

$$Lv = \lambda v, \qquad V_t = Mv,$$

leads to a PDE solvable by IST that can be written in Lax's form,

(3.7.14) $\qquad [L, M] + L_t = 0.$

(ii) The stationary solutions of (3.7.14), which include its N-soliton and its N–phase quasiperiodic solutions, satisfy a commutator relation of the form

(3.7.15) $\qquad [L, B] = 0.$

(iii) The self-similar solutions of (3.7.14) satisfy a different commutator relation,

(3.7.16) $$[L, B] = L.$$

The practical value of (3.7.15) is that it can be reduced to an algebraic equation, and finding its explicit solutions reduces to function theory on an algebraic curve. Whether the algebraic nature of (3.7.16) can be exploited as efficiently is not known. Later in this section we briefly discuss how Flaschka and Newell (1980) use the concepts of monodromy preserving deformations to examine P_{II}.

3.7.c. Singular point analysis. Given a nonlinear ODE, how does one determine whether it has movable critical points? If the equation happens to be of second order and of the form (3.7.4), one may consult the list of the 50 equations found by Painlevé et al., in Ince (1956, Chapt. 14). If the equation is on the list, then it is of P-type. If not, then it is still possible that a simple transformation will put it on the list, so we recommend determining the nature of the singularities admitted by the equation (see Exercise 8).

If the equation is third order or higher, there is no alternative to analyzing the local structure of movable singularities. Two methods are avilable. The α-method of Painlevé is described in detail by Ince. Ablowitz, Ramani and Segur (1980a) describe an alternative method, which is similar to that of Kovalevskaya, and which we illustrate below. We should note that both of these methods may miss essential singularities, which require a separate analysis.

Example 1. Consider the family of ODE's

(3.7.17) $$w'' = z^m w + 2w^3.$$

(If $m = 0$, (3.7.17) defines elliptic functions. If $m = 1$, it is P_{II}. For $m \neq 0, 1$, we will find movable critical points.) There are three main steps to the method. The first is to find the dominant behavior of the solution in the neighborhood of a movable singularity, at z_0. Thus we assume that, as $z \to z_0$,

$$w(z; z_0) \sim \frac{a}{(z-z_0)^p}.$$

In this case the dominant terms in (3.7.17) are the first and last, and $p = 1$, $a^2 = 1$. If we take $a = 1$, then as $z \to z_0$

(3.7.18) $$w \sim (z - z_0)^{-1} + o(|z - z_0|^{-1}).$$

If p had not been an integer, the dominant behavior would have been that of a (movable) algebraic branch point, and the equation would not be of P-type. (Even so, we recommend continuing the procedure since a transformation to bring the equation into P-type may be suggested.) If p had two or more roots, a separate analysis would be required for each root.

Because (3.7.17) is second order, its general solution has two constants of integration. One of these is z_0. It is necessary to carry the expansion in (3.7.18) to higher order, until the other constant of integration appears. The second step is to find the power of $(z - z_0)$ at which this arbitrary constant may enter, called a *resonance*. To do so, set $\xi = z - z_0$ and substitute

$$w(z) \sim \xi^{-1} + \beta \xi^{-1+r} \tag{3.7.19}$$

into the dominant terms of (3.7.17). To leading order in β, this gives

$$\beta[(r-1)(r-2)-6]\xi^{r-3} \sim 0,$$

which amounts to an algebraic equation for r. One root is always (-1), corresponding to the arbitrariness of z_0. In this case, the other root is $r = 4$. (If r had not been a real integer, it would have indicated a movable branch point at z_0.) This tells how high we must carry the expansion:

$$w(z) \sim \xi^{-1} + a_0 + a_1 \xi + a_2 \xi^2 + a_3 \xi^3 + \cdots. \tag{3.7.20}$$

We can expect (a_0, a_1, a_2) to be determined, and the second constant to appear when we get to a_3.

The last step is to find the coefficients in (3.7.20), by substituting it into (3.7.17) and equating powers of ξ (note that $z^m = (z_0 + \xi)^m$). The result is that

$$a_0 = 0, \quad a_1 = \frac{-z_0^m}{6}, \quad a_2 = \frac{m z_0^{m-1}}{4}. \tag{3.7.21}$$

At $O(\xi^3)$ we find

$$0 = 0 \cdot a_3 = \frac{1}{2} m(m-1) z_0^{m-2}. \tag{3.7.22}$$

There are two possibilities.

(i) If $m = 0$ or 1, (3.7.22) is identically satisfied for any value of a_3, which is therefore the second arbitrary constant. Using a method of Painlevé (cf. Ince (1956, § 14.41)) one can then show that (3.7.20, 21) does indeed represent the beginning of the Laurent series of the general solution of (3.7.17) in the neighborhood of a movable pole. Because no other algebraic singularity is possible, there are no movable algebraic branch points.

(ii) If $m \neq 0, 1$, (3.7.22) cannot be satisfied for any choice of a_3. In this case (3.7.20) must be augmented by logarithmic terms:

$$w(z) \sim \xi^{-1} + a_0 + a_1 \xi + a_2 \xi^2 + (a_3 \xi^3 + b_3 \xi^3 \log \xi) + \cdots. \tag{3.7.23}$$

Now (3.7.21) is valid, but at $O(\xi^3)$ b_3 is determined while a_3 is arbitrary. The expansion in (3.7.23) is indicative of a logarithmic branch point at the movable point z_0. Thus, the equation is not of P-type unless $m = 0, 1$. (Note that higher terms contain higher powers of both ξ and $\log \xi$.)

Example 2. A nonlinear Schrödinger equation in $(n+1)$ dimensions (cf. § 4.3) is

(3.7.24) $$i\phi_t + \nabla^2 \phi - 2|\phi|^2 \phi = 0.$$

An exact reduction is obtained by setting

(3.7.25) $$r^2 = \sum_{j=1}^{n} x_j^2, \quad \phi = R(r) \exp(i\lambda t),$$

where

(3.7.26) $$R'' + \frac{n-1}{r} R' = 2|R|^2 R + \lambda R.$$

(If we also require real R for real r, then the nonlinear term in (3.7.26) becomes $2R^3$, and we may search for (3.7.26) on Ince's list. We consider the more general case here, to show how to analyze complex equations.) The nonlinear term in (3.7.26) precludes our considering $R(r)$ as an analytic function. However, we may replace (3.7.26) with

(3.7.27) $$R'' + \frac{n-1}{r} R' = 2R^2 S + \lambda R,$$
$$S'' + \frac{n-1}{r} S = 2RS^2 + \lambda S.$$

If λ is real, and $S = R^*$ for real r, then (3.7.27) includes (3.7.26). In any case, (3.7.27) is a fourth order system of ODE's, whose singular points we may analyze.

Step 1. At leading order, the dominant behavior of any algebraic singularities of (3.7.27) is

(3.7.28) $$R \sim \frac{\alpha}{r - r_0}, \quad S \sim \frac{1}{\alpha(r - r_0)}.$$

Two of the four arbitrary constants of integration are (r_0, α).

Step 2. To find the resonances at which the other two constants of integration enter, set $x = r - r_0$,

$$R \sim \alpha x^{-1} + C_1 x^{-1+p}, \quad S = \frac{1}{\alpha} x^{-1} + C_2 x^{-1+p},$$

and substitute this into the dominant terms in (3.7.27), keeping only linear terms in (C_1, C_2). The result is a fourth order polynomial for p, with roots $(-1, 0, 3, 4)$. The first two roots represent the two free constants (r_0, α), respectively.

The last two roots determine the powers at which the other two constants may enter.

Step 3. If (3.7.28) represent the first terms in a Laurent series about a movable pole, then

(3.7.29)
$$\frac{R}{\alpha} \sim x^{-1} + a_0 + a_1 x + a_2 x^2 + a_3 x^3,$$
$$\alpha S \sim x^{-1} + b_0 + b_1 x + b_2 x^2 + b_3 x^3.$$

Substitute this into (3.7.27) and solve recursively:

$$a_0 = b_0 = -\frac{n-1}{6r_0},$$

$$a_1 = b_1 = -\frac{\lambda}{6} - \frac{(n-1)(n-7)}{36r_0^2},$$

$$a_2 + b_2 = \frac{\lambda(n-1)}{6r_0} - \frac{(n-1)(4n^2 - 35n + 85)}{108r_0^3},$$

with $(a_2 - b_2)$ free. This is the third constant of integration. (So far, so good!) However, at the next order we find

(3.7.30) $$0 \cdot (a_3 + b_3) = (n-1)(*),$$

where $(*) \neq 0$. If $n = 1$ (where we already know that (3.7.24) can be solved by IST), (3.7.30) is an identity, $(a_3 + b_3)$ is the fourth constant of integration, and (3.7.27) has no movable branch points. If $n \neq 1$, (3.7.30) is a contradiction and (3.7.29) must be augmented with a logarithmic term at $O(x^3)$. This logarithmic term then generates an infinite sequence of increasingly complicated terms at higher order. Equation (3.7.27) is not of P-type if $n \neq 1$.

If the Painlevé conjecture is correct, it follows that the nonlinear Schrödinger equation (3.7.24) can be solved by IST *only* in (1 + 1) dimensions.

These two examples do not exhaust the possible nuances of singular point analyses. The reader may consult Ablowitz, Ramani and Segur (1980a) for more details.

3.7.d. Global properties of Painlevé transcendents. Apart from its value in testing PDE's, the connection between IST and ODE's of P-type can be used to obtain global information about Painlevé transcendents. For example, we saw in § 3.6 that if $K(x, y)$ satisfies

(3.7.31)
$$K(x, y) = r\text{Ai}\left(\frac{x+y}{2}\right)$$
$$+ \sigma \frac{r^2}{4} \int_x^\infty \int_x^\infty K(x, z) \text{Ai}\left(\frac{z+s}{2}\right) \text{Ai}\left(\frac{s+y}{2}\right) dz\, ds, \quad y \geq x,$$

where Ai(z) is the Airy function, $\sigma = \pm 1$, and r is a parameter, then $w(z;r) = K(z,z;r)$ satisfies a restricted form of P_{II}:

$$\text{(3.7.32)} \qquad \frac{d^2 w}{dz^2} = zw + 2\sigma w^3, \qquad \sigma = \pm 1,$$

with the boundary condition that as $z \to +\infty$,

$$\text{(3.7.33)} \qquad w(z;r) \sim r \text{ Ai}(z).$$

Thus (3.7.32) is irreducible as an ODE, but (3.7.31) is an exact linearization of a one-parameter family of its solutions. This family includes all of the bounded solutions of (3.7.32).

The global existence of this family of solutions may be proved directly from (3.7.31), as follows. If we write (3.7.31) schematically as

$$\text{(3.7.34)} \qquad [I - \sigma r^2 \tilde{A}(z)] K = r \text{ Ai},$$

then the existence of a bounded solution of (3.7.32) follows directly from the boundedness of $[I - \sigma r^2 \tilde{A}(z)]^{-1}$ in (3.7.34). Here are the results (for details, see Ablowitz and Segur (1977b), Hastings and McLeod (1980), Ablowitz, Ramani and Segur (1980b)).

(i) $\tilde{A}(z)$ is a positive operator. Therefore, if $\sigma = -1$, (3.7.32, 33) has a unique bounded solution for all real z and for all real r.

(ii) The L_2-norm of $\tilde{A}(z)$ does not exceed 1 for any real z. Therefore if $\sigma = +1$, (3.7.32, 33) has a unique bounded solution for all real z if $-1 < r < 1$.

(iii) If $\sigma = +1, |r| = 1$, one obtains a critical branch of (3.7.32), which vanishes as $z \to +\infty$, and grows algebraically ($2w^2 + z \sim 0$) as $z \to -\infty$.

(iv) If $\sigma = +1, r > 1$, there is a real $z_0(r)$ such that $[I - \sigma r^2 \tilde{A}(z)]^{-1}$ exists only if $z > z_0$. We supect (but have not proven) that $w(z, r)$ has a pole at z_0.

(v) According to the argument given below (3.7.5a), the only singularities in the complex z-plane of these families of solutions are poles. (This result, of course, was first obtained by Painlevé.)

The story does not end with existence proofs. Just as Airy's equation,

$$\frac{d^2 w}{dz^2} = zw,$$

is representative of a simple linear turning point, so (3.7.32) is representative of a class of simple nonlinear turning points. For example, Haberman (1977) showed that the weakly nonlinear solution of

$$\text{(3.7.35)} \qquad \frac{d^2 u}{dz^2} + k(\varepsilon z) u = \varepsilon \beta(\varepsilon z) u^3, \qquad \varepsilon \ll 1$$

is approximated (asymptotically) by the solution of (3.7.32) near a simple zero of $k(\varepsilon z)$. Thus, the qualitative behavior of this solution of (3.7.35) is weakly nonlinear and exponentially decaying for $k<0$, weakly nonlinear and oscillatory for $k>0$, but fully nonlinear and represented by the solution of (3.7.32) in the transition region. In this context, we are forced to consider the solution of (3.7.32), because it connects two regions in which the solution of (3.7.35) has different qualitative behavior. Thus we come to the *connection problem*:

> Given the complete asymptotic behavior of a bounded real solution of (3.7.32) as $z \to +\infty$, find the complete asymptotic behavior of the same solution as $z \to -\infty$.

This information is sufficient to connect the two weakly nonlinear regions of the solution of (3.7.35).

It is evident that this connection problem is global, and cannot be solved by any local analysis of (3.7.32). However, recall from § 1.7 that given almost any smooth, rapidly decaying initial data for mKdV (3.7.9), its asymptotic ($t \to \infty$) solution exhibits three regions, with qualitatively different behavior in each region.

(i) For $x \gg t^{1/3}$, the solution is exponentially decaying (in x).

(ii) For $|x| = O(t^{1/3})$, the solution is approximately self-similar and governed by (3.7.32).

(iii) For $-x \gg t^{1/3}$, the solution is oscillatory.

Segu and Ablowitz (1981) found the solution of the connection problem for (3.7.32) simply by taking the appropriate limits of the asymptotic solution of the mKdV equation in regions (i) and (iii). The results are as follows.

There is a one-parameter (r) family of bounded real solutions of (3.7.32). These are exponentially decaying (as z increases) for $(2\sigma w^2 + z) > 0$, and oscillatory for $(2\sigma w^2 + z) < 0$. A typical solution is shown in Fig. 3.2. As $z \to +\infty$, all of these solutions satisfy (3.7.33), with $-1 < r < 1$ for $\sigma = +1$ and with any real r for $\sigma = -1$. We may assume $r \geq 0$ without loss. As $z \to -\infty$, these solutions have formally asymptotic expansions

(3.7.36) $$w(z) \sim d(-z)^{-1/4} \sin \theta + O(|z|^{-7/4}),$$

where

$$\theta \sim \tfrac{2}{3}(-z)^{3/2} - \tfrac{3}{2}\sigma d^2 \log(-z) + \bar{\theta} + O(|z|^{-3/2}).$$

The constants $d(\geq 0)$ and $\bar{\theta}$ are given by

(3.7.37)
$$d^2 = -\frac{\sigma}{\pi} \log\{1 - \sigma r^2\},$$

$$\bar{\theta} = \frac{\pi}{4} - \sigma \arg\left\{\Gamma\left(1 - \frac{id^2}{2}\right)\right\} - \frac{3}{2}\sigma d^2 \log 2,$$

shown in Fig. 3.3. There are no other bounded real solutions of (3.7.32).

Next we consider the general case of P_{II},

$$\frac{d^2 w}{dz^2} = zw + 2w^3 + \alpha.$$

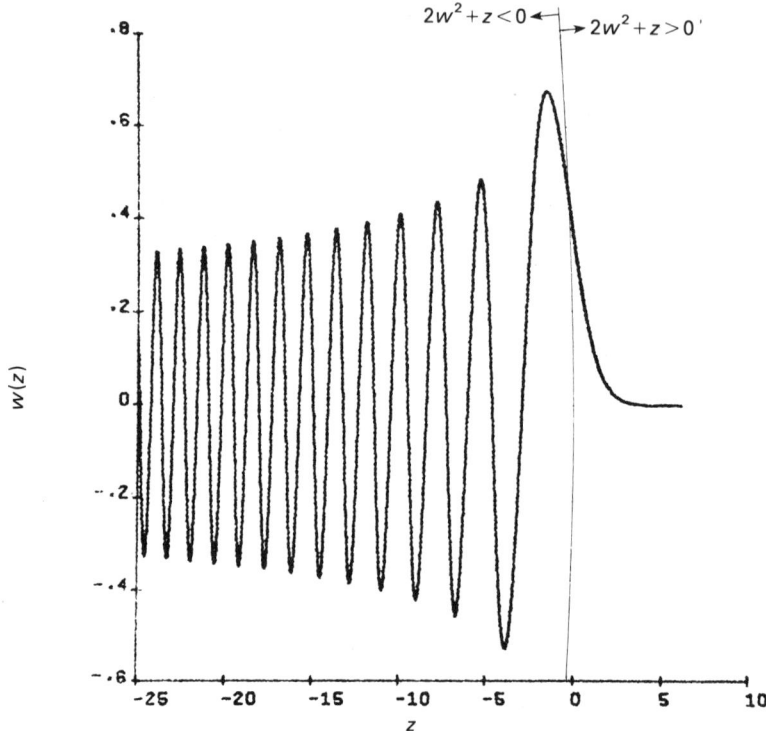

FIG. 3.2. *Typical solution of* (3.7.32, 33). *Here* $\sigma = +1$, $r = 0.9$. *The dividing parabola* $(2w^2 + z = 0)$ *also is shown.*

Airault (1979) and Boiti and Pempinelli (1979) put the Bäcklund transformation for KdV into self-similar form, and deduced a recursion relation among solutions of P_{II}:

$$(3.7.38) \qquad \tilde{w}(z, \alpha + 1) = -w(z, \alpha) - \frac{2\alpha + 1}{2w^2(z, \alpha) + z + 2w_z(z, \alpha)}.$$

This relation already had been found by other means by Lukashevich (1971). In addition, P_{II} admits a symmetry transformation

$$(3.7.39) \qquad w(z, \alpha) \to -w(z, -\alpha).$$

If this is used with (3.7.38), then any explicit solution of P_{II}, $w(z, \alpha)$, may be used to generate a doubly infinite sequence of other solutions, $w(z, \alpha \pm n)$, $n = 0, 1, 2, \cdots$. Beginning with the trivial solution ($w(z, 0) = 0$), Airault (1979) constructed in this way all of the rational solutions of the KdV equation (cf. § 3.4).

More generally, the one-parameter family of bounded real solutions of P_{II} at $\alpha = 0$ generates a one-parameter family of real solutions of P_{II} at every

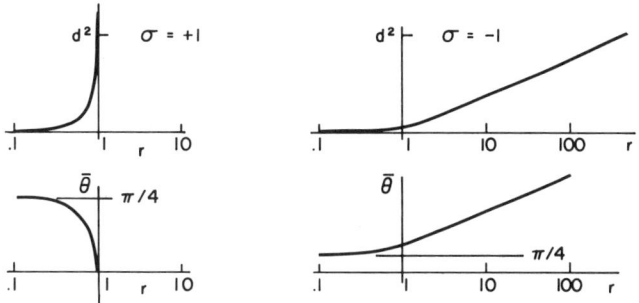

FIG. 3.3. *Asymptotic* ($z \to -\infty$) *amplitude* (d) *and phase constant* ($\bar{\theta}$) *for* P_{II} *as functions of initial* ($z \to +\infty$) *amplitude, r. Taken from* (3.7.37).

integer α. All of these solutions are bounded as $z \to +\infty$, but we show next that none of them are bounded for all real z. It will follow that for integer α P_{II} has *no* bounded real solutions except those at $\alpha = 0$.

We begin with $\alpha = 0$ and take $0 \leq r < 1$, so that all the bounded real solutions are qualitatively similar to that shown in Fig. 3.2. It is apparent from the figure that the denominator of (3.7.38) is negative where $(2w^2 + z) = 0$, is positive as $z \to +\infty$, and must vanish somewhere in between. This zero is a pole of $w(z, 1)$. (A similar argument applies if $r < 0$.) Thus every real solution of P_{II} at $\alpha = 1$ that is bounded as $z \to +\infty$ has at least one pole at some finite real z.

Now let α be any positive integer. We show next that if $w(z, \alpha)$ has a pole, then so does $w(z, \alpha + 1)$, generated by (3.7.38). Let z_0 denote the rightmost pole of $w(z, \alpha)$. Near z_0, the Laurent series for $w(z, \alpha)$ begins as

(3.7.40a) $\quad w(z, \alpha) = w_+ \sim (z - z_0)^{-1} - \frac{z_0}{6}(z - z_0) - \frac{\alpha + 1}{4}(z - z_0)^2 + \cdots,$

or as

(3.7.40b) $\quad w(z, \alpha) = w_- \sim -(z - z_0)^{-1} + \frac{z_0}{6}(z - z_0) - \frac{\alpha - 1}{4}(z - z_0)^2 + \cdots.$

In these cases the denominator of (3.7.38) takes the form

(3.7.41a) $\quad\quad\quad 2w_+^2 + z + 2\partial_z(w_+) \sim -(2\alpha + 1)(z - z_0)$

or

(3.7.41b) $\quad\quad\quad 2w_-^2 + z + 2\partial_z(w_-) \sim 4(z - z_0)^{-2}.$

For $w_-(z, \alpha)$, the first term on the right of (3.7.38) has a pole at z_0, while the second term vanishes there. Thus, z_0 is also a pole of $w_-(z, \alpha + 1)$.

For $w_+(z, \alpha)$ with $\alpha > 0$, the denominator in (3.7.38) is negative for $z = z_0 + \varepsilon$ ($0 < \varepsilon \ll 1$), positive as $z \to +\infty$ and must vanish somewhere in between. But z_0 is the rightmost pole of $w(z, \alpha)$ by hypothesis, so this zero is a pole of $w_+(z, \alpha + 1)$. Thus for $\alpha > 0$, if $w(z, \alpha)$ has a pole at some finite real z_0 and vanishes at $z \to +\infty$, then so does $w(z, \alpha+1)$. Because $w(z, 1)$ has a pole, so does $w(z, n)$ for every integer $n > 0$. By (3.7.39), $w(z, -n)$ has poles at the same locations.

We have concentrated on P_{II} up to this point, but it is clear that these methods can be applied to the other Painlevé transcendents as well. For example, Bäcklund transformations have been found by Boiti and Pempinelli (1980) for P_I, by Airault (1979) for P_{III}, P_{IV} and P_V and by Fokas and Yortsos (1980) for P_{VI}. As with (3.7.38) for P_{II}, some of these results had been found earlier by Soviet mathematicians, including Gromak, Erugin, Lukashevich and Yablonsky, using other methods. (See Erugin (1976) for a review of this Soviet work, with an extensive bibliography.)

In another direction, Ablowitz, Ramani and Segur (1980b) identified a linear integral equation whose solution also solves P_{III}, and another such equation for P_{IV}. They also found a convergent series expansion for a family of solutions of P_{III} via the integral equation. This series expansion turns out to be equivalent to that found by McCoy, Tracy and Wu (1977) by an entirely different approach. The latter authors also found connection formulae for P_{III} corresponding to (3.7.37) for P_{II}.

The original motivation of Wu, McCoy, Tracy and Barouch (1976) was that P_{III} arises in the scaling limit of the spin-spin correlation function of the two-dimensional Ising model. Their work makes no reference to IST, but aspects of it show some similarity to an IST approach. Thus the question arises: Is there some connection between the two-dimensional Ising model and IST?

A series of important papers by Sato, Miwa and Jimbo (1977), (1978), $\cdots$ have dealt with connections between: (i) monodromy-preserving deformations of linear differential equations; (ii) holonomic quantum field theory; (iii) the scaling limit of the two-dimensional Ising model; (iv) theory of Clifford groups; and (v) IST (see also Kashiwara and Kawai (1978), Ueno (1981) and a comprehensive review by Jimbo, Miwa, Mōri and Sato (1979)). The description of these ideas which follows is based on the closely related work of Flaschka and Newell (1980), who emphasized the connection with Painlevé transcendents.

We need some more definitions. Consider the (matrix) system of linear ordinary differential equations

$$(3.7.42) \qquad \frac{dy}{dx} = y \sum_{j}^{N} \frac{A_j}{x - a_j},$$

where A_j are constant $m \times m$ matrices and x is complex. The solutions of (3.7.42) at x are ordinarily multi-valued, and we denote by $Y(x)$ a single-valued fundamental solution. If x moves along some contour that encircles one of the singular points (a_j), $Y(x)$ ordinarily does not return to its original value, but may be expressed as a linear combination of the original solutions

(3.7.43) $$\tilde{Y}(x) = M_j Y(x).$$

M_j is the *monodromy matrix* at a_j (cf. § 2.3). The question of monodromy-preserving deformations is this. If the singularities (a_j) are allowed to move in the complex plane, how must the matrices (A_j) change in order that the monodromy matrices (M_j) remain fixed? In the simplest nontrivial case, where A_j is a 2×2 matrix, $N = 4$ and only one singularity is permitted to move, the problem eventually reduces to solving P_{VI}! In other words, P_{VI} may be viewed as a condition which constrains the deformation of the coefficients of a linear differential equation so that its monodromy is invariant.

In this example the ODE has only regular singular points. The emphasis of the work by Flaschka and Newell is on irregular singular points, where the monodromy matrix is effectively replaced by the *Stokes multipliers*.

In a sense, we have been dealing with the theory of deformations of linear equations all along, although we have not called it that. For example, in Chapter 1 we considered the linear ordinary differential operator

$$L = \frac{d^2}{dx^2} + q(x),$$

and asked: How may we deform the coefficient $q(x)$ as a function of an external parameter (t) so that the eigenvalues of L are invariant? The answer, of course, is that $q(x, t)$ should satisfy the KdV equation or one of the higher order analogues in (1.5.21).

Thus we have yet another way to view IST and Painlevé functions, in terms of the theory of deformations of linear differential equations. This perspective focuses attention on the direct scattering part of IST (rather than the inverse scattering part), and suggests that the scattering problem (1.2.7) for mKdV should be put in self-similar form. The transformations

$$z = \frac{x}{(3t)^{1/3}}, \quad \chi = \zeta(3t)^{1/3}, \quad q(x, t) = (3t)^{-1/3} w(z),$$

$$V_i(x, t, \zeta) = \phi_i(z, \chi)$$

finally lead to the system of linear ODE's

(3.7.44)
$$\partial_\chi \phi_1 = -i(4\chi^2 + z + 2w^2)\phi_1 + (4\chi w + 2iw_z)\phi_2,$$
$$\partial_\chi \phi_2 = (4\chi w - 2iw_z)\phi_1 + i(4\chi^2 + z + 2w^2)\phi_2.$$

The Stokes multipliers of the solutions of (3.7.44) near an irregular singular point are independent of z only if $w(z)$ satisfies a form of P_{II}. Knowing the properties of the monodromy matrices and the locations of the singular points allows Flaschka and Newell to recast the problem in terms of a system of singular linear integral equations. The authors do not examine the question of the existence of solutions. In special cases, the integral equation reproduces known results.

It may be worthwhile to summarize the results of this section. There is a very close connection between PDE's solvable by IST and ODE's of P-type. This connection may be used profitably to study either the PDE's or the ODE's. In particular, it has turned out in studying the ODE's (especially P_{II}) that almost every aspect of IST (the direct problem, the inverse problem, Bäcklund transformations) is of value when specialized to the ODE's.

3.8. Perturbations and transverse stability of solitons and solitary waves. In this section we shall briefly discuss the influence weak perturbations (including transverse perturbations) have on solitons and solitary waves. This is, of course, a very natural question to ask, and a great deal of work has been devoted to these questions in the literature.

Here we shall consider three prototype problems: (a) an example of a soliton undergoing a dissipative perturbation; (b) a solitary wave undergoing a dissipative perturbation; and (c) the transverse stability of a soliton (similar ideas apply to transverse stability of solitary waves). Each of these subsections can be read independent of the others.

For those problems giving rise to solitons, there are various methods which have been developed that use the techniques of IST (see, for example, Kaup (1976a), Kaup and Newell (1978a), Karpman and Maslov (1978), Keener and McLaughlin (1977)). The methods of Kaup and Newell (1978a) and Karpman and Maslov (1978) develop the perturbed equations of the scattering data by using perturbation theory on the associated linear scattering problem. The field variable is recovered via the inverse equations (e.g., the linear Gel'fand–Levitan equation). From a somewhat different point of view Keener and McLaughlin (1977) develop a perturbation theory using a Green's function to solve the associated linearized equation in the higher order problems. In order to calculate the Green's function, information from IST is needed.

On the other hand, it is well known that there exist very general perturbation techniques that are applicable to nonlinear problems where the leading order problem has a well defined solution (e.g., solitons, breathers, solitary waves, periodic solutions). These ideas have been applied to a wide variety of problems (see, for example, Taniuti and Wei (1968), Ablowitz (1971), Johnson (1973), Whitham (1974), Zakharov and Rubenchik (1974), Kadomtsev and Petviashvili (1970), Yajima (1974), Ko and Kuehl (1978), Kodama and Taniuti (1979), Kodama and Ablowitz (1980). In this section we will illustrate how these ideas can be applied to problems with solitons and solitary waves. It is not our intent to survey completely the various perturbation approaches for solitons and/or solitary waves.

The basic idea of the perturbation approach can be explained as follows. We study the solution of a perturbed nonlinear dispersive wave equation of a fairly general form:

(3.8.1) $$K(q, q_t, q_x, \cdots) = \varepsilon F(q, q_x, \cdots), \qquad 0 < \varepsilon \ll 1,$$

where K and F are nonlinear functions of $q, q_x, \cdots$. The unperturbed equation ($\varepsilon = 0$)

(3.8.2) $$K(q^{(0)}, q_t^{(0)}, q_x^{(0)}, \cdots) = 0$$

has as its solution $q^{(0)}$. $q^{(0)}$ is to be taken as a solitary wave or soliton solution (or perhaps a breather or a more complicated soliton state). We write this solution in terms of certain natural fast and slow variables:

(3.8.3) $$q^{(0)} = \hat{q}^{(0)}(\theta_1, \theta_2, \cdots, \theta_m, T: P_1, P_2, \cdots, P_N).$$

In (3.8.3), $\theta_i (i = 1, \cdots, m)$ are so-called "fast" variables, $T = \varepsilon t$ is a "slow" variable and the $P_l (l = 1, \cdots, N)$ are parameters which depend on the slow variable (in some problems, one might need to also introduce a slow variable $X = \varepsilon x$; see, for example, Whitham (1974), Ablowitz (1971)). In many problems we need only one fast variable, such as $\theta = x - P_1 t$ in the unperturbed problem. We generalize θ to satisfy $\partial \theta / \partial x = 1$, and $\partial \theta / \partial t = -P_1$ and use $P_1 = P_1(T)$ to remove secular terms. With this, we call such a solution (3.8.3) a quasi-stationary solution and write $q = \hat{q}(\theta, T, \varepsilon)$. It is necessary that we develop equations for the $P_1, \cdots, P_N$ by using appropriate conditions, such as secularity conditions (there must be N such independent conditions). Some of these conditions are formed from Green's identity, as follows. We assume an expression for $\hat{q}$ of the form

$$\hat{q} = \hat{q}^{(0)} + \varepsilon \hat{q}^{(1)} + \cdots$$

(after introducing appropriate variables θ_i, T, etc.). Then (3.8.2) is the leading order problem, and (if we assume K has only first order in time derivatives)

(3.8.4) $$L(\partial_{\theta_i}, \hat{q}^{(0)}) \hat{q}^{(1)} = F(\hat{q}^{(0)}) - \frac{\partial K}{\partial q_t}\bigg|_{q=\hat{q}^{(0)}} \cdot q_T \equiv \hat{F}$$

is the first order equation. Here $L(\partial_{\theta_i}, \hat{q}^{(0)})$, $u = 0$ is the linearized equation of $K(q, q_t, q_x, \cdots) = 0$ after (x, t) is transformed to the appropriate coordinate θ_i. Calling v_i ($i = 1, \cdots, M$) the M solutions of the homogeneous adjoint problem satisfying the necessary boundary conditions (e.g., $v_i \to 0$ as $|\theta| \to \infty$),

$$L^A v_i = 0, \qquad i = 1, \cdots, M, \qquad M \le N,$$

where L^A is the adjoint operator to L, we form

(3.8.5) $$(L\hat{q}^{(1)}) \cdot v_i - (L^A v_i) \cdot \hat{q}^{(1)} = \hat{F} v_i.$$

The left-hand side of (3.8.5) is always a divergence (Green's theorem). It may be integrated to give the secularity conditions. These secularity conditions allow us to be able to compute a solution $\hat{q}^{(1)}$ to (3.8.4) which satisfies suitable boundary conditions (e.g., $\hat{q}^{(1)}$ is bounded as $|\theta| \to \infty$). However, as is standard in perturbation problems, there is still freedom in the solution. This is due to the fact that some terms in the solution $\hat{q}^{(1)}$ can be absorbed in the leading order solution $\hat{q}^{(0)}$ by shifting the other parameters. The solution $\hat{q}^{(1)}$ can be made unique by imposing additional conditions which reflect specific initial conditions or other normalizations. Continuation to higher order $\hat{q}^{(N)}$ is straightforward.

Consequences of this method are the following.

(i) The expansion obtained is generally not uniformly valid on $|x| < \infty$ (see also Ablowitz (1971), Kaup and Newell (1978a), Karpman and Moslov (1978), Kodama and Ablowitz (1980)).

(ii) In its region of validity we find a quasi-stationary solution; i.e., the solution depends on the θ_i and T only.,

(iii) In order to develop a uniformly valid expansion, one must match the solution obtained via the method above to a nonstationary solution for large $|\theta_i|$ (e.g., $|\theta| \sim O(1/\varepsilon)$).

As examples of the general scheme, we study the KdV and higher "nonlinear" KdV equations with small dissipation-like perturbations. Physically, the former corresponds to the evolution of a soliton in a slowly varying medium (Johnson (1973)). An interesting feature of the former equation is the appearance of a *shelf* behind the perturbed soliton due to the dissipative perturbation (see also Leibovich and Randall (1973), Kaup and Newell (1978a), Karpman and Maslov (1978)).

3.8.a. KdV with a dissipative perturbation. Let the perturbed KdV equation be of the form

(3.8.6) $$q_t + 6qq_x + q_{xxx} = -\varepsilon \gamma q,$$

with γ, ε constant, $\gamma > 0$, γ order one, $0 < \varepsilon \ll 1$. The soliton solution to the unperturbed equation, $\varepsilon = 0$, can be written

(3.8.7) $$q^{(0)} = 2\eta^2 \operatorname{sech}^2 \eta(\theta - \theta_0), \quad \frac{\partial \theta}{\partial x} = 1, \quad \frac{\partial \theta}{\partial t} = -4\eta^2.$$

(Although we shall present results for (3.8.6), one should consider (3.8.6) only a prototype equation. The analysis applies in much wider generality.) Here η and θ_0 are arbitrary parameters which may depend on a long time scale $T = \varepsilon t$. Under the assumption of quasi-stationarity, (3.8.6) becomes

(3.8.8) $$-4\eta^2 \hat{q}_\theta + 6\hat{q}\hat{q}_\theta + \hat{q}_{\theta\theta\theta} = -\varepsilon \gamma \hat{q} - \varepsilon \hat{q}_T.$$

Expanding $\hat{q}$ in terms of ε, at leading order, we have

(3.8.9) $$-4\eta^1 \hat{q}^{(0)}_\theta + 6q^{(0)}_\theta \hat{q}^{(0)}_\theta + \hat{q}^{(0)}_{\theta\theta\theta} = 0,$$

and we take the solution (KdV soliton)

$$\hat{q}^{(0)} = 2\eta^2 \operatorname{sech}^2 \eta(\theta - \theta^{(0)}).$$

At order ε, we have

(3.8.10)
$$\hat{L}\hat{q}^{(1)} \equiv -4\eta^2 \hat{q}^{(1)}_\theta + 6(\hat{q}^{(0)} \hat{q}^{(1)})_\theta + \hat{q}^{(1)}_{\theta\theta\theta} = \hat{F}^{(1)},$$

$$\hat{F}^{(1)} \equiv -\gamma \hat{q}^{(0)} - \hat{q}^{(0)}_T = -\gamma q^{(0)} - \frac{1}{\eta} \eta_T \{2\hat{q}^{(0)} + (\theta - \theta^{(0)})q^{(0)}_\theta\} + \theta^{(0)}_T q^{(0)}_\theta.$$

From (3.8.8) we find that $\hat{q}^{(0)}$ is a proper solution of the adjoint problem of $\hat{L}u = 0$, i.e.,

(3.8.11) $$\hat{L}^A \hat{q}^{(0)} = 0, \qquad \hat{L}^A = 4\eta^2 \partial_\theta - 6q^{(0)} \partial_\theta - \partial_\theta^3,$$

decaying rapidly as $|\theta| \to \infty$. Then the compatibility condition

(3.8.12) $$\int_{-\infty}^\infty \hat{q}^{(0)} \hat{F}^{(1)} \, d\theta = 0$$

leads to

(3.8.13) $$\frac{1}{\eta} \frac{\partial \eta}{\partial T} = -\frac{2}{3}\gamma \quad \text{or} \quad \eta(T) = \eta(0) \exp\left(-\frac{2}{3} \int_0^T \gamma \, dT'\right).$$

This implies that the amplitude and speed of the soliton are decreasing ($\gamma > 0$) adiabatically simply by dissipation. (3.8.13) is the most important result in this problem. Taking (3.8.13) into account, we may solve (3.8.10) to obtain the solution

(3.8.14) $$\hat{q}^{(1)} = \frac{\gamma}{6\eta}\left[-1 + \tanh\phi + 3\left(1 + \frac{\eta}{\gamma}\theta^{(0)}_T\right)(1 - \phi \tanh\phi)\operatorname{sech}^2\phi \right.$$
$$\left. + \phi(2 - \phi \tanh\phi)\operatorname{sech}^2\phi\right],$$

$$|\phi| \ll O(\varepsilon^{-1/2}),$$

where $\phi = \eta(\theta - \theta_0)$ (See the exercises at the end of this chapter. We note that $\theta^{(0)}_T$ will be found later. Higher order calculations indicate that the expansion breaks down when $|\phi| = O(\varepsilon^{-1/2})$. It should be remarked that this order is related to the breakdown of the expansion in Kaup and Newell (1978a), Karpman and Maslov (1978), for time $t \sim O(\varepsilon^{-1/2})$.) From (3.8.14), one can see that there is a shelf introduced by the dissipation; i.e., asymptotically

(3.8.15) $$\hat{q}^{(1)} \to \begin{cases} -\dfrac{\gamma}{3\eta}\{1 - 2\phi^2 \exp(2\phi)\} & \text{for } 1 \ll -\phi \ll O(\varepsilon^{-1/2}), \\ -\dfrac{2\gamma}{3\eta}\phi^2 \exp(-2\phi) & \text{for } 1 \ll \phi \ll O(\varepsilon^{-1/2}), \end{cases}$$

which agrees with the results via the inverse method. We also notice at this point that the parameter $\theta^{(0)}$ can be taken arbitrary, since the term $\theta_T^{(0)}(1 - \phi \tanh \phi) \operatorname{sech}^2 \phi$ can be absorbed into the leading order solution $\hat{q}^{(0)}$ by shifting η to $\eta - \varepsilon \theta_T^{(0)}/(8\eta)$. It should be noted that the results obtained so far are the most crucial from the point of view of perturbation theory. For those not interested in further details, it is possible to proceed directly to § 3.8.b.

However, for η to be given by certain initial data, one can determine the evolution equation of $\theta^{(0)}$ by the following. Let us consider an initial value problem with the initial value in the form of an unperturbed solitary wave, i.e.,

(3.8.16) $$q(x, 0) = 2\eta^2 \operatorname{sech}^2 \eta x.$$

From (3.8.6) we have the following global relation (rate of change of energy):

(3.8.17) $$\frac{d}{dt} \int_{-\infty}^{\infty} q^2 \, dx = -2\varepsilon\gamma \int_{-\infty}^{\infty} q^2 \, dx.$$

Moreover, let us assume that q takes the form $q_s + \delta q$, where q_s expresses the soliton part, i.e., (3.8.7), and δq the correction to the soliton. Note that in order to use (3.8.17) we will need to use the results of asymptotic analysis far away from the soliton described below. Taking (3.8.13) into account (i.e., at leading order (3.8.17) is equivalent to (3.8.13)), we have

(3.8.18) $$\frac{d}{dt} \Delta(t) = -2\varepsilon\gamma \Delta(t),$$

where $\Delta(t) = \int_{-\infty}^{\infty} \{q_s \delta q + (\delta q)^2/2\} \, dx$. From $\delta q(x, 0) = 0$, we obtain

(3.8.19) $$\Delta(t) = 0.$$

It turns out that the length of the shelf is $O(\varepsilon^{-1})$ for $t \sim O(\varepsilon^{-1})$. Hence for these times the order of the second term in $\Delta(t)$ is the same as the first one. We argue that even though for short times the problem is not stationary, the nonstationary portion of the wave quickly moves to the tail of the soliton. For times $t \sim O(\varepsilon^{-1})$ the region near the soliton, $|\theta| \ll O(\varepsilon^{-1/2})$, is quasi-stationary, and in this region $\delta q = \delta q(\theta, T)$. Hence, in order to determine the evolution equation of the parameter $\theta^{(0)}$, we require the following relation as an additional condition:

(3.8.20) $$\int_{-\infty}^{\infty} \hat{q}^{(0)}(\theta) \hat{q}^{(1)}(\theta) \, d\theta + \frac{1}{2} \int_{-\infty}^{\infty} (\delta q)^2 \, dx = 0.$$

The condition (3.8.20) gives

(3.8.21) $$\frac{\partial \theta^{(0)}}{\partial T} = -\frac{\gamma}{3\eta} - \frac{1}{2\eta} \int_{-\infty}^{\infty} (\delta q)^2 \, dx.$$

Here we notice that for the range of time $O(1) \ll t \ll O(\varepsilon^{-1})$ the second term in (3.8.21) (i.e., $\int_{-\infty}^{\infty} (\delta q)^2 \, dx$ in $\Delta(t)$) can be ignored. For this range of time, (3.8.21) gives the same results as Karpman and Maslov (1978). Also, in

(3.8.20) and (3.8.21) we must use the results of the computation of δq for $-\theta \ll O(\varepsilon^{-1/2})$ which now follows.

In the region $|\theta| \ll O(\varepsilon^{-1/2})$ the solution is quasi-stationary; however, in the other regions the solution depends on x and t strongly. For times $t \sim O(\varepsilon^{-1})$, i.e., $O(\varepsilon^{-1/2}) \ll (-x) < \infty$, behind the soliton, the expansion is nonuniform due to the shelf. Following Knickerbocker and Newell (1980) the equation (3.8.6) is approximated in this region by $q \sim \tilde{q}$, where $\tilde{q}$ satisfies

(3.8.22) $$\tilde{q}_t + \tilde{q}_{xxx} = -\varepsilon \gamma \tilde{q},$$

with the boundary condition

(3.8.23) $$\tilde{q}(x, t) \to \begin{cases} -\dfrac{\varepsilon \gamma}{3\eta(T)} & \text{as } x \to \dfrac{1}{\varepsilon} \int_0^T 4\eta^2(T')\, dT', \\ 0 & \text{as } x \to -\infty. \end{cases}$$

If we take the moving boundary condition into account, a solution to (3.8.22) is given by

(3.8.24) $$\tilde{q}(x, t) = -\frac{\varepsilon \gamma(T_0(\varepsilon x))}{3\eta(T_0(\varepsilon x))} \exp\left(\int_T^{T_0(\varepsilon x)} \gamma\, dT'\right) \int_{-\infty}^z \text{Ai}(y)\, dy,$$

where $\text{Ai}(z)$ is the Airy function, $z = x/(3t)^{1/3}$ and $T_0(\varepsilon x)$ is given by inverting the relation $\varepsilon x = \int_0^{T_0} 4\eta(T')\, dT'$. Note that δq in (3.8.20–21) may be replaced by $\tilde{q}$.

In the region $\theta \gg O(\varepsilon^{-1/2})$, i.e., ahead of the soliton, the expansion is also nonuniform. Again the linear equation (3.8.22) applies, due now to the exponentially small solution. Using a WKB method the solution may be given asymptotically in the form $\bar{q}$ where

$$\bar{q} \equiv 8\eta^2 \exp\left(\frac{\phi(Y, T)}{\varepsilon}\right), \qquad \theta \geq O(\varepsilon^{-1/2}),$$

(3.8.25) $$\phi_T - 4\eta^2 \phi_Y + \phi_Y^3 = 0,$$

$$Y = \varepsilon(\theta - \theta_0), \qquad T = \varepsilon t.$$

Thus the "uniform" solution is obtained from

(3.8.26) $$q(x, t) = \begin{cases} \tilde{q}, & -\theta \gg O(\varepsilon^{-1/2}), \\ \hat{q}^{(0)}(\theta, T) + \varepsilon \hat{q}^{(1)}(\theta, T), & |\theta| \ll O(\varepsilon^{-1/2}), \\ \bar{q}, & \theta \gg O(\varepsilon^{-1/2}). \end{cases}$$

Although we shall not go into any details here, it is instructive to consider the conservation laws. Besides verifying them via the above formulae, one can actually consider the conservation laws as the starting point of the analysis. One must be very careful, but if the work is done correctly, all of these results may be obtained. See, for example, Ott and Sudan (1970) (note that their results must be modified due to the shelf), Kaup and Newell (1978a), Knickerbocker and Newell (1980).

Another example, a dissipatively perturbed NLS equation, is given in the exercises.

3.8.b. Higher nonlinear KdV with a dissipative perturbation. Next we briefly discuss certain perturbed "higher" nonlinear KdV and NLS equations as examples of (presumably) nonintegrable systems. We find that if the order of nonlinearity is sufficiently large then the perturbation method suggests that the perturbed solitary wave is undergoing a focusing singularity in an analogous manner to that of the higher nonlinear NLS equation (Zakharov and Synakh (1976)).

Let us consider the following "higher" nonlinear KdV equation:

$$q_t + A q^p q_x + q_{xxx} = -\varepsilon \gamma q, \quad p \geq 1 \tag{3.8.27}$$

in which the unperturbed solitary wave may be written

$$q^{(0)} = \alpha \operatorname{sech}^{2/p} \eta(\theta - \theta^{(0)}), \quad \frac{\partial \theta}{\partial x} = 1, \quad \frac{\partial \theta}{\partial t} = -4 \frac{\eta^2}{p^2}, \tag{3.2.28}$$

where α is given by $A\alpha^p = 2(p+1)(p+2)\eta^2/p^2$. By assuming a quasi-stationary solution we have

$$-4\frac{\eta^2}{p^2} \hat{q}_\theta + A\hat{q}^p \hat{q}_\theta + \hat{q}_{\theta\theta\theta} = \varepsilon \hat{F}(\hat{q}),$$
$$F(\hat{q}) = -\gamma \hat{q} - \hat{q}_T, \tag{3.8.29}$$

from which, at order ε, we obtain

$$\hat{L}\hat{q}_1 \equiv -4\frac{\eta^2}{p^2}\hat{q}_\theta^{(1)} + A((\hat{q}^{(0)})^p \hat{q}^{(1)})_\theta + \hat{q}_{\theta\theta\theta}^{(1)} = \hat{F}^{(1)}(\hat{q}^{(0)}),$$
$$\hat{F}^{(1)}(\hat{q}^{(0)}) = -\gamma \hat{q}^{(0)} - \frac{1}{\eta}\frac{\partial \eta}{\partial T}\left\{\frac{2}{p}\hat{q}^{(0)} + (\theta - \theta^{(0)})\hat{q}_\theta^{(0)}\right\} + \frac{\partial \theta^{(0)}}{\partial T}\hat{q}_\theta^{(0)}. \tag{3.8.30}$$

Using the fact that $\hat{L}^A \hat{q}^{(0)} = 0$, the compatibility condition is given by

$$\int_{-\infty}^{\infty} \hat{q}^{(0)} \hat{F}^{(1)}(\hat{q}^{(0)}) \, d\theta = 0, \tag{3.8.31}$$

which leads to

$$\frac{1}{\eta}\frac{\partial \eta}{\partial T} = -\frac{2p}{4-p}\gamma. \tag{3.8.32}$$

From (3.8.30) and (3.8.32), we find that there is a shelf which is given by

$$\hat{q}^{(1)} \to -\frac{p^3 \gamma}{4\eta^2(4-p)}\int_{-\infty}^{\infty} \hat{q}^{(0)} \, d\theta, \quad \text{for } \theta \to -\infty. \tag{3.8.33}$$

This result (found by reduction of order of (3.8.30)) is consistent with KdV. Uniform results (for $p < 4$) can be obtained following the previous ideas. From (3.8.32), one can see that the perturbation scheme breaks down at $p = 4$. This implies that the assumption of quasi-stationarity is not valid for this problem; that is, the effects of the perturbation are not adiabatic. For the case $p \geq 4$, this result suggests that the equation admits a self-focusing singularity. Whereas we have not proven the existence of the singularity, we can show that a similar situation occurs for "higher" NLS equations (where the existence of a singularity is provable).

In this regard, consider

$$(3.8.34) \qquad iq_t + q_{xx} + A|q|^{2p}q = -i\varepsilon\gamma q, \qquad p \geq 2,$$

which has an unperturbed solitary wave of the form

$$(3.8.35) \qquad q_0 = \alpha \operatorname{sech}^{1/p} \eta(\theta - \theta^{(0)}) \exp i(\sigma - \sigma^{(0)}),$$

where $A\alpha^{2p} = (p+1)\eta^2/p^2$. Here, for simplicity, we have taken the solitary wave in the rest frame, i.e.,

$$(3.8.36) \qquad \frac{\partial \theta}{\partial x} = 1, \quad \frac{\partial \theta}{\partial t} = 0, \quad \frac{\partial \sigma}{\partial x} = 0, \quad \frac{\partial \sigma}{\partial t} = \frac{\eta^2}{p^2}.$$

Under the assumption of quasi-stationarity for the solution, $q = \hat{q}(\theta, T, \varepsilon) \exp i(\sigma - \sigma^{(0)})$, we obtain

$$(3.8.37) \qquad \begin{aligned} -(\eta^2/p^2)\hat{q} + \hat{q}_{\theta\theta} + A|\hat{q}|^{2p}\hat{q} &= \varepsilon \hat{F}(\hat{q}), \\ \hat{F}(\hat{q}) &= -i\gamma\hat{q} - i\hat{q}_T - \sigma_T^{(0)}\hat{q}. \end{aligned}$$

At order ε, we have

$$(3.8.38) \qquad -\left(\frac{\eta^2}{p^2}\right)\hat{q}^{(1)} + \hat{q}_{\theta\theta}^{(1)} + (p+1)A(\hat{q}^{(0)})^{2p}\hat{q}^{(1)} + nA(\hat{q}^{(0)})^{2p}\hat{q}^{(1)*} = \hat{F}^{(1)}(\hat{q}^{(0)}),$$

in which, setting $\hat{q}^{(1)} = \hat{\phi}^{(1)} + i\hat{\psi}^{(1)}$, we obtain

$$(3.8.39a) \qquad -\left(\frac{\eta^2}{p^2}\right)\hat{\phi}^{(1)} + \hat{\phi}_{\theta\theta}^{(1)} + A(2p+1)(\hat{q}^{(0)})^{2p}\hat{\phi}^{(1)} = \operatorname{Re}[\hat{F}^{(1)}(\hat{q}^{(0)})],$$

$$(3.8.39b) \qquad -\left(\frac{\eta^2}{p^2}\right)\hat{\psi}^{(1)} + \hat{\psi}_{\theta\theta}^{(1)} + A(\hat{q}^{(0)})^{2p}\hat{\psi}^{(1)} = \operatorname{Im}[\hat{F}^{(1)}(\hat{q}^{(0)})].$$

The compatibility condition

$$(3.8.40) \qquad \int_{-\infty}^{\infty} q^{(0)} \operatorname{Im}[\hat{F}^{(1)}(\hat{q}^{(0)})] \, d\theta = 0$$

gives

$$(3.8.41) \qquad \frac{1}{\eta}\frac{\partial \eta}{\partial T} = -\frac{2p}{2-p}\gamma,$$

At $p = 2$, the perturbation scheme breaks down! Thus, if the degree of nonlinearity is greater than or equal to 5, the effects of the perturbation change the solitary wave drastically. However, this effect is not really due to the perturbation; rather it is inherent in the equation itself. By using the conservation laws Zakharov and Synakh (1976) proved that (3.8.34) admits a self-focusing singularity, as we now show.

Consider the evolution of the following quantity (moment of inertia):

$$J = \int_{-\infty}^{\infty} x^2 |q|^2 \, dx. \tag{3.8.42}$$

From (3.8.34) we obtain

$$\frac{d^2 J}{dt^2} + 4\varepsilon\gamma \frac{dJ}{dt} + 4\varepsilon^2 \gamma^2 J = 8 \int \left\{ |q_x|^2 - \frac{pA}{2(p+1)} |q|^{2p+2} \right\} dx. \tag{3.8.43}$$

For $p = 2$, this becomes

$$\frac{d^2 J}{dt^2} = 8 I_3 + O(\varepsilon), \tag{3.8.44}$$

where I_3 is one of the conserved quantities (if $\gamma = 0$)

$$I_3 = \int \left\{ |q_x|^2 - \frac{A}{3} |q|^6 \right\} dx. \tag{3.8.45}$$

This implies that if $I_3 < 0$, J goes to zero in a finite time, and the equation has a focusing singularity. No such argument has yet been given for (3.8.27) for $p \geq 4$.

We also note that in a similar manner it may be shown that the two-dimensional NLS equation

$$iA_t + A_{xx} + A_{yy} + 2A^2 A^* = 0 \tag{3.8.46}$$

has a focusing singularity. Specifically, (3.8.46) has the conserved quantities

$$I_1 = \iint |A|^2 \, dx \, dy,$$

$$I_2 = \iint (|A_x|^2 + |A_y|^2 - |A|^4) \, dx \, dy. \tag{3.8.47}$$

By direct calculation we may verify (Zakharov and Synakh (1976), see also Talanov (1965)) that

$$\frac{\partial^2}{\partial t^2} \iint (x^2 + y^2) |A|^2 \, dx \, dy = 8 I_2, \tag{3.8.48}$$

whereupon, if the initial values are such that $I_2 < 0$, then we have a focusing singularity.

3.8.c. Transverse stability of a soliton for the Kadomtsev–Petviashvili equation.
Once we have a mode such as a soliton or solitary wave, another natural question to ask is whether such solutions are stable. Ordinary stability (not transverse stability) of solitons is not in doubt when we have IST. However, it does turn out that solitons are often unstable to transverse perturbations. We shall consider the Kadomtsev–Petviashvili (K–P) equation (i.e., a two-dimensional KdV equation) as an example).

Consider the K–P equation

$$\partial_x(u_t + 6uu_x + u_{xxx}) = -3\beta^2 \sigma u_{yy}, \tag{3.49}$$

with $\sigma = \pm 1$. We shall assume long wave perturbations in the y-direction, i.e., $|\beta| \ll 1$.

The unperturbed equation and solution dictate the variables we shall use:

$$N(u^{(0)}) = \partial_x(u_t^{(0)} + 6u^{(0)}u_x^{(0)} + u_{xxx}^{(0)}) = 0. \tag{3.8.50a}$$

A soliton solution is given by

$$\begin{aligned} u_0 &= 2\eta^2 \operatorname{sech}^2 \eta(\theta - \theta^{(0)}), \\ \theta &= x - 4\eta^2 t, \quad \theta^{(0)} = \theta^{(0)}(T, y), \quad \eta = \text{const.} \end{aligned} \tag{3.8.50b}$$

Employing the usual multiple scaling ideas and using θ, T, y transforms the K–P equation (3.8.49) to

$$\partial_\theta(-4\eta^2 u_\theta + 6uu_\theta + u_{\theta\theta\theta}) = -u_{\theta T} - 3\beta^2 \sigma u_{yy}. \tag{3.8.51}$$

Expanding $u = u^{(0)} + \beta u^{(1)} + \beta^2 u^{(2)} + \cdots$ and equating powers of β yields a sequence of problems to be solved:

$$N(u^{(0)}) = \partial_\theta(-4\eta^2 u_\theta^{(0)} + 6u^{(0)}u_\theta^{(0)} + u_{\theta\theta\theta}^{(0)}) = 0, \tag{3.8.52a}$$

$$\partial_\theta L(u^{(n)}) = F^{(n)}, \tag{3.8.52b}$$

with

$$\partial_\theta L(u^{(n)}) \equiv \partial_\theta(-4\eta^2 u_\theta^{(n)} + 6(u^{(0)}u^{(n)})_\theta + u_{\theta\theta\theta}^{(n)}). \tag{3.8.52c}$$

It is somewhat more convenient to work with an integrated form of (3.8.52b),

$$Lu^{(n)} = -4\eta^2 u_\theta^{(n)} + 6(u^{(0)}u^{(n)})_\theta + u_{\theta\theta\theta}^{(n)} = \mathscr{F}^{(n)} = \int^\theta F^{(n)} d\theta'; \tag{3.8.53}$$

the adjoint problem to L in (3.8.53) is

$$L^A v = 4\eta^2 v_\theta - 6u^{(0)}v_\theta - v_{\theta\theta\theta}. \tag{3.8.54}$$

Clearly $v = u^{(0)}$ solves $L^A v = 0$ (this is a "proper" homogeneous solution). An application of Green's identity, $vLu^{(n)} - u^{(n)}L^A v = \mathcal{F}^{(n)}$, shows that (by integration)

$$\int_{-\infty}^{\infty} \mathcal{F}^{(n)} u^{(0)} \, d\theta = 0 \tag{3.8.55}$$

must be satisfied to have a bounded solution $u^{(n)}$.

When $n = 1$, we have $\mathcal{F}^{(1)} = \theta_T^{(0)} u_\theta^{(0)}$ and (3.8.55) is automatically satisfied. A forced solution $u^{(1)}$ is given by

$$u^{(1)} = \frac{1}{8\eta^2} \theta_T^{(0)} (2u^{(0)} + \theta u_\theta^{(0)}), \tag{3.8.56}$$

homogeneous solutions having been absorbed into $u^{(0)}$. When $n = 2$, we have

$$\mathcal{F}^{(2)} = -\partial_\theta (6u^{(1)}) u_\theta^{(1)} - \partial_\theta u_T^{(1)} - 3\sigma \int^\theta u_{yy}^{(0)} \, d\theta'. \tag{3.8.57}$$

The only terms remaining after applying (3.8.55) are those in $\mathcal{F}^{(n)}$ which are even in $(\theta - \theta^{(0)})$:

$$-\frac{1}{8\eta^2} \theta_{TT}^{(0)} \int_{-\infty}^{\infty} (2(u^{(0)})^2 + \theta u_\theta^{(0)} u^{(0)}) \, d\theta + 3\sigma \theta_{yy}^{(0)} \int_{-\infty}^{\infty} (u^{(0)})^2 \, d\theta = 0 \tag{3.8.58a}$$

or

$$\theta_{TT}^{(0)} - 48\sigma \eta^2 \theta_{yy}^{(0)} = 0. \tag{3.8.58b}$$

Thus, when $\sigma = +1$ we have that the soliton is unstable to transverse perturbations; otherwise, the soliton appears to be neutrally stable (although we note this has yet to be proven). Physically, in water waves, the case of instability occurs when we have sufficient surface tension (see § 4.1).

In the exercises the example of transverse instability of solitons in the two-dimensional NLS equation is discussed. We note that in this problem the one-dimensional soliton is always unstable to transverse perturbations.

It should be remarked that:

(i) In most of these cases either a multiple scale or a more standard stability analysis (e.g., the method or the method of Zakharov and Rubenchik (1974)) can be used to obtain the result (see the exercises).

(ii) These ideas apply also to multidimensional SIT (see § 4.4 and Ablowitz and Kodama (1979)), and to two-dimensional water wave packets in finite depth. (Again the solitons are unstable to transverse perturbations; see Ablowitz and Segur (1979).) Here, too, the soliton as well as the breather mode is unstable to long transverse perturbations.

EXERCISES

Section 3.1

1. Let $C(k^2/4)$ be an entire function of k^2. Show that (3.1.1) maps each higher order mKdV

$$v_t + C\left(-\frac{1}{4}\frac{\partial^2}{\partial x^2} - v^2 + v_x \int_x^\infty dy\, v\right) v_x = 0,$$

into the corresponding higher order KdV

$$u_t + C\left(-\frac{1}{4}\frac{\partial^2}{\partial x^2} - u + \frac{1}{2}u_x \int_x^\infty dy\right) u_x = 0,$$

where $C(k^2/4)$ is the linearized phase speed in each case. (This can be done in a variety of ways. An elegant one is due to A. Ramani, who uses the linear integral equations (1.3.27) and (1.3.37), and identifies $(K_1 + \bar{K}_1)$ in (1.3.27) with K in (1.3.37).)

2. Clearly, one may pose any $D(u) = 0$, then substituting in (3.1.2) determines $E(\theta) = 0$, which maps into $D(u) = 0$ by (3.1.2). Conversely, suppose $E(\theta) = 0$ is linear with a given dispersion relation, $\omega(k)$, i.e., a generalization of (3.1.3). What is $D(u) = 0$?

3. Consider

$$\left(\frac{u+v}{2}\right)_x = \lambda,$$

$$\left(\frac{u+v}{2}\right)_t = -\left(\frac{u-v}{2}\right)_{xx}\lambda' - \frac{\alpha}{4}\lambda^4 + \frac{1}{2}\left[\left(\frac{u-v}{2}\right)_x\right]^2 \lambda'',$$

where α is constant and $\lambda = \lambda((u-v)/2)$ satisfies

$$\lambda''' + 2\alpha\lambda\lambda' = 0.$$

Rund (1976) proposed these relations as a BT from

$$u_t + \frac{\alpha}{4}(u_x)^4 + u_{xxx} = 0$$

to itself. The latter equation is of interest because if we set $u_x = \phi$, then

$$\phi_t + \alpha\phi^3\phi_x + \phi_{xxx} = 0,$$

which has no solitons and only three polynomial conservation laws. Show that the proposed BT is not a BT.

4. Let

$$V_x = c_1 e^{\beta(u)} + u_x[\beta'(u)V + \alpha(u)],$$
$$V_t = c_2 e^{\beta(u)} + u_t[\beta'(u)V + \alpha(u)],$$

where α, β are arbitrary functions and c_1, c_2 are constant.

(a) Show that this is not a BT.
(b) Show that it is a point transformation; i.e., $v = V(x, t, u(x, t))$. Find V explicitly.

5. (a) The scattering problem for the sine-Gordon equation is (from 1.2.17 and 1.2.18)

$$v_{1x} - i\zeta v_1 = qv_2, \qquad v_{1t} = \frac{a}{\zeta}v_1 + \frac{b}{\zeta}v_2,$$

$$v_{2x} - i\zeta v_2 = rv_1, \qquad v_{2t} = \frac{c}{\zeta}v_1 - \frac{a}{\zeta}v_2.$$

Find $D(\mathbf{u}) = 0$ and $E(\mathbf{v}) = 0$. What is the relation between $D(\mathbf{u}) = 0$ and the sine-Gordon equation?

(b) Define $V = v_1/v_2$, $U = q = -r$. Show that the scattering problem also gives a BT between $D(U) = 0$ and $E(V) = 0$. Derive (3.1.7) from this BT.

6. Zakharov (1974) writes the Boussinesq equation as

$$u_t = A_x, \qquad A_t = (u + u^2 + \tfrac{1}{4}u_{xx})_x.$$

The appropriate scattering problem is

$$\psi_{xxx} + (u_x + i(\tfrac{4}{3})^{1/2}A)\psi + (1 + 2u)\psi_x = \lambda\psi,$$

$$i(\tfrac{4}{3})^{1/2}\psi_t = \psi_{xx} + \tfrac{4}{3}u\psi.$$

Show that this a Bäcklund transformation.

7. Show that the scattering problem for the three-wave interaction (§ 2.1) is a Bäcklund transformation.

8. (a) Find a Bäcklund transformation between the fifth order KdV and fifth order mKdV equations.
(b) Find a BT between the fifth order KdV equation and itself.

Section 3.2

1. Show that the BT between KdV and itself (3.1.19, 20) may be put in the form of (3.2.5) with $N = 1$. Thus, this BT is also a pseudopotential.

2. (a) Show that

$$v = -2\frac{u_x}{u + \lambda} + u$$

is a BT from (3.2.6) to itself, with an arbitrary parameter (λ).

(b) Find the general traveling wave solution of (3.2.6) by assuming $u = u(x - ct)$.

(c) Use this traveling wave solution in the BT to find a second exact solution of (3.2.6). Does it represent two solitons in any sense?

(d) This BT may be deduced from the Cole–Hopf transformation (3.1.2) as follows. (i) If θ satisfies the heat equation and u is related to θ through (3.1.2), then u satisfies (3.2.6). (ii) $\psi = \theta_x$ also satisfies the heat equation. (iii) If V is related to ψ through (3.1.2), then V also satisfies (3.2.6). (iv) The relation between V and u is in (a) (M. Kruskal, private communication).

3. Find a matrix representation of a solution of (3.2.19). Construct the linear pseudopotential. Is it a BT?

4. (a) Show that (3.2.20) has a pseudopotential if and only if (3.2.21) has a nontrivial solution. Show that the pseudopoential is trivial if $\alpha = \gamma = 0$.
(b) Show that the only solution of (3.2.21) with $N = 1$ has $\alpha = \gamma = 0$.
(c) Equation (3.2.20) has no apparent conservation laws. To what does an Abelian solution of (3.2.21) correspond?

5. Suppose that in (3.2.30) $\underline{\alpha}_1$ has the Jordan canonical form

$$\underline{\alpha}_1 = \begin{pmatrix} \lambda & 1 & 0 & & \\ 0 & \lambda & 1 & 0 & \cdots \\ 0 & 0 & \lambda & 1 & \cdots \\ & & \cdots & & \\ & & & 0 & \lambda \end{pmatrix}$$

and only one eigenvector, v_n. Let $(v_1, \cdots, v_n)$ be an orthonormal basis such that

$$\underline{\alpha}_1 v_n = \lambda v_n,$$
$$\underline{\alpha}_1 v_{n-1} = \lambda v_{n-1} + v_n,$$
$$\underline{\alpha}_1 v_{n-2} = \lambda v_{n-2} + v_{n-1}, \quad \text{etc.}$$

(a) Show that $(v_m, v_{m+1}, \cdots, v_n)$ spans an invariant subspace of $\underline{\alpha}_1$, for each $m = 1, \cdots, n$. Show that two commuting matrices have the same invariant subspaces.

(b) Show that v_n, the eigenvector of $\underline{\alpha}_1$, corresponds to a conservation law involving $(u, u_x, \cdots, u_{(p-1)x})$, that v_{n-1} corresponds to a conservation law involving $(v_n, u, u_x, \cdots, u_{(p-1)x})$, etc.

(c) Show that every Abelian Lie algebra corresponds to a set of conservation laws for the original problem, and that these may or may not be trivial.

6. Show that

$$u_t + u_{xxx} + f(u)u_x = 0$$

has a pseudopotential depending on (u, u_x, u_{xx}) if and only if

$$f(u) = c_0 + c_1 u + c_2 u^2$$

(Wahlquist and Estabrook (1975), (1976)).

7. Prove that (3.2.21) has no non-Abelian solutions. Taking α, β, γ, δ as $N \times N$ constant matrices, choose a coordinate system for q that puts α in Jordan canonical form.

(a) Suppose α is diagonal. Show from (3.2.21a) that $\alpha \equiv 0$. Show that the resulting Lie algebra is Abelian.

(b) Therefore assume that α cannot be diagonalized. Suppose α has only one eigenvalue (λ), and one eigenvector. Compute the trace of (3.2.21a) and show that $\lambda = 0$. Hence

$$\alpha_{i,i+i} = 1, \qquad \alpha_{ij} = 0 \quad \text{otherwise.}$$

Show from (3.2.21a) that

$$\gamma_{i,j} = 0 \quad \text{if } i > j,$$
$$\gamma_{i,i} = i + M \quad \text{for some } M, \quad i = 1, \cdots, N.$$

Compute the trace of (3.2.21d) and show that

$$M = -\frac{N+1}{2}.$$

Show from (3.2.21d) that

$$\beta_{i+1,i} = \frac{i(N-i)}{2} > 0, \qquad i = 1, \cdots, N-1.$$

In (3.2.21b) call $\mathbf{Q} = [\alpha, \delta] + [\beta, \gamma] - \alpha$. Compute $\sum_{i=1}^{N-1} (\mathbf{Q})_{i+1,i}$ in (3.2.21b) and show that there is a contradiction for $N > 1$.

(c) The only alternative is that α has more than one but less than N eigenvectors. Thus, α has two or more Jordan blocks along its diagonal, with zero elsewhere. These may be arranged so that the upper left block is the largest. Compute the partial trace of (3.2.21a) *within* each block, and show that *every* eigenvalue of α must vanish. γ may be partitioned into the same size blocks as α, but its off-diagonal blocks (which need not be square) need not vanish. Show from (3.2.21a) that the diagonal blocks of γ have the same form as in (b), and that each off-diagonal block is upper triangular. The rest of the proof parallels that in (b), except that one must also show from (3.2.21d) that each off-diagonal block of γ has only zeros along its own diagonal. Again define $\mathbf{R} = [\alpha, \beta] - \gamma$ from (3.2.21d). Then $\sum_i (\mathbf{R})_{i+1,i}$ in (3.2.21d) leads to a contradiction unless $N = 1$, as before.

(d) Conclude that (3.2.21) has no solution at all unless α is diagonalizable, and in that case (3.2.21) has only an Abelian solution.

8. Consider

$$u_t + u^2 u_x - u_{xxx} = u_{xx}.$$

(a) Show that the equation has bounded traveling wave solutions ($u = f(x - ct)$) on $-\infty < x < \infty$, but that none of these are periodic. Sketch the aperiodic solutions.

(b) A more delicate question is whether the equation has bounded traveling wave solutions that do not oscillate.

(c) Corones and Testa (1976) showed that the equation has a non-Abelian pseudopotential. They did not determine whether the pseudopotential yields a BT or a scattering problem, or whether the solutions in (a) interact like solitons, or whether the problem can be solved exactly. Does it have a pseudopotential?

Section 3.3

1. Investigate the phase shift of the two-soliton solution for the mKdV equation. Do the same for the sine-Gordon equation and the nonlinear Schrödinger equation.

2. Use the Bäcklund transformation in bilinear form (3.3.72) to generate a two-soliton (3.3.76) solution from a one-soliton solution (3.3.74).

3. Use the soliton permutation formula and the first few soliton solutions to evaluate the arbitrary constant C in (3.3.86).

4. Use (3.3.87) to evaluate the eigenfunctions ψ from the first two soliton solutions.

Section 3.4

1. Show that (3.4.25) along with either the KdV equation (3.4.1) or the limiting form of the Bäcklund transformation (3.4.26) yield the same rational solutions.

2. What is the difference between the rational solutions of the self-similar form of the KdV equation obtained by setting $u = (3t)^{-2/3} w(z)$, $z = x/(3t)^{1/3}$ and the full class of rational solutions of the KdV equation.

3. Obtain the next rational solution for the Boussinesq equation (3.4.34) (corresponding to F_3) and to the mKdV equation (3.4.43) with nonzero background (corresponding to F_3, G_3).

Section 3.5

1. Investigate the structure of a two-soliton solution to (3.5.1). What is the phase shift of interaction? What would it be for an N-soliton solution?

2. Show how to obtain (3.5.38) from (3.5.37).

3. Verify that (3.5.39) indeed yields (3.5.41) as $\delta \to 0$ and (3.5.42) as $\delta \to \infty$.

4. Given (3.5.48), suppose we take $k_1 \to 0$. Show that unless $\delta k \sim \pi$ we have a trivial result.

5. In what sense is (3.5.63) a Riemann–Hilbert problem? How may it be viewed as a scattering problem?

Section 3.6

1. In what sense is the result that (3.6.8) with $L_i F = 0$, L_1, L_2 given in (3.6.12), (3.6.16) more general than the inverse scattering results of Chapter 1? In what sense do we get less information?

2. Show that the "factorized" equation (3.6.51) yields the Gel'fand–Levitan equations (3.6.53).

3. Let $\hat{M}_1, \hat{M}_2$ be given by (3.6.61). Show that $[\hat{M}_1, \hat{M}_2] = 0$ indeed yields the Kadomtsev–Petviashvili equations (3.6.63).

4. When does the resonant solution described by Miles (1977b) arise in the soliton representation given in (3.6.68)?

Section 3.7

1. (a) Show that $f = z/2$ is an exact solution of (3.7.1c). What is the corresponding solution of (3.7.1a)?

(b) Set $f(z) = z/2 + g(z)$. What equation does $g(z)$ satisfy? Show that g itself is an integrating factor, and integrate the equation once. The second order ODE is equation XXXIV on Ince's list (1956, Chapt. 14). Show that his "solution" of this equation is Miura's (1968) transformation.

2. A "partial proof" of the Painlevé conjecture is sketched below (3.7.5a). In what sense is this proof partial?

3. Burgers' equation,

$$u_t + u u_x = u_{xx},$$

is linearized exactly by the Cole–Hopf transformation (cf. § 3.1, 3.2). Show that it has both a traveling wave solution and a similarity solution. Show that in either case, the ODE can be integrated once, and the first-order ODE is a generalized Riccati equation (3.7.3), and therefore is of P-type.

4. Fisher's equation,

$$u_t = u - u^2 + u_{xx},$$

has no pseudopotential (cf. § 3.2). What ODE is satisfied by its traveling wave solution, $u = u(x - ct)$? Show that it is not of P-type unless $c^2 = \frac{25}{6}$.

For these special values of c, Ablowitz and Zeppetella (1979) integrated the ODE explicitly, and found what are apparently the only solutions of the equation known in closed form.

5. Generalized KdV equations are

$$u_t + u^n u_x + u_{xxx} = 0.$$

What family of ODE's do the traveling waves satisfy? For what n are these of P-type? (Compare this result with that in Exercise 6, § 3.2.)

6. (a) What is the self-similar form of (3.7.10)? Show that this ODE has a movable logarithmic branch point. This shows that the linear integral equation in IST cannot give *directly* the solution of the sine-Gordon equation.

(b) Show that the movable branch point disappears either by differentiation or by the exponential transformation below (3.7.11). Recall from (1.2.18) that the IST formulation actually gives u_x, rather than u. This example shows that if the movable critical point is simple enough, it can be removed by a suitable transformation.

7. There is strong evidence that the "double sine-Gordon equation,"

$$u_{xt} = \sin u + \lambda \sin 2u$$

cannot be solved by IST (cf. § 4.4). Find its similarity solution. The ODE has movable logarithmic branch points, just as the sine-Gordon equation does. Show that after an exponential transformation, the ODE still has movable critical points.

8. Mikhailov (1980) showed that

$$u_{xt} = a\, e^u + b\, e^{-2u}$$

could be solved by IST. Show that the self-similar form of the equation is not compatible with (3.7.4), but can be put in this form by an exponential transformation. Show that this equation is of P-type. (This ODE is essentially P_{III}, but is *not* on Ince's list! This shows the danger of using that list superficially. The equations on it are "canonical," and some other equations equivalent to them are not mentioned. Therefore we recommend actually finding the singular point structure for ODE's not on the list.)

9. Zakharov (1981) has reported numerical experiments that suggest that

$$u_{xx} - u_{tt} = u^3$$

may have special properties. Find its self-similar solutions. Show that the ODE is not of P-type.

10. The "Boussinesq equations" (note the plural)

$$h_t + (uh)_x = 0,$$

$$u_t + uu_x + gh_x + \frac{H}{3} h_{xtt} = 0$$

describe the evolution of long water waves that travel in two directions (cf. Whithem (1974). Find the traveling wave solutions, and eliminate u. Show that the equation for h is not of P-type.

11. Consider
$$i\phi_t + \phi_{xx} - \phi_{yy} - 2|\phi|^2\phi = 0,$$
which governs the evolution of packets of deep water waves without surface tension (cf., § 4.3). Show that this equation has an exact reduction to (3.7.26). Can it be solved by IST?

12. The Kadomtsev–Petviashvili equation,
$$(u_t + 6uu_x + u_{xxx})_x + 3\sigma u_{yy} = 0, \qquad \sigma = \pm 1,$$
can be solved by IST (cf. § 2.1). Show the following:
   (a) The y-independent solutions satisfy KdV.
   (b) The traveling waves $u(x - ct, y)$ satisfy the Boussinesq equation.
   (c) By setting $c = 0$, this gives the steady solutions.
   (d) A similarity variable is
$$z = (3t)^{-1/3}x + \sigma(3t)^{-4/3}y^2.$$
Show that the ODE obtained for $F(z)$ can be integrated once. Multiply by F, and show that (F^2) satisfies P_{II} (Redekopp (1980)).
   (e) A "modified K–P" equation is
$$(v_t - 6v^2 v_x + v_{xxx})_x + \sigma v_{yy} = 0,$$
which may be thought of as a $(2+1)$-dimensional analogue of the mKdV equation (cf. § 4.1). What ODE is satisfied by the self-similar solutions of the time-independent problem? Show that this ODE is *not* of P-type.

13. Here we show that the Painlevé conjecture must be modified from its original formulation by Ablowitz, Ramani and Segur (1978). The Benjamin–Ono equation
$$u_t + uu_x + H(u_{xx}) = 0,$$
where H is the Hilbert transform, seems to be solvable by IST (cf. § 3.5, 4.1). A reduction to a system of ODE's is obtained by finding its rational solutions, as discussed in § 3.5. In this way we obtain Calogero–Moser systems such as
$$\ddot{x}_1 = \frac{2}{(x_1 - x_2)^3}, \qquad \ddot{x}_2 = -\frac{2}{(x_1 - x_2)^3}.$$
Alternatively we may view these ODE's as nonlinear evolution equations solvable by a variant of IST. However, the solution of Calogero–Moser systems does not rely on a linear integral equation like (3.7.5a).
   (a) Show that these ODE's have movable algebraic branch points; i.e., as $t \to t_0$,
$$x_1 \sim (-1)^{1/4}(t-t_0)^{1/2}, \qquad x_2 \sim -(-1)^{1/4}(t-t_0)^{1/2}.$$
   (b) Obtain the same result from the general solution of these ODE's given in § 3.5. This shows the importance of an exact reduction to an ODE in a variable related to the inherent linear integral equation.

(c) The difficulty here cannot be blamed on the fact that the B–O equation is not a PDE. Show that (3.5.36), the ODE's obtained by seeking rational solutions of the KdV equation, also exhibit movable algebraic branch points.

(d) More trivially, the reduction of (3.7.1a) to (3.7.1c) may be written as

$$u(x,t) = g(t)f(z), \qquad z = \frac{x}{(3t)^{1/3}}.$$

The ODE for $f(z)$ is (3.7.1c), and is of P-type. Show that the ODE for $g(t)$ is not of P-type. Note again that (t) plays no role in the linear integral equation for KdV.

(e) Note, however, that in all the above cases the relevant equations may be transformed to Painlevé type. Such transformations must be allowed in the conjecture.

Section 3.8

1. (a) Show that, if we set $y = \tanh \eta(\theta - \theta^{(0)})$, then (3.8.10) can be written

$$\hat{L}\hat{q}^{(1)} = \frac{d}{dy}(1-y^2)\frac{d}{dy}\hat{q}^{(1)} + \left(12 - \frac{4}{1-y^2}\right)\hat{q}^{(1)} = \hat{F}^{(1)},$$

where $\hat{F}^{(1)}$ is given by

$$\hat{F}^{(1)} = \frac{2\gamma}{3\eta}\frac{1}{1+y} + \frac{2\gamma}{3\eta}\log\frac{1+y}{1-y} + 2\frac{\partial\theta^{(0)}}{\partial T}.$$

(b) Taking into account that $\hat{L}v = 0$ is a Legendre equation, show that $v = P_3^2(y) = 15y(1-y^2)$ is the only proper solution. Then by using the variation of constant method, i.e.,

$$\hat{q}^{(1)}(y) = A(y)P_3^2(y),$$

obtain the equations for $B(y) = dA/dy$:

$$\frac{dB}{dy} + \frac{2(1-4y^2)}{y(1-y^2)}B = G,$$

$$G = \frac{2\gamma}{45\eta}\frac{1-y}{y(1-y^2)^3} + \frac{2\gamma}{45\eta}\frac{1}{y(1-y^2)^2}\log\frac{1+y}{1-y}$$

$$+ \frac{2}{15}\frac{\partial\theta^{(0)}}{\partial T}\frac{1}{y(1-y^2)^2}.$$

(c) Since this is just a first order ordinary differential equation, obtain the solution

$$\hat{q}^{(1)} = \frac{\gamma}{6\eta}\left[-1 + y + \frac{3}{2}\left(1 + \frac{\eta}{\gamma}\frac{\partial\theta^{(0)}}{\partial T}\right)(1-y^2)\left(2 - y\log\frac{1+y}{1-y}\right)\right.$$

$$\left. + (1-y^2)\left(1 - \frac{1}{4}y\log\frac{1+y}{1-y}\right)\log\frac{1+y}{1-y}\right].$$

2. Consider a prototype perturbed NLS equation in the form

(1) $$iq_t + q_{xx} + 2q^2 q^* = -i\varepsilon\gamma q.$$

(a) Show the unperturbed soliton solution is given by

$$q_0 = \eta \, \text{sech}\, \eta(\theta - \theta^{(0)}) \exp[i\xi(\theta - \theta^{(0)}) + i(\sigma - \sigma^{(0)})],$$

where

$$\frac{\partial \theta}{\partial t} = -2\xi, \quad \frac{\partial \theta}{\partial x} = 1,$$

$$\frac{\partial \sigma}{\partial t} = \eta^2 + \xi^2, \quad \frac{\partial \sigma}{\partial x} = 0.$$

Here, ξ, η, $\theta^{(0)}$ and $\sigma^{(0)}$ are arbitrary functions of the long time scale $T = \varepsilon t$.

(b) Assume that the quasi-stationary solution of (3.1.67) takes the form

$$q = \hat{q}(\theta, T, \varepsilon) \exp[i\xi(\theta - \theta^{(0)}) + i(\sigma - \sigma^{(0)})].$$

Substituting this into (1), show

$$-\eta^2 \hat{q} + \hat{q}_{\theta\theta} + 2\hat{q}^2 \hat{q}^* = \varepsilon \hat{F}(q),$$

$$\hat{F}(q) \equiv -i\gamma \hat{q} - i\hat{q}_T + (\theta - \theta^{(0)})\xi_T \hat{q} - (\xi \theta_T^{(0)} + \sigma_T^{(0)})\hat{q}.$$

Assume that $\hat{q}$ can be expanded by

$$\hat{q}(\theta, T, \varepsilon) = \hat{q}^{(0)}(\theta, T) + \varepsilon \hat{q}^{(1)}(\theta, T) + \cdots,$$

where $\hat{q}^{(0)}$ is the leading order solution

$$\hat{q}^{(0)} = \eta \, \text{sech}\, \eta(\theta - \theta^{(0)}).$$

Show that, at order ε, we have

$$-\eta^2 \hat{q}^{(1)} + \hat{q}^{(1)}_{\theta\theta} + 4(\hat{q}^{(0)})^2 \hat{q}^{(1)} + 2(\hat{q}^{(0)})^2 \hat{q}^{(1)*} = \hat{F}^{(1)},$$

where $\hat{F}^{(1)} = \hat{F}(\hat{q}^{(0)})$ and hence setting $\hat{q}^{(1)} = \hat{\phi}^{(1)} + i\hat{\psi}^{(1)}$, where $\hat{\phi}^{(1)}$ and $\hat{\psi}^{(1)}$ are real-valued functions, obtain the system of equations

$$\hat{L}\hat{\phi}^{(1)} \equiv -\eta^2 \hat{\phi}^{(1)} + \hat{\phi}^{(1)}_{\theta\theta} + 6(\hat{q}^{(0)})^2 \hat{\phi}^{(1)} = \text{Re}[\hat{F}^{(1)}],$$

$$\hat{M}\hat{\psi}^{(1)} \equiv -\eta^2 \hat{\psi}^{(1)} + \hat{\psi}^{(1)}_{\theta\theta} + 2(\hat{q}^{(0)})^2 \hat{\psi}^{(1)} = \text{Im}[\hat{F}^{(1)}].$$

(c) By noting that the operators $\hat{L}$ and $\hat{M}$ are self-adjoint and $\hat{L}\hat{q}^{(0)}_\theta = 0$, $\hat{M}\hat{q}^{(0)} = 0$, show that the conditions for solvability are given by the secularity conditions

$$\int_{-\infty}^{\infty} \hat{q}^{(0)}_\theta \, \text{Re}[\hat{F}^{(1)}] \, d\theta = 0,$$

$$\int_{-\infty}^{\infty} \hat{q}^{(0)} \, \text{Im}[\hat{F}^{(1)}] \, d\theta = 0.$$

Use these conditions to obtain the evolution equations

$$\frac{\partial \xi}{\partial T} = 0 \quad \text{and} \quad \frac{1}{\eta}\frac{\partial \eta}{\partial T} = -2\gamma,$$

which show that the amplitude of the soliton is decreasing ($\gamma > 0$) but the velocity is constant. Then obtain the solutions

$$\hat{\phi}^{(1)} = -\frac{1}{2\eta}(\xi\theta_T^{(0)} + \sigma_T^{(0)})\{1 - (\theta - \theta^{(0)}) \tanh \eta(\theta - \theta^{(0)})\} \operatorname{sech} \eta(\theta - \theta^{(0)}),$$

$$\hat{\psi}^{(1)} = \frac{\eta}{2}(\theta - \theta^{(0)})\{\theta_T^{(0)} + \gamma(\theta - \theta^{(0)})\} \operatorname{sech} \eta(\theta - \theta^{(0)})$$

(the expansions must be modified for sufficiently large $|\theta - \theta^{(0)}|$ in a manner similar to those for KdV). Here we have two arbitrary parameters $\theta^{(0)}$ and $\sigma^{(0)}$ which provide the shift of the location and phase of the soliton. For the initial value problem, show that we must take the orthogonality conditions

$$\int_{-\infty}^{\infty} \hat{q}_\theta^{(0)} \hat{\psi}^{(1)} \, d\theta = 0, \quad \int_{-\infty}^{\infty} \hat{q}^{(0)} \hat{\phi}^{(1)} \, d\theta = 0,$$

which give

$$\frac{\partial \theta^{(0)}}{\partial T} = 0, \quad \frac{\partial \sigma^{(0)}}{\partial T} = 0.$$

Show that these conditions may be obtained by using the following relations (modified conserved quantities) in a similar way as in the KdV case:

$$\frac{d}{dt} \int_{-\infty}^{\infty} qq^* \, dx = -2\varepsilon\gamma \int_{-\infty}^{\infty} qq^* \, dx,$$

$$\frac{d}{dt} \int_{-\infty}^{\infty} (q_x q^* - q_x^* q) \, dx = -2\varepsilon\gamma \int_{-\infty}^{\infty} (q_x q^* - q_x^* q) \, dx.$$

3. Consider the multidimensional NLS equation (see Zakharov and Rubenchik (1974))

$$iq_t + q_{xx} + 2|q|^2 q = \varepsilon^2 \alpha q_{yy}.$$

(a) Show that the unperturbed soliton solution is given by

$$q^{(0)} = \hat{q}^{(0)}(x) e^{i\lambda^2(t + \sigma^{(0)})},$$

$$\hat{q}^{(0)}(x) = \lambda \operatorname{sech} \lambda(x + \theta^{(0)}).$$

(b) Let $T = \varepsilon t$, $\sigma^{(0)} = \sigma^{(0)}(T, y)$, $\theta^{(0)} = \theta^{(0)}(T, y)$, and find

$$iq_t + q_{xx} + 2|q|^2 q = \varepsilon^2 \alpha q_{yy} - \varepsilon i q_T.$$

Calling

$$q = \hat{q}(x) e^{i\lambda^2(t + \sigma^{(0)})},$$

show that
$$-\lambda^2 \hat{q} + \hat{q}_{xx} + 2|\hat{q}|^2 \hat{q} = \varepsilon \lambda^2 \sigma_T^{(0)} \hat{q} - i\varepsilon \hat{q}_T$$
$$+ \varepsilon^2 \lambda (\hat{q}_{yy} + 2i\lambda^2 \sigma_y^{(0)} \hat{q}_y + i\lambda^2 \sigma_{yy} \hat{q} - \lambda^4 (\sigma_y^{(0)})^2 \hat{q}),$$

and hence, expanding $\hat{q} = \hat{q}^{(0)} + \varepsilon \hat{q}^{(1)} + \cdots$, show that we have
$$-\lambda^2 \hat{q}^{(0)} + \hat{q}_{xx}^{(0)} + 2|\hat{q}^{(0)}|^2 \hat{q}^{(0)} = 0$$

with the above soliton solution. Show that at $O(\varepsilon)$
$$-\lambda^2 \hat{q}^{(1)} + \hat{q}_{xx}^{(1)} + 4(\hat{q}^{(0)})^2 \hat{q}^{(1)} + 2(\hat{q}^{(0)})^2 \hat{q}^{(1)*} = \lambda^2 \sigma_T^{(0)} \hat{q}^{(0)} - i\hat{q}_T^{(0)}.$$

(c) Split this equation into real and imaginary parts via $\hat{q}^{(1)} = \phi^{(1)} + i\psi^{(1)}$ and find
$$L_0 \psi_1 = -\hat{q}_T^{(0)} = -\theta_T^{(0)} \hat{q}_x^{(0)} = F_0^{(1)},$$
$$L_1 \phi_1 = \lambda^2 \sigma_T^{(0)} \hat{q}^{(0)} = F_1^{(1)},$$

where L_0, L_1 are the self-adjoint operators
$$L_0 \equiv \frac{d^2}{dx^2} + 2(\hat{q}^{(0)})^2 - \lambda^2,$$
$$L_1 \equiv \frac{d^2}{dx^2} + 6(\hat{q}^{(0)})^2 - \lambda^2.$$

(d) Noting that $L_0 \hat{q}_0 = 0$, $L_1 \hat{q}_x^{(0)} = 0$, find the compatibility conditions
$$\int_{-\infty}^{\infty} \hat{q}^{(0)} F_0^{(n)} \, dx = 0,$$
$$\int_{-\infty}^{\infty} \hat{q}_x^{(0)} F_1^{(n)} \, dx = 0,$$

and verify that ψ_1, ϕ_1 are given by
$$\psi_1 = -\frac{\theta_T^{(0)}}{2} x \hat{q}^{(0)},$$
$$\phi_1 = \frac{1}{2} \lambda \sigma_T^{(0)} \frac{\partial}{\partial x} \hat{q}^{(0)}.$$

(e) Show that at $O(\varepsilon^2)$ if $\hat{q}^{(2)} = \phi^{(2)} + i\psi^{(2)}$. Then ϕ_2, ψ_2 satisfy
$$L_0 \psi^{(2)} = \lambda^2 \sigma_T^{(0)} \psi^{(1)} - \phi_T^{(1)} + 2\alpha\lambda^2 \sigma_y^{(0)} \hat{q}_y^{(0)} + \alpha\lambda^2 \sigma_{yy}^{(0)} \hat{q}^{(0)} - 4\phi^{(1)} \psi^{(1)} \hat{q}^{(0)} = F_0^{(2)},$$
$$L_1 \phi^{(2)} = \lambda^2 \sigma_T^{(0)} \phi^{(1)} + \psi^{(1)} + \alpha \hat{q}_{yy}^{(0)} - \alpha\lambda^4 (\sigma_y^{(0)})^2 \hat{q}^{(0)}$$
$$- 2((\phi^{(1)})^2 - (\psi^{(1)})^2) \hat{q}^{(0)} - 4\hat{q}^{(0)}((\phi^{(1)})^2 + (\psi^{(1)})^2) = F_1^{(2)}.$$

(f) Show that the compatibility conditions give

$$\int \hat{q}^{(0)}(-\phi_T^{(1)} + \alpha\lambda^2 \sigma_{yy}^{(0)} \hat{q}^{(0)}) \, dx = 0,$$

$$\sigma_{TT}^{(0)} - 4\alpha\lambda^2 \sigma_{yy}^{(0)} = 0$$

and

$$\int \hat{q}_x^{(0)}(\psi_T^{(1)} + \alpha q_{yy}^{(0)}) \, dx = 0,$$

or

$$\theta_{TT}^{(0)} + \frac{4\lambda^2 \alpha}{3} \theta_{yy}^{(0)} = 0.$$

Hence, regardless of the sign of α, we have that the NLS equation is unstable to transverse perturbations. As we have mentioned it turns out that when $\alpha = -1$ the equation (3.8.46) has a focusing singularity (see, for example, Zakharov and Synakh (1976)).

4. In this example we show that via a slightly different analysis the result in Exercise 3 above may be obtained.

(a) First linearize the equation in Exercise 3 by letting $q = q^{(0)} + q^{(1)}$, $|q^{(0)}| \gg |q^{(1)}|$, and find

$$iq_t^{(1)} + q_{xx}^{(1)} + 4|q^{(0)}|^2 q^{(1)} + 2q^{(0)2} q^{(1)*} = \alpha q_{yy}^{(1)}.$$

(b) Define

$$q^{(0)} = \lambda \operatorname{sech} \lambda x \, e^{i\lambda^2 t} \equiv \hat{q}^{(0)}(x) \, e^{i\lambda^2 t},$$

$$q^{(1)} = (\phi(x) + i\psi(x)) \, e^{i\lambda^2 t} e^{\Omega t} \sin ky,$$

and show that ϕ, ψ satisfy

$$(L_0 + \alpha k^2)\psi = -\Omega\phi,$$

$$(L_1 + \alpha k^2)\phi = \Omega\psi,$$

where L_0, L_1 are given in Exercise 3.

(c) Expand, in a long limit $k \to 0$

$$\Omega = k\Omega^{(1)} + k^2 \Omega^{(2)} + \cdots,$$

$$\phi = \phi^{(0)} + k\phi^{(1)} + \cdots,$$

$$\psi = \psi^{(0)} + k\psi^{(1)} + \cdots.$$

Then we have at $O(1)$:

$$L_0 \psi^{(0)} = 0,$$

$$L_1 \phi^{(0)} = 0,$$

and take the neutrally stable solutions
$$\psi^{(0)} = A^+ \hat{q}^{(0)},$$
$$\phi^{(0)} = A^- \hat{q}_x^{(0)}$$
(+, − refer to the parity of the solutions) at $O(k)$:
$$L_0 \psi^{(1)} = -\Omega^{(1)} \phi^{(0)} = F_0^{(1)},$$
$$L_1 \phi^{(1)} = \Omega^{(1)} \psi^{(0)} = F_1^{(1)}.$$

Show that we have the solutions
$$\psi^{(1)} = -\tfrac{1}{2}\Omega^{(1)} A^- \hat{q}_x^{(0)},$$
$$\phi^{(1)} = \frac{1}{2\lambda} \Omega^{(1)} A^+ \frac{\partial}{\partial \lambda} \hat{q}^{(0)}.$$

(d) Show that the compatibility conditions
$$\int_{-\infty}^{\infty} \hat{q}^{(0)} F_0^{(n)} \, dx = 0,$$
$$\int_{-\infty}^{\infty} \hat{q}_x^{(0)} F_1^{(n)} \, dx = 0,$$
are automatically satisfied when $n = 1$. However show that at $O(k^2)$ we have
$$L_0 \psi^{(2)} = -\Omega^{(2)} \phi^{(1)} - \alpha \psi^{(0)} - \Omega^{(1)} \phi^{(1)} = F_0^{(2)},$$
$$L_1 \phi^{(2)} = \Omega^{(2)} \psi^{(0)} - \alpha \phi^{(0)} + \Omega^{(1)} \psi^{(1)} = F_1^{(2)}.$$

Show that the compatibility conditions now give
$$\alpha \int_{-\infty}^{\infty} \hat{q}^{(0)} \psi^{(0)} \, dx + \Omega^{(1)} \int_{-\infty}^{\infty} \hat{q}^{(0)} \phi^{(1)} \, dx = 0$$
or
$$\Omega^{(1)2} = 4\lambda^2 \alpha,$$
and
$$\alpha \int_{-\infty}^{\infty} \hat{q}_x^{(0)} \phi^{(0)} \, dx - \Omega^{(1)} \int_{-\infty}^{\infty} \hat{q}_x^{(0)} \psi^{(1)} \, dx = 0$$
or
$$\Omega^{(1)2} = \tfrac{4}{3}\alpha \lambda^2,$$
which agrees with the result of Exercise 3.

Chapter 4

Applications

In this chapter we consider certain completely integrable equations as models of physical phenomena. The equations that arise as physical models typically do so in many physical systems, but almost always as a result of the same kind of assumptions. For each of these equations we will concentrate primarily on two or three physical problems in order to demonstrate in some detail the "typical derivation" of that equation. Other applications of the same equation usually can be worked out analogously. A consequence of this approach is that we have omitted some important applications, but a comprehensive survey of all nonlinear phenomena in physics is not possible here. Discussions of other applications may be found in Karpman (1975), Yajima and Kakutani (1975), Makhankov (1978) and Yajima and Ichikawa (1979).

Many of these derivations follow a more or less standard pattern, in which one is interested in relatively small deviations of a certain type from some equilibrium state of a physical system. To leading order, this amounts simply to linearizing the problem about the equilibrium state. One hopes to develop an asymptotic expansion in this way, but usually secular terms arise when the expansion is carried to some higher order. The method of multiple time scales (cf. Cole (1968)) is then used to suppress secular terms, in order to extend the range of validity of the expansion. This is the most common physical meaning of the nonlinear evolution equations that we have been studying: they are the consequence of suppressing secular terms in a formally asymptotic expansion.

The fact that these equations play this particular role has some consequences that are useful in relating soliton theories to physical phenomena.

(i) The evolution equations are fully nonlinear, and their solutions are not necessarily small. Even so, solutions that are order one represent relatively small deviations from the equilibrium state in the physical problem.

(ii) What one gains ordinarily by solving these nonlinear evolution equations is not primarily the ability to discuss large deviations from the equilibrium state, but rather the ability to discuss relatively small deviations over a long time scale.

(iii) The nonlinear evolution equations covers the next time scale beyond that of the appropriate linear problem. In interpreting observations of physical phenomena, there is no need to mention solitons unless the time scale of the observations exceeds that of the linear problem.

(iv) The solutions of these completely integrable equations exhibit no stochastic behavior. The corresponding conclusion about the physical system is restricted to relatively small deviations from equilibrium, and to the appropriate time scales. Stochastic behavior of the same physical system may still occur for large deviations from equilibrium, or over longer time scales.

One other general comment is germane before we discuss specific examples. The world we inhabit has 3 (space) + 1 (time) dimensions, whereas most of the equations we have been studying have $(1+1)$ dimensions. In what sense can these equations model physical phenomena? Depending on the particular application, one may find that the linearized problem admits wave propagation in $(1+1)$, $(2+1)$ or $(3+1)$ dimensions. There is no difficulty in the first case; the nonlinear evolution equation describes a phenomenon that is intrinsically $(1+1)$-dimensional. In the other two cases, these equations arise only after the solutions have been restricted from $(2+1)$ or $(3+1)$ down to $(1+1)$ dimensions. Then two interpretations are possible.

(i) The equation in $(1+1)$ dimensions is a (nontrivial) toy problem, whose solution may provide insight into some of the phenomena that occur in the higher dimensional problem. There is no intention to compare any solutions of the equation directly with physical observations.

(ii) The restriction to $(1+1)$ dimensions has physical meaning. There are realistic circumstances under which the solution of the higher dimensional problem might evolve in this restricted manner, at least approximately. This interpretation is possible only if the $(1+1)$-dimensional solutions are stable with respect to perturbations in the other dimensions. Otherwise this lower dimensional evolution is theoretically possible but is unlikely to occur.

All of these possibilities occur in the applications we will discuss.

4.1. KdV problems and their cousins. The prototype here is the Korteweg–de Vries equation

(4.1.1) $$u_t + 6uu_x + u_{xxx} = 0,$$

but several other exactly solvable equations arise under variations of the simplifying assumptions. These include the modified KdV equation

(4.1.2) $$u_t + 6\sigma u^2 u_x + u_{xxx} = 0, \quad \sigma = \pm 1,$$

the Benjamin–Ono equation

(4.1.3) $$u_t + 2uu_x + H[u_{xx}] = 0,$$

where H is the Hilbert transform,
$$H[f] = \frac{1}{\pi} \int_{-\infty}^{\infty} \frac{f(y)\,dy}{y-x},$$
and the Kadomtsev–Petviashvili equation

(4.1.4) $$(u_t + 6uu_x + \sigma u_{xxx})_x + u_{yy} = 0.$$

Each of these equations arises as the condition required to eliminate secular terms in a formally asymptotic, small amplitude expansion. What distinguishes these equations from those in §§ 4.2 and 4.3 is that these equations arise in problems that are nondispersive at leading order, so that the lowest order approximation is the wave equation

$$\phi_{\tau\tau} = c^2 \phi_{xx}.$$

Ordinarily this occurs in the long wave limit, but there are exceptions.

The general nature of the derivation we now present was recognized by several people, including Benney (1966), Gardner and Su (1969), Karpman (1975). The method was formalized by Taniuti, Wei, et al. (e.g., (1968)) and is called the reductive perturbation method by them. A variety of applications beyond those discussed here may be found in the references cited in Karpman (1975), Yajima and Kakutani (1975), Makhankov (1978) and Yajima and Ichikawa (1979).

4.1.a. Water waves. The problem of long water waves dates back to the early experimental work of Russell (1838), (1844) and the (conflicting) theory of Airy (1845). In fact, the original objective of Korteweg and de Vries (1895) was to provide an alternative theory to Airy's that was in closer agreement with the observations of Russell. Extensive expositions of this subject have been given by Lamb (1932), Stoker (1957), Wehausen and Laitone (1960), Whitham (1974) and Miles (1980), but from slightly different perspectives from that presented here.

The classical problem of water waves is to find the irrotational motion of an inviscid, incompressible, homogeneous fluid with density ρ, subject to a constant gravitational force g. The fluid rests on a horizontal and impermeable bed of infinite extent at $z = -h$, and has a free surface at $z = \zeta(x, y, t)$, where there is a finite surface tension T. The surface tension is unimportant for many applications, and one may set $T = 0$ without essential loss. We carry it here because its inclusion is important in some circumstances.

The fluid has a velocity potential ϕ which satisfies

(4.1.5) $$\nabla^2 \phi = 0, \quad -h < z < \zeta(x, y, t)$$

(irrotational motion of an incompressible fluid). It is subject to boundary conditions on the bottom ($z = -h$)

(4.1.6) $$\phi_z = 0 \quad \text{(impermeable bed)},$$

and along the free surface $(z = \zeta)$

(4.1.7a) $\quad \dfrac{D\zeta}{Dt} \equiv \zeta_t + \phi_x \zeta_x + \phi_y \zeta_y = \phi_z \quad$ (kinematic condition),

(4.1.7b) $\quad \phi_t + g\zeta + \dfrac{1}{2}|\nabla \phi|^2 = \dfrac{T}{\rho} \dfrac{\zeta_{xx}(1+\zeta_y^2) + \zeta_{yy}(1+\zeta_x^2) - 2\zeta_{xy}\zeta_x\zeta_y}{(1+\zeta_x^2 + \zeta_y^2)^{3/2}}$

(dynamic condition).

Boundary conditions in (x, y) and initial conditions are also required. If the waves in question are isolated, then $|\nabla \phi|$ and ζ should vanish as $(x^2 + y^2) \to \infty$. In other problems, periodic boundary conditions in x and in y may be relevant.

We may linearize (4.1.5–7) about $|\nabla \phi| = 0$, $\zeta = 0$, and seek solutions of the linearized equations proportional to $\exp\{i(kx + my - \omega t)\}$ (see, e.g., Lamb (1932, § 228)). The result is the linearized dispersion relation

(4.1.8) $\quad\quad\quad\quad\quad \omega^2 = (g\kappa + \kappa^3 T) \tanh \kappa h,$

where $\kappa^2 = k^2 + m^2$. (These concepts from linear theory are discussed in the Appendix.) From this one computes the group velocity and shows that the linearized problem is dispersive at most wave numbers, but not at $\kappa = 0$ (i.e., long waves), where it is only weakly dispersive. This is an essential point in the derivation of the KdV equation; we focus our attention on those waves for which the linearized problem is only weakly dispersive. For many problems this occurs at $\kappa = 0$.

To derive KdV, we assume that

(A) The motion is strictly two-dimensional, $m = 0$.

(B) The relevant length scale in the x-direction is much longer than the fluid depth,

$$(kh)^2 \ll 1.$$

(C) Wave amplitudes are small,

$$\varepsilon = \dfrac{|\zeta|_{\max}}{h} \ll 1.$$

(D) These last two effects approximately balance,

$$(kh)^2 = O(\varepsilon).$$

Note. Under these additional assumptions, (4.1.5–7) reduce to a simplified model, which will turn out to be (4.1.1). The model is consistent if its solution satisfies these assumptions for $t > 0$ whenever the initial data satisfied them. To be practical, its solution also should approximately satisfy these assumptions whenever the initial data approximately satisfied them. The fact that a model is derived in an apparently rational manner does not guarantee that it is either consistent or practical.

Note. Assumption (A) may be weakened. It is necessary only that the waves be nearly one-dimensional, so that $(m/k)^2 \ll 1$. We could replace (A) with

$$\left(\frac{m}{k}\right)^2 \ll \varepsilon$$

and still derive (4.1.1); but if we assumed

$$\left(\frac{m}{k}\right)^2 = O(\varepsilon)$$

instead, the same derivation would lead to (4.1.4).

Note. Assumption (D) is an example of Kruskal's (1963) "principle of maximal balance", which states that in a perturbation expansion involving two or more small parameters a scaling which reduces the problem as little as possible is of interest.

The assumptions (A)–(D) suggest the following dimensionless variables (marked *):

(4.1.9)
$$z^* = \frac{z}{h}, \quad x^* = \sqrt{\varepsilon}\,\frac{x}{h}, \quad t^* = \sqrt{\frac{\varepsilon g}{h}}\,t,$$
$$\phi = h\sqrt{\varepsilon g h}\,\phi^*, \quad \zeta = \varepsilon h \zeta^*.$$

If ϕ is analytic at $z = -h$, it has a convergent power series expansion

$$\phi = h\sqrt{\varepsilon g h}\sum_{n=0}^{\infty}(z^*+1)^n \phi_n^*(x^*, t^*).$$

Rayleigh (1876) observed that for long waves this expansion is also asymptotic. Substituting it into (4.1.5) and equating powers of z^* yields

(4.1.10)
$$\phi_{n+2}^* = -\frac{\varepsilon}{(n+2)(n+1)}\frac{\partial^2 \phi_n^*}{\partial (x^*)^2}.$$

From (4.1.6) and (4.1.10), $\phi_{2n+1}^* \equiv 0$. Thus

(4.1.11) $$\phi = h\sqrt{ghe}\left[\phi_0^* - \frac{\varepsilon}{2!}(z^*+1)^2 \frac{\partial^2 \phi_0^*}{\partial (x^*)^2} + \frac{\varepsilon^2}{4!}(z^*+1)^4 \frac{\partial^4 \phi_0^*}{\partial (x^*)^4} + \cdots\right].$$

No approximations have been made to this point.

To leading order, the conditions at the free surface can be put in the form

(4.1.12)
$$\frac{\partial \zeta^*}{\partial t^*} + \frac{\partial}{\partial x^*}\frac{\partial \phi_0^*}{\partial x^*} = O(\varepsilon),$$

$$\frac{\partial}{\partial t^*}\frac{\partial \phi_0^*}{\partial x^*} + \frac{\partial \zeta^*}{\partial x^*} = O(\varepsilon).$$

To solve (4.1.12), we assume

(4.1.13) $$\zeta^* = \zeta_0 + \varepsilon\zeta_1 + \cdots, \qquad \frac{\partial \phi_0^*}{\partial x^*} = u_0 + \varepsilon u_1 + \cdots.$$

Anticipating the secular terms that will arise at $O(\varepsilon)$, we also introduce a slower time variable (cf. Cole (1968)),

(4.1.14a) $$\tau = \varepsilon t^*,$$

so that

(4.1.14b) $$\frac{\partial}{\partial t^*} \to \frac{\partial}{\partial t^*} + \varepsilon \frac{\partial}{\partial \tau}.$$

Now (4.1.12) becomes simply the linear wave equation for (ζ_0, u_0); the general solution is

(4.1.15) $$\begin{aligned}\zeta_0 &= f(x^* - t^*; \tau) + g(x^* + t^*; \tau), \\ u_0 &= f(x^* - t^*; \tau) - g(x^* + t^*; \tau).\end{aligned}$$

This may also be written in terms of the (linear) characteristic coordinates

(4.1.16) $$r = x^* - t^* = \frac{\sqrt{\varepsilon}}{h}(x - \sqrt{gh}\, t), \qquad l = x^* + t^* = \frac{\sqrt{\varepsilon}}{h}(x + \sqrt{gh}\, t).$$

Thus, on a short time scale ($t^* = O(1)$), arbitrary initial data split into right and left running waves. Solitons cannot be distinguished on this time scale, because *every* solution consists of a superposition of (two) waves of permanent form. There is no interaction because the nonlinear terms are too weak to exert any influence this quickly.

At the next order, (4.1.12) becomes

(4.1.17) $$\begin{aligned}\frac{\partial \zeta_1}{\partial t^*} + \frac{\partial u_1}{\partial x^*} &= -\left[\frac{\partial \zeta_0}{\partial \tau} + \frac{\partial}{\partial x^*}(u_0 \zeta_0) - \frac{1}{6}\frac{\partial^3 u_0}{\partial x^{*3}}\right], \\ \frac{\partial u_1}{\partial t^*} + \frac{\partial \zeta_1}{\partial x^*} &= -\left[\frac{\partial u_0}{\partial \tau} + \frac{\partial}{\partial x^*}\left(\frac{u_0^2}{2}\right) - \frac{1}{2}\frac{\partial^3 u_0}{\partial x^{*2} \partial t^*} + \hat{T}\frac{\partial^3 \zeta_0}{\partial x^{*3}}\right],\end{aligned}$$

where $\hat{T} = T/\rho g h^2$ is the dimensionless surface tension. These equations may be integrated rather easily by using characteristic coordinates (4.1.16). The solution so obtained contains terms which grow linearly in l, and others which grow linearly in r. These terms are secular; they render the expansion in (4.1.13) nonuniform. We eliminate them by forcing the coefficients of l, r in these terms to vanish:

(4.1.18) $$\begin{aligned}2\frac{\partial f}{\partial \tau} + 3f\frac{\partial f}{\partial r} + \left(\frac{1}{3} - \hat{T}\right)\frac{\partial^3 f}{\partial r^3} &= 0, \\ -2\frac{\partial g}{\partial \tau} + 3g\frac{\partial g}{\partial l} + \left(\frac{1}{3} - \hat{T}\right)\frac{\partial^3 g}{\partial l^3} &= 0.\end{aligned}$$

Thus, the left and right running waves each evolve according to their own KdV equations, which describe how the two sets of waves each interact with themselves over a long time scale ($\tau = \varepsilon t^* = O(1)$).

For an air-water interface, $(\frac{1}{3} - \hat{T}) > 0$ if $h > 0.5$ cm. Usually $\hat{T} \ll \frac{1}{3}$ and we may neglect $\hat{T}$ altogether. Then the dimensional surface displacement due to a single soliton becomes

$$(4.1.19) \quad \frac{z}{h}(x, t) = \frac{4}{3}\varepsilon a^2 \operatorname{sech}^2\left\{\frac{\sqrt{\varepsilon}\, a}{h}\left[x \pm \sqrt{gh}\left(1 + \frac{2\varepsilon a^2}{3}\right)t + x_0\right]\right\}.$$

Note that:

(i) Our arbitrary choice of ε drops out, because only the combination (εa^2) appears.

(ii) A soliton raises the free surface, regardless of its direction of propagation.

(iii) The speed of every soliton exceeds $\sqrt{gh}$.

(iv) To leading order, the speed of a soliton with amplitude α is $c = \sqrt{g(h + \alpha)}$. This relation was obtained empirically by Russell (1838).

(v) For thin sheets of water $\hat{T} > \frac{1}{3}$, and the conclusions in (ii) and (iii) must be reversed.

The interaction between the left and right running waves also may be found by integrating (4.1.17):

$$(4.1.20) \quad 4\zeta_1(l, r; \tau) = [\partial_l g(l) - \partial_l g(l_0)]\int_{r_0}^{r} f\, dr + [\partial_r f(r) - \partial_r f(r_0)]\int_{l_0}^{l} g\, dl$$
$$+ 2[g(l) - g(l_0)][f(r) - f(r_0)],$$

with a similar expression for u_1. To prevent these terms from becoming secular, we also require that

$$(4.1.21) \quad |f|, |\partial_r f|, \left|\int f\, dr\right|, |g|, |\partial_l g|, \left|\int g\, dl\right| < \infty.$$

These conditions assure that the interaction between the left and right waves is both weak and localized. For the KdV model to be consistent (4.1.21) must hold for all time (τ) if the initial data for (4.1.18) satisfied them.

Under the assumptions (A)–(D), we now have replaced (4.1.5–7) with a simpler problem whose solution formally approximates that of (4.1.5–7) to $O(\varepsilon^2)$. This is the sense in which the KdV equation models long water waves of moderate amplitude. Before comparing the predictions of this model with experimental data, some comments on the derivation may be in order.

(i) The "KdV model" of water waves actually consists of the linear wave equation (4.1.12) on a short time scale and two KdV equations (4.1.18) on a longer time scale. The model is intended to be valid asymptotically as $\varepsilon \to 0$, and it is important that both (4.1.12) and (4.1.18) remain nontrivial in the limit

$\varepsilon \to 0$. As pointed out by Kruskal (1975), the KdV model is preferable in this respect to alternative models of long water waves of moderate amplitude, such as the Boussinesq equation

(4.1.22) $$u_{tt} = u_{xx} + \varepsilon[(u^2)_{xx} + u_{xxxx}] + O(\varepsilon^2).$$

In the limit $\varepsilon \to 0$, (4.1.22) reduces to (4.1.12) and (4.1.18).

(ii) In the derivation given here, τ represents a slow time scale, and a solution of KdV at fixed τ corresponds to a snapshot of a water wave. But x^* and t^* are somewhat interchangeable in light of (4.1.16), and we could have introduced a slow space scale, χ, instead of τ. We would still get two KdV equations, but now a KdV solution at fixed χ would correspond to the signature of a water wave as it passed a probe at a fixed location. Most wave measurements are made this way.

(iii) This derivation is formal, and does not actually prove that the solutions of KdV are asymptotic to those of (4.1.5-7). No such proof is available. In fact, it was proved only recently that (4.1.5-7) admits solitary waves with angles of the water surface up to $\pi/6$, i.e., up to the limiting case of Stokes (Amick and Toland (1979)).

(iv) The fluid is assumed here to be inviscid, and the model admits no dissipation. But water has a finite viscosity, and one may estimate a dissipative time scale based on either laminar or turbulent boundary layers (e.g., Keulegan (1948)). The validity of the KdV model requires that the dissipative time scale greatly exceed the KdV time scale ($\tau = O(1)$).

(v) The original equations of motion, (4.1.5-7), are Galilean invariant. The KdV equations maintain this invariance if we interpret f (or g) as a horizontal velocity, because of (4.1.15b). Then the transformation

$$\tau \to \tau, \quad r \to r - c\tau, \quad f \to f + c$$

leaves (4.1.18a) invariant.

(vi) This derivation suggests that irrotationality is fundamental to derivation of KdV. This is false, as shown by Benney (1966); see also Benjamin (1966).

(vii) The dynamic condition at the free surface (4.1.7b) states that the pressure must vanish there. This is only approximate for most water waves. A more precise statement is that the pressure must match to the pressure of the air above; i.e., the surface wave is actually an internal wave. This subtle point is usually neglected because the ratio of air to water densities is approximately 10^{-3}. However, if the waves were *very* long ($kh \ll 10^{-3}$) this effect would become important, and the appropriate model would be (4.1.3), rather than (4.1.1). Similar considerations apply to the rotation of the earth and its curvature.

The KdV equation was tested as a model of water waves by Zabusky and Galvin (1971), Hammack and Segur (1974), (1978) and Weidman and Maxworthy (1978); the original experimental work of Russell (1838), (1845) is still

APPLICATIONS

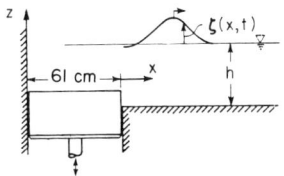

FIG. 4.1. *Schematic diagram of wave maker. (Hammack and Segur (1974)).*

worth consideration as well. An extensive comparison was made by Hammack and Segur, whose work we now discuss. These experiments were conducted in a wave tank 31.6 m long, 61 cm deep and 39.4 cm wide. As shown schematically in Fig. 4.1, the wave generator consisted of a rectangular piston located in the tank bed adjacent to the upstream end wall of the tank. The piston spanned the tank width, and was 61 cm long for the experiments we will discuss. The time history of its vertical displacement was prescribed for each experiment.

Wave measurements were made during each experiment at several positions down the tank using parallel-wire resistance gauges. In the first set of experiments (1974) the fluid depth h was 5 cm, and the waves were measured at $x/h = 0$, 20, 180 and 400, where $x = 0$ at the downstream edge of the piston. In the second set (1978) $h = 10$ cm, and waves were measured at $x/h = 0$, 50, 100, 150 and 200.

Figure 4.2 shows the wave generated simply by raising the piston. The piston motion was fast enough that the shape of the wave at $x/h = 0$ is effectively the shape of the piston (because of the reflecting wall at the upstream end of the piston, the wave at $x = 0$ was actually twice as long as the piston and half as high as its displacement).

On a short time scale, according to (4.1.15), this wave should simply translate with speed $\sqrt{gh}$. The wave measured at $x/h = 20$ (Fig. 4.2b) fits this description approximately; its shape is basically that of the wave at $x = 0$. (The front of the wave is to the left in these figures, and a wave which translates with speed $\sqrt{gh}$ shows no horizontal displacement in succeeding frames).

That solitons emerge on a long time scale may be seen in Fig. 4.2c, d. Solving the Schrödinger eigenvalue problem (1.3.33) with the wave measured at $x = 0$ as the potential yields 3 discrete eigenvalues, representing 3 solitons. These correspond to the 3 positive, more or less permanent waves seen at $x/h = 180$ and $x/h = 400$. According to (4.1.19), these waves all should move to the left in these figures, since their speeds all exceed $\sqrt{gh}$. That they do not is a measure of the effect of viscosity in these experiments.

Even so, we assert that these waves are solitons on the basis of their shapes. The entire profile of a single soliton is determined from (4.1.19) once its amplitude is known. The peak amplitudes of the first two waves in Fig. 4.2d were measured, and the dots in that figure represent evaluations of (4.1.19)

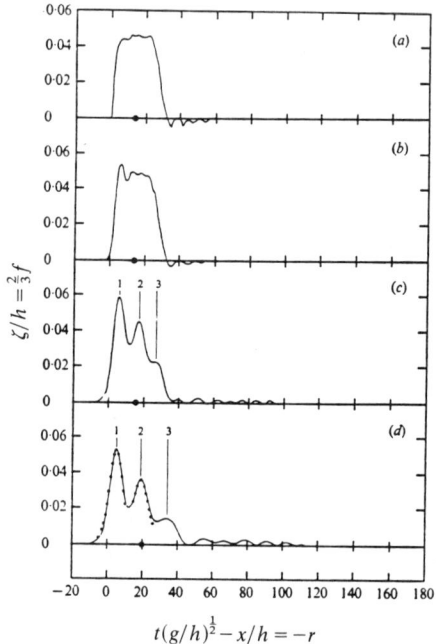

FIG. 4.2. *Evolution of a long, positive water wave into* 3 KdV *solitons.* $h = 5$ cm. —, *measured profiles;* ···, *soliton profiles computed using* (4.1.19). (a) $x/h = 0$, (b) $x/h = 20$, (c) $x/h = 180$, (d) $x/h = 400$. (*Hammack and Segur* (1974)).

based on those amplitudes. The agreement with the measured wave shapes is striking.

The results shown in Fig. 4.2 suggest the following picture of long water waves of moderate amplitude.

(i) There is a short (linear) time scale, during which the left and right running waves separate from each other.

(ii) There is a long (KdV) time scale, during which the right (or left) running waves evolve into N solitons plus radiation.

(iii) There is an even longer viscous time scale, during which the energy in these solitons is gradually dissipated. Because the KdV time scale is shorter, however, the solitons continually readjust their shapes and speeds as they lose energy, so that *locally*, as in Fig. 4.2d, they look and act like solitons.

Note that the mean water level is positive in the oscillatory waves in Fig. 4.2c,d. Recall from § 1.7 that the *mean* of the oscillatory waves in the long time KdV solution is negative (cf. (1.7.51)). The discrepancy is a viscous effect, and may be explained as follows. The solitons slowly lose energy due to viscous effects, but the total mass of the waves is conserved. Therefore, as mass is forced out of the solitons, it builds a "shelf" behind them, which raises the mean water level there. This is analogous to the shelf discussed in § 3.8.

APPLICATIONS   285

Several other experiments involving solitons were also made in this series. In some experiments the initial wave amplitudes were very small and the solitons had not yet emerged by $x/h = 400$. In every experiment in which the number of solitons observed at $x/h = 400$ was unambiguous, it agreed with the number predicted by using the initial data for that experiment as the potential for the Schrödinger eigenvalue problem and counting the number of discrete eigenvalues (i.e., the number of solitons). The amplitude of the leading soliton at $x/h = 400$ also was predicted with reasonable accuracy in two steps: (i) by using the eigenvalue problem to determine its (inviscid) amplitude; (ii) by using a formula due to Keulegan (1948) to determine the viscous decay of a solitary wave of this amplitude as it propagates over 400 depths. More details may be found in Hammack and Segur (1974).

The piston displacement in Fig. 4.3 was exactly reversed from that of Fig. 4.2. If the wave evolution were linear, then each wave record in Fig. 4.3 would be the reverse of the corresponding record in Fig. 4.2. This is approximately the case for the first two records, which are measured on the short, linear time scale. On a longer time scale, however, the wave records are much different. An entirely negative initial wave, such as that in Fig. 4.3a can produce no solitons; all of the energy must go to the continuous spectrum. Fig. 4.3 shows a representative example of the radiation part of the KdV solution.

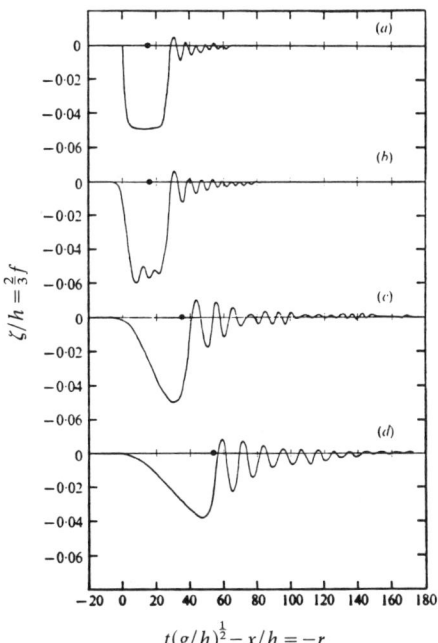

FIG. 4.3. *Evolution of a long, negative water wave into oscillatory waves without solitons.* $h = 5$ cm. —, *measured profiles.* (a) $x/h = 0$, (b) $x/h = 20$, (c) $x/h = 180$, (d) $x/h = 400$. (*Hammack and Segur* (1974)).

A more detailed test of how well the KdV model predicts this radiation may be made in Fig. 4.4, which shows a larger amplitude wave in somewhat deeper water. Recall from § 1.7 that the asymptotic solution of (4.1.1) without solitons consists of four regions.

(i) For $x \gg (3t)^{1/3}$, the solution is exponentially small.

(ii) For $|x| \leq O((3t)^{1/3})$ the solution is approximately self-similar, and approaches the special solution

$$u = \frac{x + x_0}{6t}.$$

(iii) There is a relatively thin collisionless shock layer near $-x = [O((3t)^{1/3}(\log 3t)^{2/3+p})]$, $1 \gg p \geq 0$.

(iv) For $(-x) \gg (3t)^{1/3}(\log 3t)^{2/3+p}$, the solution consists of decaying oscillations. These oscillations form into wave groups, with nodes given by the zeros of the reflection coefficient, $\rho(k)$. The wavenumbers within each group are

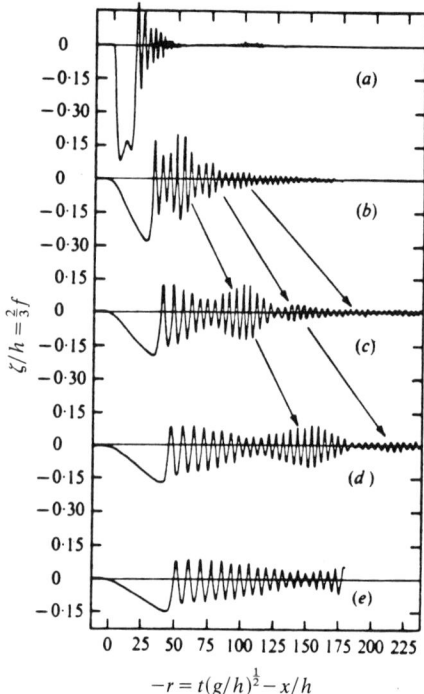

FIG. 4.4. *Evolution of a long, negative water wave into oscillatory waves, with clearly defined wave groups. The initial amplitude is larger than in Fig. 4.3.* $h = 10$ cm. —, *measured profiles;* →, *trajectory of wave group, based on average measured wave frequency and linearized dispersion relation* (4.1.8). (a) $x/h = 0$, (b) $x/h = 50$, (c) $x/h = 100$, (d) $x/h = 150$, (e) $x/h = 200$. (*Hammack and Segur* (1978)).

fixed, and the group travels with the group velocity of the linearized problem evaluated at the dominant wavenumber of the group.

Certainly this description is in qualitative agreement with the wave records shown in Fig. 4.4. A quantitative comparison of regions (i) and (ii) is shown in Fig. 4.5, where the leading portion of the wave record in Fig. 4.4e is compared with the asymptotic KdV solution evaluated at the appropriate time. For comparison, the asymptotic solution of the linearized equation, (A.1.49), with the same initial data is also plotted. In this experiment, the KdV prediction of this leading wave is surprisingly accurate. Moreover, there is evidence that the agreement is even closer when corrections for the finite viscosity of the fluid are made (Hammack and Segur (1978)).

The KdV prediction of the oscillatory region is not as accurate in this experiment, largely because the waves generated in this region are not long in the sense of our original assumption (B), p. 278. In other words, the initial data consisted primarily of long waves, but the long waves generated short waves, and their evolution is not modeled well by the KdV equation. Even so, the qualitative picture from the KdV equation is correct. Wave groups are clearly identifiable in Fig. 4.4b, c, d and a detailed analysis of these records shows that the dominant wavenumbers remain essentially unchanged as the waves evolve. Moreover, the trajectories plotted in Fig. 4.4, which clearly coincide with the observed trajectories of the wave groups, actually were obtained by evaluating the linearized group velocity, (4.1.8), at the wave frequencies measured.

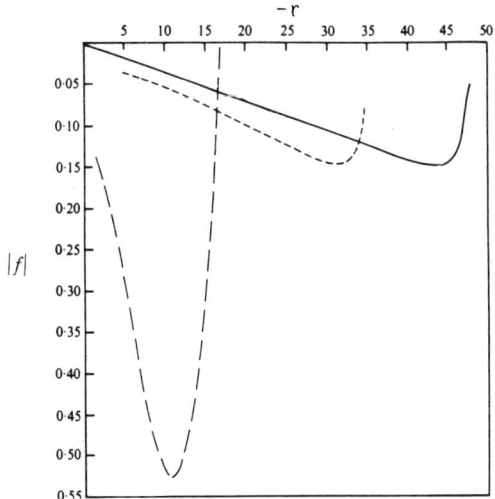

FIG. 4.5. *Theoretical and experimental profiles for wave front at $x/h = 200$ in Fig. 4.4.* —, *measured profile;* — —, *linear asymptotic theory* (A.1.58); - - - -, *KdV asymptotic theory* (1.7.41–45) (*Hammack and Segur* (1978)).

The fact that realistic initial data for (4.1.5–7) usually contain some high frequency waves raises the question of whether the presence of these short waves might invalidate the KdV model. That is, if the evolution of the long waves changes significantly due to the presence of short waves, and if the KdV model is valid only when all short waves are absent, then the KdV equation would not be a practical model of (4.1.5–7). The results of a series of experiments designed to test this possibility are shown in Fig. 4.6. In three separate experiments (each shown vertically in this figure), the piston was displaced upward with the same mean motion, but with increasing amounts of superposed high frequency motion. The first experiment was qualitatively like that in Fig. 4.2: the initial wave was roughly the shape of the piston, and evolved into four separate solitons. The early wave records of the other two experiments look different because the long and short waves have not yet separated. However, small amplitude water waves are strongly dispersive (cf.

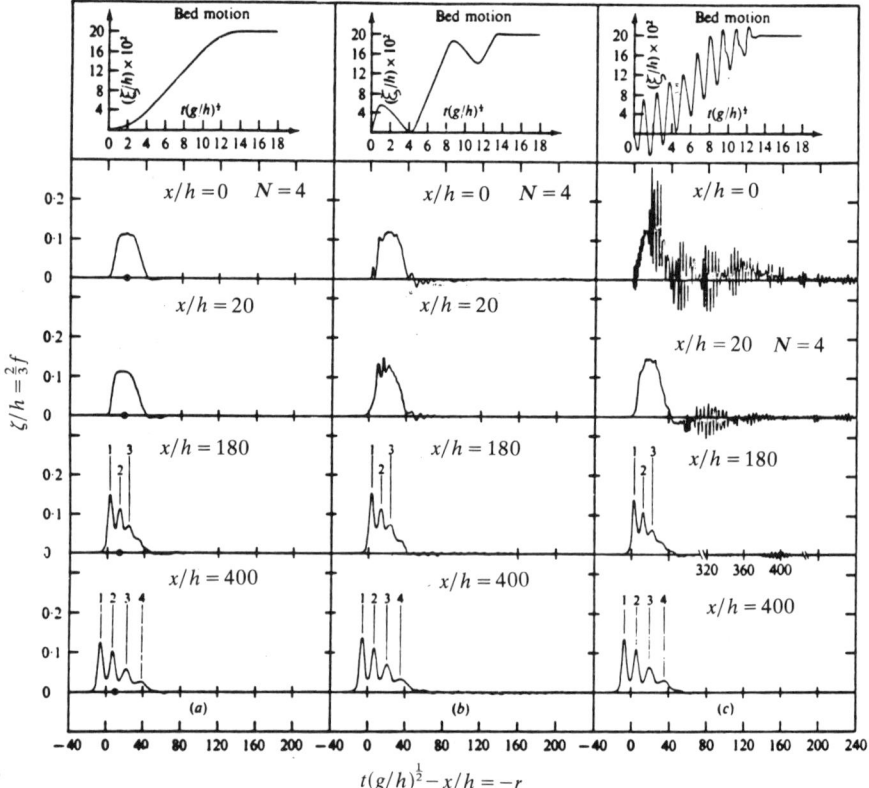

FIG. 4.6. *Time-displacement histories of three piston motions, and the water waves they generate.* $h = 5$ cm. (a) *Mean piston motion*, (b), (c) *mean motion with superposed oscillation. N represents the number of discrete eigenvalues obtained in each experiment by using the measured wave profiles at* $x/h = 0$ *in the scattering problem* (1.3.33). (Hammack and Segur (1974)).

4.1.8), and by $x/h = 180$ (i.e., the KdV time-scale), the long and short waves are physically separated. The solitons that emerge at $x/h = 400$ are indistinguishable in the different experiments.

To summarize, the KdV equation is part of a model of how two-dimensional long water waves of moderate amplitude evolve over a relatively long time scale. It predicts the evolution of the long waves fairly well, especially if its predictions are corrected to account for viscous effects. It predicts the evolution of short waves incorrectly, but the presence of these short waves does not seriously impair the accuracy of the model for the long waves, because of the dispersive nature of water waves.

Several alternatives arise when the water waves are not truly two-dimensional (cf. Miles (1980).) One possibility is that the waves are nearly two-dimensional. In this case (4.1.18) should be replaced with

$$(4.1.23) \qquad (2f_\tau + 3ff_r + (\tfrac{1}{3} - \hat{T})f_{rrr})_r + f_{\eta\eta} = 0,$$

where $\eta = \varepsilon y/h$. Here is a summary of what is known about the solution of (4.1.23).

(i) Every solution of (4.1.18) also solves (4.1.23).

(ii) Based on the results of Kadomtsev and Petviashvili (1970), one-dimensional solitons are unstable with respect to long transverse perturbations if $(\tfrac{1}{3} - \hat{T}) < 0$ (i.e., very thin sheets of water), but not in the usual case, where $(\tfrac{1}{3} - \hat{T}) > 0$. The experimental results in Figs. 4.2 and 4.6 indicate that solitons are not unstable with respect to short transverse perturbations, either.

(iii) If $\hat{T} < \tfrac{1}{3}$ (the usual case), (4.1.23) has N-soliton solutions, with the solitons interacting obliquely (3.3.90, 91). The interaction of two oblique solitons, which is stationary in the appropriate coordinate system, is shown in Fig. 4.7a.

This kind of interaction is very suggestive of the ocean waves shown in Fig. 4.7b. These waves were photographed off the coast of Oregon. According to the photographer, the water was about $\tfrac{1}{2}$ meter deep where the interaction occurred. The figure shows that the two waves each belong to periodic wave trains (presumably coming in from deep water), but each wave is so long relative to the (locally) shallow water depth that the waves may be considered solitary. No more quantitative information is available about these waves, but even so the similarity between this interaction and that predicted by (4.1.23) is striking.

(iv) Surface tension dominates gravity for very thin sheets of water, where $\hat{T} > \tfrac{1}{3}$. Here one-dimensional solitons are unstable, but "lump" solutions of (4.1.23) exist; cf. § 3.4. These $(2+1)$-dimensional analogues of solitons have not yet been observed experimentally.

For the experimentalist interested in finding lumps, we should note that (4.1.23) with $\hat{T} > \tfrac{1}{3}$ also applies if we disregard gravity altogether, remove the horizontal bed (and the viscous boundary layer it creates) on which the thin

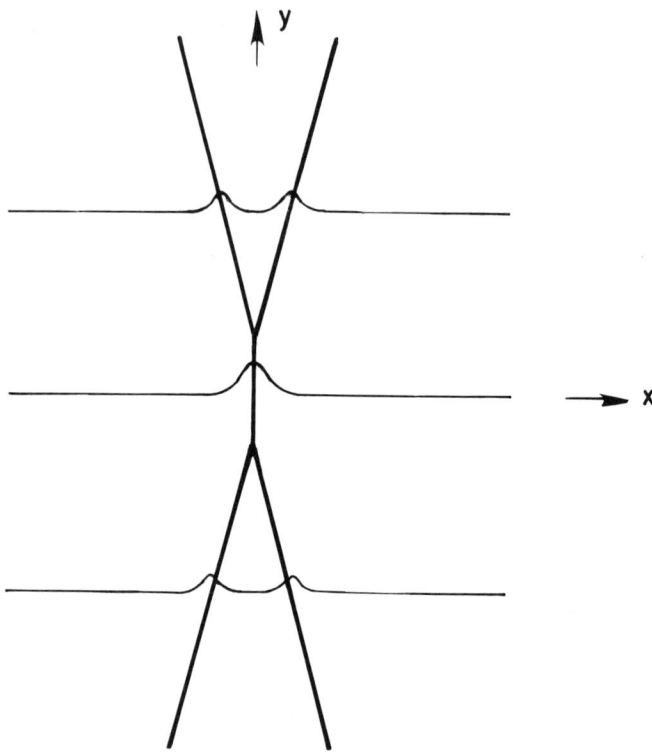

FIG. 4.7a. *Sketch of two-soliton solution of Kadomtsev-Petviashvili equation from (3.3.91). In this symmetric case, $k_1 = k_2$, $p_1 = -p_2 > \sqrt{3}k_1$. This pattern moves in the x-direction with speed $(k_1^2 + p_1^2)$.*

sheet of water lies, and restrict our attention to the symmetric modes (so that (4.1.6) holds). Now we are considering waves on a freely suspended sheet of water, such as that studied by Taylor (1959). The main advantage of this configuration is that viscous effects are much less important.

(v) If $\hat{T} > \frac{1}{3}$ (dominant surface tension), Zakharov and Manakov (1979) have solved (4.1.23) exactly by a generalization of IST. They require boundary conditions restrictive enough that lumps are excluded a priori. The asymptotic $(t \to \infty)$ behavior of the solution with these boundary conditions was given by Manakov, Santini and Takhtadzhyan (1980).

4.1.b. Internal waves. The internal oscillations due to gravity of a stably stratified fluid are known as internal waves. Both the oceans and the atmosphere usually are stratified, and they support rich spectra of these waves (e.g., Phillips (1977)). In fact, the waves at the air-water interface that we have just discussed may be thought of as an extreme case of internal waves (caused by an extremely large density gradient at the interface).

FIG. 4.7b. *Oblique interaction of two shallow water waves. (Photograph courtesy of T. Toedtemeier)*

Under appropriate circumstances, long internal waves of moderate amplitude evolve according to the KdV equation, just as surface waves do. However, the KdV equation does not play the ubiquitous role for internal waves that it does for surface waves; depending on circumstances, long internal waves might evolve according to the mKdV equation (4.1.2), the Benjamin–Ono equation (4.1.3) or a model intermediate between KdV and (4.1.3). In this subsection, we analyze one fairly simple (two-layer) model of internal waves in order to see how these various equations arise, and in what sense they model the evolution of long internal waves.

Consider two incompressible, immiscible fluids, with densities $\rho_1 < \rho_2$ and depths h_1, h_2 ($H = h_1 + h_2$), as shown in Fig. 4.8. The lower, heavier fluid rests

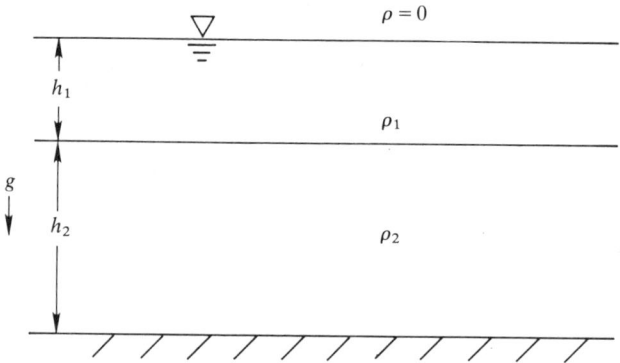

FIG. 4.8. *Two-layer configuration. The surface waves have their maximum displacements at the free surface; the internal waves have theirs at the interface.*

on a horizontal impermeable bed, while the upper fluid is bounded above by a free surface. Both fluids are subject to a vertical gravitational force. We will neglect surface tension, both at the fluid interface and at the free surface. We also neglect the earth's rotation (but see Gibbon, James and Moroz (1979)). If we assume that each fluid is also irrotational, then each has a velocity potential (ϕ_1, ϕ_2) that satisfies the Laplace equation in its own domain. The appropriate boundary conditions are as follows: at the bottom ($z = -H$), the vertical velocity must vanish; at the interface ($z = -h_1 + \eta(x, y, t)$), the vertical velocity and pressure are continuous; at the free surface ($z = \zeta(x, y, t)$), the conditions are those in (4.1.7), with $T = 0$; as $x^2 + y^2 \to \infty$, we require that ($\nabla\phi_1, \nabla\phi_2, \zeta, \eta$) all vanish. Then, once initial conditions are given (at $t = 0$), we may ask for the subsequent evolution of $\eta(x, y, t)$, $\zeta(x, y, t)$, $\phi_1(x, y, z, t)$ and $\phi_2(x, y, z, t)$ for $t > 0$. We will concentrate primarily on two-dimensional motion ($\partial_y \equiv 0$).

APPLICATIONS

We begin by linearizing the problem, and finding the linearized dispersion relation. This is given by Lamb (1932, § 231). Let

(4.1.24) $$\Delta = \frac{\rho_1 - \rho_2}{\rho_2}.$$

Then

(4.1.25) $$\omega^4[1+(1-\Delta)\tanh kh_1 \tanh kh_2] - \omega^2 gk[\tanh kh_1 + \tanh kh_2] + \Delta g^2 k^2 \tanh kh_1 \tanh kh_2 = 0.$$

The following points are of interest.

(i) For each k, (4.1.25) is a quartic equation for $\omega(k)$, whereas the linearized dispersion relation for the surface wave (4.1.8) is quadratic.

(ii) (4.1.25) reduces to (4.1.8) with $T=0$ if $\Delta = 0$, or if $h_2 = 0$, or if $h_1 = 0$, or if $\Delta = 1$ (note that $\Delta \to 1$ can be interpreted either as $\rho_1 \to 0$ or as $\rho_2 \to \infty$).

(iii) For Δ small enough, one may show from a comparison of ζ (surface) and η (interface) that, for fixed k, the larger (in magnitude) pair of roots of (4.1.25) represent two waves whose maximum amplitudes occur at the free surface. These two modes are called surface waves, and reduce to the waves in a homogeneous fluid if $\Delta \to 0$.

(iv) The smaller (in magnitude) pair of roots of (4.1.25) represent two waves whose maximum amplitudes occur at the interface if Δ is small enough. These two modes are called internal waves, and disappear if $\Delta = 0$. Our primary interest here is in the long time evolution of the internal waves.

(v) For either the surface waves or the internal waves,

(4.1.26)
$$\omega^2 = O(k^2) \quad \text{as } kh_1, kh_2 \to 0,$$
$$\frac{\partial \omega}{\partial k} > 0, \quad \frac{\partial^2 \omega}{\partial k^2} < 0 \quad \text{for } k > 0,$$
$$\omega^2 = O(|k|) \quad \text{as } kh_1, kh_2 \to \infty.$$

From these results it follows that long internal waves travel faster than other internal waves, but that there are both faster and slower surface waves (in terms of group velocity).

Next we want to solve (4.1.25) in the long wave limit. However, as noted by Benjamin (1967) (see also Davis and Acrivos (1967) and Lamb (1932, § 231)), (4.1.25) has several limits that might qualify as "the long wave limit". Each of these possibilities is physically meaningful.

(i) If the waves in question are longer than all vertical depths, then kh_1, kh_2, $kH \ll 1$. In this limit, (4.1.25) reduces to

(4.1.27) $$\left(\frac{\omega}{k}\right)^4 - gH\left(\frac{\omega}{k}\right)^2 + g^2 \Delta h_1 h_2 = O(k^2),$$

and all four roots have the form

(4.1.28) $$\omega \sim Ck + ak^3 \quad \text{for } kH \ll 1.$$

This case is analogous to the surface wave problem that we have already discussed. An initial disturbance of small amplitude consisting only of long waves (in the sense that $kH \ll 1$), splits into two surface waves and two internal waves on a short (linear and nondispersive) time scale (see Lamb (1932, §§ 231–234)). If all four wave speeds are distinct, then the four waves separate in space on this short time scale, and each interacts only with itself on the next (nonlinear and dispersive) time scale. As we will show below, the consequence of (4.1.28) is that the governing equation on the long time scale is either the KdV (4.1.1) or the mKdV (4.1.2) equation.

(ii) If $h_1 \ll h_2$, then the waves in question may be long in comparison with the thin layer ($kh_1 \ll 1$) but still short in comparison with the total depth ($kH \gg 1$). This "long wave limit" often is relevant for oceanic internal waves. For example, the depth of the main seasonal thermocline is typically 50–100 m in summer, whereas the depth of the ocean may be several thousand meters. As discussed in detail by Benjamin (1967), in this limit (4.1.25) reduces to

$$\omega^2 = g|k| \quad \text{(short surface waves)}$$

and

(4.1.29) $$\left(\frac{\omega}{k}\right)^2 - g\,\Delta h_1 = O(|k|h_1) \quad \text{(long internal waves)}.$$

As before, an initial disturbance of small amplitude breaks into four linear wave modes (two surface, two internal) on a short time scale. On the next time scale, each of the internal waves is governed by the Benjamin–Ono equation (4.1.3); see also Ono (1975) and Ablowitz and Segur (1980). We repeat that (4.1.3) appears to be solvable by some version of IST, although the theory is incomplete at this time.

(iii) Another possibility is that $kh_1 \ll 1$, $kH = O(1)$. In this version of the "long wave limit," Joseph (1977) and Kubota, Ko and Dobbs (1978) have derived (3.5.39), an equation intermediate between (4.1.1) and (4.1.3). This intermediate equation contains an arbitrary parameter (which is effectively kH); it reduces to (4.1.1) if $kH \to 0$ and to (4.1.3) if $kH \to \infty$. It also is exactly solvable (see § 3.5).

Now let us return to the case in which $kH \ll 1$. For simplicity, we also require $\Delta \ll 1$ (in the ocean, typically $\Delta < 0.02$). From (4.1.27), the linear long wave speeds are

$$\left(\frac{\omega}{k}\right)^2 \sim \begin{cases} C_s^2 = gH & \text{(surface waves)}, \\ C_i^2 = \dfrac{g\,\Delta h_1 h_2}{H} & \text{(internal waves)}. \end{cases}$$

Because $\Delta \ll 1$, these speeds are distinct. Hence, if the initial disturbance is of finite extent and consists only of long waves, then the surface and internal waves separate on the linear time scale.

To follow the internal waves on the next time scale, we define

(4.1.30)
$$\tau = \frac{1}{6}\left(\frac{g\Delta}{H}\right)^{1/2} t, \qquad \chi = \frac{x - C_i t}{(h_1 h_2)^{1/2}},$$

$$f = \frac{3(h_1 - h_2)}{2 h_1 h_2} \eta_i,$$

where η_i is the dimensional deflection of the interface due to the internal wave mode with speed C_i. Then one can show that this internal wave evolves according to the KdV equation

(4.1.31) $$f_\tau + 6 f f_\chi + f_{\chi\chi\chi} = 0.$$

Because a soliton is an intrinsically positive solution of (4.1.31), it follows from (4.1.30) that *an internal soliton is always a wave in which the thin layer thickens.* That is, the interface deflects downward if the upper layer is smaller and upward if the lower layer is smaller. These internal solitons were first discussed by Keulegan (1953).

An example of such an internal soliton is shown in Fig. 4.9, which was generated in the same wave tank used in Figs. 4.1–4.6, but with a density-stratified fluid. The figure shows both the initial internal wave and the wave to which it evolves. As in Fig. 4.2, the dots mark the shape of the exact, one-soliton solution of (4.1.27) with the measured peak amplitude. For more details, see Segur and Hammack (1981). Similar experiments were done by Koop and Butler (1981).

As with surface waves, it is necessary to correct for viscous dissipation in order to compare this theory directly with data. It turns out that this dissipation is stronger for long internal waves than it is for long surface waves. The reason is that in addition to all of the viscous boundary layers that form for surface waves, internal waves also have a boundary layer at the interface, where the velocity shear is the greatest. In fact, this additional boundary layer is often the most important energy sink in the problem.

There is no conceptual difficulty in generalizing Keulegan's (1948) theory to include this extra boundary layer (Leone (1974)). This generalized theory, coupled with the inviscid KdV model, then predicts the evolution of long internal wave solitons (Leone and Segur (1981)).

Note from (4.1.30) that f vanishes if the two layers have equal depth and their density difference is small. In this special configuration, the problem attains a symmetry such that the coefficient of the nonlinear term vanishes. In this case, the scaling implicit in assumption (D), p. 278, is not appropriate, and KdV is not the correct asymptotic equation. The correct scaling here is

$$kh = O(\varepsilon),$$

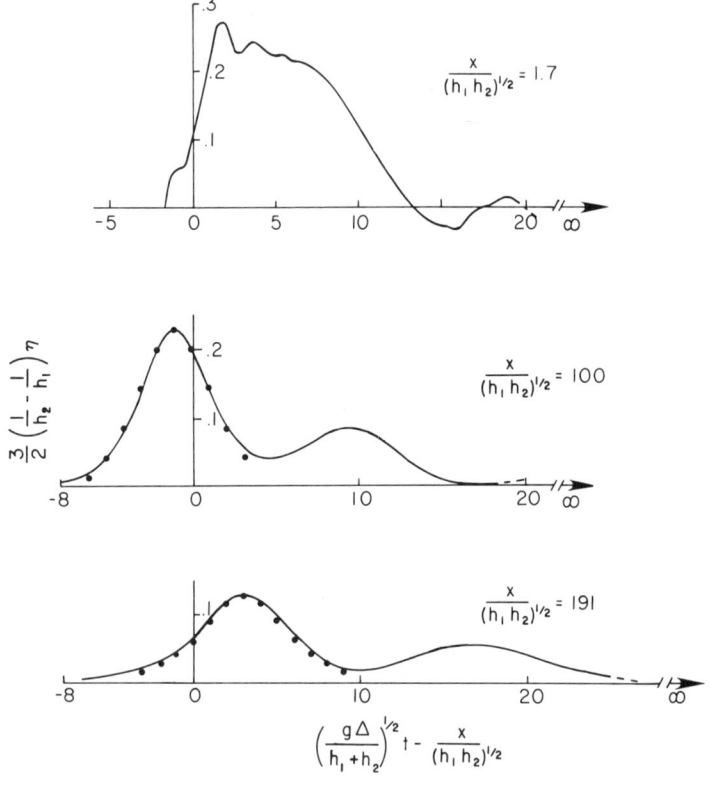

FIG. 4.9. *Evolution of a long internal wave from its initial shape into two solitons.* ——, *measured profile;* • • •, *KdV soliton.*

and the asymptotic equation is the mKdV equation, (4.1.2), which governs the evolution of the wave on the *next* time scale ($\varepsilon^2 t = O(1)$). Typically, the mKdV equation arises when $\omega(k) \sim C_0 k + ak^3$ in the long wave limit, but the problem has a symmetry that makes positive and negative waves dynamically indistinguishable, as they are here when $h_1 = h_2$.

One additional anomaly may be mentioned here. When the appropriate one-dimensional equation is mKdV because of some symmetry in the problem, then nearly one-dimensional waves are governed not by (4.1.4) but by

(4.1.32) $\quad (u_t + au^2 u_x + bu_{xxx})_x + u_{yy} = 0,$

where a, b are constants. As noted in § 3.7, this equation lacks the Painlevé property. Presumably it lacks all of the features that make these equations special: solitons, extra conservation laws, complete integrability, etc.

Finally, we note one more variation on the equations governing the evolution of long internal waves of moderate amplitude. Long internal waves can couple resonantly with a packet of short surface waves whose group velocity matches the internal wave speed. The result is not (4.1.1, 2 or 3) but a coupled pair of nonlinear evolution equations (Grimshaw (1975), Djordjevic and Redekopp (1977), Ma (1978), Ma and Redekopp (1979)). Because of this coupling, long internal waves sometimes may be observed in the form of resonantly excited surface waves. Descriptions of field observations of this phenomenon are discussed by Phillips (1974) and by Osborne and Burch (1980).

4.1.c. Rossby waves. In the atmosphere of a rotating planet, a fluid particle is endowed with a certain rotation rate, determined by its latitude. It follows that its motion in the north-south direction is inhibited by conservation of angular momentum, just as gravity inhibits the vertical motion of a density-stratified fluid. The large scale atmospheric waves caused by the variation of rotation rate with latitude are known as Rossby waves (Rossby (1939)).

As one might suspect, there is an analogy between internal waves and Rossby waves under appropriate assumptions. Because the KdV equation models internal waves, one might suspect that a KdV model of Rossby waves could be constructed as well. This was demonstrated by Benney (1966) and by Long (1964); see also Miles (1980) and the references cited therein. Under the simplest assumptions (long waves, incompressible fluid, β-plane approximation, etc.), the derivation of the KdV equation as a model of Rossby waves in the presence of a steady east-west zonal flow follows the relatively standard pattern given by Benney (1966), to which the reader is referred for more details. This particular application of the KdV equation warrants special attention, however, because of the conjecture of Maxworthy and Redekopp (1976) that Jupiter's Great Red Spot might be a solitary Rossby wave.

Figure 4.10 (see insert following page 310) shows a photo of the planet Jupiter, taken during the recent Voyager project. The cloud patterns show that the atmospheric motion on Jupiter is dominated by a number of east-west zonal currents, corresponding to the jet streams in our own atmosphere. Several oval-shaped spots also may be seen, including the prominent Great Red Spot in the southern hemisphere. The Great Red Spot has been seen at approximately this latitude for hundreds of years; it is known to migrate slowly to the west, and to maintain its integrity despite interactions with other atmospheric objects. A number of models have been proposed over the years to explain this intriguing feature of the Jovian atmosphere, including the model of the Red Spot as a solitary wave.

Following a recent version of this model described by Redekopp and Weidman (1978), we begin with the quasigeostrophic form of the potential vorticity equation for an incompressible fluid (Pedlosky (1971)):

(4.1.33)
$$\{(\partial_t + U \partial_x) + \varepsilon \psi_y (\partial_x - \psi_x \partial_y)\}\{\mu^2 \partial_x^2 + \partial_y^2 + \partial_z (K^2 \partial_z)\}\psi + (\beta - U'')\psi_x = 0.$$

Here (x, y, z) represent the (east, north, vertical) directions, and we have written the total horizontal stream function Ψ as

$$\Psi(x, y, z, t) = \int_{\bar{y}}^{y} U(\eta) \, d\eta + \varepsilon \psi(x, y, z, t)$$

in terms of a zonal shear flow and a perturbation. Moreover, we have used the β-plane approximation, so that the Coriolis parameter is approximated by

(4.1.34) $$f = 2\Omega \sin \theta_0 + \beta y.$$

Also,

(4.1.35) $$K(z) = \frac{2\Omega \sin \theta_0 l_2}{N(z) \, d},$$

where $N(z)$ is the Brunt–Väisälä frequency and l_2, d are characteristic length scales in the north-south and vertical directions, respectively. K compares the effects of rotation and density variation (i.e., centrifugal and gravitational forces). Finally,

$$\mu = \frac{l_2}{l_1}$$

represents the ratio of length scales in the north-south and east-west directions.

In the linear ($\varepsilon \to 0$), long wave ($\mu \to 0$) limit, ψ has the form

(4.1.36) $$\psi = \sum_n A_n(x - c_n t) \phi_n(y) p_n(z),$$

where c_n is the end result of *two* eigenvalue problems:

(4.1.37)
$$(K^2 p_n')' + k_n^2 p_n = 0, \qquad p_n(0) = p_n(1) = 0,$$
$$\phi_n'' - k_n^2 \phi_n + \frac{\beta - U''}{U - c_n} \phi = 0, \qquad \phi_n(y_S) = \phi_n(y_N) = 0.$$

Here the atmosphere is assumed to be confined between horizontal, rigid lids (at $z = 0, 1$), and the zonal shear flow lies between y_S and y_N. From (4.1.37), wave propagation is possible only in the east-west direction, i.e., this problem is intrinsically $(1+1)$-dimensional, so the question of transverse stability does not arise here.

If the various modes in (4.1.36) separate on a short time scale, then one may derive an evolution equation to describe how an individual mode interacts with itself on a longer timescale. As in the previous derivations, this is done by eliminating secular terms that arise at higher order in the expansion. Depending on the nature and existence of a stable density stratification, characterized

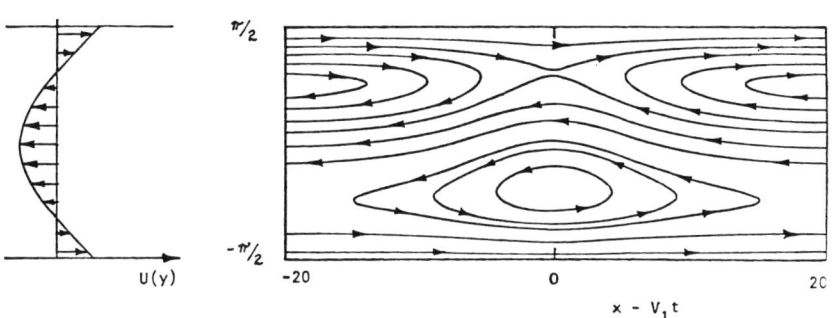

FIG. 4.11. *Streamline patterns for a Rossby wave soliton in a homogeneous atmosphere. The assumed background velocity profile is shown on the left. (Redekopp and Weidman (1978))*.

by $N(z)$, either KdV or mKdV is possible for a given mode. Coupled evolution equations also are possible if two or more modes have nearly equal linear phase speeds. Unfortunately, so little is known about any details of the Jovian atmosphere that the model cannot be made very precise or tested quantitatively at this time.

Figure 4.11 shows a horizontal streamline pattern corresponding to a single soliton in a coordinate system traveling with the wave, for a homogeneous atmosphere and an assumed zonal shear flow. The shape of the recirculating region is tantalizingly similar to that of the Great Red Spot, shown in Fig. 4.10. The region just above the recirculating region in Fig. 4.11 corresponds to the Hollow, which is observed to persist just north of the Red Spot. However, we reiterate that this comparison should be considered only suggestive at this time.

A separate test of the soliton model of the Red Spot and Hollow is that the combination should interact with other waves as a soliton. As pointed out by Maxworthy and Redekopp (1976), there was another large disturbance in the southern hemisphere, known as the South Tropical Disturbance, that first appeared early in this century and lasted for several decades. These two disturbances had different propagation speeds, and went through each other about 9 times during the lifetime of the South Tropical Disturbance. One interaction was described in detail by Peek (1958):

> During the six weeks which would have been required for the *p* end of the Disturbance to pass from one end of the Hollow to the other, ··· there was no sign whatever of any encroachment upon the region; instead, within a few days of its arrival at the *f* end of the Hollow, a facsimile of the *p* end of the Disturbance was seen ··· to be forming at the other end of the Hollow, ···. The new development ··· proved to be a true *p* end of the Disturbance which drew away from the Red Spot at approximately the same rate (at which it had approached), ···. Thus its passage through ··· the Red Spot, which would have taken 3 months at its normal rate of progress, must have been accomplished in a matter of fourteen days.

Maxworthy and Redekopp (1976) interpret this interaction as that of two solitons, complete with the required phase shift! A more detailed analysis of some of these interactions is given by Maxworthy, Redekopp and Weidman (1978).

For the sake of simplicity, several technical but important aspects of this model have been glossed over in this presentation. One of these is that the amplitude of the Great Red Spot is not small; i.e., $\varepsilon = O(1)$ in (4.1.33), so the ε-expansion used here must be considered purely formal. Another is that the Red Spot has lasted so long that any viable model of it must include some mechanism by which the wave extracts energy from the zonal shear flow at a rate sufficient to balance the loss of energy through dissipation. Discussion of these and other points may be found in Maxworthy and Redekopp (1976), Redekopp (1977), Redekopp and Weidman (1978), Maxworthy, Redekopp and Weidman (1978).

4.2. Three-wave interactions. The resonant interaction of three waves is perhaps the simplest nonlinear interaction of waves that are dispersive in the linear limit. Ordinarily, these equations arise in the study of a dispersive system in the weakly nonlinear limit. The system should have two ingredients.

(i) At the lowest order in a small amplitude expansion, one obtains the linearized dispersion relation $\omega(\mathbf{k})$. It must admit a resonant triad, i.e., three linear waves that satisfy a resonance condition:

(4.2.1) $$\mathbf{k}_1 + \mathbf{k}_2 + \mathbf{k}_3 = 0, \qquad \omega_1 + \omega_2 + \omega_3 = 0.$$

(We assume in this section that $\omega(\mathbf{k}) = -\omega(-\mathbf{k})$; otherwise (4.2.1) is incomplete.)

(ii) At next order in this expansion, quadratic interactions must occur. Then secular terms may arise at this order unless the slowly varying amplitudes of these three waves satisfy

(4.2.2)
$$\partial_\tau a_1 + (\mathbf{C}_1 \cdot \nabla) a_1 = i\gamma_1 a_2^* a_3^*,$$
$$\partial_\tau a_2 + (\mathbf{C}_2 \cdot \nabla) a_2 = i\gamma_2 a_3^* a_1^*,$$
$$\partial_\tau a_3 + (\mathbf{C}_3 \cdot \nabla) a_3 = i\gamma_3 a_1^* a_2^*,$$

where $\mathbf{C}_j$ is the linearized group velocity evaluated at $\mathbf{k}_j$. If the original system is conservative, then $(\gamma_1, \gamma_2, \gamma_3)$ are real and $\gamma_1 \gamma_2 \gamma_3 \leq 0$.

The general nature of these equations was noted explicitly by Benney and Newell (1967) among others, although special cases had been derived earlier. Some of the contexts in which these resonant triads arise, along with the appropriate references, may be found in Davidson (1972), Kaup (1976b) and Phillips (1974).

A special case of (4.2.2) that has received a good deal of attention occurs when the complex wave amplitudes depend on only one independent variable.

These are unmodulated, uniformly evolving wavetrains, and (4.2.2) reduces to

(4.2.2′)
$$\dot{a}_1 = i\gamma_1 a_2^* a_3^*, \quad \dot{a}_2 = i\gamma_2 a_3^* a_1^*,$$
$$\dot{a}_3 = i\gamma_3 a_1^* a_2^*,$$

where either $\dot{a}_i = \partial_t a_i$ or $\dot{a}_i = c_i \partial_x a_i$. Just as (4.2.2) can be solved exactly by IST, (4.2.2′) can be solved exactly in terms of elliptic functions (Ball (1964), Bretherton (1964)).

4.2.a. Nonlinear optics. An ideal dielectric medium may be thought of as one in which the electrons in each molecule are tightly bound to the nucleus. Applying an electric field to such a medium does not create a current, but rather displaces each bound electron by some finite amount. The macroscopic effect, obtained by summing over all the displacements in a unit volume, is the *polarization*, **P**.

The relation between the polarization and the electric field is one of the constitutive relations that defines the medium. In the simplest case, the relation is linear,

(4.2.3) $$P_i = \chi_{ij} E_j;$$

χ_{ij} is the (linear) *susceptibility* of the material. For isotropic materials, χ_{ij} reduces to a scalar. A simple generalization of (4.2.3) is to consider the field to be a plane wave with frequency ω, and to allow χ_{ij} to depend on ω.

This model was adequate for most problems in optics prior to the invention of the laser. However, lasers are capable of producing such high field intensities that nonlinear corrections to the susceptibility become important. "Nonlinear optics" usually denotes the study of phenomena caused by the nonlinear corrections to the susceptibility of a dielectric material. It is among the most active areas in applied physics, and our discussion of it necessarily will be incomplete. The discussion which follows is based on those in Akhmanov and Khokhlov (1972), Whitham (1974) and Yariv (1975); see also Kleinman (1972).

Consider an ideal dielectric medium that is nonmagnetic and homogeneous. Maxwell's equations (in MKS units) are

(4.2.4a) $$\nabla \times \mathbf{H} = \partial_t \mathbf{D}, \quad \nabla \times \mathbf{E} = -\partial_t \mathbf{B},$$

(4.2.4b) $$\nabla \cdot \mathbf{D} = 0, \quad \nabla \cdot \mathbf{B} = 0,$$

where

(4.2.4c) $$\mathbf{D}(\mathbf{E}) = \frac{1}{\mu_0 c^2} \mathbf{E} + \mathbf{P}, \quad \mathbf{B} = \mu_0 \mathbf{H},$$

and c is the speed of light in vacuo. These may be combined to

(4.2.5) $$\frac{1}{c^2} \partial_t^2 \mathbf{E} + \partial_t^2 \mathbf{P} + \nabla \times \nabla \times \mathbf{E} = 0.$$

The electron displacement, **Z**, in the lossless medium may be modeled as a forced anharmonic oscillator:

(4.2.6) $$m\mathbf{Z}_{tt} + \nabla_z U(\mathbf{Z}) = \mathbf{l}q\tilde{E}$$

where m is the effective mass of the oscillator, U denotes its potential energy, $\mathbf{l}(t)$ is directed along the electric dipole moment of the oscillation, q is the charge on an electron, and $\tilde{E}$ denotes the local field strength. In the simplest case, every molecule is identical and

(4.2.7) $$\mathbf{P} = Nq\mathbf{Z},$$

where N is the number of such oscillators per unit volume. The final result is that **P** satisfies

(4.2.8) $$\partial_t^2 \mathbf{P} + \nabla_P V(\mathbf{P}) = \frac{Nq^2}{m}\mathbf{E} = \frac{\Omega^2}{c^2}\mathbf{E},$$

where V includes both the sum over the individual potential energies and the difference between the local and macroscopic fields. Once V is specified, (4.2.5) and (4.2.8), along with boundary and initial conditions, determine the field. Note that both the nonlinearity and the anisotropy of the material are related to $V(\mathbf{P})$.

Many lossless dielectric materials are isotropic in the limit of weak fields, and (4.2.8) may be approximated by

(4.2.9) $$\partial_t^2 P_i + \omega_0^2 P_i + d_{ijk} P_j P_k \sim \frac{\Omega^2}{c^2} E_i, \qquad i = 1, 2, 3.$$

From symmetry considerations in (4.2.9), $d_{ijk} = d_{ikj}$. Yariv (1975) gives a list of values of (d_{ijk}) for several crystals. For crystalline quartz, $d_{111} \neq 0$, and all three interacting polarization vectors are parallel. For ammonium dihydrogen phosphate (ADP) and potassium dihydrogen phosphate (KDP), $d_{ijk} = 0$ unless all of the subscripts are different; i.e., P_1 and P_2 excite P_3, etc. (These two crystals are also anisotropic in the linear limit.)

If the material is isotropic, or is crystalline with a center of symmetry, then $d_{ijk} \equiv 0$ and the first nonlinear effect comes at the next order:

(4.2.10) $$\partial_t^2 P_i + \omega_0^2 P_i + c_{ijkl} P_j P_k P_l \sim \frac{\Omega^2}{c^2} E_i, \qquad i = 1, 2, 3.$$

For such a material there are no resonant triads, but resonant quartets become important.

In the linear limit, the solution of (4.2.5, 9) may be represented by transverse plane waves:

(4.2.11) $$\mathbf{E} = \mathbf{E}_0 e^{i\theta}, \quad \mathbf{P} = \mathbf{P}_0 e^{i\theta}, \quad \theta = \mathbf{k} \cdot \mathbf{x} - \omega t,$$

with $\mathbf{k} \cdot \mathbf{E}_0 = 0$. The result of this substitution is the linearized dispersion relation

(4.2.12) $$c^2 k^2 = \omega^2 - \frac{\omega^2 \Omega^2}{\omega^2 - \omega_0^2},$$

which is sketched in Fig. 4.12. Where the initial value problem has been relevant, we have specified dispersion relations in the form $\omega(\mathbf{k})$. For optical experiments it is more natural to consider $\mathbf{k}(\omega)$; i.e., ω is given and $\mathbf{k}$ is determined by the physics. Moreover, it is also standard in this field to define the index of refraction of the medium,

(4.2.13) $$\mathbf{n} = \mathbf{k} \frac{c}{\omega},$$

and write the dispersion relation as ($n^2 = \mathbf{n} \cdot \mathbf{n}$)

(4.2.14) $$n^2 = 1 - \frac{\Omega^2}{\omega^2 - \omega_0^2}.$$

Clearly this linear lossless model is inadequate for ω near ω_0, where the frequency of the applied field nearly matches the (linearized) atomic frequency

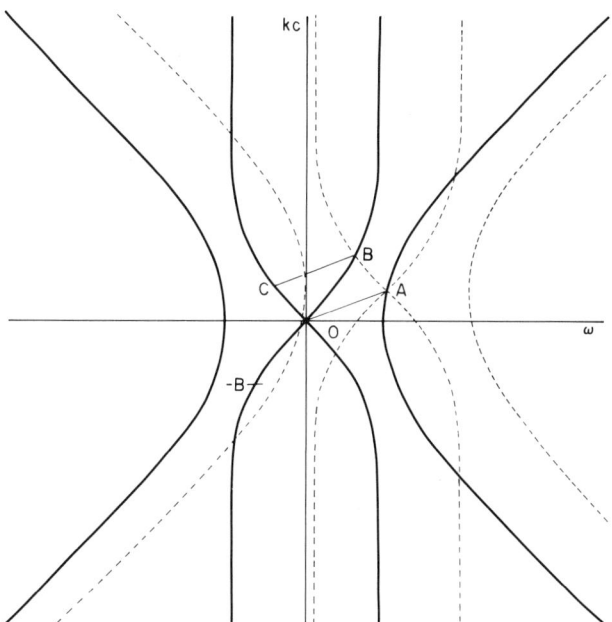

FIG. 4.12. *Linearized dispersion relation* (4.2.12) *for isotropic dielectric material*. A, $(-B)$ *and* C *form a resonant triad*: $\vec{A} + (-\vec{B}) + \vec{C} = 0$.

of the medium. We consider this resonant case more carefully in § 4.4. Here we consider only the nonresonant case.

We now illustrate in Fig. 4.12 a graphical method to find the resonant triads admitted by a linearized dispersion relation such as (4.2.12). This procedure has been discovered several times (e.g., Ziman (1960), Ball (1964)). First, pick any point A on one of the branches of the dispersion curve. Then reproduce all branches of the dispersion relation, with the origin translated from O to A. In Fig. 4.12, this is shown with dashed lines. Each intersection of a dashed and a solid curve (e.g., the point B) represents a second wave that can participate with A in a resonant triad. From B, draw a vector parallel and equal to $\overrightarrow{AO}$. By construction, this vector ends on a dispersion curve, at C. Therefore, the points A, B and C all lie on a dispersion curve, and therefore all represent solutions of (4.2.12). Moreover, it is evident that $\overrightarrow{OA} + \overrightarrow{OC} = \overrightarrow{OB}$, so that the waves represented by $(A, -B, C)$ also satisfy (4.2.1).

The triads obtained in this way are one-dimensional ($\mathbf{k}_1$, $\mathbf{k}_2$, $\mathbf{k}_3$ are collinear). More generally, if $\mathbf{k}_1$, $\mathbf{k}_2$, and $\mathbf{k}_3$ are only coplanar, the solutions of (4.2.12) lie on surfaces obtained by rotating Fig. 4.12 about the ω-axis. However, the geometric method just described to find resonant triads still applies in the higher dimensional problem.

If the dielectric material in question remains anisotropic in the linear limit, the dispersion relation necessarily is higher dimensional. Again, the geometric method applies, although it is apparent that it can become rather involved.

Next, let us find a weakly nonlinear solution of (4.2.5, 9). An appropriate measure of nonlinearity is

$$\varepsilon = \max\left\{\frac{d_{ijk}P_j}{\omega_0^2}\right\};$$

from (4.2.9), the field is weakly nonlinear if $\varepsilon \ll 1$. We seek a solution in the form

$$\mathbf{E}(\mathbf{x}, t, \varepsilon) = \varepsilon \mathbf{E}_1(\mathbf{x}, t, \varepsilon) + \varepsilon^2 \mathbf{E}_2 + O(\varepsilon^3),$$

(4.2.15a)
$$\mathbf{P}(\mathbf{x}, t, \varepsilon) = \varepsilon \mathbf{P}_1 + \varepsilon^2 \mathbf{P}_2 + O(\varepsilon^3),$$

where

$$\mathbf{E}_1 = \sum_{m=1}^{N} \{\mathbf{A}_m(\mathbf{y}, \tau) \exp(i\theta_m) + \mathbf{A}_m^* \exp(-i\theta_m)\},$$

(4.2.15b)
$$\tau = \varepsilon t, \quad \mathbf{y} = \varepsilon \mathbf{x}, \quad \theta_m = \mathbf{k}_m \cdot \mathbf{x} - \omega_m t.$$

Some interpretive statements follow.

(i) In laboratory situations, where the incoming waves are controlled precisely, representing the field by a discrete set of N waves is quite realistic.

APPLICATIONS   305

However, if the incoming waves contain two waves of a resonant triad, then the third wave of the triad should be included in $\mathbf{E}_1$ as well. It may have zero amplitude initially.

(ii) Time and space reverse their usual roles in this problem. One imposes an electric field at the boundary of the material for all time, and this field evolves in space as it progresses through the medium. Possible slow modulations of the incoming wave packet are specified by variations of $\mathbf{A}_m$ with respect to τ.

(iii) The notion of "slow" modulations should be placed in context here. The period of a wave emitted from a ruby laser is about 2×10^{-15} seconds. In order to achieve higher field intensities, laser beams often are pulsed (Q-switched) with an on-time so short it may be measured in picoseconds (i.e., 10^{-12} sec). Each such pulse contains about a thousand waves, and may be considered a slow modulation!

At the lowest order in the expansion in (4.2.15) we simply reproduce the solution of the linear problem,

$$\mathbf{P}_1 = \sum_{m=1}^{N} \frac{\Omega^2 c^{-2}}{\omega_0^2 - \omega_m^2} \{\mathbf{A}_m \exp(i\theta_m) + \mathbf{A}_m^* \exp(-i\theta_m)\},$$

where $(\mathbf{k}_m, \omega_m)$ are related by (4.2.12).

Let us suppose that this linearized solution contains a single resonant triad, in which $(\mathbf{k}, \omega)$ satisfy (4.2.1), and that the field vectors have the constant orientation required by d_{ijk}. For example, in a quartz crystal ($d_{111} \neq 0$) we may take

$$\mathbf{A}_m = \mathbf{v} a_m(\mathbf{y}, \tau), \quad m = 1, 2, 3,$$

where $\mathbf{v}$ is a constant unit vector and each a_m is a scalar. This solution at the lowest order generates secular terms at the next order unless the three resonant waves interact over a long space scale according to

(4.2.16a)
$$\partial_\tau a_1 + (\mathbf{c}_1 \cdot \nabla_y) a_1 = i\gamma_1 a_2^* a_3^*,$$
$$\partial_\tau a_2 + (\mathbf{c}_2 \cdot \nabla_y) a_2 = i\gamma_2 a_3^* a_1^*,$$
$$\partial_\tau a_3 + (\mathbf{c}_3 \cdot \nabla_y) a_3 = i\gamma_3 a_1^* a_2^*,$$

where

(4.2.16b)
$$\mathbf{c}_1 \cdot \nabla_y = \sum_{j=1}^{3} \left.\frac{\partial \omega}{\partial k_j}\right|_{\mathbf{k}=\mathbf{k}_1} \frac{\partial}{\partial y_j},$$

$$\gamma_i = \frac{c^2 d_{ijk} \omega_i}{\left[1 + \frac{\omega_0^2 \Omega^2}{(\omega_i^2 - \omega_0^2)^2}\right] \prod_{m=1}^{3} (\omega_m^2 - \omega_0^2)}.$$

Because $d_{ijk} \geq 0$ and $(\omega_1, \omega_2, \omega_3)$ cannot all have the same sign (from (4.2.1)), $\gamma_1, \gamma_2, \gamma_3$) cannot either. Thus, we have the decaying instability discussed in §2.1.

Second harmonic generation is a special case of a resonant triad in which $\omega_3 = \omega_1$, $\omega_2 = -2\omega_1$. Furthermore, we may identify a_3 with a_1, and scale the equations to

(4.2.17)
$$\partial_\tau a_1 + (\mathbf{c}_1 \cdot \nabla_y) a_1 = i a_1^* a_2^*,$$
$$\partial_\tau a_2 + (\mathbf{c}_2 \cdot \nabla_y) a_2 = -2i(a_1^*)^2.$$

Then the second harmonic (a_2) is generated by the fundamental (a_1), even if it vanished initially. This special case has received a great deal of experimental attention in nonlinear optics. It happens to be a singular limit in the IST formulation (see Kaup (1978)).

Let us now restrict our attention to the steady one-dimensional problem, where all three waves propagate along one axis; i.e., $(\mathbf{k}_1, \mathbf{k}_2, \mathbf{k}_3)$ are parallel and $(\mathbf{A}_1, \mathbf{A}_2, \mathbf{A}_3)$ are oriented in accord with d_{ijk}. We may denote the significant spatial coordinate by (y). Then (4.2.16) reduce to coupled ordinary differential equations:

(4.2.18)
$$\dot{a}_1 = \frac{i\gamma_1}{c_1} a_2^* a_3^*,$$
$$\dot{a}_2 = \frac{i\gamma_2}{c_2} a_3^* a_1^*,$$
$$\dot{a}_3 = \frac{i\gamma_3}{c_3} a_1^* a_2^* m$$

where $\dot{a} = \partial_y(a)$.

Two integrals of these equations, sometimes known as the Manley–Rowe relations, are

(4.2.19)
$$\frac{c_1|a_1|^2}{\gamma_1} - \frac{c_2|a_2|^2}{\gamma_2} = \text{const.},$$
$$\frac{c_1|a_1|^2}{\gamma_1} - \frac{c_3|a_3|^2}{\gamma_3} = \text{const.}$$

Conservation of energy follows from these:

(4.2.20)
$$\sum_{j=1}^{3} \frac{k_j|a_j|^2}{\omega_j} = \text{const.},$$

where we have used (4.2.16b), (4.2.12) and (4.2.1). Thus the total energy of the incoming waves is shared by the three interacting waves. These relations also have a quantum-mechanical interpretation, discussed by Akhmanov and Khokhlov (1972). Ball (1964) noted that the complete solution of (4.2.18) may be given in terms of elliptic functions.

What about the experimental evidence? The field of nonlinear optics may be said to have started with the experimental demonstration of second harmonic generation by Franken, Hill, Peters and Weinreich (1961).

They focused a "steady" ruby laser beam (0.6943 μm) on the front surface of a quartz crystal, and detected radiation emitted from the back surface at twice the frequency of the incoming beam, i.e., blue light at 0.347 μm. (In this context, a pulse whose duration exceeds 10^{-6} sec may be considered steady.) The fraction of power converted to the second harmonic in this steady, one-dimensional experiment was only about 10^{-8}. More recent experiments which are much more sophisticated are capable of converting a significant fraction of the total power to the second harmonic. A striking demonstration of this effect is shown in Fig. 4.13 (see insert following page 310). Further discussion may be found in Yariv (1975) or in Kleinman (1972).

It should be noted that this experiment has a very practical consequence. One obtains in this way a source of coherent light at twice the frequency of a ruby laser; i.e., one may build a "blue laser." Obviously, for this application it is desirable to convert as much of the power as possible to the harmonic mode.

This is one application of a three-wave interaction in nonlinear optics. In fact, several rather clever applications of (4.2.1) have been made in this field, and are discussed by Akhmanov and Khokhlov (1972) and by Yariv (1975). We mention only two here.

(i) *Parametric oscillation* occurs when a laser beam at a high frequency $(-\omega_3)$ is used to "pump" up the amplitudes of two lower frequency signals at ω_1, ω_2 ($\omega_1 + \omega_2 = -\omega_3$) in a resonator. The nonlinear interaction must be strong enough that the waves at ω_1, ω_2 receive energy at least as fast as they lose it to imperfect mirrors, etc. The practical benefit here is that, although ω_3 is fixed by the resonant transition frequencies of the laser that produced it, ω_1 and ω_2 are defined only by (4.2.1) and the linear dispersion relation of the medium. Thus, at some cost in efficiency, one gains the ability to "tune" the frequencies (ω_1, ω_2) of the pumped waves over a fairly large range. That is, one obtains a source of coherent light with a variable frequency.

(ii) *Frequency-up conversion* refers to an interaction in which an optical signal at a low frequency ω_1 is pumped by a strong laser beam at ω_2 to produce a signal at a high frequency $(-\omega_3 = \omega_1 + \omega_2)$. The practical value of this conversion is that it permits one to detect infrared radiation, for which existing detectors are limited, by converting the signal up towards the visible part of the spectrum where much better detectors are available.

For all of these applications, we should reiterate that even if the problem were one-dimensional, (4.2.18) is appropriate only if the incoming beams are steady. If the beams consist of sufficiently short pulses, then (4.2.16) is the appropriate model.

4.2.b. Internal waves. One of the fundamental goals of physical oceanography is to explain the dynamic origins of the spectra of internal waves measured in the oceans, such as those discussed by Garrett and Munk (1972), (1975). One aspect of this problem is to explain how internal waves are generated ab initio. Another is to explain how energy is transferred among the various internal wave modes. Resonant triads are thought to play a role in both of these processes, but here we discuss only the first (see also Phillips (1974), Watson, West and Cohen (1976), McComas and Bretherton (1977) and Olbers and Herterich (1979)).

One possible mechanism for generating internal waves is a resonant triad involving a combination of surface and internal waves. In particular, if the surface wave(s) are originally energetic and the internal wave(s) originally quiescent, then this triad may initiate internal wave motion.

The simple two-layer system discussed in § 4.1 admits resonant triads involving surface and internal waves (Ball (1964)). For simplicity, therefore, we will use this system to analyze the interaction. However, the reader should note that this system admits only one set of internal waves in addition to surface waves. Therefore, it can demonstrate how energy is transferred into the first internal wave mode, but not how the energy is transferred from this mode to other internal waves. This limitation is important if the system is to be used to study ocean waves.

The particular configuration in question was shown in Fig. 4.8. The linearized dispersion relation for waves in this system was given in (4.1.25); see Fig. 4.14. Using the geometric method of Ball (1964) (see also Segur (1980)), it is

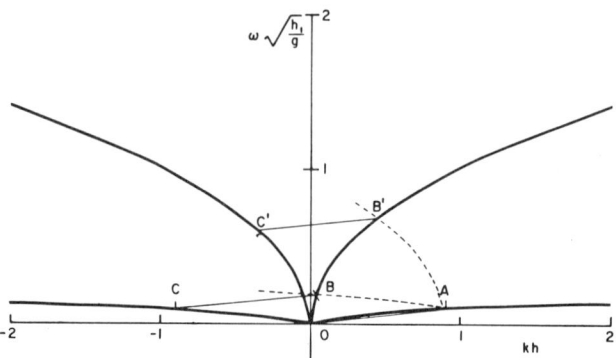

FIG. 4.14. *Linearized dispersion relation for two layer system, (4.1.25). $h_2 = \infty$ in the case drawn. Two resonant triads are shown, one involving two surface waves and the other involving two internal waves.*

easy to show that (4.1.25) admits several resonant triads, involving either two surface waves and one internal wave, or two internal waves and one surface wave.

Next, it is necessary to show that the interaction coefficients in (4.2.2) do not vanish. This calculation may be done in several ways, one of which is a standard application of the method of multiple time scales (see, e.g., Thorpe (1966), Joyce (1974)). In order to avoid some of the rather tedious algebra that one encounters by this approach, Simmons (1969) devised a variational method, which is computationally more efficient. Other derivations may be found in the references cited in Phillips (1974). Typically, these derivations involve uniform wave trains, so that the evolution equations are (4.2.2′) rather than (4.2.2). It is a straightforward extension of the analysis to include both spatial and temporal modulations, however. What is important for this discussion is that the interaction coefficients for resonant triads in the two-layer model do not vanish identically.

What are the consequences of these triads? For simplicity, consider first the case of spatially uniform wave trains. By renumbering the linearized wave modes, the kinematic condition (4.2.1) may be rewritten as

(4.2.21) $$\mathbf{k}_1 + \mathbf{k}_2 = \mathbf{k}_3, \qquad \omega_1 + \omega_2 = \omega_3,$$

with $\omega_i > 0$. Hasselmann (1967) noted that for a conservative system without spatial variation, (4.2.2a) may be written as

(4.2.22) $$\partial_\tau a_1 = i a_2^* a_3^*,$$
$$\partial_\tau a_2 = i a_3^* a_1^*,$$
$$\partial_\tau a_3 = -i a_1^* a_2^*;$$

i.e., the negative interaction coefficient in (4.2.22) is associated with the highest frequency in (4.2.21). This association is *independent* of the details of the interaction coefficients, provided only that they do not vanish. Now consider a uniform wavetrain of frequency ω_3 and dimensionless amplitude a_3 (of order unity). If a_1 and a_2 are infinitesimal initially, then for early times (4.2.22) implies that a_3 is constant, and that

(4.2.23) $$\partial_\tau^2 a_i \sim |a_3|^2 a_i, \qquad i = 1, 2.$$

Thus, the infinitesimal modes grow exponentially at the expense of a_3; i.e., a_3 is nonlinearly unstable with respect to perturbations in the triad defined by (4.2.21). (Bear in mind, however, that this instability is a short term effect; the general solution of (4.2.22) is periodic.)

When the resonant triads associated with the two-layer system are numbered to fit (4.2.21) with $\omega > 0$, ω_3 always is a surface wave (see, e.g., Fig. 4.14). Because the interaction coefficients do not vanish, it follows from Hasselmann's theorem that surface waves in a two-fluid system are unstable. The initial growth rate of the instability is proportional to the amplitude, $|a_3|$, from (4.2.23). We should note that this growth rate is *faster* than the growth rate of the Benjamin–Feir instability, which is proportional to $|a_3|^2$ (cf. § 4.3).

What is the experimental evidence of this instability? Lewis, Lake and Ko (1974) regard (4.2.22) as a stability problem, and are concerned only with the early growth of an internal wave mode at the expense of two forced surface waves. Their perturbation solution of (4.2.22) for early times predicts their observed results (growth rate, etc.) relatively well, but the experiment stops before the periodic nature of the solution becomes apparent.

Joyce (1974) also studied the growth of an internal wave at the expense of two surface waves. Viscous damping was comparable to the nonlinear growth in his experiments, so linear damping terms were added to (4.2.22). One of the primary objectives of his experiments was to observe a transient approach to a steady state, in which energy is continually fed to the surface waves at the same rate that it is dissipated by all three waves. Clearly, the periodic solutions of (4.2.22) have no bearing on these experiments.

McEwan (1971) noted that when linear damping terms are added to (4.2.22), there is a minimum amplitude of a_3, below which the instability disappears. He also demonstrated this amplitude cutoff experimentally (see also McEwan, Mander and Smith (1972)).

Note that all of these experiments examine modulations of uniform wave trains in time *or* space; i.e., they relate to (4.2.2′), rather than (4.2.2). Note also that the exact periodic solutions of (4.2.2′) were not observed in these experiments, either because of the short duration of the experiment because of strong damping effects that were omitted from (4.2.2′).

Now let us consider resonant triads of wave packets, modulated both in time and space, so that the governing equation is (4.2.2). This model is undoubtedly more relevant for interacting ocean waves. Because the kinematic condition (4.2.21) does not depend on the group velocities of the waves, three resonant wave packets (of finite spatial extent) can separate in space. In fact, they ordinarily do (cf. § 2.1). Thus, out of all the resonant triads admitted by the linearized dispersion relation (4.1.25), special attention should be given to any in which the group velocities also match, because their effective interaction time is longer.

There is a family of such triads admitted by (4.1.25) in which k_1 represents a long internal wave and k_2, k_3 represent short surface waves, with $k_1 = (k_3 - k_2) \ll k_3$. By (4.2.21),

$$(4.2.24) \qquad \frac{\omega_3 - \omega_2}{k_3 - k_2} = \frac{\omega_1}{k_1}.$$

For $(k_3 - k_2)$ sufficiently small, the left side of (4.2.24) is the group velocity of a surface wave at $k_3 (\approx k_2)$. The right side is the phase velocity of the internal wave. But because the internal wave is long, this is also its group velocity. Thus

FIG. 4.10. *Jupiter. The Great Red Spot is the large oval-shaped swirl below the equator. The Hollow lies just north of it. (Courtesy of NASA).*

FIG. 4.13. *Optical demonstration of second harmonic generation: an incoming beam of red light generates on outgoing beam of blue light in a crystal of ammonium dihydrogen phosphate (Photograph courtesy of R. W. Terhune).*

the long internal wave and the short surface wave propagate together. This is sometimes known as a long wave–short wave interaction (Gargett and Hughes (1972), Phillips (1974), Benney (1977), Rizk and Ko (1978), Ma and Redekopp (1979)).

Finally, we should note that we have considered no background shear in this discussion. Cairns (1979) and Craik and Adam (1979) demonstrated that if the equilibrium configuration consists of a stably stratified fluid with shear (i.e., a horizontal velocity, $U(z)$, with $U'(z) \neq 0$) then the interaction coefficients all may have the same sign. This is the "explosive instability," previously known only in plasmas (cf. Davidson (1972), Sugaya, Sugawa and Nomoto (1977)). In this case the three resonant waves extract energy from the background to produce a singularity in a finite time (in this approximation).

4.2.c. Resonant quartets. We have seen that resonant triads often represent the first nonlinear corrections to a linear theory in a small amplitude limit. The governing equations, (4.2.2) or (4.2.2)', are required to suppress secular terms at $O(\varepsilon^2)$. However, if no secular terms arise at this order, then these equations need not be satisfied, and the first nontrivial nonlinear correction arises at $O(\varepsilon^3)$. In nonlinear optics, this occurs if the dielectric is isotropic or is crystalline with a center of symmetry. In internal and surface waves, this occurs if one considers only interactions among different waves in the same vertical mode. In particular, surface waves in a homogeneous fluid admit no resonant triads (Phillips (1960)).

Resonant quartets arise at the next order if the appropriate interaction coefficients do not vanish. The resonance condition, corresponding to (4.2.1), is

(4.2.25) $\quad \mathbf{k}_1 + \mathbf{k}_2 + \mathbf{k}_3 + \mathbf{k}_4 = 0, \qquad \omega_1 + \omega_2 + \omega_3 + \omega_4 = 0,$

where $\omega_i = \omega(\mathbf{k}_i)$ from the linearized dispersion relation. Because we have assumed that $\omega(\mathbf{k}) = -\omega(-\mathbf{k})$, (4.2.25) always has solutions of the form

(4.2.26) $\quad \mathbf{k}_1 = -\mathbf{k}_2, \qquad \mathbf{k}_3 = -\mathbf{k}_4.$

Thus resonant quartets always are admitted by the dispersion relation. (This holds even if $\omega(\mathbf{k}) + \omega(-\mathbf{k}) \neq 0$.)

For resonant triads, the appropriate slow variables are $(\varepsilon \mathbf{x}, \varepsilon t)$. Because resonant quartets arise at the next order, their slow variables are $(\varepsilon^2 \mathbf{x}, \varepsilon^2 t)$; the nonlinear coupling is weaker, and the evolution is slower. The governing equations are:

(4.2.27)
$$\partial_\tau a_m + (\mathbf{c}_m \cdot \nabla) a_m = i \sum_{p=1}^{4} \beta_{mp} a_m a_p a_p^* + i \sum_{q,r,s \neq m} \gamma_{mqrs} a_q^* a_r^* a_s^*,$$

where $m = 1, 2, 3, 4$ and c_m is the linearized group velocity, evaluated at k_m (e.g., see Benney and Newell (1967)).

They also noted that, if (4.2.26) are the *only* solutions of (4.2.25), then the last sum in (4.2.27) vanishes and the simplified equations may be solved in closed form:

(4.2.28)
$$a_m(\mathbf{x}, \tau) = f_m(\mathbf{x} - \mathbf{c}_m \tau) \exp i \left\{ \int_0^\tau \sum_p \beta_{mp} |f_p(\mathbf{x} - \mathbf{c}_m \tau + (\mathbf{c}_m - \mathbf{c}_p)T)|^2 \, dT \right\},$$

where $a_m(\mathbf{x}, 0) = f_m(\mathbf{x})$ are the initial complex amplitudes. We are aware of no other general results about (4.2.27). In particular, the question of whether (4.2.27) is completely integrable seems to be open.

A quite different picture of four-wave interactions has been developed by Hasselmann (1962), (1963a, b) in the context of surface waves; see also Hasselmann et al. (1973), West, Thomson and Watson (1974) and Willebrand (1975). Whereas the effects of randomness are assumed to be weak relative to the effects of nonlinearity in deriving (4.2.27), exactly the opposite is true in the model of Hasselmann et al. The result is not (4.2.27), but a Boltzmann-type transport equation. Presumably the solutions of these two models do not agree; their basic assumptions are different. What is needed is a careful discussion of the regions of validity of each model. To our knowledge, this has not yet been given.

4.3. The nonlinear Schrödinger equation and generalizations. We saw in § 4.2 that one basic derivation can lead to the equations for either three-wave or four-wave (or higher) interactions, depending on details of the problem. The (cubic) nonlinear Schrödinger equation

(4.3.1) $$i\psi_t + \psi_{xx} + 2\sigma |\psi|^2 \psi = 0, \quad \sigma = \pm 1.$$

also follows from this kind of derivation in a linearly dispersive system; it corresponds to a different balance of terms.

We may make this assertion more explicit. To leading order, let some physical quantity be composed of N plane waves, with N complex scalar amplitudes, a_i. As shown in § 4.2, the equations for resonant triads or quartets arise if the N envelopes vary slowly in time and space so that

$$(\partial_t + \mathbf{c}_i \cdot \nabla) a_i \sim \text{nonlinear terms},$$

where $\mathbf{c}_i$ is the group velocity of the ith wave. A different balance is achieved if the nonlinear terms are much weaker than this, so that

$$(\partial_t + \mathbf{c}_i \cdot \nabla) a_i \sim 0.$$

Now each wave packet travels with its own group velocity and without interaction on this time scale. If none of the group velocities coincide, and if each wave packet is localized, then the packets separate in space. Therefore, the nonlinear interactions that occur on the next time scale can involve only the interactions of each wave packet with itself. In $(1+1)$ dimensions, (4.3.1) often is the governing equation on this next time scale. Subject to some restrictions, this conclusion also applies if the initial data were generated by a localized disturbance, rather than consisting of N plane waves.

The reader may have noticed in § 4.2 that if a linearized group velocity admits one resonant triad, it usually admits several (e.g., see Figs. 4.12 and 4.14). Additional justification is required to select a single "dominant" triad, whose governing equations are (4.2.2). There is no corresponding difficulty with (4.3.1) so long as no two linear group velocities coincide. In this case, only self-interactions are included because all other wave packets have dispersed away on an earlier time scale. However, because of transverse instabilities, the $(1+1)$-dimensional nature of (4.3.1) limits its physical relevance, as we shall see.

4.3.a. Nonlinear optics. Consider an isotropic dielectric material whose index of refraction exhibits a nonlinear correction in the presence of a moderately strong electric field:

$$(4.3.2) \qquad n(\omega, |\mathbf{E}|) = \frac{kc}{\omega} \sim n_0(\omega) + n_2(\omega) |\mathbf{E}|^2.$$

For definiteness, we take $n_0 > 0$. Akhmanov, Khokhlov and Sukhorukov (1972) discuss several mechanisms that contribute to such a nonlinear correction, and give appropriate references; see also Yariv (1975) and Karpman (1975). Among these mechanisms are the following.

(i) The "orientational (or high frequency) Kerr effect" refers to the tendency for anisotropic molecules in a liquid to align themselves with a strong electric field. Naturally, the medium becomes anisotropic in the presence of the field. This effect often produces the dominant contribution to n_2.

(ii) "Electrostriction" refers to the compression of a dielectric material by an applied field. The compression then changes the refractive index. This is usually the dominant effect in a liquid with isotropic molecules, in gases and in isotropic solids.

(iii) If the medium absorbs any of the light, the absorbed energy heats the material and expands it. Landauer (1967) has shown that $n_2 > 0$ in the absence of absorption. We will see that $n_2 > 0$ corresponds to self-focusing and $n_2 < 0$ to self-defocusing. Thus, self-defocusing can be realized only by mechanisms involving light absorption.

Whatever the mechanism, (4.3.2) assumes that the response of the medium is quasistatic for each frequency and quasilocal. Akhmanov, Khokhlov and Sukhorukov (1972) note that electrostriction and thermal effects are usually nonlocal, and that the latter are often nonstatic as well. We present next the simplest derivation of (4.3.1), which assumes the validity of (4.3.2). However, some nonlocal effects also can be included, as will be seen in the discussion of water waves (§ 4.3b).

In the simplest case, a steady plane wave at a fixed frequency traveling in the x_1-direction shines on the dielectric medium. The complex amplitude of the field may vary slowly in space, but not in time (by hypothesis). From (4.3.2), an appropriate measure of the nonlinearity is

$$\varepsilon^2 = \max_{x,t} \left| \frac{n_2 |\mathbf{E}|^2}{n_0} \right|;$$

we assume $\varepsilon \ll 1$. It is consistent with (4.3.2) to assume

(4.3.3) $\qquad c^2 \mathbf{P} = (n^2 - 1)\mathbf{E} \sim (n_0^2 - 1 + 2n_0 n_2 |\mathbf{E}|^2)\mathbf{E},$

so that (4.2.5) becomes

(4.3.4) $\qquad c^{-2} \partial_t^2 \{(n_0^2 + 2n_0 n_2 |\mathbf{E}|^2)\mathbf{E}\} + \nabla \times \nabla \times \mathbf{E} \sim 0.$

Introduce slow space scales,

(4.3.5) $\qquad y_1 = \varepsilon x_1, \quad y_2 = \varepsilon x_2, \quad y_3 = \varepsilon x_3, \quad \chi = \varepsilon^2 x_1.$

To leading order ($O(\varepsilon)$), the electric field take the form

(4.3.6) $\qquad \mathbf{E} = \hat{e} \varepsilon \{\psi(\mathbf{y}; \chi) \exp\{ikx_1 - i\omega t\} + (*)\} + O(\varepsilon^2),$

where $\hat{e}$ is a unit vector orthogonal to $\mathbf{k}$, and $\omega(k)$ is given by the linear dispersion relation (i.e., by $n_0(\omega)$, (4.2.14)). If we omit unnecessary homogeneous terms at the next order, we obtain

(4.3.7) $\qquad \dfrac{\partial \psi}{\partial y_1} = 0;$

i.e., no modulation of the wave occurs in the direction of propagation over distances that are $O(\varepsilon^{-1})$.

Secular terms arise at $O(\varepsilon^3)$ unless

(4.3.8) $\qquad 2ik \, \partial_\chi \psi + \nabla_\perp^2 \psi + \{2\omega^2 n_0^2 c^{-2} \operatorname{sgn}(n_2)\} |\psi|^2 \psi = 0,$

where $\nabla_\perp^2 = \partial_{y_2}^2 + \partial_{y_3}^2$ is the transverse Laplacian operator. Equation (4.3.8) was derived by Kelley (1965), and in a different form by Talanov (1965). In the steady problem under consideration, the shape of the incoming beam is specified by $\psi(y_2, y_3; \chi = 0)$. Then (4.3.8), along with boundary conditions in (y_2, y_3), determines its spatial evolution as it propagates across the dielectric medium in the x_1-direction.

Note that (4.3.8) is ε-independent, as are (4.3.7) and (4.2.14). Thus, these equations do not degenerate in the limit $\varepsilon \to 0$. This is a desirable feature, inasmuch as they were derived as part of an asymptotic expansion in small ε.

APPLICATIONS   315

We will discuss two experimental configurations. In the less conventional one, the incoming beam is modulated in only one transverse direction; say

$$\frac{\partial \psi}{\partial y_3} = 0.$$

In this case (4.3.8) describes evolution in $(1+1)$ dimensions. It may be rescaled to (4.3.1), and we may apply the results of inverse scattering theory (cf. Chapter 1). In particular, if $\psi \to 0$ rapidly as $|y_2| \to \infty$ and if $n_2 > 0$, then an incoming beam of sufficient strength will "self-focus" into N solitons plus some residual radiation (associated with the continuous spectrum). In this context, the solitons are seen as N straight lines (in the (y_2, χ)-plane) along which the intensity of the beam becomes constant as $\chi \to \infty$. These solitons are sometimes called "waveguides," and the parameter that usually gives the speed of the soliton gives the direction of propagation of the waveguide in this setting. The phenomenon is sometimes referred to as "self-trapping" to distinguish it from the more violent self-focusing that occurs in $(2+1)$ dimensions.

The notion of a self-trapped waveguide in a dielectric material is rather enticing, since the beam suffers no dispersion (or diffraction, as it is called in this context). Unfortunately, these $(1+1)$-dimensional waveguides are unstable with respect to long transverse perturbations (i.e., in the y_3-direction), as discussed in § 3.8. Presumably this instability is the reason for the lack of experimental evidence about them.

It may be appropriate to note here that the corresponding envelope solitons in water waves also are unstable to long transverse perturbations. Even so, they have been produced experimentally without undue difficulty; a wave record of one is shown in Fig. 4.16. The trick is to use relatively narrow wave tanks, so that the unstable transverse modes are excluded by the geometry. In principle, the same concept could be exploited to produce stable, self-trapped waveguides in dielectric media. However, we are aware of no experimental work along these lines.

We now consider the more conventional configuration associated with (4.3.8) in which the incoming beam has an approximately circular cross section. A fundamental question of interest is whether (4.3.8) can be solved by IST in $(2+1)$ dimensions. Apparently the answer is negative: at least (4.3.8) does not have the Painlevé property (cf. § 3.7). To see this, note that the ansatz

(4.3.9)
$$r = \lambda (y_2^2 + y_3^2)^{1/2},$$

$$\psi(y_2, y_3; \chi) = \lambda |2\omega^2 n_0^2 c^{-2}|^{-1/2} R(r) \exp\left(\frac{i\lambda^2 \chi}{2k}\right)$$

provides an exact reduction of (4.3.8) to the ordinary differential equation

(4.3.10)
$$R'' + \frac{1}{r} R' + \text{sgn}(n_2) R^3 - R = 0.$$

This equation has severe movable logarithmic singularities, so it is not of P-type. Based on the conjecture in § 3.7, therefore, we do not expect (4.3.8) to be solvable by IST, or to have a complete set of action-angle variables, or to show recurrence, etc.

Another fundamental question regards the nature of self-focusing in $(2+1)$ dimensions. Zakharov and Synakh (1976) proved that a class of solutions of (4.3.8) with $n_2 > 0$ must focus in a finite time; see § 3.8 for the proof. The existence of a blowup type of singularity in the solution of (4.3.8) indicates only the breakdown of the perturbation expansion that led to (4.3.8) as the approximate governing equation. It does not necessarily correspond to a singularity in the original (unperturbed) problem. However, the precise nature of the singularity in (4.3.8) may be used to guide the separate analysis that is required in this region. Estimates of the strength of the singularity have been made by Kelley (1965) and by Zakharov and Synakh (1976). Neither of these seems to be satisfactory, as discussed by Ablowitz and Segur (1979). The precise nature of the self-focusing singularity in the solution of (4.3.8) should be regarded as open at this time.

Akhmanov, Khokhlov and Sukhorukov (1972) discuss many of the experimental observations of self-focusing in $(2+1)$ dimensions. In some cases the focused beam was so intense that the material was physically damaged. As noted by Yariv (1975), the "phenomenon is of great concern to experimentalists working with very high power laser pulses since such damage can occur within the laser source itself."

Up to this point, we have considered only steady incoming beams. This restriction is physically undesirable because the high intensities required for (4.3.2) often are produced by Q-switched lasers with short or ultrashort pulses. It is also unnecessary. To remove it, we may define a slow time scale

$$(4.3.11) \qquad \tau = \varepsilon t$$

in addition to the scales in (4.3.5), and generalize (4.3.6) so that $\psi = \psi(\mathbf{y}, \tau; \chi)$. Then at $O(\varepsilon^2)$ (4.3.7) must be replaced by

$$(4.3.12) \qquad (\partial_\tau + c_1 \partial_{y_1})\psi = 0.$$

where c_1 is the one-dimensional linearized group velocity; i.e., the pulse propagates without distortion at the linearized group velocity of the carrier wave. At the next order, (4.3.8) becomes

$$(4.3.13) \quad 2ik\, \partial_x \psi + \alpha\, \partial_\eta^2 \psi + \nabla_\perp^2 \psi + \{2\omega^2 n_0^2 c^{-2} \text{ sgn }(n_2)\}|\psi|^2 \psi = 0,$$

where $\alpha = -k(\partial^2 k)/(\partial \omega^2)$ and $\eta^{\cdot} = \tau - y_1/c_1$. Here the evolution occurs in $(3+1)$ dimensions, but is qualitatively similar to that in $(2+1)$ dimensions. Equation (4.3.13) does not have the Painlevé property, and presumably cannot

APPLICATIONS   317

be solved by IST. The existence of a self-focusing singularity was proved by Zakharov and Synakh (1976) and by Glassey (1977) for $\alpha > 0$. The nature of the singularity seems to be less delicate in this case, and Zakharov and Synakh (1976) reasoned that as $\chi \to \bar{\chi}$,

(4.3.14) $$\psi \sim (\bar{\chi} - \chi)^{-1/2}.$$

4.3.b. Water waves. The nonlinear Schrödinger equation models the evolution of a one-dimensional packet of surface waves on sufficiently deep water. Various derivations have been given by Zakharov (1968), Benney and Roskes (1969), Hasimoto and Ono (1972), Davey and Stewartson (1974), Yuen and Lake (1975), Freeman and Davey (1975), Djordjevic and Redekopp (1977); see also Ablowitz and Segur (1979), Yuen and Lake (1980).

There are two significant differences between the problem of water waves and the nonlinear optics problems that we have discussed. The first is that in water of finite depth, the oscillatory waves induce a mean flow, which is a nonlocal effect. (Nonlocal effects also may arise in optical problems, but they were omitted from our earlier discussion.) The second difference is the interpretation of the equation, and the boundary conditions attached to it. Here are some of the contexts in which (4.3.1) or generalizations of it in $(2+1)$ dimensions arise.

(i) Waves at a single frequency may be produced by an oscillating paddle at one end of a long wave tank. The evolution of these waves is very similar to the evolution of the EM-waves discussed above.

(ii) A localized storm at sea generates a wide spectrum of waves, which propagate away from the source region in all horizontal directions. If the propagating waves have small amplitudes and encounter no wind away from the source region, then because of their dispersive nature they eventually sort themselves into nearly one-dimensional packets of nearly monochromatic waves. If the scales are chosen properly, a generalization of (4.3.1) in $(2+1)$ dimensions governs the long time evolution of each of these packets. If the packet is sufficiently localized, it is appropriate to require that the waves vanish far from the center of the packet.

(iii) Nearly monochromatic, nearly one-dimensional waves could cover a broad range of the sea as a result of a steady wind of long duration and fetch. The same generalization of (4.3.1) in $(2+1)$ dimensions can govern the evolution of these waves after the wind stops. In this case, periodic boundary conditions in the horizontal directions would be appropriate (but see Exercise 5).

Whatever the cirumstances, we are interested in a solution of (4.1.5)–(4.1.7) that consists primarily of a small amplitude, nearly monochromatic, nearly one-dimensional wave train. This wave train travels in the x-direction with an identifiable (mean) wavenumber, $\boldsymbol{\kappa} = (k, l)$. Denote by a the characteristic amplitude of the disturbance and by δk the characteristic variation in k. The nonlinear Schrödinger equation in $(2+1)$ dimensions is the consequence of assuming (with $\kappa^2 = k^2 + l^2$):

(i) small amplitudes,

(4.3.15a) $$\varepsilon \equiv \kappa a \ll 1;$$

(ii) slowly varying modulations,

(4.3.15b) $$\frac{\delta k}{\kappa} \ll 1;$$

(iii) nearly one-dimensional waves,

(4.3.15c) $$\frac{|l|}{\kappa} \ll 1;$$

(iv) balance of all three effects,

(4.3.15d) $$\frac{\delta k}{\kappa} = O(\varepsilon),$$

(4.3.15e) $$\frac{|l|}{\kappa} = O(\varepsilon).$$

The dimensionless depth, kh, can be finite or infinite, but to avoid the shallow water limit (and KdV), we need

(4.3.16) $$(kh)^2 \gg \varepsilon.$$

In this limit, the solution of the lowest order (linear) problem is

(4.3.17a) $$\phi \sim \varepsilon \left(\frac{\cosh k(z+h)}{\cosh kh} [\tilde{A} \exp(i\theta) + (*)] + \text{const.} \right)$$

where (*) denotes complex conjugate,

(4.3.17b) $$\theta = kx - \omega(k)t$$

and $\omega(k)$ is given by (4.1.8). To go to higher order, we introduce slow (dimensional) variables (again, using the method of multiple scales),

(4.3.18) $$x_1 = \varepsilon x, \quad y_1 = \varepsilon y, \quad t_1 = \varepsilon t, \quad t_2 = \varepsilon^2 t,$$

and expand ϕ and ζ:

$$\phi \sim \varepsilon \left\{ \Phi(x_1, y_1, t_1, t_2) + \frac{\cosh k(z+h)}{\cosh kh} [\tilde{A}(x_1, y_1, t_1, t_2) \exp(i\theta) + (*)] \right\} + O(\varepsilon^2),$$

(4.3.19) $$\zeta = \varepsilon \{\tilde{\zeta}_{11} \exp(i\theta) + (*)\} + O(\varepsilon^2), \quad \tilde{\zeta}_{11} = \frac{i\omega}{g + k^2 T} \tilde{A}.$$

These expansions must be carried out to $O(\varepsilon^3)$. The variations allowed in $\tilde{A}$ reflect the fact that this is a wave packet, rather than a uniform wave train, and $\tilde{\Phi}$ provides a mean motion generated by the packet. In what follows we shall only discuss the secular effects that the higher order terms have on $\tilde{\Phi}$, and $\tilde{A}$; details can be found in Benney and Roskes (1969), Davey and Stewartson (1974) or Djordjevic and Redekopp (1977).

At the next order of approximation, a secular condition requires that the wave packet travel with its linear group velocity,

(4.3.20) $$\frac{\partial \tilde{A}}{\partial t_1} + C_g(k)\frac{\partial \tilde{A}}{\partial x_1} = 0,$$

where $C_g = d\omega/d\kappa$. On this same time scale, $\tilde{\Phi}$ satisfies a forced wave equation,

(4.3.21) $$\frac{\partial^2 \tilde{\Phi}}{\partial t_1^2} - gh\left\{\frac{\partial^2 \tilde{\Phi}}{\partial x_1^2} + \frac{\partial^2 \tilde{\Phi}}{\partial y_1^2}\right\} = k\omega\beta_1 \frac{\partial}{\partial x_1}|\tilde{A}|^2,$$

where

$$\beta_1 = \frac{kC_g}{\omega}\,\text{sech}^2\,kh + \frac{2}{1+\tilde{T}},$$

$$\tilde{T} = \frac{k^2 T}{g} = (kh)^2 \hat{T}.$$

The solution of (4.3.21) changes dramatically, depending on whether or not

(4.3.22) $$gh > C_g^2.$$

If the ratio $C_g/\sqrt{gh}$ is interpreted as the "Mach number" of the wave packet, then (4.3.22) is the condition for "subsonic" flow. In this case, if $\tilde{A}$ has compact support, then $\tilde{\Phi}$ has a forced component that travels with speed C_g (i.e., it satisfies (4.3.20)), and a free component that radiates outward with speed $\sqrt{gh}$, and is $O(t_1^{-1/2})$ as $t_1 \to \infty$. Hence with (4.3.22), as $t_1 \to \infty$, we find that $\tilde{\Phi}$ satisfies both (4.3.20) and

(4.3.23) $$\alpha\frac{\partial^2 \tilde{\Phi}}{\partial x_1^2} + \frac{\partial^2 \tilde{\Phi}}{\partial y_1^2} = -\frac{k\omega}{gh}\beta_1 \frac{\partial}{\partial x_1}|\tilde{A}|^2,$$

where

$$\alpha = \frac{gh - C_g^2}{gh},$$

along with the boundary condition that $\tilde{\Phi}$ vanishes as $(x_1^2 + y_1^2) \to \infty$. These are the boundary conditions prescribed by Davey and Stewartson (1974), and they are correct without surface tension.

If the effects of surface tension are strong enough, (4.3.22) fails and the flow is "supersonic." Now, even if $\tilde{A}$ has compact support, $\tilde{\Phi}$ and its derivatives are nonzero along "Mach lines" that emanate from the support of $\tilde{A}$. In the limit $t_1 \to \infty$, $\tilde{\Phi}$ satisfies both (4.3.20, 23) as before. However, the appropriate boundary conditions for (4.3.23) now are that $\tilde{\Phi}$ and its derivatives vanish ahead of the support of $\tilde{A}$ (e.g., as $x_1 \to \infty$), and *no* conditions as $x_1 \to -\infty$. Hence, in general, we can not expect that global integrals involving $\tilde{\Phi}$ will converge.

The limit $t_1 \to \infty$ is of interest because the nonlinear Schrödinger equation appears when one eliminates secular terms on the *next* time scale, $t = O(\varepsilon^{-2})$. Carrying this out, and putting the result in dimensionless form, we define

(4.3.24)
$$\xi = \varepsilon k(x - C_g t), \qquad \eta = \varepsilon k y,$$
$$\tau = \varepsilon^2 (gk)^{1/2} t,$$
$$A = k^2 (gk)^{-1/2} \tilde{A}, \qquad \Phi = k^2 (gk)^{-1/2} \tilde{\Phi},$$

and find that A and Φ satisfy

(4.3.25a) $\qquad iA_\tau + \lambda A_{\xi\xi} + \mu A_{\eta\eta} = \chi |A|^2 A + \chi_1 A \Phi_\xi,$

(4.3.25b) $\qquad \alpha \Phi_{\xi\xi} + \Phi_{\eta\eta} = -\beta(|A|^2)_\xi,$

where

$$\sigma = \tanh kh, \quad \tilde{T} = \frac{k^2 T}{g}, \quad \kappa = \sqrt{k^2 + l^2},$$

$$\omega^2 = gh\sigma(1 + \tilde{T}) \geq 0,$$

$$\omega_0^2 = g\kappa.$$

$$\lambda = \frac{\kappa^2 \left(\frac{\partial^2 \omega}{\partial \kappa^2}\right)}{2\omega_0},$$

$$\mu = \frac{\kappa^2 \left(\frac{\partial^2 \omega}{\partial l^2}\right)}{2\omega_0} = \frac{\kappa C_g}{2\omega_0} \geq 0,$$

(4.3.26) $\qquad \chi = \left(\frac{\omega_0}{4\omega}\right) \left\{ \frac{(1-\sigma^2)(9-\sigma^2) + \tilde{T}(2-\sigma^2)(7-\sigma^2)}{\sigma^2 - \tilde{T}(3-\sigma^2)} + 8\sigma^2 \right.$

$$\left. - 2(1-\sigma^2)^2 (1+\tilde{T}) - \frac{3\sigma^2 \tilde{T}}{1+\tilde{T}} \right\}$$

$$\chi_1 = 1 + \frac{\kappa C_g}{2\omega}(1-\sigma^2)(1+\tilde{T}) \geq 0,$$

$$\alpha = \frac{(gh - C_g^2)}{gh},$$

$$\beta = \left(\frac{\omega}{\omega_0 kh}\right) \left\{ \frac{\kappa C_g}{\omega}(1-\sigma^2) + \frac{2}{1+\tilde{T}} \right\} \geq 0,$$

$$\nu = \chi - \frac{\chi_1 \beta}{\alpha}.$$

In the above formulae, all functions are evaluated at $l = 0$, since we are considering our underlying wave train to be propagating purely in the x-direction.

The coupled pair of equations (4.3.25) describe the evolution of the complex wave amplitude A and the mean flow ($\nabla \Phi$) induced by it. If the wave packet is local, it is appropriate to require that A vanishes as $\xi^2 + \eta^2 \to \infty$. As discussed above, the appropriate boundary conditions for Φ depend on the sign of α.

The character of the solution of (4.3.25) depends fundamentally on the signs of the coefficients in the equations. Fig. 4.15 is a map of parameter space,

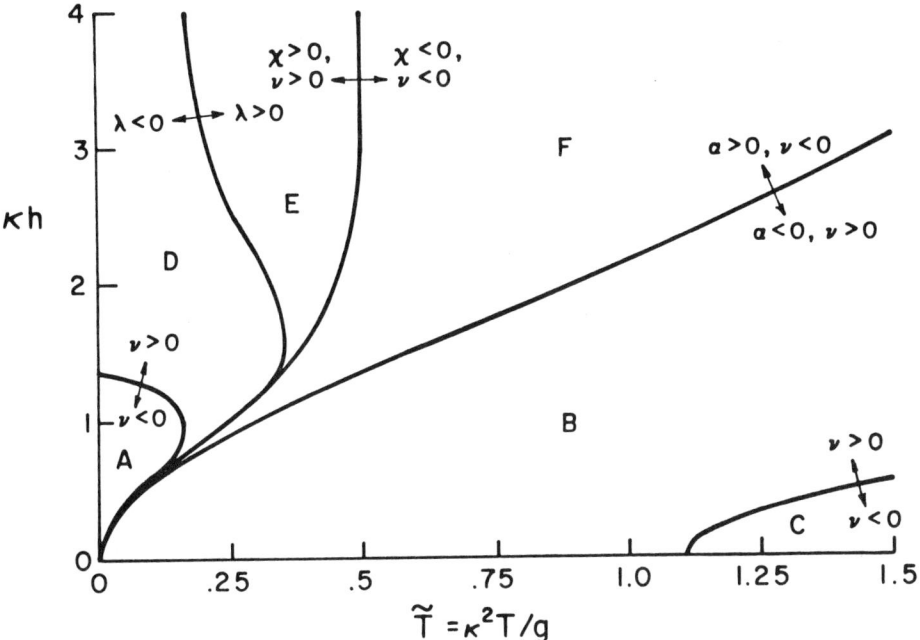

FIG. 4.15. *Map of parameter space for packets of oscillatory water waves, showing where the coefficients in (4.3.25) change sign. The dynamics of wave evolution is different in each region. Self-focusing is possible in region F. (Ablowitz and Segur (1979)).*

showing where these signs change. Each boundary line corresponds to a simple zero of a coefficient, as shown, except for the two curves bounding region F. These two curves denote singularities of ν. In a neighborhood of each of these two curves phenomena occur on a shorter time scale than the $O(\varepsilon^{-2})$ scale required elsewhere; cf. Djordjevic and Redekopp (1977).

This completes the derivation of the governing equations (4.3.27). However, two additional comments may help to place these equations in context before discussing their consequences.

(i) Recent work by Longuet-Higgins (1975), Cokelet (1977) and others has clarified some of the dynamics of water waves that are very close to breaking. There is apparently no overlap between their work and (4.3.25); they are concerned with large amplitude waves, whereas (4.3.25) describes waves of small amplitude.

(ii) Zakharov's (1972) equations of plasma physics are very closely related to (4.3.21, 25a). In $(1+1)$ dimensions, we may identify the oscillatory water wave with a high frequency Langmuir wave in the plasma, and identify Φ_ξ with the ion density. Then (4.3.21) describes the ion-acoustic mode, with the right-hand side representing the ponderomotive force, while (4.3.25a) with $\chi = 0$ describes the evolution of the Langmuir wave.

Now let us consider the solvability of (4.3.25) in various limiting cases. First, in the deep water limit $(kh \to \infty)$ the mean flow vanishes and (4.3.25) reduce to the nonlinear Schrödinger equation in $(2+1)$ dimensions,

(4.3.27)
$$iA_\tau + \lambda_\infty A_{\xi\xi} + \mu_\infty A_{\eta\eta} = \chi_\infty |A|^2 A,$$

where

$$\lambda_\infty = -\frac{\omega_0}{8\omega}\left(\frac{1-6\tilde{T}-3\tilde{T}^2}{1+\tilde{T}}\right),$$

$$\mu_\infty = \frac{\omega_0}{4\omega}(1+3\tilde{T}),$$

$$\chi_\infty = \frac{\omega_0}{4\omega}\frac{8+\tilde{T}+2\tilde{T}^2}{(1-2\tilde{T})(1+\tilde{T})}.$$

The appropriate boundary conditions for localized initial data are that A vanishes as $\xi^2 + \eta^2 \to \infty$.

As discussed above, this equation does not have the Painlevé property, and presumably cannot be solved by IST. Moreover, with periodic boundary conditions we do not expect to observe recurrence over arbitrarily long times. Yuen and Ferguson (1978) report approximate recurrence of a numerically integrated solution of (4.3.27) over a relatively short time. We expect that integration over a longer time scale will show that this recurrence is only approximately valid for short times. Subsequent work by Martin and Yuen (1980) seems to be consistent with this expectation.

In region F of Fig. 4.15 (i.e., sufficiently strong surface tension in sufficiently deep water), (4.3.27) has solutions that focus in a finite time, just as (4.3.8) does. Focusing is not limited to waves in infinitely deep water; solutions of (4.3.25) may also focus in region F. The phenomenon has not yet been observed in water waves, as it has been in optics. More details may be found in Ablowitz and Segur (1979).

The situation is much different in the shallow water limit of (4.3.25); i.e., $kh \to 0$, but $\varepsilon \ll (kh)^2$. In this limit we obtain, after rescaling,

(4.3.28)
$$iA_t - \sigma A_{xx} + A_{yy} = \sigma|A|^2 A + A\Phi_x,$$
$$\sigma \Phi_{xx} + \Phi_{yy} = -2(|A|^2)_x,$$

where $\sigma = \text{sgn}(\tfrac{1}{3} - \hat{T})$. These equations are of IST type, and explicit N-soliton solutions have been found (Ablowitz and Haberman (1975a), Anker and Freeman (1978)). Lump-type envelope hole solitons have been found by Satsuma and Ablowitz (1979). Thus, the solutions of (4.3.25) apparently are well behaved in the shallow water limit, but may be badly behaved in sufficiently deep water.

Finally let us restrict (4.3.25) to (1+1) dimensions. As noted by Ablowitz and Segur (1979) and by Hui and Hamilton (1979), many such restrictions are possible, corresponding to different directions in which modulations of the wave envelope are permitted. However, the only experimental evidence available is for waves that are modulated in the direction of wave propagation, so we consider only this case. If $\partial_\eta = 0$ in (4.3.25), the second equation may be integrated once, and the first equation becomes

(4.3.29)
$$iA_\tau + \lambda A_{\xi\xi} = \nu |A|^2 A,$$

with λ, ν defined by (4.3.26). Initial data can be created experimentally by modulating (in time) the stroke of an oscillating paddle at one end of a one-dimensional wave tank. If $\lambda\nu > 0$, as it is in regions A, B and E of Fig. 4.15, there are no solitons. Arbitrary initial data that vanish as $|\xi| \to \infty$ and are smooth evolve into a field of radiation that decays as $\tau^{-1/2}$; see § 1.7.

Envelope solitons are possible in regions C, D and F, where $\lambda\nu < 0$. The one-soliton solution of (4.3.29) is

(4.3.30) $$A = a\left|\frac{2\lambda}{\nu}\right|^{1/2} \text{sech}\{a(\xi - 2b\tau)\} \exp\{ib\xi + i\lambda(a^2 - b^2)\tau\}.$$

(Recall from (4.3.19) that the shape of the free surface is proportional to $[A\exp(i\theta) - (*)]$.) Fig. 4.16 shows the experimental measurements of such a wave at two downstream locations in a wave tank. We have superposed on these measurements the soliton solution of (4.3.29) with the appropriate peak amplitude at each location. Note that the wave amplitude has decreased in the second measurement, an indication of viscous effects. Even so, because the viscous time scale is longer than that in (4.3.29), the wave readjusts its shape to appear locally like a soliton.

Recall from § 1.1 that solitons are defined in terms of their ability to retain their identities despite interactions. Yuen and Lake (1975) gave an interesting experimental demonstration of this property, shown in Fig. 4.17. The first column shows an envelope soliton propagating without change over 9.15 m. The second column shows another wave at a somewhat different carrier-wave

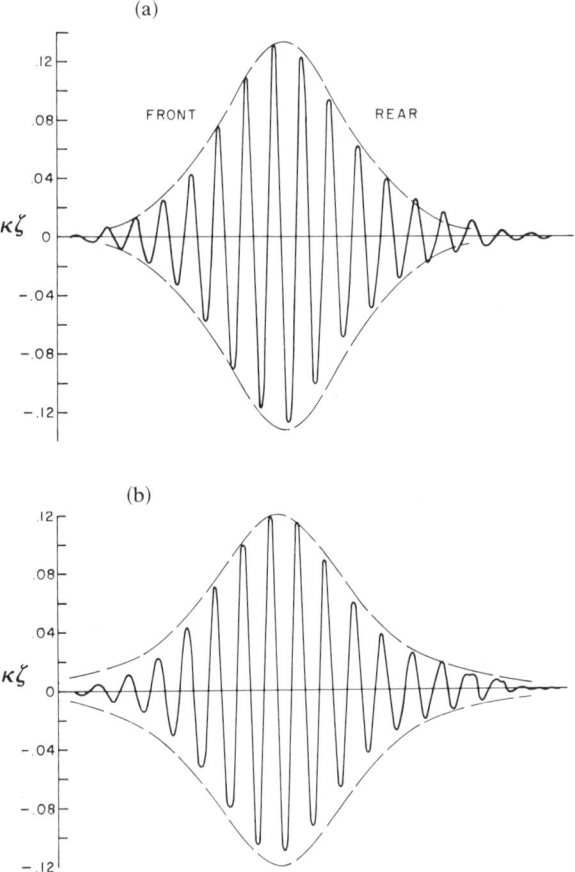

FIG. 4.16. *Measured displacement of water surface showing evolution of envelope soliton at two downstream locations.* $h = 1$ m, $kh = 4.0$, $\omega = 1$ HZ, $T = 1.0 \times 10^{-4}$. ———, *measured history of surface displacement*; – – –, *theoretical envelope shape*; $k\zeta = ka \operatorname{sech}(z)$, $z = [ag/\omega](\nu/8\lambda)^{1/2}(C_g t - x)$. (a) 6 m *downstream of wave maker*, $ka = 0.132$, (b) 30 m *downstream of wave maker*, $ka = 0.116$. (*Ablowitz and Segur* (1979)).

frequency. This wave is not a soliton, and some evolution of the wave packet is evident as it propagates over the same distance. The last column shows the interaction of these two wave packets. Even though the interaction is complicated, the final waveform measured is essentially a juxtaposition of the two waves recorded earlier; i.e., the waves regained their separate identities after the interaction.

Lake, Yuen, Rungaldier and Ferguson (1977) also consider (4.3.29) with periodic boundary conditions. Here the theory predicts recurrence, which they observe experimentally. They also relate the recurrence time to the initial (Benjamin–Feir) instability of the Stokes wave train at this frequency. Yuen and Lake (1980) give a more complete description of this work.

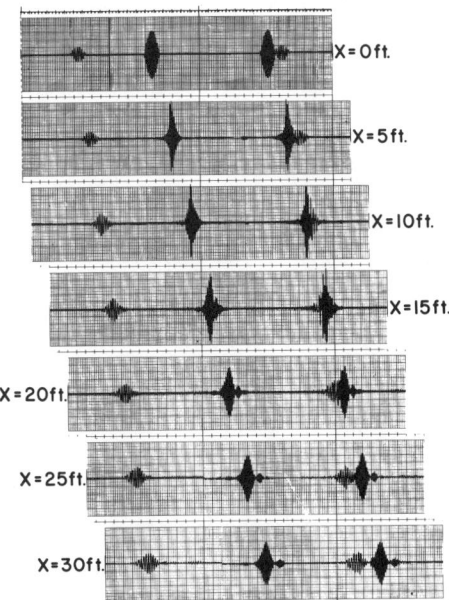

FIG. 4.17. *One water wave pulse overtaking and passing through another wave pulse. Left-hand trace: first pulse alone, $\omega_0 = 1.5$ HZ, initial $(ka)_{max} \simeq 0.10$, 6-cycle pulse. Center trace: second pulse alone, $\omega_0 = 3$ HZ, initial $(ka)_{max} \simeq 0.2$, 12-cycle pulse which disintegrates into two solitons. Right-hand traces: interaction of the two pulses. (Yuen and Lake (1975)).*

Every solution of (4.3.29) is also a solution of (4.3.25), of course, but the solitons are unstable with respect to long transverse perturbations, as discussed in § 3.8. The solitons shown in Figs. 4.16, 4.17 happened to be measured in relatively narrow tanks, which excluded the destabilizing modes. Thus, these same experiments could not have been run successfully in significantly wider tanks! Experimental evidence of this instability may be seen by comparing the waves in Figs. 4.16 and 4.18, both measured by Hammack (1979). The initial and ambient conditions for these two waves were nearly identical, except that Fig. 4.16 was recorded in a relatively narrow tank (86.4 cm wide), whereas Fig. 4.18 was recorded in a wider tank (244 cm). In particular, the wide tank admitted the destabilizing transverse modes that were excluded in the narrow tank. The unstable nature of the envelope soliton is evident in Fig. 4.18.

In summary, (4.3.25) governs the evolution of localized packets of oscillatory water waves of relatively small amplitude in $(2+1)$ dimensions. When the equations are restricted to $(1+1)$ dimensions they can be solved exactly by IST. This theory predicts with reasonable accuracy the experimental observations of waves that are also constrained to $(1+1)$ dimensions. However, because of the instability of the solitons to long transverse perturbations,

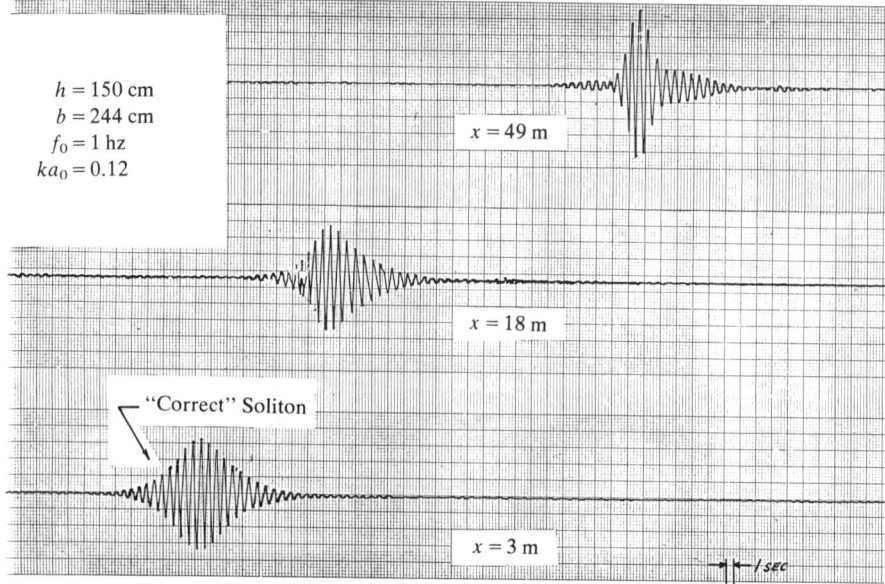

FIG. 4.18. *Evolution of water wave packet in a wide tank, showing the transverse instability that was absent in Fig. 4.16. (Courtesy of J. L. Hammack).*

neither the theory nor the experiments in (1 + 1) dimensions can be used to predict the evolution of the waves in (2 + 1) dimensions. No adequate theory in (2 + 1) dimensions is available at this time.

4.4. Equations of the sine-Gordon type. The sine-Gordon equation in (1 + 1) dimensions may be written as

(4.4.1a) $$\phi_{xx} - \phi_{tt} = \sin \phi$$

or as

(4.4.1b) $$\phi_{x\tau} = \sin \phi,$$

depending on the underlying coordinate system. Like the other equations we have studied in this chapter, it arises in a wide variety of applications; Scott, Chu and McLaughlin (1973) list several of these, with extensive references. We will see that in the sharp line limit of self-induced transparency, the equation arises when one eliminates secular terms in a perturbation expansion involving small amplitudes. Gibbon, James and Moroz (1979) have shown that a similar perturbation analysis of baroclinic instability of a rotating, two-layer fluid system also leads to (4.4.1). This seems to be the "typical" derivation of (4.4.1). It corresponds to those given in §§ 4.1, 4.2 and 4.3 in the sense that the evolution equation is a secular condition of a perturbation expansion carried to higher order. However, there are other applications, such as the description of pseudospherical surfaces in differential geometry and the propagation of magnetic flux in an infinitely long Josephson junction, in which the sine-Gordon equation appears without a perturbation scheme. The applications presented here have been chosen primarily to indicate the diversity of the phenomena described by the sine-Gordon equation.

4.4.a. Differential geometry. We begin with the earliest application known of the sine-Gordon equation. Equation (4.4.1) describes two-dimensional surfaces of constant negative curvature. In this case the model is exact, and the problem of interest has $(1+1)$ dimensions, so (4.4.1) is not a restriction of a higher dimensional problem. The important work of Bäcklund on transformations of surfaces (see § 3.1) was motivated by this application. The presentation given here, which assumes some knowledge of differential geometry, summarizes the more extensive discussion in Eisenhart (1909).

Consider a smooth two-dimensional surface embedded in a three-dimensional Euclidean space. Let y^i ($i = 1, 2, 3$) denote orthogonal Cartesian coordinates in the three-dimensional space, and let u^α ($\alpha = 1, 2$) denote some intrinsic coordinates on the surface. Three equations of the form

$$y^i = y^i(u^1, u^2)$$

define the surface. Let $\mathbf{r}(y^i)$ denote the position vector of a point P on the surface relative to the coordinate system of the enveloping space. If P is displaced infinitesimally to a new position on the surface, the position vector is changed by

$$d\mathbf{r} = \frac{\partial \mathbf{r}}{\partial u^\alpha} du^\alpha$$

(summation over repeated indices is implied in this section).

The element of arc length is defined by the first fundamental quadratic form,

(4.4.2) $$\mathrm{I} \equiv d\mathbf{r} \cdot d\mathbf{r} = g_{\alpha\beta} du^\alpha du^\beta,$$

where

$$g_{\alpha\beta} = \frac{\partial \mathbf{r}}{\partial u^\alpha} \cdot \frac{\partial \mathbf{r}}{\partial u^\beta}$$

is the covariant metric tensor of the surface. Similarly, let $\hat{\mathbf{n}}$ denote the unit normal vector to the surface at P. When P is displaced infinitesimally, $\hat{\mathbf{n}}$ changes by

$$d\hat{\mathbf{n}} = \frac{\partial \hat{\mathbf{n}}}{\partial u^\alpha} du^\alpha.$$

The second fundamental quadratic form is

(4.4.3) $$\text{II} \equiv -d\hat{\mathbf{n}} \cdot d\mathbf{r} = h_{\alpha\beta}\, du^{\alpha}\, du^{\beta};$$

$h_{\alpha\beta}$ is the extrinsic curvature tensor. Any smooth curve on the surface through a point P has a radius of curvature (in Euclidean space) at that point. By varying the direction of the curve through P, one finds a maximum (ρ_1) and a minimum (ρ_2) radius of curvature, corresponding to the two *principal directions* at P; these directions are orthogonal to each other. The *total (or Gaussian) curvature* of the surface at P is

(4.4.4) $$K = \frac{1}{\rho_1 \rho_2}.$$

The surface has negative total curvature ($K < 0$) at a point if the principal radii of curvature lie on opposite sides of the tangent plane at P. The bell of a trumpet, a saddle, and an ordinary potato chip all are examples of surfaces with negative curvature.

Let (λ^1, λ^2) denote the unit tangent vector to a curve C that passes through P. C is an *asymptotic line* if

(4.4.5) $$h_{\alpha\beta} \lambda^{\alpha} \lambda^{\beta} = 0$$

along C. There are two real, distinct asymptotic lines through a point on a surface if and only if the total curvature is negative there. If the entire surface has negative curvature, we may use the system of asymptotic lines to define the intrinsic coordinates on the surface.

Consider a surface of constant negative curvature ($K = -1/a^2$) with intrinsic coordinates defined by its asymptotic lines. By a suitable choice of scales, the first fundamental quadratic form may be written as

(4.4.6) $$\text{I} = a^2 (du^2 + 2\cos\phi\, du\, dv + dv^2),$$

where ϕ denotes the angle between the asymptotic lines. In terms of these coordinates, the equation of Gauss becomes the sine-Gordon equation

$$\frac{\partial^2 \phi}{\partial u\, \partial v} = \sin\phi.$$

(The equation of Gauss is a compatibility condition that an arbitrary tensor $h_{\alpha\beta}$ must satisfy to be the extrinsic curvature tensor of a surface.) Thus every solution of this equation defines a surface of constant negative curvature ($-1/a^2$), with its first fundamental form defined by (4.4.6). These surfaces are called *pseudospherical*. Some special cases are discussed by Eisenhart (1909, §§ 116, 117).

This application is our first example of an equation solvable by IST that arises as an exact model. No asymptotic expansion is even suggested in this derivation. Thus, the question arises as to whether the sine-Gordon equation is different from the other equations solvable by IST in some fundamental way. The answer is negative. Sasaki (1979a,b) showed that *every* equation of the form (1.5.16) describes surfaces with constant negative curvature. Different equations of this class simply represent different metrics. That the sine-Gordon equation, rather than the mKdV equation, should be associated with this geometric property is a matter of history, not mathematics.

It is natural to ask what equation(s) define pseudospherical surfaces in 3 or more dimensions, to obtain a generalization of (4.4.1) to higher dimensions. Recent work along these lines has been reported by Chern and Terng (1980). Whether the generalized sine-Gordon equation that they obtain can be solved by IST is unknown.

4.4.b. Self-induced transparency (SIT). Recall from § 4.2 that the linear index of refraction (i.e., the linearized dispersion relation) of an ideal dielectric material is singular at any resonant atomic frequency of the medium. Self-induced transparency is one of a variety of phenomena that may occur when a dielectric material is irradiated by an electric field at a frequency near a resonant frequency of the medium.

The phenomenon was discovered by McCall and Hahn (1965), (1967), (1969), (1970). By now there is an extensive literature on the subject; we mention specifically the papers by McCall and Hahn (1969), Lamb (1971), Slusher and Gibbs (1972), Courtens (1972) and Kaup (1977a), all of which were more or less reviews of the subject at the time they were written.

We begin with a physical description of self-induced transparency. In the simplest version of SIT, the dielectric material consists of two-level atoms, each of which has a ground state and an excited state. We assume that these two states cannot be further subdivided in terms of angular momentum; i.e., there is no *level degeneracy*. Further, the atoms are in their ground state initially; i.e., the medium is an *attenuator* rather than an *amplifier*. The incident electric field is tuned to the resonant frequency of these atoms, and excites the irradiated atoms. This transfer of energy from the electric field to the medium is usually irreversible, and it eventually depletes the energy of an electric pulse. The rate of energy absorption by the medium is given by Beer's law (1852).

SIT occurs when a sufficiently strong and very short incident pulse is properly shaped (in time) so that the front of the pulse loses energy (coherently) to the medium, which stores it for a time before returning it (coherently) to the back of the pulse. For a such a special pulse, the medium is left in its ground state, there is no net energy transfer, and the pulse propagates with a fixed reduced speed through the medium, which has become effectively transparent. This is self-induced transparency.

Next we give the governing (Maxwell–Bloch) equations. Maxwell's equations in an ideal dielectric material reduce to

(4.4.7) $$c^{-2} \partial_t^2 \mathbf{E} + \partial_t^2 \mathbf{P} + \nabla \times \nabla \times \mathbf{E} = 0.$$

In this equation, $\mathbf{P}$ represents the total polarization of the material, due to both resonant and nonresonant dipoles. (For example, McCall and Hahn's experiments were conducted in ruby, in which only the very small fraction of Cr^{+3} ions resonate. The "host medium", Al_2O_3, is nonresonant.) In this discussion of SIT, it will be convenient to let $\mathbf{P}$ represent only the polarization due to resonant or nearly resonant dipoles. This interpretation is permissible if we let c denote the phase speed of light in the medium when the (nearly) resonant dipoles are removed. Thus for ruby, c now denotes the phase speed of light in Al_2O_3, and $\mathbf{P}$ now represents the polarization due only to the Cr^{+3} ions.

In SIT problems, the resonant dipoles are assumed to be so sparsely distributed that they interact with the imposed electric field, but not with each other. Under these circumstances, we let $\mathbf{p}(\mathbf{x}, t; \omega)$ represent the polarization of an individual two-level dipole with transition frequency ω, and let $\tilde{\eta}(\mathbf{x}, t; \omega)$ represent the difference in normalized population densities between the excited and ground states. Thus, $|\tilde{\eta}| \leq 1$ and $\tilde{\eta} = -1$ if all the dipoles with frequency ω reside in the ground state (as they do as $t \to -\infty$, by assumption). Lamb (1971) has shown from quantum mechanical considerations that $\mathbf{p}$, $\tilde{\eta}$ and $\mathbf{E}$ are related by Bloch-type equations

(4.4.8a) $$\partial_t^2 \mathbf{p} + \omega^2 \mathbf{p} = -\frac{1}{3}\left(\frac{2\omega \not{p}^2}{\not{h}}\right) \mathbf{E} \tilde{\eta},$$

(4.4.8b) $$\partial_t \tilde{\eta} = \left(\frac{2}{\not{h}\omega}\right) \mathbf{E} \cdot \partial_t \mathbf{p},$$

where $\not{h}$ is Planck's constant and $\not{p}$ is the dipole matrix element for such a transition;

$$\not{p} = O(q\bar{r}),$$

where q is the charge on an electron and $\bar{r}$ is the average radius of the dipole. In the $(1+1)$-dimensional problem, the factor of $(\frac{1}{3})$ should be omitted from (4.4.8a). Note that, if we neglect the right-hand side of (4.4.8b), then (4.4.8a) is equivalent to the linearized version of (4.2.6).

Various slow relaxation and damping terms may be included in (4.4.8) as well (e.g., see Slusher and Gibbs (1972)). We omit them from this presentation, but we must then accept that our model will fail if individual dipoles remain excited for times comparable to these relaxation times: for Rb vapor, the shortest of these times is about 3×10^{-8} sec (Slusher and Gibbs (1972)).

The medium is said to be *inhomogeneously broadened* if the transition frequencies of the resonant dipoles are not identical, but merely clustered about the resonant frequency, ω_0. This broadening of the spectral line is due to Doppler frequency shifts in gases, and due to static crystalline electric and magnetic fields in solids (McCall and Hahn (1969)). In any case, we require

(4.4.9) $$|\omega - \omega_0| \ll \omega_0$$

for approximate resonance. Then if there are N_0 (= constant) resonant dipoles per unit volume, the total polarization is

(4.4.10) $$\mathbf{P} = N_0 \int_{-\infty}^{\infty} \mathbf{p}(\mathbf{x}, t, \omega) g(\omega) \, d\omega \equiv N_0 \langle \mathbf{p} \rangle,$$

where $g(\omega)$ is the probability density, representing the inhomogeneous broadening, normalized so that

$$\int g \, d\omega = 1.$$

Equations (4.4.7, 8, 10) are sometimes known as the Maxwell–Bloch equations without damping.

An essential ingredient of SIT is that the resonant dipoles are so sparsely distributed that their total polarization is weak, i.e.,

(4.4.11) $$|c^{-2} \partial_t^2 \mathbf{E}| \gg |\partial_t^2 \mathbf{P}|$$

in (4.4.7), so that backscattering may be neglected in that equation. Eilbeck (1972) showed that an appropriate measure of this sparseness is

(4.4.12) $$\delta^2 = \frac{N_0 \hat{p}^2 c^2}{2\hbar\omega_0} \ll 1,$$

which may be interpreted as a ratio of energies. δ is the small parameter required to derive the SIT equations from the Maxwell–Bloch equations.

Like all of the dynamical models previously discussed, the SIT equations arise in the limit of weak but finite fields. In this case, we require

$$|\mathbf{E}| = O\left(\delta \frac{\hbar\omega_0}{\hat{p}}\right).$$

Thus, we assume that the electric field takes the form of a weak transverse wave at frequency ω_0, with a slowly varying envelope, traveling in the x-direction. In the $(1+1)$-dimensional problem we have

(4.4.13a) $$\mathbf{E} \sim \frac{\hbar\omega_0}{2\hat{p}}[\delta\hat{\mathbf{j}}\{E(\chi, \tau) e^{i\theta} + E^* e^{-i\theta}\} + \delta^2 \mathbf{E}_1],$$

where

$$\theta = k_0 x - \omega_0 t, \quad \chi = \delta k_0 x, \quad \tau = \delta\omega_0 t.$$

We treat here the case of a linearly polarized field, but no essential changes are required for circular polarization. It is consistent with (4.4.7, 8, 10) to take

(4.4.13b) $\quad \omega = \omega_0(1 + 2\delta\alpha),$

(4.4.13c) $\quad \tilde{\eta} \sim \eta_0(\chi, \tau; \alpha) + \delta\eta_1,$

(4.4.13d) $\quad \mathbf{p} \sim \dfrac{\hbar}{2}\hat{\mathbf{j}}\{p(\chi, \tau; \alpha)\, e^{i\theta - i\pi/2} + p^*\, e^{-i\theta + i\pi/2}\} + \hbar\,\delta\mathbf{p},$

(4.4.13e) $\quad \mathbf{P} = \delta^2 \dfrac{2\hbar\omega_0}{c^2 \hbar^2} \langle\mathbf{p}\rangle.$

Then, at leading order, (4.4.7) yields

(4.4.14) $\quad k_0^2 = \dfrac{\omega_0^2}{c^2}$

so that the wavenumber of the carrier wave is determined by the host medium, with the resonant atoms absent. At the next order, secular terms arise (in $\hat{\mathbf{j}} \cdot \mathbf{E}_1$) unless

(4.4.15) $\quad \dfrac{\partial E}{\partial \chi} + \dfrac{\partial E}{\partial \tau} = \langle p \rangle.$

Similarly, removal of secular terms in (4.4.8) yields

(4.4.16a) $\quad \partial_\tau p + 2i\alpha p = E\eta,$

(4.4.16b) $\quad \partial_\tau \eta = -\tfrac{1}{2}(Ep^* + E^*p).$

In terms of a characteristic coordinate

$$X = \chi, \quad T = \tau - \chi,$$

we obtain the SIT equations in the form given by Lamb (1973):

(4.4.17)
$$E_X = \langle p \rangle,$$
$$p_T + 2i\alpha p = E\eta,$$
$$\eta_T = -\tfrac{1}{2}(Ep^* + E^*p).$$

The appropriate initial-boundary conditions are that for all $\chi > 0$, $E \to 0$ as $T \to \pm\infty$, $p \to 0$, $\eta \to -1$ as $T \to -\infty$, and that $E(\chi = 0, T)$ is given and vanishes rapidly as $T \to \pm\infty$. Aspects of these equations were analyzed by Eilbeck, Gibbon, Caudrey and Bullough (1973).

The sine-Gordon equation is a special case of (4.4.17) that arises in the sharp line limit, in which inhomogeneous broadening is neglected. In this limit,

$$g(\omega) = \delta(\omega - \omega_0)$$

in (4.4.10) and (4.4.17) becomes

(4.4.18)
$$E_x = p,$$
$$p_T = E\eta,$$
$$\eta_T = -\tfrac{1}{2}(Ep^* + E^*p).$$

The last two of these give

(4.4.19)
$$\eta^2 + |p|^2 = 1,$$

which suggests the substitution

(4.4.20a)
$$\eta = \cos\theta, \qquad p = \exp(i\psi)\sin\theta.$$

Then it follows that, if E has constant phase initially,

(4.4.20b)
$$\psi = \text{const.}, \qquad E = \exp(i\psi)\,\partial_T\theta,$$

and

(4.4.21)
$$\theta_{xT} = \sin\theta.$$

However, both (4.4.21) and (4.4.17) may be solved by IST, so that there is no compelling reason to restrict our attention to (4.4.21).

Following the important work of Lamb (1973), Ablowitz, Kaup and Newell (1974) considered the scattering problem,

(4.4.22a)
$$\partial_T v_1 + i\zeta v_1 = \tfrac{1}{2} E v_2,$$
$$\partial_T v_2 - i\zeta v_2 = -\tfrac{1}{2} E^* v_1,$$

along with the "time" dependence,

(4.4.22b)
$$\partial_x v_1 = A v_1 + B v_2,$$
$$\partial_x v_2 = C v_1 - A v_2.$$

Integrability of (4.4.22) requires

(4.4.23)
$$A_T = \tfrac{1}{2}(EC + E^*B),$$
$$B_T + 2i\zeta B = \tfrac{1}{2} E_x - AE,$$
$$C_T - 2i\zeta C = -\tfrac{1}{2} E_x^* - AE^*,$$

corresponding to (1.2.8). By comparing these equations to (4.4.17), they noted that a solution of (4.4.23) is

(4.4.24)
$$A(\chi, T; \zeta) = \frac{i}{4}\left\langle \frac{\eta}{\zeta - a} \right\rangle = \frac{i}{4}\int_{-\infty}^{\infty} \frac{\eta(\chi, T; \alpha)g}{\zeta - \alpha}\,d\omega(\alpha),$$
$$B = -\frac{i}{4}\left\langle \frac{p}{\zeta - \alpha} \right\rangle, \qquad C = -B^*.$$

It follows that (4.4.17) may be solved by IST, although some generalizations of the method are required for this problem. Aspects of the solution are discussed in detail by Lamb (1973), Ablowitz, Kaup and Newell (1974) and Kaup (1977a). Here are some of the main points.

(1) A soliton is often referred to in the literature as a "2π-pulse". The electric field envelope for an isolated soliton is given by

(4.4.25a) $$E(\chi, T) = 4\zeta_i \exp(-i\phi) \operatorname{sech} \psi,$$

where $\zeta = \zeta_r + i\zeta_i$ is the discrete eigenvalue for this soliton,

(4.4.25b) $$\psi = \Omega_i \chi - 2\zeta_i T + \psi_0,$$

(4.4.25c) $$\phi = \Omega_r \chi - 2\zeta_r T + \phi_0,$$

and

(4.425d) $$\Omega_r + i\Omega_i = -\frac{1}{2} \int_{-\infty}^{\infty} \frac{g \, d\omega(\alpha)}{\zeta_r - \alpha + i\zeta_i}.$$

This reduces to the solution first found by McCall and Hahn (1967) if $g(\omega)$ is symmetric about ω_0 and $\zeta_r = 0$. The dimensional speed of propagation of a 2π-pulse is

(4.4.26) $$v = c\left(1 + \frac{\Omega_i}{2\zeta_i}\right)^{-1},$$

so that the pulse always travels *slower* than the speed of light in the host medium. The resulting delay in travel time across an attenuator is a principal means of identifying SIT experimentally (Patel (1970)).

(2) A "0π-pulse" is the analogue of a breather. The electric field envelope in this case is given by

(4.4.27) $$E = 8\zeta_r \zeta_i \frac{\zeta_r \cosh \psi \sin \phi + \zeta_i \sinh \psi \cos \phi}{\zeta_r^2 \cosh^2 \psi + \zeta_i^2 \cos^2 \phi},$$

where ψ, ϕ are given by (4.4.25). These have special practical interest because there are corresponding solutions in the presence of level degeneracy (see Lamb (1971) and Kaup and Scacca (1980)).

(3) As discussed in detail in § A.2, the linearized version of (4.4.17) has no dispersion relation, and exhibits Landau-type damping. Correspondingly, the solution of (4.4.17) related to the continuous spectrum does not decay algebraically, as described in § 1.7, but exhibits (exponential) Landau-type damping as well. As shown by Ablowitz, Kaup and Newell (1974), the decay rate of $p(\chi, T; \alpha)$ is proportional to $g(\omega_0[1 + 2\delta\alpha])$. Thus, all modes damp exponentially if and only if $g(\omega) > 0$. The decay rate in Beer's law is usually based on $g(\omega_0)$.

(4) According to (1.3.16), there are no discrete eigenvalues, and therefore no self-induced transparency if

$$\int_{-\infty}^{\infty} |E|\, d\tau \leq 0.904.$$

Kaup (1977) discussed the relation of this result to McCall and Hahn's (1969) famous area theorem; see also Hmurcik and Kaup (1979).

(5) An arbitrary initial pulse of sufficient strength will break up into a finite number of 2π-pulses and 0π-pulses, plus radiation (or "ringing") that usually decays exponentially. Gibbs and Slusher (1970) demonstrated this behavior experimentally, as shown in Fig. 4.19. The left figure shows the results of a series of experiments in which the intensity of the initial pulse was varied. As shown, the final pulses show absorption as in (a), reshaping into a 2π-pulse as in (c), and breakup into two or three 2π-pulses as in (d), (e).

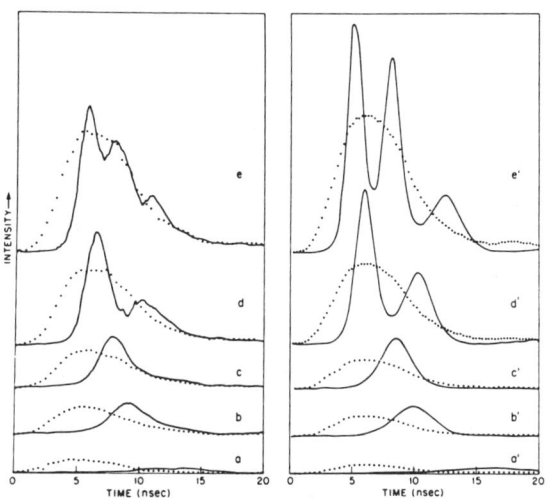

FIG. 4.19. *Evolution of optical pulses of different intensities in a nearly resonant medium, showing their input* ($\cdots$) *and output* (———) *shapes. The weak pulse in* (a) *is absorbed, whereas those in* (c), (d) *and* (e) *demonstrate SIT. Computer generated output pulses are shown in* (a') *through* (e'). (*Gibbs and Slusher* (1970)).

(6) The SIT equations, (4.4.17), contradict a great many general statements one would like to make about problems solvable by IST. Specifically, we should note that: (i) (4.4.17) has no linearized dispersion relation; (ii) except in special cases the process being described is irreversible; (iii) there are infinitely many local conservation laws, but (4.4.19) may be the only global constant of the motion; (iv) the transmission coefficient, $a(\zeta)$, is time dependent, and therefore does not generate action-type variables. These statements are all related, of

course. In particular, they all require some inhomogeneous broadening $(g(\omega) \neq \delta(\omega - \omega_0))$. Even so, they show what an anomalous problem SIT is.

This completes our discussion of the $(1+1)$ dimensional problem without damping or level degeneracy. As usual, in $(3+1)$ dimensions the question arises of whether the 2π-pulses and 0π-pulses are stable to transverse perturbations. Kodama and Ablowitz (1980), (1981) have shown that both kinds of pulses are unstable. This lateral instability had been observed experimentally by McCall and Hahn (1969), who described it in terms of "pulse stripping;" see also numerical work by Mattar and Newstein (1977). Even so, Slusher and Gibbs (1972) report that apertures at the beginning and end of the attenuator can be used effectively to control the instability in practice.

The problem of level degeneracy seems to be more serious. For most attenuators, suitable transitions show level degeneracy at one level or the other (or both). Lamb (1971) gives an extensive discussion of this problem. For our purposes it is sufficient to note that in the sharp-line limit $(g(\omega) = \delta(\omega - \omega_0))$, (4.4.20) must be replaced by

$$\theta_{xt} = \sin \theta + \lambda \sin 2\theta \qquad (4.4.28)$$

in the simplest case with level degeneracy. It is straightforward to show that the self-similar solution of this equation does not have the Painlevé property, and presumably cannot be solved by IST. Dodd, Bullough and Duckworth (1975) reached this same conclusion, because (4.4.28) does not have a Bäcklund transformation of a certain type. Numerical experiments by Ablowitz, Kruskal and Ladik (1979) confirm that two solitary wave solutions of (4.4.28) generate radiation during their interaction for $0 < \lambda < \infty$, although the radiation may be very weak under some conditions. Salamo, Gibbs and Churchill (1974) emphasize the weakness of the radiation on the basis of their experiments on sodium vapor, and Bullough and Caudrey (1978) also note the weakness of the radiation in their numerical experiments. The situation seems to be that (4.4.28) and its generalization to include inhomogeneous broadening are well *approximated* by exactly solvable problems. This property may be sufficient for practical applications. However, there is no reason to expect that either of these problems can be solved exactly by IST, or that they exhibit true self-induced transparency in the general sense.

4.4.c. General relativity. One of the most exciting potential applications of IST is that it may provide a new class of nontrivial solutions of Einstein's equations of general relativity. The significant advances in this direction have been made by Maison (1978), Belinskii and Zakharov (1978), and Harrison (1978); see also Maison (1979) and Neugebauer (1979). Because of the preliminary nature of their results, we will only outline the main points. We expect that this discussion soon will be obsolete.

APPLICATIONS

In four-dimensional space-time, both Maison (1978) and Belinskii and Zakharov (1978) use the convention in which the metric takes the form

$$-ds^2 = g_{ij} dx^i dx^j,$$

where g_{ij} has signature $(-+++)$. Maison is interested in a model of a steady rotating star (with no angular dependence), so he assumes

(4.4.29a) $\qquad ds^2 = \lambda_{ij} dx^i dx^j - h(dz^2 + dr^2), \qquad i, j = 1, 2,$

where both λ_{ij} and h depend only on (z, r), and $\det \lambda < 0$. Belinskii and Zakharov have cosmological questions in mind, and assume

(4.4.29b) $\qquad ds^2 = \lambda_{ij} dx^i dx^j - h(dz^2 + dt^2), \qquad i, j = 1, 2,$

where both λ_{ij} and h depend on (z, t), and $\det \lambda < 0$. However, one of these can be changed into the other by a complex change of variables, and their analyses are quite similar up to a point. In either case, we define

(4.4.30) $\qquad\qquad \tau^2 = -\det \lambda > 0.$

Einstein's equations in a vacuum are that the Ricci tensor should vanish,

(4.4.31) $\qquad\qquad R_{ij} = 0.$

For (4.4.29a), one component of (4.4.31) amounts to the (2×2) matrix equation,

(4.4.32) $\qquad\qquad \partial^i(\tau \lambda^{-1} \partial_i \lambda) = 0,$

where $i = 3, 4$ and $x^3 = z$, $x^4 = r$. We will omit the equations for h that are implicit in (4.4.31), because they may be integrated by quadratures once λ is known.

Maison introduces new coordinates

(4.4.33) $\qquad\qquad \xi = z + ir, \qquad \xi^* = z - ir,$

and notes that the trace of (4.4.32) becomes

(4.4.34) $\qquad\qquad \tau_{\xi\xi^*} = 0,$

which is trivially integrated. Up to this point, there is no essential difference between the work of Maison and that of Belinskii and Zakharov. Their works diverge beyond this point.

Maison defines new variables, α (real) and A (complex), and shows that (4.4.32) can be written as (4.4.34) plus

(4.4.35)
$$2A_{\xi^*} + \tau^{-1}\tau_\xi A^* \cos\alpha + \tau^{-1}\tau_{\xi^*} A = 0,$$
$$2A^*_\xi + \tau^{-1}\tau_{\xi^*} A \cos\alpha + \tau^{-1}\tau_\xi A^* = 0,$$
$$\alpha_{\xi\xi^*} + |A|^2 \sin\alpha - \mathrm{Re}\left\{\tau^{-1}\tau_\xi\left(\frac{A^*}{A}\right)\sin\alpha\right\}_\xi = 0,$$

which are the equations to be solved. These may be viewed as a generalization of the (Euclidean) sine-Gordon equation, to which they reduce if $\tau = A = 1$.

Maison then postulates a (2×2) scattering problem, and shows that its compatibility condition is (4.4.35). This is the crucial first step in solving these equations by IST. Moreover, if $\tau = A = 1$, his scattering problem reduces to the Euclidean version of (1.2.7) for the usual sine-Gordon equation.

The nature of his scattering problem is elliptic (rather than hyperbolic), and it prevents him from applying the usual IST formalism to solve (4.4.35). Even so, the scattering problem can be viewed as a Bäcklund transformation, as discussed in § 3.1. In fact, it led Harrison (1978) to a Bäcklund transformation for the Ernst equations, which are essentially equivalent to (4.4.32).

Next, we compare these results to those obtained by Belinskii and Zakharov (1978). They also find a scattering problem, whose integrability conditions are equations corresponding to (4.4.35). Their scattering problem seems to be much different from that of Maison, however. In fact, it seems to be a new kind of scattering problem altogether, inasmuch as it involves differentiation with respect to the eigenvalue. Even so, they use it to construct explicitly one- and two-soliton solutions by the method of reduction to a Riemann–Hilbert problem. They remark that the stationary solution of Kerr can be obtained from their solitons by a complex transformation.

4.5. Quantum field theory. Zabusky and Kruskal (1965) originally invented the word "soliton" to describe nonlinear waves that interact like particles. By now, solitons have become well-defined mathematical objects, and one can reverse this comparison. Here we discuss some of the work that indicates the extent to which physical particles can be described by solitons. This question has been the focus of a great deal of recent research in quantum field theory. The brief description given here is intended primarily as a guide to some of this literature.

Quantum field theory differs in several respects from the other applications we have discussed. One difference is that in many cases the equations with solitons (sine-Gordon, nonlinear Schrödinger, etc.) are not approximations of a larger set of governing equations. Rather, the soliton-equations are taken as models of the governing equations. More precisely, in many aspects of quantum field theory there are no "governing equations;" there are only principles of symmetry (such as Galilean or Lorentz invariance), and any dynamic model that satisfies these principles is considered legitimate. Consequently, there has been a good deal of interest in the quantized version of the nonlinear Schrödinger equation,

(4.5.1a) $$i\phi_t = -\phi_{xx} - \alpha |\phi|^2 \phi,$$

and the sine-Gordon equation

(4.5.1b) $$u_{xx} - u_{tt} = m^2 \sin u;$$

APPLICATIONS

(4.5.1a) is the simplest (nontrivial) infinite dimensional nonlinear Hamiltonian system that is completely integrable, while (4.5.1b) is the simplest completely integrable, nonlinear relativistic Hamiltonian system.

As always, the equations that have been solved are $(1+1)$-dimensional. Quantum results in $(1+1)$ dimensions have physical meaning in some aspects of solid state physics, but none in high energy physics. Physically relevant models of elementary particles have $(3+1)$ dimensions. Thus, most of the results obtained by quantizing equations like those in (4.5.1) should be considered suggestive of what might happen in more dimensions.

As discussed in § 1.6, most of the equations solvable by IST can be viewed as infinite dimensional Hamiltonian systems that are completely integrable. This viewpoint of the classical (i.e., nonquantum) equations is quite natural for the purpose of quantization. Thus the Hamiltonians for (4.5.1a,b) are

(4.5.2a) $$H_{\text{NLS}} = -i \int_{-\infty}^{\infty} \left\{ \phi_x^* \phi_x - \frac{\alpha}{2} \phi^* \phi^* \phi \phi \right\} dx,$$

(4.5.2b) $$H_{\text{SG}} = \int_{-\infty}^{\infty} \left\{ \frac{1}{2}(u_x^2 + p^2) + m^2(1 - \cos u) \right\} dx,$$

respectively. In these two cases the conjugate variables are (ϕ, ϕ^*) and (u, p). Hamilton's equations and Poisson brackets were defined in (1.6.22) and (1.6.29), respectively, and we should note that Hamilton's equations also may be written in terms of these Poisson brackets as

(4.5.3) $$q_t = -\langle H, q \rangle, \qquad p_t = -\langle H, p \rangle.$$

What does it mean to quantize these equations? First, it means that there is a Hilbert space $\mathcal{H}$, and the conjugate variables of the classical theory are now to be interpreted as operators which act on $\mathcal{H}$. Second, the Poisson brackets are to be replaced with "equal time commutation relations",

(4.5.4a) $$[\phi(x, t), \phi^*(y, t)] = \delta(x - y),$$

(4.5.4b) $$[p(x, t), u(y, t)] = \delta(x - y),$$

respectively, where $\hbar = 1$ and

$$[a, b] \equiv ab - ba.$$

More precisely, (4.5.4) means that for any two elements of $\mathcal{H}$ (denoted by α, β),

$$(\alpha, [p, q]\beta) = (\alpha, \delta(x - y)\beta),$$

where $(\cdot, \cdot)$ is the inner product on $\mathcal{H}$. In fact, all operator equations are to be interpreted in this way. Third, the dynamical equations, which now define the

evolution of operators, have the form
(4.5.5) $$p_t = -[H, p], \qquad q_t = -[H, q],$$
rather than (4.5.3).

There are some rather delicate questions in this process. One of them is to define $\mathcal{H}$ explicitly in such a way that all of the quantities in the theory are meaningful. This question is sometimes considered pedantic, but Oxford's (1979) result that only a few of the infinite set of motion constants for (4.5.1a) exist in the quantized version suggests that the question should be taken seriously.

Another delicate question relates to factor ordering in the quantum problem. Thacker and Wilkinson (1979) used (4.5.2a) as the quantized Hamiltonian corresponding to (4.5.1a), whereas Kaup (1975) used

(4.5.6) $$H = -i \int \left\{ \phi_x^* \phi_x - \frac{\alpha}{2} \phi^* \phi \phi^* \phi \right\} dx.$$

These are equivalent in the classical problem, but not in the quantum problem because ϕ and ϕ^* no longer commute. In fact, their final results differ, presumably because of differences in their ordering of factors.

Discrepancies such as this indicate that a complete solution of the quantum problem is not obvious, even when the classical problem is well understood. Historically, quantization of these completely integrable problems has proceeded in steps of increasing sophistication. The first step (i.e., the lowest level of approximation) is called quasiclassical. For the nonlinear Schrödinger equation, this consists of transforming the classical problem to its action-angle variables (see § 1.6), and then quantizing these variables (Kaup (1975), Kulish, Manakov and Faddeev (1976)). As noted by the latter authors, there is no guarantee a priori that transforming and then quantizing is the same as quantizing and then transforming. However, in this case the results obtained by quasiclassical quantization are equivalent (up to factor ordering) to those obtained by using the Bethe ansatz (1931) to solve the fully quantized problem, as had been done by Lieb and Lininger (1963), Berezin, Pokhil and Finkel'berg (1964), and McGuire (1964). In other words, in this case the quasiclassical approximation gives results that happen to be exact. (The works of Thacker (1978) and Oxford (1979) are also of interest here.)

For the sine-Gordon equation, quasiclassical (or semi-classical) quantization is somewhat more complicated. Even so, ingenious methods were developed by Dashen, Hasslacher and Neveu (1974), (1975) and by Korepin and Faddeev (1975) to take explicit solutions of the classical problem and to calculate quantum corrections to them. We should emphasize that these results do not rely on the usual perturbation theory used in this field, which corresponds roughly to a small amplitude expansion in the classical problem. Dashen et al. also applied their methods to solutions of other models, including the Gross–Neveu model.

APPLICATIONS 341

The second level of refinement has been to develop a fully quantized version of IST completely parallel to that developed in Chapter 1, but in which all functions (potentials, eigenfunctions, scattering data) are replaced by operators. When this version is completed, the classical problem will be used only as a guide. All variables will be quantum mechanical. Recent work on several quantum models along these lines has been reported by Sklyanin and Faddeev (1978), Sklyanin (1979), Thacker and Wilkinson (1979), Bergknoff and Thacker (1979), Honerkamp, Weber and Wiesler (1981) and Takhtadzhyan (1981). Some of this work relies heavily on the important contributions of Baxter (1972). At this time it appears that a fully quantized version of the direct scattering problem is available for several quantum models, including those in (4.5.2). The inverse scattering problem has not yet been quantized.

How to quantize solitons is a question that is being answered now, and a comprehensive review of the subject will likely be given in the next few years. Our objective here has been simply to identify some of the important work in the subject. In addition to the papers already mentioned, we should also note:

(i) the introductory articles of Flaschka and McLaughlin, Hasslacher and Neveu, Campbell, Noll, and Sutherland in the Conference Proceedings edited by Flaschka and McLaughlin (1978);

(ii) the article by Rebbi (1979) in Scientific American (in this article, we interpret "solitary wave" where the author writes "soliton");

(iii) the survey of integrable quantum systems by Ruijsenaars (1980), which contains an extensive bibliography;

(iv) the application of the usual perturbation theory to quantum solitons (e.g., Callan and Gross (1975)); and

(v) the work of Korepin, Kulish and Sokolov in the Proceedings edited by Zakharov and Manakov (1981).

EXERCISES

Section 4.1

Physical applications of these completely integrable, long wave models are virtually endless, and each application requires a new description of the physical problem, the terminology, the relevant physical questions, etc. All of the exercises given here are related to one application, chosen because it is common to almost everyone's experience: the transverse vibrations of a stretched string, such as a guitar string or a telephone cord. Background material may be found in Love (1944), Fung (1965) or Mott (1973).

1. Consider an infinitesimal element of a stretched string, whose undeformed cross section is uniform, as in Fig. 4.20.

(a) Justify

$$\rho \frac{\partial^2 w}{\partial t^2} = \frac{\partial}{\partial x}(\tilde{T} \sin \alpha + F \cos \alpha),$$

342   CHAPTER 4

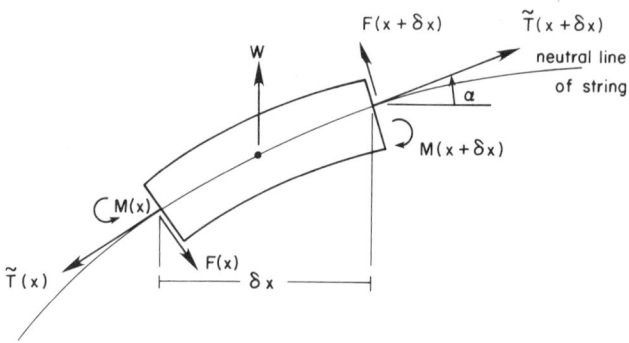

FIG. 4.20

where ρ is the (constant) linear density of the material and

$$\frac{\partial w}{\partial x} = \tan \alpha.$$

(b) Show that, if rotary inertia may be neglected,

$$0 = -\frac{\partial M}{\partial x} + \frac{F}{\cos \alpha}.$$

(c) The Euler–Lagrange hypothesis is

$$M = \frac{EI}{R} = -EI\frac{\partial}{\partial x}(\sin \alpha),$$

where E is Young's modulus, I is the moment of inertia of the cross section about an axis on the neutral line, and R is the radius of curvature of the neutral axis. Combine these to obtain

$$\rho\frac{\partial^2}{\partial t^2}\frac{\partial w}{\partial x} = \frac{\partial^2}{\partial x^2}\left\{\tilde{T}\sin\alpha - EI\cos^2\alpha\frac{\partial^2}{\partial x^2}\sin\alpha\right\}.$$

What is the physical meaning of each term in this equation?

(d) Let T denote the constant tension of the undeflected string. Justify

$$\tilde{T} = T\left(1 + \frac{\mu}{2}\left(\frac{\partial w}{\partial x}\right)^2\right) + O\left(\left(\frac{\partial w}{\partial x}\right)^4\right),$$

where μ is a nonnegative empirical constant. Under what conditions may variations in T be neglected? (Here it is convenient to consider both longitudinal and transverse modes.)

2. The motion of the string is determined by the equation in (c) in Exercise 1 along with those in (a) and (d). Based on these equations, what constitutes

a long wave? What is a small amplitude wave? Derive the dimensionless equation

$$\frac{\partial^2 u}{\partial \hat{t}^2} = \frac{\partial^2}{\partial \hat{x}^2}\left\{u(1-2\varepsilon^2 u^2) - \delta^2 \frac{\partial^2 u}{\partial \hat{x}^2}\right\} + O(\varepsilon^4, \delta^4, \delta^2\varepsilon^2),$$

where ε is a measure of small amplitudes and δ a measure of long waves. Show that $\delta = O(\varepsilon)$ results in minimal simplification of the equation. In the linear limit $(\varepsilon \to 0)$, this equation was derived by Mott (1973) in a much different way. We consider three possible boundary conditions for this equation.

A. A guitar string may be considered clamped at two points, a distance L apart. The wavelength of the disturbance, λ, typically is $O(L)$.

B. The string is clamped at two points a distance L apart, but $L \gg \lambda$.

C. The string is being wound under tension from one spool onto the other. The distance between the spools is L, and the speed of the string is V.

3. Take $\delta = \varepsilon$ in the equation in Exercise 2. Expand

$$u = u_0 + \varepsilon^2 u_1 + \varepsilon^4 u_2 + \cdots.$$

(a) Show that the general solution of the leading order equation is

$$u_0(\hat{x}, \hat{t}) = f(\hat{x} - \hat{t}) + g(\hat{x} + \hat{t}).$$

(b) Show that for boundary conditions A and B in Exercise 2, f and g each are periodic with period $2L\delta(T/EI)^{1/2}$, and that

$$\oint f \, d\hat{x} = 0 = \oint g \, d\hat{x}.$$

(c) Show that for boundary conditions C, the periods of f and g are $2\Lambda/|1 \pm v|$, where

$$\Lambda = L\delta\left(\frac{T}{EI}\right)^{1/2}, \qquad v = V\left(\frac{\rho}{T}\right)^{1/2}.$$

Show that, if $V^2 > T/\rho$, then one of these periods is smaller than Λ, and that arbitrary initial data cannot be prescribed over the length of the string. What is the physical meaning of this contradiction? Which of the assumptions is invalid when $V^2 > T/\rho$?

4. Show that if $V^2 < T/\rho$ secular terms arise at $O(\varepsilon^2)$ in the solution of the equation in Exercise 2 unless

$$f_\tau - 6(C_g)f_r - 6f^2 f_r - f_{rrr} = 0,$$
$$g_\tau + 6(C_f)g_l + 6g^2 g_l + g_{lll} = 0,$$

where

$$C_f = \frac{1+v}{2\Lambda}\oint f^2(\xi, \tau)\, d\xi, \qquad C_g = \frac{1-v}{2\Lambda}\oint g^2(l, \tau)\, dl.$$

Show that both C_f and C_g are constants of the motion for these equations, so that the evolution equations for the left and right running waves are effectively uncoupled. What is the physical reason that the transverse vibrations of a string are modeled by mKdV, whereas the longitudinal vibrations are modeled by KdV? If the string had a round cross section, would you expect the torsional vibrations to be modeled by KdV or mKdV?

5. (a) Show that each of the equations in Exercise 4 has a periodic solution of the form $f = bk \operatorname{cn}\{b(r + U\tau); k\}$, where b is arbitrary and $\operatorname{cn}(\theta; k)$ is the Jacobian elliptic function (cf. Byrd and Friedman (1971)). Write this solution in dimensional terms for w_x.

(b) If a guitar string were given the appropriate initial deflection with no momentum, then $f = g$ initially, and this solution represents the nonlinear motion of the string. Let ω_0 denote the frequency of the string in the linear limit. Show that the nonlinear frequency is given by

$$\omega = \omega_0\left\{1 - (b\delta)^2\left[3\left(\frac{E}{K} - 1 + k^2\right) + \left(k^2 - \frac{1}{2}\right)\right]\right\},$$

where $K(k)$, $E(k)$ are the complete elliptic integrals of the first and second kinds. Thus, there are two contributions to the frequency shift. The first term represents the change in speed of the right running wave due to the presence of the left running wave. It always *reduces* the frequency of the string. The second term is due to the self-interaction of the right running wave, and reduces the frequency only if $k^2 > \frac{1}{2}$.

(c) This solution also can be interpreted in terms of boundary condition C in Exercise 2. If $V^2 < T/\rho$ all long infinitesimal waves travel faster than the string speed. How big must the finite amplitude wave in (a) be for its speed to match the string speed? What are the physical implications of this possibility? Is the asymptotic expansion valid for a wave this big?

6. Under condition B in Exercise 2, we may solve the mKdV equation on $(-\infty, \infty)$. Show that in this case, the solution that evolves from appropriate initial conditions consists of N solitons, ordered by amplitude (with the biggest one in *back*), *preceded* by a train of dispersive waves.

7. Here are estimates of the scales required to observe solitons experimentally ($g = 0$ in Exercise 4).

(a) To produce at least one soliton, we must violate (1.7.1) more than marginally. In dimensional terms,

$$\left[(1-\mu)\left(\frac{T}{EI}\right)\right]^{1/2} \int \left|\frac{\partial w}{\partial x}\right| dx > 2$$

is probably adequate.

(b) The largest soliton separates from the infinitesimal waves with a speed that does not exceed $(2|f|_{\max})^2$. If the initial dimensionless length of the wave is $\hat{\lambda}$, the time required for the soliton to emerge is on the order of

$$\tau = \frac{\hat{\lambda}}{(2|f|_{\max})^2}.$$

If the wave is to remain in the test section during this time, the ratio of test section length L to initial wave length λ must satisfy

$$\frac{L}{\lambda} > \frac{2}{1-\mu}\left(\left|\frac{\partial w}{\partial x}\right|_{\max}\right)^{-2}.$$

Note that the linearized wave speed, $\sqrt{T/\rho}$, does not enter here.

(c) Consider a nylon E-string for a guitar. Suppose $T \sim 0.1\, YA$, where Y is the yield strength. For nylon, $E/Y \sim 50$. For an E-string, $I/A \sim (0.018 \text{ cm})^2$. Thus, long waves satisfy

$$\lambda^2 \gg 10 \frac{E}{Y}\frac{I}{A} \sim (0.4 \text{ cm})^2,$$

and waves longer than 2 cm should qualify in practice. If we balance

$$\frac{EI}{T\lambda} = \delta^2 \sim \left(\frac{1-\mu}{2}\right)\left(\left|\frac{\partial w}{\partial x}\right|_{\max}\right)^2,$$

so that the maximum slope is about $\frac{1}{5}$, then the solitons should appear about 1 m away. The wave speed here is about 2 m/sec, so the whole experiment takes only a fraction of a second.

(d) A more accessible demonstration might be possible if the nylon string were replaced by a long telephone cord or a long "Slinky". In either case, the displacements are larger and the speeds are slower.

Section 4.2

1. (a) In experimental situations, (4.2.1a) should be replaced by

$$\mathbf{k}_1 + \mathbf{k}_2 + \mathbf{k}_3 = \boldsymbol{\kappa},$$

when $\delta = |\boldsymbol{\kappa}|/|\mathbf{k}_3|$ is small. How is (4.2.16) modified if $\delta = O(\varepsilon)$? If $\delta \gg \varepsilon$? If $\delta \ll \varepsilon$?

(b) How do these results change in a one-dimensional problem?

2. Add to $\mathbf{E}_1$ in (4.2.15) two more waves with frequencies (ω_4, ω_5) such that

$$\mathbf{k}_3 + \mathbf{k}_4 + \mathbf{k}_5 = 0, \qquad \omega_3 + \omega_4 + \omega_5 = 0,$$

in addition to (4.2.1).

(a) If no other resonances exist among these waves, they interact in two triads, each involving ω_3. How must (4.2.16) be modified? If the $\mathbf{A}_i$ are y-independent, the interaction is described by 5 complex ordinary differential equations. Do they have the Painlevé property? (Warning: this is a long calculation.)

(b) Define a sixth wave by

$$\mathbf{k}_6 = \mathbf{k}_1 - \mathbf{k}_5.$$

Show that if this wave is resonantly coupled, i.e., if $\omega_6 = \omega_1 - \omega_5$, then each of the 6 waves interacts in two triads. What are the governing equations in this case? Ablowitz and Haberman (1975b) showed that in (1 + 1) dimensions, these equations are of IST type. They did not perform the inverse scattering analysis for the 4×4 scattering problem that results.

3. (a) In an imperfect dielectric medium, an applied electric field produces a weak current in addition to the polarization. Show that (4.2.5) becomes

$$c^{-2} \partial_t^2 \mathbf{E} + \partial_t^2 \mathbf{P} + \frac{\sigma}{c^2} \partial_t \mathbf{E} + \nabla \times \nabla \times \nabla \mathbf{E} = 0,$$

where σ is the (small) conductivity of the medium. Suppose this current is the only loss in the system, and that $\sigma = O(\varepsilon)$. How are (4.2.16) modified?

(b) Consider a uniform wave train in such a medium, along with two or more weak parasitic waves capable of forming a resonant triad, as in Exercise 2. What effect does the current have on the instability?

4. Express (4.2.20) in terms of $(\mathbf{E} \times \mathbf{H})$, the flux of electromagnetic energy density.

5. What is the period of the waves in (4.2.18)? Estimate from this an optimal crystal size, to convert as much power as possible into the second harmonic from an incoming (fundamental) wave. This size is actually an upper bound, because other small effects also limit the size of the crystal. See Yariv (1975) for some of these considerations.

6. (a) Solve (4.1.25) for ω^2 explicitly in the limit $h_2 \to \infty$. Show that the larger pair of roots (for the surface waves) are independent of the densities in this limit.

(b) Denote the three wavenumbers in (4.2.24) by $k_1 = \delta k$, $k_2 = (1-\delta)k$, $k_3 = k$. Show that for $h_2 = \infty$, the triad in (4.2.24) occurs where $4\Delta k h_1 = 1$.

(c) Let $h_1 \to \infty$ as well. Find explicitly the wavenumber and frequencies for a triad involving two surface and one internal waves, and another involving one surface and two internal waves.

Section 4.3

1. An alternative to hypothesizing (4.3.2) is to approximate (4.2.8) with the first nonlinear correction for an isotropic medium,

$$\partial_t^2 \mathbf{P} + \omega_0^2 \mathbf{P} + d|\mathbf{P}|^2 \mathbf{P} \sim \frac{\Omega^2}{c^2} \mathbf{E},$$

and to couple this with (4.2.5).

(a) Show that if the wave amplitude is time independent, then this derivation also leads to (4.3.2). Find n_2 explicitly in terms of d, Ω^2, $(\omega^2 - \omega_0^2)$, c^2 and n_0.

(b) Derive (4.3.8) from (4.2.5) and the equation for **P** above.

2. Justify the statements following (4.3.13).

(a) Find a $(3+1)$-dimensional generalization of (4.3.9) that reduces (4.3.13) to an ordinary differential equation. Show that this ODE is not P-type.

(b) Based on (4.2.12), for what range of ω is $\alpha = -k(\partial^2 k/\partial \omega^2) > 0$?

(c) Show that (4.3.13) admits a self-focusing singularity if $\alpha > 0$ and $n_2 > 0$. What can you conclude if $\alpha < 0$?

3. The deep water approximation in water waves amounts to letting $kh \to \infty$.

(a) Show that, if $kh \gg 1$, the error introduced into the linear solution (i.e., (4.3.19)) by this approximation is of exponential order in kh. On this basis it is often asserted that effects of finite depth can be ignored as soon as the fluid depth exceeds the wavelength, because the oscillatory wave no longer "feels" the bottom.

(b) Show that as $h \to \infty$, $\nabla \Phi = O(h^{-1})$, so that the (nonlinearly) induced mean flow decays algebraically rather than exponentially in this limit.

(c) Take $\tilde{T} = 0$ in (4.3.26) and consider a one-dimensional water wave with a period of 1 second and a maximum free-surface slope of (0.1). What is the wavelength of this wave if it is propagating in an ocean whose total depth is 3 km? Estimate the mean flow induced by this wave at the sea floor. Armi (1977) uses 4 cm/sec as a "typical" value of a bottom current. Is the induced mean flow significant in the deep ocean? What if the same wave were propagating over the continental shelf, where the total depth might be 200 m?

4. Stokes (1847) noted that oscillatory water waves induce a second order (in ε) "drift velocity". Thus individual fluid particles experience a slow mean motion in the direction of the group velocity of the waves (see Phillips (1977, p. 43) for details).

(a) Show that the mean motion related to Φ in (4.3.25) is in the opposite direction, and should be subtracted from the drift velocity of Stokes.

(b) Any steady, inviscid theory of water waves in finite depth is ambiguous with regard to the total mass transport due to the waves, because the equations admit an arbitrary, uniform, horizontal, mean flow that can be superposed on the motion. One way to remove this ambiguity is by considering the fluid to be slightly viscous; see Liu and Davis (1977) for a review of work along these lines. An alternative is to solve (4.3.23) for a localized wave packet that vanishes as $(x_1^2 + y_1^2) \to \infty$, and require no motion at ∞. Show that this method makes the inviscid mass transport unambiguous. Find the total mass transport.

5. The local nature of the wave amplitude (A) was used to reduce (4.3.21) to (4.3.23). If A is periodic in (x_1, y_1), show that this reduction follows only by assuming that Φ also satisfies (4.3.20). If this assumption is not made, what replaces (4.3.25)?

6. The experiment shown in Fig. 4.17 fully demonstrates the interaction of solitons only if the two carrier wave frequencies are close enough that the interaction of the two wave packets occurs on a long enough ($\varepsilon^2 t \sim 1$) time scale. Based on the measured waves, define an appropriate ε and estimate the time scale of the interaction. To what extent does this experiment confirm the theory?

7. (a) In $(1+1)$ dimensions, we may define the mass of a localized wave by

$$M = \rho \int_{-\infty}^{\infty} \zeta\, dx,$$

where ρ is the fluid density. Define the horizontal momentum and the total (potential + kinetic) energy of the wave in the same way. These three quantities are conserved exactly by (4.1.5–7).

(b) Using the expansions required to derive (4.3.25), expand the mass, horizontal momentum, and energy to $O(\varepsilon^3)$. Show that

$$M = \varepsilon I_1 + \varepsilon^2 I_2 + \varepsilon^3 I_3 + O(\varepsilon^4),$$

where I_1, I_2, I_3 are proportional to the first three conservation laws for (4.3.29). What are the corresponding expansions of the horizontal momentum and energy? This gives some insight into the question of why (4.3.29) has infinitely many constants of the motion: they are the coefficients in an asymptotic expansion of an exactly conserved quantity, such as M. It leaves open the question of why other approximate equations have only a finite number of such constants.

Section 4.4

1. Pseudospherical surfaces.

(a) Every solution of the sine-Gordon equation defines a (family of) pseudospherical surfaces. ϕ denotes the angle between the asymptotic lines, which coalesce if $\phi = n\pi$. But they must be distinct if the total curvature is negative, so lines on which $\phi = n\pi$ are singular lines of the surface. According to a theorem of Hilbert, every pseudospherical surface contains at least one singular line; this is a consequence of the Gauss–Bonnet theorem.

(b) A vast array of solutions of the sine-Gordon equation are now available, including the explicit soliton solutions (or "kinks", as they are called for (4.4.1)). Each such solution defines a pseudospherical surface, pieced together on its singular lines. It might be interesting to build geometric objects that correspond to one kink, two kinks, a kink–antikink pair, a breather, etc.

2. SIT has a simple mechanical analogue, due to McCall and Hahn. Consider a set of ideal, identical pendula, well separated and hanging in a line just above a horizontal plane.

(a) If a ball whose mass exceeds that of a pendulum ($m_b > m_p$) rolls along the plane, it imparts some of its momentum to each pendulum it strikes, until it eventually loses all of its momentum to the medium (of pendula). Show that the velocity of the ball after the nth collision is $V_n = V_0 \exp(-\alpha n)$, where $\alpha = \log((m_b + m_p)/(m_b - m_p))$. This is the analogue of Beer's law.

(b) If the mass of the ball equals the mass of a pendulum, and if its initial velocity is sufficient to swing a pendulum all the way around, show that the ball gives all of its momentum to a pendulum, waits while the pendulum swings around once, regains all of its momentum when the pendulum returns to its original position, and then travels with its original speed to the next pendulum. This is the analogue of SIT. What is the minimum initial velocity required for SIT? What is the average velocity of the ball if this minimal velocity is exceeded? What if the initial velocity is met exactly? What happens if the initial velocity is too small?

(c) What happens if the mass of the pendulum exceeds that of the ball?

(d) These results assume that the ball and pendula are point masses. What happens if both have finite diameters?

3. Show from (4.4.8) that for *any* electric field,

$$\partial_t \left\{ |\partial_t \mathbf{p}|^2 + \omega^2 |\mathbf{p}|^2 + \frac{\omega^2 \not{h}^2}{3} \tilde{\eta}^2 \right\} = 0.$$

What is the physical meaning of this identity? Show from this relation that $|\mathbf{p}| = O(\not{h})$.

4. Show that any other terms added to those in (4.4.13) are nonsecular, and do not change (4.4.14, 15, 16).

5. (a) Show that increasing the amplitude of a 2π-pulse in SIT increases its dimensional speed of propagation.

(b) A realization of (4.4.10) may be obtained by assuming

$$g(\omega) = \frac{1}{\pi} \frac{\Gamma}{(\omega - \omega_0)^2 + \Gamma^2}.$$

Compute the speed of a 2π-pulse explicitly in this case.

Appendix

Linear Problems

A.1. Fourier transforms. The objective of this section is to outline the use of Fourier transform methods for solving certain linear equations, and to demonstrate the method with several examples. More precisely, the method outlined is separation of variables, the end result of which is a representation of the solution in terms of its Fourier transform. The method also may be described as looking for "normal modes," or for "solutions in the form $e^{ikx-i\omega t}$". This approach is *not* necessarily equivalent to taking a Fourier–Laplace transform, and we shall discuss the difference between these two approaches in the next section.

The simplest types of evolution equation for which Fourier transform methods are useful have the following form:

(A.1.1) $$u_t = F(u, u_x, u_{xx}, \cdots),$$

where F is linear in its arguments, homogeneous and with constant coefficients. It is desired to solve (A.1.1) for $-\infty < x < \infty$, $t > 0$, subject to the constraints that u vanish as $|x| \to \infty$, and that as $t \to 0$, $u(x, t) \to U(x)$, a prescribed function ("initial data").

Even within the context of linear evolution equations, Fourier transform methods apply to a wider class of problems than (A.1.1). We may consider equations involving more than one time derivative or, more generally, replacing the scalar, u, in (A.1.1) with a vector $(u_1, u_2, \cdots, u_N)$, and making appropriate changes in F. Or the equation may be of interest only in a finite interval, $a < x < b$, perhaps with periodic boundary conditions. Some of the possible generalizations are explored in the example problems and in the exercises at the end of the chapter.

Here are the basic steps in the method.
1. Does the problem have a unique solution?

Useful information often can be obtained by looking for a few "conservation laws," i.e., relationships of the form

$$\frac{\partial}{\partial t}T + \frac{\partial}{\partial x}F = 0, \tag{A.1.2}$$

where T and F may depend on x, t, u and its derivatives. In many cases, the equation itself is in this form. Other conservation laws sometimes can be obtained by multiplying the equation by some function (e.g., u, u_x, etc.) and integrating by parts. In any case, if the boundary conditions require that F vanish as $|x| \to \infty$, and if $\int_{-\infty}^{\infty} T\,dx$ is defined (at $t = 0$), then this integral is time independent. Of particular value is any conservation law such that T is positive definite; i.e.,

$$T \geq 0 \text{ and } T = 0 \Rightarrow u = 0. \tag{A.1.3}$$

Then $\int T\,dx$ may define a norm for a particular space of functions, and the uniqueness of the solution (within this particular set of functions) follows directly from the conservation law. In these cases, we shall identify $\int T\,dx$ as the "energy integral" of the system, whether or not it represents any physical energy. Moreover, this energy integral need not be related to the Hamiltonian of the system, if one exists. (Friedrichs (1958) made extensive use of "energy methods" to establish the uniqueness of solutions of symmetric, positive systems of differential equations. Closely related methods may be used to establish the stability of a finite difference scheme, as discussed by Richtmeyer and Morton (1967). The concept is particularly appropriate for the equations solvable by IST, because often one of the exactly conserved quantities may be identified as an energy.)

Not all problems have energy integrals, and in some cases the function space is not large enough for the application in mind. In these cases, the question of uniqueness must be answered by other means.

2. Is there a dispersion relation?

Substitute a trial solution of the form

$$u(x, t) \sim A \exp(ikx - i\omega t) \tag{A.1.4}$$

into the differential equation. Here k is real, and ω is some complex number. For this method to work, the differential equation must reduce to

$$D(\omega, k)A \exp(ikx - i\omega t) = 0, \tag{A.1.5}$$

for all (x, t). The exponential (or even its real or imaginary part) does not vanish for all (x, t), and $A = 0$ represents only a trivial solution. It follows that (A.1.4) satisfies the partial differential equation nontrivially only if ω and k are related through the dispersion relation

$$D(\omega, k) = 0. \tag{A.1.6}$$

There are several variants of this step.

(i) If the problem is defined on $a<x<b$, rather than $-\infty<x<\infty$, k typically is restricted to a countable set of real values.

(ii) For an nth order system of equations, replace (A.1.4) with

(A.1.7) $$\mathbf{v}(x, t) \sim \mathbf{A} \exp(ikx - i\omega t).$$

Then (A.1.5) becomes

$$\underline{\underline{M}} \mathbf{A} \exp(ikx - i\omega t) = 0,$$

where $\underline{\underline{M}}$ is an $n \times n$ matrix. The dispersion relation is defined by

(A.1.8) $$\det(\underline{\underline{M}}) = 0.$$

(iii) For a differential-difference equation (discrete in space, continuous in time), replace (A.1.4) with

(A.1.9) $$u_n(t) \sim A z^n \exp(-i\omega t),$$

where z is a complex number on the unit circle and n is an integer. (The analogy with (A.1.4) is more apparent if we set $z = \exp(i\theta)$, with θ real.) Then the dispersion relation takes the form

(A.1.10) $$D(\omega, z) = 0.$$

(iv) For a finite difference scheme (discrete in space and time), replace (A.1.4) with

(A.1.11) $$u_n^m \sim A \Omega^m z^n,$$

where z lies on the unit circle, Ω is a complex number, and m, n are integers. The dispersion relation has the form

(A.1.12) $$D(\Omega, z) = 0.$$

In numerical analysis, Ω is often called the *amplification factor*, following von Neumann's (1944) work on stability of difference schemes.

Now we return to the continuous problem, and (A.1.6).

3. Are the normal modes complete?

Solve (A.1.6) for $\omega(k)$. For each k, the number of solutions should equal the order (in t-derivatives) of the differential equation. At any fixed t, each $[k, \omega(k)]$ in (A.1.4) represents one "mode" in a Fourier integral (or sum), and a sum over the modes represents a (formal) solution of the differential equations. The sum is taken over all real k (for $-\infty < x < \infty$):

(A.1.13) $$u(x, t) = \frac{1}{2\pi} \int_{-\infty}^{\infty} A(k) \exp(ikx - i\omega t)\, dk,$$

unless further restrictions on k already have been imposed. The method is successful if the sum is general enough to represent arbitrary initial conditions on u, u_t, etc., at $t = 0$. (Here "arbitrary" means arbitrary in some space such as L_2, the space of square-integrable functions.)

We say that the above method fails either if there is no dispersion relation (i.e., for fixed k, ω is unrestricted), or if the set of normal modes is not complete and so cannot represent the initial data. Some examples are given in the next section.

When the method works, one obtains an integral representation of the solution. The sense in which this "solution" actually solves the equation is sometimes delicate, and will be discussed in the context of the examples.

4. What is the character of the solution?

The real advantage of Fourier transforms lies here. Once we have determined that we have the general solution of the initial value problem, much of the relevant information in the problem can be obtained directly from the dispersion relation.

The growth rate of any particular mode is given by Im (ω).

(i) If Im $(\omega) > 0$ for some real k (i.e., for one of the possible modes of the system), this mode grows exponentially in time, and the problem is *unstable*. The most unstable mode is the one that maximizes Im (ω), if one exists. Unless the initial amplitude of this mode were *exactly* zero, it would dominate the solution after a sufficiently long time. If the initial data in the problem were known only to within a certain tolerance (which might depend on a method of measurement, for example), one would expect to observe predominantly the most unstable mode eventually, regardless of the initial conditions.

(ii) If Im $(\omega) \to \infty$ in any limit (e.g., $k \to \infty$), there is no bound on the growth rate and the problem is *ill-posed* (in the sense of Hadamard). Here any uncertainty in the initial data precludes virtually all predictions about the solution for $t > 0$. If a model of a physical problem is ill-posed, it may not be properly formulated.

(iii) If Im $(\omega) < 0$ for all real k, the problem is *asymptotically stable* (or *dissipative*), because every mode decays exponentially for $t > 0$. After a long enough time, the predominant mode is the one that maximizes Im (ω), except for very special initial conditions.

(iv) If the solution of (A.1.1) can be represented in the form (A.1.13) with $\omega(k)$ uniquely defined, then its energy integral is $\int |u|^2 \, dx$. By Parseval's relation, we have

$$\int |u|^2 \, dx = \frac{1}{2\pi} \int |A(k)|^2 \exp\{2 \text{ Im } (\omega(k))t\} \, dk.$$

Thus the problem has a time-independent energy integral only if Im $(\omega) = 0$ for real k; i.e., the dispersion relation is real. Then the dominant feature of the solution is neither exponential decay nor growth, but *wave propagation*. Most of the problems discussed in this book, when linearized, fit this description.

(v) By way of comparison, a finite difference scheme is *unstable* if $|\Omega| > 1$ for some z on the unit circle, and *stable* if $|z| = 1 \Rightarrow |\Omega| \leq 1$. A real dispersion relation corresponds to $|z| = 1 \Rightarrow |\Omega| = 1$.

These definitions are consistent with the theory of ordinary differential equations (e.g., Birkhoff and Rota (1969)). The connection may be seen by writing the solution of (A.1.1) in the form

$$u(x, t) = \frac{1}{2\pi} \int \hat{u}(k, t) e^{ikx} \, dk;$$

then $\hat{u}(k, t)$ formally satisfies the ordinary differential equation

$$\frac{d}{dt} \hat{u} = -i\omega(k) \hat{u},$$

whose stability is determined by Im (ω).

5. What is the long time behavior of the solution in problems with real dispersion relations?

For a problem on a finite interval in one spatial dimension, a representative solution has the form

(A.1.14) $$u(x, t) = \sum_{n=-\infty}^{\infty} A_n \exp(ik_n x - i\omega_n t).$$

Here the $\{k_n\}$ depend on the length of the interval, $\omega_n = \omega(k_n)$ on the dispersion relation, and $\{A_n\}$ on the initial data; k_n and ω_n are real, and we assume that ω_n is uniquely defined. Then the energy integral is given by

(A.1.15) $$\int |u|^2 \, dx = \sum_n |A_n|^2,$$

provided the sum exists.

On intuitive grounds, one feels that if a fixed amount of energy is confined to a finite interval in modes that cannot transfer energy among themselves, then there ought to be no asymptotic $(t \to \infty)$ state. This notion is correct; instead of tending to an asymptotic state, the system almost returns to its initial conditions after a finite time ("Poincaré recurrence").

The validity of this assertion may be seen by approximating the solution in (A.1.14) to any desired accuracy by a finite set of modes:

$$u_N(x, t) = \sum_{-N}^{N} A_n \exp(i\theta_n), \qquad \theta_n = k_n x - \omega_n t.$$

It is not difficult to show (e.g., see Arnold (1978)) that this partial sum corresponds to the solution of a Hamiltonian system with $(2N + 1)$ degrees of freedom. (The Hamiltonian is $H = i \sum_{-N}^{N} |A_n|^2 \omega_n$, the action variables are $|A_n|^2$ and the angle variables are $i\theta_n$.) Then Liouville's theorem regarding conservation of volume in phase space states that there can be no asymptotic state, and Poincaré's recurrence theorem states that almost every initial state recurs in a finite time.

On an infinite interval, a typical solution is

$$(A.1.16) \qquad u(x, t) = \frac{1}{2\pi} \int_{-\infty}^{\infty} A(k) \exp(ikx - i\omega t) \, dk,$$

and again $\omega(k)$ is real. Approximate evaluation of such integrals is a subject in itself. We present here only the basic ideas; for a more thorough exposition, the reader may consult the texts by Copson (1965), Olver (1974) or Bleistein and Handlesman (1975). These problems also have fixed amounts of energy, but because the energy can spread over an infinite interval, asymptotic states are possible. For each mode (or "wave"), points of constant phase travel with the *phase velocity*,

$$(A.1.17) \qquad c_p(k) = \frac{\omega}{k}.$$

If every wave has the same phase velocity, (i.e., $\omega = c_0 k$), the solution at any time $t > 0$ is simply the initial function, translated through space by an amount $c_0 t$:

$$(A.1.18) \qquad u(x, t) = \frac{1}{2\pi} \int_{-\infty}^{\infty} A(k) \exp(ik(x - c_0 t)) \, dk = u(x - c_0 t, 0).$$

The problem is *dispersive* if

$$(A.1.19) \qquad \frac{d^2 \omega}{dk^2} \neq 0.$$

Here different waves have different phase speeds, and the behavior of the solution depends on how the waves reinforce or interfere with each other. The important propagation velocity here is the group velocity,

$$(A.1.20) \qquad c_g(k) = \frac{d\omega}{dk}.$$

The significance of this velocity is that after a sufficiently long time, each wave number k dominates the solution in a region defined by

$$(A.1.21) \qquad x \sim c_g(k) t + o(t).$$

The precise form of the solution in this region may be found by evaluating (A.1.16) by either of two related methods, stationary phase or steepest descents.

Some partial differential equations admit self-similar solutions in the form

$$(A.1.22) \qquad u(x, t) = t^{-p} f\left(\frac{x}{t^q}\right),$$

where p and q are constant, and f satisfies an ordinary differential equation. (Said differently, the equation is invariant under the transformation

$$T_b: \quad t \to bt, \quad x \to b^q x, \quad u \to b^{-p} u,$$

where b is a scalar. The set of all such transformations (T_b) form a group; e.g., see Hall (1959).) These solutions often lie outside the function space of interest (e.g., they need not be square-integrable in x). Nevertheless, in many problems, the asymptotic $(t \to \infty)$ solution becomes approximately self-similar locally, but is modulated by a "slowly varying" function that depends on the initial data. The advantages of this representation, when it is available, are that: (i) it clearly identifies what part of the solution is controlled by the differential equation, and what is controlled by the initial conditions; (ii) it may be uniformly valid in x (as $t \to \infty$), even if the representation obtained by stationary phase is not.

This completes the outline of the method. Next, we demonstrate its application in some example problems.

Example 1: The Schrödinger equation

(A.1.23) $\qquad i\psi_t + \psi_{xx} = 0, \qquad -\infty < x < \infty, \quad t > 0,$

(A.1.24) $\qquad \psi \to 0 \text{ as } x \to \infty,$

(A.1.25) $\qquad \psi(x, t = 0) = \Psi(x) \quad \text{with} \int |\Psi|^2 \, dx = 1.$

In quantum mechanics, $\psi(x, t)$ represents the (complex) wave function of a free particle, $|\psi|^2(x, t)$ represents the probability density at time t of locating the particle at position x, and $\int |\Psi|^2 \, dx = 1$ allows this interpretation at $t = 0$.

A conservation law of interest is

(A.1.26) $\qquad i(|\psi|^2)_t + (\psi^* \psi_x - \psi \psi_x^*)_x = 0,$

obtained by multiplying (A.1.23) by ψ^* (the complex conjugate of ψ) and subtracting the complex conjugate of the resulting equation. Integrating (A.1.26) in x yields the time-independent "energy integral",

(A.1.27) $\qquad \int |\psi|^2 \, dx = 1.$

This integral identifies L_2 as a natural space for the problem, and this meshes nicely with the probabilistic interpretation. Next we prove that the problem cannot have two different solutions in L_2. Suppose that there are two such solutions in L_2. Then their difference, $\Delta(x, t)$, is in L_2, satisfies (A.1.23, 24) and vanishes identically at $t = 0$. It follows that

(A.1.28) $\qquad \int |\Delta|^2 \, dx = 0$

for all time, so that $\Delta(x, t)$ must be the zero function.

To find the dispersion relation, let

(A.1.29) $$\psi \sim \psi_0 \exp(ikx - i\omega t),$$

and find

(A.1.30) $$\omega(k) = k^2.$$

Here ω is real for real k, as we anticipate from the existence of the energy integral.

Some notation is necessary before reconstructing the solution of the problem. If $\phi(x)$ is in L_2, we define its Fourier transform by

(A.1.31) $$\hat{\phi}(k) = \int_{-\infty}^{\infty} \phi(x) e^{-ikx} dx.$$

Then the inverse Fourier transform is

(A.1.32) $$\phi(x) = \frac{1}{2\pi} \int_{-\infty}^{\infty} \hat{\phi}(k) e^{ikx} dk.$$

We use this convention throughout the book. Now sum over the modes in (A.1.29) to obtain the formal solution of (A.1.23–25),

(A.1.33) $$\psi(x, t) = \frac{1}{2\pi} \int_{-\infty}^{\infty} \hat{\Psi}(k) e^{ikx - ik^2 t} dk,$$

where $\hat{\Psi}(k)$ is the Fourier transform of the initial data.

In what sense does this integral solve the problem? Certainly it reproduces $\Psi(x)$ at $t = 0$, by construction. If $\hat{\Psi}$ is absolutely integrable on $-\infty < x < \infty$ (i.e., an element of L_1), then the integral satisfies (A.1.24) by the Riemann–Lebesgue lemma (this famous theorem is in most books on real analysis, such as Hewitt and Stromberg (1969)). If $\hat{\Psi}$ converges rapidly enough as $|k| \to \infty$ to permit two differentiations under the integral sign, then the integral also satisfies (A.1.23), and is a (pointwise) solution of the problem.

If either $\psi(x)$ or its derivative is discontinuous, differentiation under the integral is not permitted. An alternative procedure may be used if $\hat{\Psi}(k)$ can be continued analytically into the complex k-plane. In the simplest case, the contour of k-integration may be rotated from the real axis through an angle $(-\delta)$, $0 < \delta < \pi/2$, without encountering any singularities of $\hat{\Psi}(k)$ in the finite plane, as shown in Fig. A.1. We also need that $\hat{\Psi}(k) \to 0$ rapidly as $|k| \to \infty$ in the sectors through which the contour is deformed. Under these circumstances, integration along the deformed contour gives the same value as integration along the original contour, by Cauchy's integral theorem. But under the transformation

$$k \to r e^{-i\delta}, \quad r \text{ real}, \quad 0 < \delta \leq \frac{\pi}{2},$$

(A.1.34)
$$e^{-ik^2 t} \to \exp\left(r^2 t \cos\left(\frac{\pi}{2} + 2\delta\right)\right) \exp\left(-ir^2 t \sin\left(\frac{\pi}{2} + 2\delta\right)\right),$$

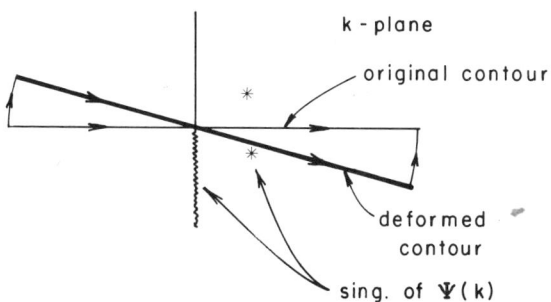

FIG. A.1. *Contours of integration for* (A.1.32).

and for $t > 0$ the integral converges exponentially fast as $|r| \to \infty$. It follows that even though $\psi(x)$ might have been discontinuous, the solution of (A.1.23) that evolves from it for $t > 0$ is not only continuous, but infinitely differentiable! This shows the smoothing effect of the Schrödinger operator. A specific example is examined in Exercise 1.

If $\hat{\Psi}$ permits neither differentiation under the integral nor continuation into the complex plane, consideration of "weak solutions" becomes necessary (see Lax (1954) or (1973)). However, one or the other of the methods discussed ordinarily is adequate for applications. To summarize, we have found that the problem has at most one solution in L_2, and that (A.1.33) is a representation of it. It remains to describe the asymptotic behavior of the solution as $t \to \infty$. To do so, note that (A.1.33) has the form

(A.1.35a)
$$\int \hat{\Psi}(k) e^{i\phi(k)t} dk,$$

where

(A.1.35b)
$$\phi(k) = k\frac{x}{t} - \omega(k) = k\frac{x}{t} - k^2,$$

if we hold (x/t) fixed as $t \to \infty$. The intuitive notion behind Kelvin's method of stationary phase (cf. Copson (1965)) is the following. This integral represents a superposition of infinitely many wavetrains, but for sufficiently large t the phases of wave trains represented by k and $(k + \delta k)$ will be very different unless $\phi'(k)$ vanishes. Thus, one expects destructive interference from most of the wave trains, and that the dominant contributions to the the integral should come from small neighborhoods of points where $\phi'(k)$ vanishes (i.e., where the phase $\phi(k)t$ is stationary). Following this reasoning, we would expect a particular wavenumber k to dominate the solution where

(A.1.36)
$$\phi'(k) = \frac{x}{t} - 2k = \frac{x}{t} - c_g(k) = 0.$$

This shows the significance of the group velocity: as $t \to \infty$, each wavenumber k dominates the solution in a region described approximately by (A.1.36).

Thus, $(k - \delta k)$ dominates along one straight line in (x, t)-space, and $(k + \delta k)$ along a slightly different straight line. It follows that the (time-independent) contribution to the energy integral from the packet of wavenumbers between them,

$$\frac{1}{2\pi} \int_{k-\delta k}^{k+\delta k} |\hat{\Psi}|^2 \, dk,$$

is spread over a region of space that is increasing linearly in time. This suggests that $|\psi|^2$ should decrease at t^{-1} (to maintain the invariance of $\int |\psi|^2 \, dx$), so that

(A.1.37) $$|\psi| = O(t^{-1/2}) \quad \text{as } t \to \infty.$$

This heuristic argument turns out to be correct provided $\phi''(k) \neq 0$. If $\phi''(k) = 0$, the trajectories of $(k \pm \delta k)$ separate more slowly, and the decay rate is correspondingly slower than the rate given by (A.1.37).

The method of stationary phase provides explicit formulas for the dominant behavior of ψ, but the easiest way to justify these formulas is to use Debye's method of steepest descents (cf. Copson (1965)). For a fixed value of (x/t), this consists of extending $\hat{\Psi}(k)$ into the complex k-plane, and deforming the path of k-integration so as to:

(i) pass through a zero of $\phi'(k)$;
(ii) keep the real part of $\phi(k)$ constant along the path;
(iii) maximize the imaginary part of $\phi(k)$ at the zero of $\phi'(k)$.

This can get rather complicated, but for (A.1.35) the only zero of $\phi'(k)$ lies at $k = (x/2t)$, and the entire deformed path is defined by

$$k = \frac{x}{2t} + re^{-i\pi/4},$$

$-\infty < r < \infty$. With this change of variables, (A.1.33) becomes

(A.1.38) $$\psi(x, t) = \frac{1}{2\pi} \exp\left(it\left(\frac{x}{2t}\right)^2 - i\left(\frac{\pi}{4}\right)\right) \int_{-\infty}^{\infty} \hat{\Psi}\left(\frac{x}{2t} + re^{-i\pi/4}\right) e^{-r^2 t} \, dr,$$

where again we have assumed that the rotation does not encounter any singularities of $\hat{\Psi}$ (but see Exercise 2). If $\hat{\Psi}$ is well behaved, the dominant contribution to the integral as $t \to \infty$ comes from the neighborhood of $r = 0$. Thus, we expand $\hat{\Psi}$ in a Taylor series about $k = (x/2t)$, and evaluate each of the resulting integrals. The result is

(A.1.39) $$\psi(x, t) \sim \frac{1}{2\sqrt{\pi t}} \exp\left(it\left(\frac{x}{2t}\right)^2 - i\left(\frac{\pi}{4}\right)\right) \left[\hat{\Psi}\left(\frac{x}{2t}\right) + \sum_{n=1}^{\infty} \frac{\hat{\Psi}^{(2n)}(x/2t)}{(4it)^n n!}\right].$$

As anticipated by the previous discussion, the role of the group velocity

($k = x/2t$) is apparent, and the amplitude decays as $t^{-1/2}$. If $\hat{\Psi}$ is sufficiently well behaved, (A.1.39) is uniformly valid for all (x/t) as $t \to \infty$.

The final point in the discussion of this problem is to relate (A.1.39) to a "slowly varying similarity solution" of (A.1.23). To this end, we look for a special solution of (A.1.23) in the form

$$\psi(x, t) = t^{-p} f(\eta), \qquad \eta = \frac{x}{t^q},$$

and find that $q = \frac{1}{2}$, and that $f(\eta; p)$ satisfies

(A.1.40) $$f'' - \frac{i}{2}\eta f' - ipf = 0.$$

Under the transformation $z = i\eta^2/4$, f can be identified with the confluent hypergeometric function, but for our purposes it is sufficient to observe that one solution of (A.1.40) for $p = \frac{1}{2}$ is

(A.1.41) $$f(\eta; \tfrac{1}{2}) = A\, e^{i\eta^2/4} = e^{ix^2/(4t)},$$

where A is constant. Thus, as $t \to \infty$, the solution of (A.1.23–25) tends to a solution that is locally self-similar, but modulated by a slowly varying function (i.e., A is now to be thought of as a slowly varying function) that depends on the initial conditions:

(A.1.42) $$\psi(x, t) \sim [t^{-1/2} e^{ix^2/(4t)}] \left[\frac{1}{2\sqrt{\pi}} e^{-i\pi/4} \hat{\Psi}\left(\frac{x}{2t}\right)\right].$$

Example 2: The heat equation

(A.1.43)
$$T_t = \kappa T_{xx}, \qquad -\infty < x < \infty, \quad \kappa, t > 0,$$
$$T \to 0 \quad \text{as } |x| \to \infty,$$
$$T(x, t = 0) = T_0(x).$$

If $T_0(x)$ is real, T can be interpreted (for example) as the temperature in a long tube containing a monatomic gas with no mean motion. The tube should have insulated sides, with no variation across it. This temperature is measured relative to some reference temperature, $\bar{T} > 0$, so that $(T + \bar{T})$ is the absolute temperature. The heat flux is $(-\kappa T_x)$.

The equation is already in the form of a conservation law,

(A.1.44) $$\frac{\partial}{\partial t} \int_a^b T\, dx = \kappa T_x \Big|_a^b,$$

which states that any change in the average temperature in an interval is due to a net heat flux across the ends of the interval.

The temperature of the gas measures the (random) kinetic energy of the molecules, and (A.1.44) is the remnant of the law of conservation of energy in (A.1.43). However, T need not be positive, and (A.1.44) is not useful in establishing uniqueness.

The appropriate "energy integral," which does *not* represent the physical energy, is obtained by multiplying (A.1.43) by T and integrating by parts:

$$\text{(A.1.45)} \qquad \frac{1}{2}\frac{\partial}{\partial t}\int_a^b T^2\,dx = \kappa T_x T\Big|_a^b - \int_a^b (T_x)^2\,dx.$$

This is not a conservation law ($\int T^2\,dx$ is not conserved), but it still can be used to show uniqueness. If $T_0(x)\in L_2$, then $\int_{-\infty}^{\infty} T^2\,dx$ exists initially and is positive definite. Then (A.1.45) shows that this integral does not grow if the boundary terms vanish, so that the solution remains in L_2 for $t>0$. As in the previous problem, uniqueness is established by computing $\int \Delta^2\,dx$, for the difference of two solutions that evolve from the same initial data and showing from (A.1.45) that it remains zero for $t>0$.

The dispersion relation for (A.1.43) is

$$\text{(A.1.46)} \qquad \omega = -i\kappa k^2.$$

Thus, Im $(\omega) \leq 0$, and the problem is asymptotically stable; this is consistent with (A.1.45) on $(-\infty, \infty)$, which shows that the energy integral must decrease in time from any finite positive value. In fact, Im $(\omega) = 0$ only for $k = 0$; the fact that $\omega = 0$ for $k = 0$ is simply a restatement of (A.1.44), that

$$\hat{T}(k=0) = \int T\,dx$$

is time independent.

The solution of (A.1.43) is

$$\text{(A.1.47)} \qquad T(x,t) = \frac{1}{2\pi}\int_{-\infty}^{\infty} \hat{T}_0(k)\, e^{-\kappa k^2 t}\, e^{ikx}\,dk,$$

where $\hat{T}_0(-k) = \hat{T}_0^*(k)$, because $T_0(x)$ is real. There is no difficulty establishing the validity of (A.1.47); differentiation under the integral (any number of times) is permitted for $t>0$.

As $t \to \infty$, the dominant contribution to (A.1.47) comes from the neighborhood of $k = 0$, where Im $(\omega) = 0$. If $\hat{T}_0(k)$ is analytic there, we may expand it in a Taylor series about $k = 0$, and evaluate the separate integrals. Using the identities

$$\hat{T}_0(0) = \int T_0(x)\,dx,$$

$$i\hat{T}_0'(0) = \int xT_0(x)\,dx, \quad \text{etc.}$$

we obtain

(A.1.48)
$$T(x,t) = \frac{\int T_0(\xi)\,d\xi}{2(\pi\kappa t)^{1/2}}\exp\left(-\frac{x^2}{4\kappa t}\right)$$
$$+\frac{\int \xi T_0(\xi)\,d\xi}{2\sqrt{\pi}(\kappa t)}\frac{x}{2(2\kappa t)^{1/2}}\exp\left(-\frac{x^2}{4\kappa t}\right)+\cdots.$$

Again, we observe that the solution tends to a self-similar form as $t \to \infty$. There is no slow modulation in this case, because all of the contributions come from the neighborhood of $k = 0$.

Example 3: The linearized Korteweg–de Vries equation

(A.1.49)
$$u_t + u_{xxx} = 0, \quad -\infty < x < \infty, \quad t > 0$$
$$u \to 0 \quad \text{as } |x| \to \infty,$$
$$u(x,0) = U(x).$$

The original discovery of IST followed the discovery by Miura (1968) of an explicit transformation between the Korteweg–deVries (KdV) equation

$$u_t + uu_x + u_{xxx} = 0,$$

and the modified KdV equation

$$v_t + v^2 v_x + v_{xxx} = 0.$$

In the limit of small amplitudes both of these reduce to (A.1.49). Other applications of (A.1.49) are discussed in the exercises.

If $U(x)$ is real, u remains real for $t > 0$. We consider only real solutions here. The energy integral in this problem is $\int u^2\,dx$, which is time independent. Thus L_2 is an appropriate space, and the solution of (A.1.49) is unique in L_2. Because $\int u^2\,dx$ is time independent, we expect a real dispersion relation and substituting $\exp(ikx - i\omega t)$ into (A.1.49) confirms this:

(A.1.50)
$$\omega = -k^3.$$

The (formal) Fourier transform solution of the problem is

(A.1.51)
$$u(x,t) = \frac{1}{2\pi}\int \hat{U}(k)\,e^{ikx + ik^3 t}\,dk,$$

where $\hat{U}(-k) = \hat{U}^*(k)$ because $U(x)$ is real. The sense in which this actually solves (A.1.49) may be seen, for example, in Cohen (1979), who made no assumptions about the behavior of $\hat{U}(k)$ except on the real k-axis.

In evaluating (A.1.51) as $t \to \infty$, we restrict our attention to cases in which $\hat{U}(k)$ can be continued off of the real k-axis and in which the deformed k-contours do not encounter any singularities of $\hat{U}(k)$. Then the asymptotic evaluation of (A.1.51) is closely related to the evaluation of the Airy function,

(A.1.52) $$\text{Ai}(\eta) = \frac{1}{2\pi} \int_{-\infty}^{\infty} \exp\left(ik\eta + \frac{ik^3}{3}\right) dk,$$

discussed at length by Copson (1965). The Airy function is plotted in Fig. A.2.

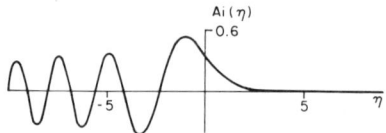

FIG. A.2. *The Airy function,* Ai(η).

The points of stationary phase in (A.1.51) occur where

$$\frac{x}{t} + 3k^2 = 0,$$

and different results obtain for

$$\frac{x}{t} < 0, \quad \frac{x}{t} > 0, \quad |x| = o(t) \quad \text{as } t \to \infty.$$

For $x/t < 0$, there are two real stationary points, at $k = \pm |x/(3t)|^{1/2}$; the path of steepest descents must touch both of these points, as shown in Fig. A.3.

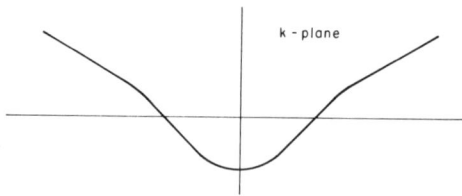

FIG. A.3. *Steepest descents path for* $x/t < 0$ *in* (A.1.51).

Near these points, the contour is described locally by

(A.1.53) $$k = \pm \left|\frac{x}{3t}\right|^{1/2} + re^{\pm i\pi/4} + \cdots,$$

from which one may obtain the dominant term in the large t-expansion of (A.1.51). For $t \to \infty$ with fixed $x/t < 0$,

(A.1.54) $$u(x, t) \sim \frac{\rho(x/t)}{(\pi t)^{1/2} |3x/t|^{1/4}} \cos\left\{-2\left|\frac{x}{3t}\right|^{3/2} t + \frac{\pi}{4} + \phi\left(\frac{x}{t}\right)\right\},$$

where

$$\rho\left(\frac{x}{t}\right) = \left|\hat{U}\left(\left|\frac{x}{3t}\right|^{1/2}\right)\right|,$$

$$\phi\left(\frac{x}{t}\right) = \arg\left\{\hat{U}\left(\left|\frac{x}{3t}\right|^{1/2}\right)\right\}.$$

Note that the decay rate (in x) of u as $x/t \to -\infty$ is faster than the apparent $(-\frac{1}{4})$ power, because $\hat{U}$ also decays.

For $x/t > 0$, there are no real stationary points. The dominant contribution as $t \to \infty$ is obtained by lifting the contour up to the stationary point in the upper k-plane,

$$k = i\left(\frac{x}{3t}\right)^{1/2} + l,$$

and expanding $\hat{U}$ near the stationary point. Thus, as $t \to \infty$, $x/t > 0$,

(A.1.55) $$u(x,t) \sim \frac{1}{2(\pi t)^{1/2}(3x/t)^{1/4}} \hat{U}\left(i\left(\frac{x}{3t}\right)^{1/2}\right) \exp\left(-2\left(\frac{x}{3t}\right)^{3/2} t\right).$$

Both (A.1.54) and (A.1.55) break down as $x/t \to 0$. To examine the behavior in this region it is convenient to change variables in (A.1.51):

(A.1.56) $$s = k(3t)^{1/3}, \quad \eta = \frac{x}{(3t)^{1/3}},$$

so that

(A.1.57) $$u(x,t) = \frac{1}{2\pi(3t)^{1/3}} \int \hat{U}\left(\frac{s}{(3t)^{1/3}}\right) \exp\left(is\eta + \frac{is^3}{3}\right) ds.$$

Then Taylor-expanding $\hat{U}$ near $s = 0$ yields the asymptotic expansion for u in this region as $t \to \infty$ in terms of the Airy function and its derivative:

(A.1.58) $$u(x,t) \sim (3t)^{-1/3} \hat{U}(0) \text{Ai}(\eta) - (3t)^{-2/3} i\hat{U}'(0) \text{Ai}'(\eta) + O((3t)^{-1}).$$

Using the asymptotic properties of Ai (η), one can show that (A.1.58) matches smoothly to (A.1.55) as $\eta \to +\infty$, and to (A.1.54) as $\eta \to -\infty$.

Thus the solution of (A.1.49) decays as $t^{-1/2}$ for fixed $x/t < 0$, as $t^{-1/3}$ near $x/t = 0$, and exponentially fast for fixed $x/t > 0$. Chester, Friedman and Ursell (1957) showed how to derive an alternative representation of the asymptotic solution that is uniformly valid in (x/t). The result is

(A.1.59)
$$u(x,t) \sim (3t)^{-1/3} \text{Ai}(\eta) \left[\frac{\hat{U}(k) + \hat{U}(-k)}{2}\right]$$
$$+ (3t)^{-2/3} \text{Ai}'(\eta) \left[\hat{U}(k) - \frac{\hat{U}(-k)}{2ik}\right],$$

where

$$\eta = \frac{x}{(3t)^{1/3}}, \qquad k = \left(\frac{-x}{3t}\right)^{1/2}.$$

This replaces the separate expansions in (A.1.54, 55, 58). It is worth noting that both

$$u = (3t)^{-1/3} \operatorname{Ai}(\eta) \quad \text{and} \quad u = (3t)^{-2/3} \operatorname{Ai}'(\eta)$$

are self-similar solutions of (A.1.49), and that (A.1.59) is in the form of a "slowly varying similarity solution," where the modulation depends on the initial data, through $\hat{U}$.

Example 4: *The Klein–Gordon equation*

(A.1.60)
$$u_{TT} - u_{XX} + u = 0, \qquad -\infty < X < \infty, \qquad T > 0,$$
$$u \to 0 \quad \text{as } |X| \to \infty,$$

$u(X, T = 0)$, $u_T(X, T = 0)$ *both given and real*. The Klein–Gordon equation arises in various contexts in relativistic quantum mechanics (cf. Morse and Feshbach (1963)). Our main interest in it lies in the fact that it is the linearized version of the sine-Gordon equation, although the converse of this statement is probably more accurate historically.

This problem is hyperbolic, and the method of characteristics can be used to advantage. Thus, we define characteristic coordinates (in terms of the original "laboratory coordinates"),

(A.1.61)
$$x = \frac{T + X}{2}, \qquad t = \frac{X - T}{2},$$

and (A.1.60) becomes

(A.1.62)
$$u_{xt} = u, \qquad x - t > 0.$$

The theory of hyperbolic equations is too broad for us to develop here in any serious way. The classical text by Courant and Freidrichs (1948) is among the best references on the subject for our purposes. Two of the important consequences of the hyperbolic nature of (A.1.60) are the following:

(i) Any disturbance propagates outward along the characteristics emanating from it. It follows that if the initial data for (A.1.60) have compact support (in laboratory coordinates), the solution of (A.1.60) has compact support for any finite T. We will restrict our discussion here to these cases.

(ii) Discontinuities in u or its derivatives propagate along characteristics. The transformation from (A.1.60) to (A.1.62) applies only where the second derivatives of u are defined.

The energy integral in this case is more complicated than in previous problems:

(A.1.63) $$\frac{\partial}{\partial T}\int_{-\infty}^{\infty}(u_T^2+u_X^2+u^2)\,dx - 2u_Xu_T\Big|_{-\infty}^{\infty} = 0.$$

Thus, it is not sufficient that $u \in L_2$; we need that u_T and $u_X \in L_2$ as well. If the initial conditions lie in this slightly smaller space, then (A.1.63) assures that the solution remains there for $T>0$. Further, there is at most one solution to the problem in this space. On the other hand, (A.1.63) also implies that if the initial conditions do not lie in this space, whatever solution evolves from them will remain outside the space for all T. (Note that this problem differs from the heat equation in this respect.) The reason is that if the integral ever exists then from (A.1.63) its time derivative must vanish for all time.

The dispersion relation for (A.1.60) is obtained by substituting $u \sim \exp(i\kappa X - i\Omega T)$:

(A.1.64) $$\Omega^2 = \kappa^2 + 1.$$

The dispersion relation is real, as suggested by (A.1.63). For each real κ, there are two roots of (A.1.64), because the problem is second order in time. Thus, there are three speeds of interest in this problem:

(i) discontinuities in u or its derivatives propagate along the characteristics with speed l ("the speed of light");

(ii) for a given κ, the phase speed is

$$\left|\frac{\Omega}{\kappa}\right| = \left|\frac{\sqrt{\kappa^2+1}}{\kappa}\right| \geq 1;$$

(iii) for a given κ, the group speed is

$$\left|\frac{d\Omega}{d\kappa}\right| = \left|\frac{\kappa}{\sqrt{\kappa^2+1}}\right| \leq 1.$$

The Fourier transform solution of the problem is

(A.1.65) $$u(X,T) = \frac{1}{2\pi}\int A(\kappa)\,e^{i\kappa X + i\sqrt{\kappa^2+1}\,T}\,d\kappa$$
$$+ \frac{1}{2\pi}\int B(\kappa)\,e^{i\kappa X - i\sqrt{\kappa^2+1}\,T}\,d\kappa.$$

If u and u_T are real at $T=0$, then for all real κ the second integral in (A.1.65) is the complex conjugate of the first. If the initial data has a finite energy integral, then

(A.1.66) $$\frac{1}{2\pi}\int (1+\kappa^2)(|A|^2+|B|^2)\,d\kappa = \int (u^2+u_X^2+u_T^2)\,dx < \infty.$$

The solution does not become smoother in time, and a classical solution exists for all X ($T>0$) only if u_{XX} and u_{TX} are defined everywhere at $T=0$; otherwise, the solutions are "weak." In particular u has continuous second derivatives if

(A.1.67) $$\int (1+\kappa^2)(|A|+|B|)\,d\kappa <\infty,$$

by the Riemann–Lebesgue lemma.

To find the long time behavior of u, we must find the points of stationary phase in (A.1.65). For fixed X/T, these occur in the first integral where

(A.1.68) $$\frac{X}{T}=-\frac{\kappa}{\sqrt{\kappa^2+1}} \Rightarrow \kappa=-\frac{X}{\sqrt{T^2-X^2}}.$$

Using the stationary phase formula, as $T\to\infty$ for fixed $|X/T|<1$,

(A.1.69) $$u(X,T)\sim \frac{1}{\sqrt{2\pi}}\frac{T}{(T^2-X^2)^{3/4}}A\left(-\frac{X}{\sqrt{T^2-X^2}}\right)$$
$$\times\exp\left(i\sqrt{T^2-X^2}+\frac{i\pi}{4}\right)+(*).$$

Thus, the asymptotic solution inside the light cone consists of oscillations, whose amplitude decays as $T^{-1/2}$. Outside the light cone u vanishes identically if it had compact support initially.

The behavior of u along the light cone ($X/T=\pm 1$) is of particular interest because this function provides the "initial data" if (A.1.62) is to be solved as an initial value problem. This can be determined from (A.1.69) if $A(\kappa)$ has a limit as $\kappa\to\pm\infty$, which we now assume. If A also satisfies (A.1.66), then

$$\kappa^{3/2}A(\kappa)=o(1) \quad \text{as } \kappa\to\pm\infty.$$

Then it follows from (A.1.68, 69) that as $|X/T|\to 1$, $T\gg 1$,

$$|u|\sim|\kappa^{3/2}A(\kappa)|(2\pi)^{-1/2}\frac{T}{|X|^{3/2}}=o(1).$$

Thus, assuming only that the initial data have a finite energy integral and that $A(\kappa)$ has a limit as $\kappa\to\pm\infty$, it follows that the solution of (A.1.60) must decay along any characteristic as $T\to\infty$, even if this solution is only a weak solution. The solution is a classical solution if (A.1.67) also holds, and then the decay rate is faster.

Finally, let us rewrite (A.1.65) in terms of the characteristic coordinates. This is done conveniently by changing the variable of integration. Let

$$\kappa=\frac{1}{2}\left(\zeta-\frac{1}{\zeta}\right),$$

and use $\zeta > 0$ for the first integral and $\zeta < 0$ for the second. Then (A.1.65) becomes

(A.1.70) $$u = \frac{1}{2\pi} \int_{-\infty}^{\infty} \mathcal{A}(\zeta) \exp\left(i\zeta x - \frac{it}{\zeta}\right) d\zeta,$$

where

$$\mathcal{A}(\zeta) = \begin{cases} \frac{1}{2}(1 + \zeta^{-2}) A(\frac{1}{2}(\zeta - \zeta^{-1})), & \zeta > 0, \\ \frac{1}{2}(1 + \zeta^{-2}) B(\frac{1}{2}(\zeta - \zeta^{-1})), & \zeta < 0. \end{cases}$$

Clearly, (A.1.70) has the form of the Fourier transform solution of (A.1.62). It is also clear that the integral has no meaning for $t \neq 0$ unless

(A.1.71) $$\mathcal{A}(0) = 0.$$

At first sight, this additional restriction on the "initial data" (i.e., along the characteristic $t = 0$) may seem unnatural. However, if the initial data in laboratory coordinates satisfies (A.1.67), so that the transformation to characteristic coordinates is *defined* everywhere, then (A.1.71) is guaranteed. Kaup and Newell (1978c) discuss the analogue of (A.1.71) for the sine-Gordon equation.

Example 5: Discrete problems.

A semidiscrete version of (A.1.23–25) is

(A.1.72)
$$-i\frac{d}{d\tau}\psi_n(\tau) = \psi_{n+1}(\tau) + \psi_{n-1}(\tau) - 2\psi_n(\tau),$$
$$n = 0, \pm 1, \pm 2, \cdots, \quad \tau > 0,$$
$$\psi_n \to 0 \quad \text{as } |n| \to \infty, \quad \tau > 0,$$
$$\psi_n(\tau = 0) = \Psi_n \quad \text{given with } \sum_{n=-\infty}^{\infty} |\Psi_n|^2 = 1.$$

Here $\psi_n(\tau)$ is the nth function of time in an infinite sequence of such functions. In this example, the subscript $(\cdot)_n$ denotes a spatial index, rather than differentiation.

Alternatively, a finite difference scheme (Crank–Nicolson) for (A.1.23–25) is

(A.1.73)
$$-i\frac{\psi_n^{m+1} - \psi_n^m}{\Delta t} = \frac{\psi_{n+1}^{m+1} + \psi_{n-1}^{m+1} - 2\psi_n^{m+1}}{2h^2} + \frac{\psi_{n+1}^m + \psi_{n-1}^m - 2\psi_n^m}{2h^2},$$
$$\psi_n^m \to 0 \quad \text{as } |n| \to \infty, \quad \psi_n^0 = \Psi_n \quad \text{given with } \sum_{n=-\infty}^{\infty} |\Psi_n|^2 = 1.$$

Superscripts denote the discrete time index, rather than exponents. (A.1.73) is a viable scheme to compute (numerically) an approximate solution of (A.1.23–25). We could also attribute meaning to (A.1.72), or we could simply think of it as a problem intermediate between (A.1.23) and (A.1.73).

The analyses of (A.1.72) and (A.1.73) are similar, and are analogous to the Fourier transform methods already discussed. We concentrate here on (A.1.72), leaving (A.1.73) for an exercise. The first step is to find the "energy integral" for (A.1.72), by multiplying it by $\psi_n^*(\tau)$ and subtracting the complex conjugate of the resulting equation. In the sum over n, some of the terms form telescoping series, so that

(A.1.74) $$i\frac{d}{d\tau} \sum_{n=-\infty}^{\infty} |\psi_n(\tau)|^2 = 0,$$

which is analogous to (A.1.27). This identifies l_2, the set of square-summable sequences, as an appropriate space for the problem. Uniqueness of the solution in l_2 also follows from (A.1.74).

The dispersion relation for (A.1.72) is obtained by substituting (A.1.9) into (A.1.72). The result is

(A.1.75) $$\omega = \frac{-(z-1)^2}{z},$$

which is real if $|z| = 1$. Again, the real dispersion relation is consistent with the existence of a time-independent energy integral.

The next step is to mimic the Fourier transform. In the simplest case, we may suppose that

(A.1.76) $$\sum_{m=-\infty}^{\infty} |\psi_m| < \infty.$$

Then the function

(A.1.77) $$\tilde{\psi}(z) = \sum_m \psi_m z^{-m}$$

is defined for complex z on the unit circle; this is the analogue of the Fourier transform. The inverse transform comes from multiplying (A.1.77) by z^{n-1}, and integrating around the unit circle:

$$\frac{1}{2\pi i} \oint \tilde{\psi}(z) z^{n-1} \, dz = \frac{1}{2\pi i} \oint \sum_m \psi_m z^{-m+n-1} \, dz.$$

The order of integration and summation may be interchanged, using Fubini's theorem, because of (A.1.76) and the fact that the integral is over continuous functions on a finite interval. After we apply Cauchy's integral theorem, the

LINEAR PROBLEMS 371

result is the inversion formula for (A.1.77):

(A.1.78) $$\psi_n = \frac{1}{2\pi i} \oint \tilde{\psi}(z) z^{n-1} \, dz.$$

Now we may construct the "Fourier transform" solution of (A.1.72):

(A.1.79a) $$\psi_n(\tau) = \frac{1}{2\pi i} \oint \tilde{\Psi}(z) \exp\left(i(z-1)^2 z^{-1} \tau\right) z^{n-1} \, dz,$$

where the integral is taken around the unit circle, and

(A.1.79b) $$\tilde{\Psi}(z) = \sum_m \Psi_m z^{-m}.$$

An alternative representation is obtained by substituting

$$z = e^{i\theta}, \qquad \tilde{\Psi}(z) = \bar{\Psi}(\theta),$$

so that (A.1.79a) becomes

(A.1.80) $$\psi_n(\tau) = \frac{1}{2\pi} \int_0^{2\pi} \bar{\Psi}(\theta) \exp\{in\theta + 2i\tau(\cos\theta - 1)\} \, d\theta.$$

In obtaining the solution of (A.1.72), we have considered n to be a discrete variable, taking on only integer values. The solution, however, is defined for any real n, even though we may be more interested in integer values. This slight change in viewpoint allows us to evaluate (A.1.80) as $\tau \to \infty$ by the usual asymptotic methods. Thus, we consider (n/τ) to be an arbitrary, fixed constant, with τ large. The phase in (A.1.80) is

$$\tau\phi\left(\theta; \frac{n}{t}\right) = \left[\frac{n}{\tau}\theta + 2(\cos\theta - 1)\right]\tau,$$

and the stationary points occur where

(A.1.81) $$\frac{n}{\tau} = 2 \sin\theta.$$

Thus, (A.1.81) defines the *group velocity* corresponding to the dispersion relation (A.1.75). It is important to note that the group velocity for (A.1.72) is bounded. This identifies a qualitative difference between (A.1.72) and (A.1.23). In the continuous problem, the group velocity is given by (A.1.36), and arbitrarily large wavenumbers travel arbitrarily fast. The effect of the spatial discretization in (A.1.72) is to bound the maximum wavenumber, which in turn bounds the group velocity, as shown in (A.1.81).

At the stationary points of $\phi(\theta)$,

$$\phi''(\theta) = -2 \cos\theta;$$

this does not vanish except at $\theta = \pi/2$ and $3\pi/2$, corresponding to $n/\tau = +2$

and -2, respectively. At these two exceptional points, $\phi'''(\theta)$ does not vanish. This information allows us to find the dominant behavior of the solution as $\tau \to \infty$, with (n/τ) fixed. Here are the major results.

(i) If $|n| \ll 2\tau$, the solution oscillates with an amplitude that decays as $\tau^{-1/2}$. The behavior of the solution of (A.1.72) in this region is qualitatively similar to that of (A.1.23).

(ii) If $|n| \gg 2\tau$, the integral has no stationary points, and the solution decays faster than τ^{-1}. If the initial data had compact support, the decay rate would be exponential in this region. As we have discussed, this relatively quiet region exists because very high wavenumbers (in x) are excluded from the solution of (A.1.72).

(iii) Near $n = \pm 2\tau$, there is a *wavefront* that has no counterpart in the continuous problem. The solution decays only as $\tau^{-1/3}$ near the wavefront, which becomes the dominant feature of the solution as $\tau \to \infty$.

In light of the qualitative difference between the asymptotic solutions of (A.1.23) and (A.1.72), it may be worthwhile to reconsider in what sense they approximate each other. We begin with (A.1.23), on $-\infty < x < \infty$. Identify uniformly spaced points by

$$x_n = nh$$

for some constant $h \ll 1$. Then

(A.1.82) $$\left.\frac{\partial^2 \psi}{\partial x^2}\right|_n = \frac{\psi_{n+1} + \psi_{n-1} - 2\psi_n}{h^2} + O(h^2),$$

and (A.1.23) becomes

(A.1.83) $$-i\frac{d}{dt}\psi_n = \frac{\psi_{n+1} + \psi_{n-1} - 2\psi_n}{h^2} + O(h^2).$$

This may be approximated by (A.1.72) until the neglected terms in (A.1.83) produce a cumulative effect that is significant. Based on (A.1.83), the time for the breakdown of (A.1.72) as an approximation to (A.1.23) may be estimated by

$$t = O(h^{-2}).$$

Thus, (A.1.72) approximates (A.1.23) only for a limited time. From this standpoint it is not so surprising that their asymptotic $(t \to \infty)$ behaviors are different. There are two limits, $(t \to \infty)$ and $(h \to 0)$; they do not commute for all space (n).

Even so, it is worth asking where we should look in the solution of (A.1.23) to find the wavefront that appears in the solution of (A.1.72). To align (A.1.83) with (A.1.72), set

$$\tau = \frac{t}{h^2}$$

and neglect the higher order terms. The wave front occurs where
$$n = \pm 2\tau;$$
i.e.,
$$\frac{x_n}{h} = \pm \frac{2t}{h^2} \quad \text{or} \quad x_n = \pm \frac{2t}{h}.$$

Comparing with (A.1.36) shows that this trajectory corresponds to a wavenumber

(A.1.84) $$k = \frac{1}{h}.$$

There are two possibilities.

(i) The initial data for (A.1.23) contained no information at such a high wavenumber. There is no important contradiction between the solutions of (A.1.23) and (A.1.72) because the slowly decaying wavefront occurs where the amplitude vanishes.

(ii) The initial data for (A.1.23) contain significant information at this wavenumber (and beyond). Then (A.1.82) is a lousy approximation, the neglected terms apparently become important relatively quickly and the asymptotic solutions become valid after (A.1.72) has ceased to approximate (A.1.23).

In this problem, we have come face to face with an approximation which has a limited range of validity in time. This sort of difficulty also surfaces in Chapter 4, where we discuss physical applications of evolution equations.

This completes our description of the "method of Fourier transforms" for solving linear evolution equations with constant coefficients. The method has the advantages that it is straightforward to apply, and that the qualitative behavior of the solution for sufficiently large time is determined directly from the dispersion relation. Its disadvantage is that it is not as general as some other methods, such as Fourier–Laplace transforms. However, in a problem where this method fails, it must fail in one or the other (or both) of two ways:

(i) there is no dispersion relation;
(ii) the set of Fourier modes is incomplete.

In the next section, we consider some problems for which this method fails. Phase mixing, algebraically growing modes and Landau damping are common features of these problems, and the linearized limit of the problem of self-induced transparency (SIT) is an example. With the exception of this problem, however, the material in § A.2 does not relate directly to IST or to problems solvable thereby.

A.2. Failure of the Fourier transform method. We consider next some problems in which this method fails. Frequently, the failure is dramatic: there is no dispersion relation, and a continuous range of ω's is permitted for each fixed k. In contrast to problems which have dispersion relations, these problems can exhibit exponential decay (in time) even though ω is necessarily real for real k. The phenomenon of "Landau damping" in plasma physics is an example of this decay.

Example 1. A model of kinetic theory in one dimension.

(A.2.1)
$$\frac{d}{dt}f(t) = -\alpha \int_{-\infty}^{\infty} U(x)g(x,t)\,dx, \qquad t > 0,$$

$$\frac{\partial g}{\partial t} + v\frac{\partial g}{\partial x} = U(x)f(t), \qquad -\infty < x < \infty, \quad t > 0,$$

$$g(x,t) \to 0 \quad \text{as } x \to -\infty,$$

$U(x), f(0), g(x,0) = G(x)$ given and real,

$U(x), G(x) \in L_2$.

This problem was proposed by Ramanathan and Sandri (1969) as a simple model to test the validity of hypotheses used in the derivation of the kinetic theory of gases. In that context, f corresponds to the departure from equilibrium of a one-particle distribution function, and g to the departure from equilibrium of a two-particle correlation function. Both v and α are positive constants, with $\alpha \ll 1$. The integral term models two-body interactions, but three-body and higher interactions are neglected. We will analyze (A.2.1) in some detail, because it is a prototype of a number of linear problems that have no dispersion relations.

A time-independent energy integral for (A.2.1) is easily found:

(A.2.2)
$$\frac{d}{dt}\left[f^2 + \alpha \int_{-\infty}^{\infty} g^2\,dx\right] = -\alpha v g^2 \bigg|_{x=-\infty}^{+\infty}$$

except that only one boundary condition (as $x \to -\infty$) is available for g. However, if $U(x)$ and $G(x)$ have compact support, then one may show $g(x,t)$ has compact support for all time, so that the right side of (A.2.2) vanishes. For the remainder of the discussion, we assume that $U(x)$ and $G(x)$ decay fast enough that the right side of (A.2.2) vanishes for all time. Then g remains in L_2 for all time, and (A.2.1) has at most one solution with $g \in L_2$.

Thus, (A.2.1) has a time-independent energy integral, but this does not prevent f from decaying to zero at an exponential rate as $t \to \infty$, provided g grows correspondingly. In fact, if $U(x)$ is symmetric and $G(x)$ is antisymmetric in x, then (A.2.1) is strictly time reversible; i.e., it is invariant under the transformation

(A.2.3) $\qquad\qquad t \to -t, \quad x \to -x, \quad g \to -g.$

Even so, f can decay as $t \to \infty$; time reversibility only means that it also decays as $t \to -\infty$. (This behavior contrasts strongly with solutions of the heat equation, which also exhibit exponential decay, but are time irreversible.)

We now show that the method of Fourier transforms fails for (A.2.1), because the problem has no dispersion relation. Some care is required in applying the usual ansatz because of the integral term in the equations. Take the Fourier transform (in x) of the equations, making use of the fact that g and U are in L_2, to obtain

(A.2.4a) $$\frac{df}{dt}(t) = -\frac{\alpha}{2\pi} \int U(-k)g(k, t)\, dk,$$

(A.2.4b) $$\frac{\partial}{\partial t}\hat{g}(k, t) + ikv\hat{g}(k, t) = \hat{U}(k)f(t).$$

Next, assume that, for each mode,

$$f \sim \tilde{f}(\omega) e^{-i\omega t}, \quad \hat{g}(k, t) \sim \tilde{g}(k, \omega) e^{-i\omega t}.$$

More precisely,

(A.2.5) $$f(t) = \int \tilde{f}(\omega) e^{-i\omega t}\, d\omega, \quad \hat{g}(k, \omega) = \int \tilde{g}(k, \omega) e^{-i\omega t}\, d\omega,$$

where $\tilde{f}$ and $\tilde{g}$ are to be interpreted as generalized functions, in order that the integrals be defined (e.g., see Lighthill (1958)). Then (A.2.4) becomes

(A.2.6a) $$-i\omega\tilde{f}(\omega) = -\frac{\alpha}{2\pi} \int \hat{U}(-k)\tilde{g}(k, \omega)\, dk,$$

(A.2.6b) $$(-i\omega + ikv)\tilde{g}(k, \omega) = \hat{U}(k)\tilde{f}(\omega).$$

If we eliminate $\tilde{f}$ from (A.2.6), multiply the resulting equation by $\tilde{g}$, use the reality of $U(x)$ and integrate over all real k, the result is

(A.2.7) $$-\omega^2 \int |\tilde{g}|^2\, dk + \omega v \int k|\tilde{g}|^2\, dk + \alpha \left|\int \hat{U}(-k)\tilde{g}(k, \omega)\, dk\right|^2 = 0.$$

This is a quadratic equation for ω, with real coefficients. Its discriminant is positive, so (A.2.7) has two real roots. Thus, if k is real, ω must be real, which is consistent with the existence of the energy integral, (A.2.2). However, (A.2.7) does not yield a dispersion relation or demonstrate its absence, because $\tilde{g}$ is still free.

Now, we solve (A.2.6b) for $\tilde{g}$. The (formal) general solution (cf. Lighthill (1958)) is

(A.2.8) $$\tilde{g}(k, \omega) = \frac{\hat{U}(k)\tilde{f}(\omega)}{i(kv - \omega)} + iC(\omega)\delta(kv - \omega).$$

In the first term, we take only the principal part of the singular function; in the second term, δ is the Dirac delta function and $C(\omega)$ is arbitrary. Special cases are $C(\omega) = \pm \pi \hat{U}(\omega/v)\tilde{f}(\omega)$, which correspond to specifying contours over or under the singularity in the complex ω-plane. Substituting (A.2.8) into (A.2.6a) yields

(A.2.9) $$\left[\omega + \frac{\alpha}{2\pi} \int \frac{|\hat{U}(k)|^2}{kv - \omega} dk\right]\tilde{f}(\omega) - \frac{\alpha}{2\pi v} \hat{U}\left(-\frac{\omega}{v}\right) C(\omega) = 0.$$

But this merely fixes $C(\omega)$ in terms of $\tilde{f}(\omega)$; it does *not* fix $\omega(k)$. This is the important distinction between (A.2.1) and the problem discussed in the preceding section: for each fixed k in (A.2.1), all real ω's are possible; there is no dispersion relation. (This distinction was clearly identified by van Kampen (1955) for the linearized Vlasov equations.)

Some readers may feel uneasy that generalized functions have suddenly appeared, although they had not been mentioned before. In fact, they have been lurking in the background all along, but it was not necessary to acknowledge them before. To see this explicitly, consider a variation of (A.2.1) that does have a dispersion relation:

(A.2.10)
$$\frac{d}{dt}f(t) = -\alpha \int g(x, t) \, dx = -\alpha \hat{g}(0, t),$$

$$\frac{\partial}{\partial t} g(x, t) + v \frac{\partial}{\partial x} g(x, t) = U(x)f(t).$$

Proceeding as before from (A.2.10) we obtain

(A.2.11a) $$-i\omega \tilde{f}(\omega) = -\alpha \tilde{g}(0, \omega),$$
(A.2.11b) $$i(kv - \omega)\tilde{g}(k, \omega) = \hat{U}(k)\tilde{f}(\omega).$$

The second of these again gives (A.2.8) at $k = 0$,

$$\tilde{g}(0, \omega) = i \frac{\hat{U}(0)\tilde{f}(\omega)}{\omega} + ic(\omega)\delta(-\omega).$$

However, when we substitute this into (A.2.11a) and multiply by ω, the result is

(A.2.12) $$[\omega^2 - \alpha \hat{U}(0)]\tilde{f}(\omega) = 0.$$

Because $f(\omega)$ is arbitrary, (A.2.12) defines ω^2. Thus, generalized functions appear in this problem as well, but it still has a well-defined dispersion relation.

To this point, all we have shown is that the method of Fourier transforms fails for (A.2.1), because there is no dispersion relation. Fourier–Laplace transforms provide another approach that often succeeds when Fourier transform methods fail. Next we will use Fourier–Laplace transforms to solve (A.2.1).

Solve (A.2.4b) for $\hat{g}(k, t)$, in terms of $f(t)$, and then Fourier transform back to $g(x, t)$. The result is

$$(A.2.13) \qquad g(x, t) = \int_0^t U(x - v(t-\tau))f(\tau)\, d\tau + G(x - vt).$$

Then (A.2.1) reduces to a single equation for f:

$$(A.2.14) \qquad \frac{df}{dt} = -\alpha \int_0^t f(\tau) K(v(t-\tau))\, d\tau - \alpha \int_{-\infty}^{\infty} U(x) G(x - vt)\, dx,$$

where

$$(A.2.15) \qquad K(y) = \int_{-\infty}^{\infty} U(x-y)U(x)\, dx = \frac{1}{2\pi} \int_{-\infty}^{\infty} |\hat{U}(k)|^2 e^{iky}\, dk.$$

(Because $K(y)$ is even, certain signs are arbitrary in the subsequent analysis. The final result does not depend on this choice of signs, so we simply make a choice and use it consistently.)

We restrict our attention to cases in which $G(x) \equiv 0$ (no pairwise correlation of particles at $t = 0$). Then (A.2.15) reduces to

$$(A.2.16) \qquad \frac{df}{dt} = -\alpha \int_0^t f(\tau) K(v(t-\tau))\, d\tau.$$

For a special family of potentials, (A.2.16) can be solved in closed form (see Exercise 1). Here we are interested in general features, which do not require special assumptions about $U(x)$. The Laplace transform of (A.2.16) yields

$$\left\{ p + \frac{\alpha}{2\pi} \int \frac{|\hat{U}(k)|^2}{p + ikv}\, dk \right\} \tilde{f}(p) = f(0), \qquad \operatorname{Re}(p) > 0,$$

after we interchange the order of two integrals. The inverse transform gives the formal solution of (A.2.16):

$$(A.2.17) \qquad f(t) = \frac{f(0)}{2\pi i} \int_C \left\{ p + \frac{\alpha}{2\pi} \int \frac{|\hat{U}|^2}{p + ikv}\, dk \right\}^{-1} e^{pt}\, dp,$$

where the Bromwich contour C is taken upward along a vertical line to the right of all singularities of the integrand in the complex p-plane. These singularities are the solutions of

$$(A.2.18) \qquad p + \frac{\alpha}{2\pi} \int_{-\infty}^{\infty} \frac{|\hat{U}|^2}{p + ikv}\, dk = 0.$$

Next we evaluate (A.2.17), and see how Landau damping appears. It is not difficult to see that the only solutions of (A.2.18) are purely imaginary. If we set $p = i\omega$, this merely states that ω must be real for real k, as we found from (A.2.7). However, some care must be exercised in moving the Bromwich

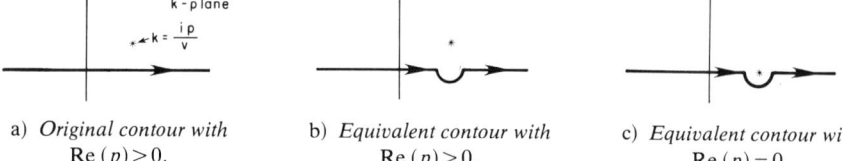

a) *Original contour with* Re $(p) > 0$.

b) *Equivalent contour with* Re $(p) > 0$.

c) *Equivalent contour with* Re $(p) = 0$.

FIG. A.4. *Contours for integration in* (A.2.18).

contour over to the imaginary axis of the *p*-plane. (This extra care was precisely the difference between Vlasov's (1945) and Landau's (1946) treatment of Vlasov's equations of a collisionless plasma.) If Re $(p) > 0$, the *k*-integral in (A.2.18) is along a contour that goes *under* the apparent singularity at $k = ip/v$, as shown in Fig. A.4a. By Cauchy's theorem, the *k*-integral is unchanged if its contour is deformed as in Fig. A.4b. Now if we let Re $(p) \to 0$, as in Fig. A.4c, this contour still goes *under* the singularity. But now k need not be real, and ω may become complex as well. Hence, (A.2.18) becomes (with $p = i\omega$)

$$\text{(A.2.19)} \quad i\omega + \frac{\alpha}{2\pi i} \oint \frac{|\hat{U}(k)|^2}{\omega + kv} dk + \frac{\alpha |\hat{U}(-\omega/v)|^2}{2v} = 0,$$

where

$$\text{(A.2.20)} \quad \oint \frac{\phi(x)}{x} dx = \lim_{\varepsilon \to 0} \left[\int_{-\infty}^{-\varepsilon} \frac{\phi(x)}{x} dx + \int_{\varepsilon}^{\infty} \frac{\phi(x)}{x} dx \right]$$

is the principal-value integral. Solving (A.2.19) recursively, and keeping only the $O(\alpha)$ terms, gives the approximate solution

$$\omega = \omega_r + i\omega_i,$$

$$\text{(A.2.21)} \quad \omega_i = \frac{\alpha}{2v} |\hat{U}(0)|^2 = \frac{\alpha}{2v} \left(\int U(x) \, dx \right)^2 \geq 0,$$

$$\omega_r = o(\alpha).$$

Finally, we substitute back into (A.2.17) to get the dominant behavior of f, as $t \to \infty$,

$$\text{(A.2.22a)} \quad f(t) \sim f(0) \exp\left\{ -\frac{\alpha}{2v} \left(\int U \, dx \right)^2 t \right\}.$$

This can also be written as

$$\text{(A.2.22b)} \quad f(t) \sim f(0) \exp\left\{ -\frac{\alpha}{v} \int_0^\infty K(y) \, dy \, t \right\},$$

which corresponds to Bogolyubov's (1962) formula for the decay to equilibrium of a nonequilibrium distribution of particles.

It follows from (A.2.22) that there are two possibilities. If $\int U\,dx = 0$, then to lowest order in α there is neither damping nor growth of $f(t)$. The asymptotic behavior of the solution of (A.2.1) requires a more accurate solution of (A.2.19) than that given by (A.2.21). To leading order in α, f is constant as is $\int g^2\,dx$ as $t \to \infty$.

If $\int U\,dx \neq 0$, then f decays to zero at an exponential rate as $t \to \infty$ (or as $t \to -\infty$). One might question whether it is appropriate to call this "damping," but certainly zero is the only stable equilibrium point for f.

Given $f(t)$ (approximately), $g(x, t)$ may be found (approximately) from (A.2.13). For large time, if $\int U\,dx \neq 0$,

(A.2.23)
$$g(x, t) \sim f(0) \int_0^\infty U(x - vt + v\tau)e^{-\tau/t_0}\,d\tau$$
$$= t_0 f(0) \sum_{n=0}^\infty U^{(n)}(x - vt)(vt_0)^n,$$

where

$$t_0^{-1} = \frac{\alpha}{2v}\left(\int U\,dx\right)^2.$$

Thus $g(x, t)$ tends toward a solitary wave of permanent form, traveling with speed v, and with $\alpha \int g^2\,dx = f^2(0)$. The shape of the wave depends on details of $U(x)$.

In summary, if

(A.2.24) $\qquad G(x) = 0, \quad U(x) = U(-x), \quad \int U\,dx \neq 0,$

then (A.2.1) allows time reversal and has a time–independent energy integral. However, there is no dispersion relation, and as $t \to \infty$ (or $t \to -\infty$), energy is transferred from f^2 to $\int g^2\,dx$ at an exponential rate. The only stable configuration of the system is $f = 0$, with g in the form of a solitary wave of permanent form.

Example 2. Self-induced transparency (linear limit).

The phenomenon of self-induced transparency is discussed in detail in Chapter 4. The equations are

(A.2.25)
$$\left.\begin{array}{l} \lambda_\tau + 2i\alpha\lambda = \varepsilon N \\ N_\tau = -\tfrac{1}{2}(\varepsilon^*\lambda + \varepsilon\lambda^*) \\ \varepsilon_x = \langle\lambda\rangle \end{array}\right\} \quad x > 0, \quad -\infty < \tau < \infty.$$

Here $\varepsilon(x, \tau)$ is the (complex) envelope of the electric field, $\lambda(x, \tau, \alpha)$ represents the (complex) induced polarization, $N(x, \tau, \alpha)$ is the (real) normalized population inversion, and

(A.2.26a) $\qquad \langle\lambda\rangle \equiv \int_{-\infty}^\infty g(\alpha)\lambda(x, \tau, \alpha)\,d\alpha,$

where $g(\alpha)$ represents the inhomogeneous broadening of the medium. We assume that g is a real, nonnegative function normalized by

(A.2.26b) $$\int g \, d\alpha = 1.$$

The usual laboratory experiment is conducted as an initial value problem in x; $x = 0$ is where the EM wave first enters the resonant medium. The appropriate initial-boundary data for (A.2.25) are

(A.2.27)
$$\lambda \to 0, \quad N \to -1 \quad \text{as } \tau \to -\infty, \quad \text{for all } x > 0,$$
$$\varepsilon(x = 0, \tau) \quad \text{given}, \quad \int |\varepsilon(0, \tau)| \, d\tau < \infty.$$

It is important to remember that (x, τ) reverse their usual roles in this problem: x is timelike, τ is spacelike.

The nonlinear problem, (A.2.25), has a time-independent energy integral; at every x,

$$\frac{\partial}{\partial \tau}(\lambda \lambda^* + N^2) = 0.$$

Then the boundary conditions $(\tau \to -\infty)$ show that for all (x, τ, α)

(A.2.28) $$\lambda \lambda^* + N^2 = 1.$$

If the electric field imposed at $x = 0$ is weak, then an approximate solution can be obtained by linearizing (A.2.25) about its unperturbed state at $\tau = -\infty$. Thus, for $\delta \ll 1$,

(A.2.29)
$$\varepsilon(x, t) \sim \delta E(x, \tau; \delta) + \cdots,$$
$$\lambda(x, t, \alpha) \sim \delta \Lambda(x, \tau, \alpha; \delta) + \cdots,$$
$$N(x, t, \alpha) \sim -1 + \delta N_{(1)}(x, \tau, \alpha; \delta) + \cdots,$$

and the linearized equations are

(A.2.30)
$$\Lambda_\tau + 2i\alpha \Lambda = -E, \quad -\infty < \tau < \infty, \quad x > 0,$$
$$E_x = \langle \Lambda \rangle,$$
$$N_{(1)} \equiv 0,$$
$$\Lambda, E \to 0 \quad \text{as } \tau \to -\infty,$$
$$E(0, \tau) \quad \text{given}, \quad E(0, \tau) \to 0 \quad \text{as } \tau \to -\infty.$$

Notice that the linearized equations involve fewer unknowns than the original problem, and that the constraint of a time-independent energy integral, (A.2.28), is lost in the linearization.

With the ansatz

$$(\Lambda, E) \sim (\tilde{\Lambda}, \tilde{E}) \, e^{ikx - i\omega \tau},$$

(A.2.30) become

(A.2.31)
$$(-i\omega + 2i\alpha)\tilde{\Lambda} = -\tilde{E},$$
$$ik\tilde{E} = \langle \tilde{\lambda} \rangle.$$

From these equations, one can deduce that

(A.2.32)
$$2k\langle \alpha |\tilde{\lambda}|^2 \rangle + k\omega \langle |\tilde{\lambda}| \rangle = |\langle \tilde{\lambda} \rangle|^2.$$

For fixed, real ω (remember to reverse (x, τ)), this is a linear equation for k with real coefficients. Thus, k must be real if ω is real, but we cannot deduce a dispersion relation from (A.2.32), since λ is still unknown.

As before, we may solve (A.2.31) for $\tilde{\Lambda}$:

(A.2.33)
$$\tilde{\Lambda} = \frac{i\tilde{E}}{2\alpha - \omega} + iC(k, \omega)\delta(2\alpha - \omega),$$

where the first term on the right is interpreted in the sense of principal part, and $C(k, \omega)$ is arbitrary. Therefore,

(A.2.34)
$$\langle \tilde{\Lambda} \rangle = i\tilde{E} \left\langle \frac{1}{2\alpha - \omega} \right\rangle + \frac{i}{2} C(k, \omega) g\left(\frac{\omega}{2}\right) = ik\tilde{E}.$$

This equation determines $C(k, \omega)$ in terms of E; no restriction is imposed on (k, ω). Thus, (A.2.30) has no dispersion relation, and we must solve these equations by other means.

Analysis of the linearized equations is simplified somewhat if we also assume

(A.2.35)
$$E(0, \tau) \equiv 0, \quad \tau < 0;$$

i.e., the imposed electric field was not turned on until $\tau = 0$. Then it is straightforward to show that both Λ and E vanish identically for all $x > 0, \tau < 0$. This permits the boundary condition at $\tau = -\infty$ to be imposed instead at $\tau = 0$.

It is convenient to eliminate Λ from (A.2.30):

(A.2.36)
$$\Lambda(x, \tau, \infty) = -\int_0^\tau E(x, T) \exp\{2i\alpha(T - \tau)\}\, dT,$$

and

$$\langle \Lambda \rangle = \int \Lambda g \, d\alpha = -\int_0^\tau E(x, T) \int_{-\infty}^\infty g(\alpha) \exp\{2i\alpha(T - \tau)\}\, d\alpha\, dT.$$

The interchange of integrals is justified for finite τ if E is bounded, because $g \in L_1$. Define

(A.2.37)
$$G(m) = \int g(\alpha) e^{-2i\alpha m}\, d\alpha,$$

which is related to the Fourier transform of g. Then (A.2.30) reduces to

$$(A.2.38) \qquad E_x(x, \tau) = -\int_0^\tau E(x, T) G(\tau - T)\, dT, \qquad x > 0, \quad \tau > 0,$$

Some insight may be obtained by studying one type of inhomogeneous broadening:

$$(A.2.39) \qquad g(\alpha) = \frac{a}{\pi} \frac{1}{a^2 + \alpha^2}, \qquad a > 0.$$

In the limit $a \to 0$, g becomes a delta function and we say that there is no broadening. For any a, (A.2.38) now implies

$$(A.2.40) \qquad E_{x\tau} = -E - 2aE_x, \qquad x > 0, \quad \tau > 0.$$

This equation has a dispersion relation:

$$\omega k = -1 - 2iak$$

or

$$(A.2.41) \qquad k = \frac{-\omega}{\omega^2 + (2a)^2} + \frac{2ia}{\omega^2 + (2a)^2}.$$

Thus, (A.2.30) does not have a dispersion relation, but if we eliminate Λ, then the resulting equation (A.2.40) does! Im $(k) > 0$ for any $a > 0$, so (A.2.40) is stable; every Fourier mode decays (in x) to zero at an exponential rate. In the limit $a \to 0$, (A.2.40) becomes the Klein–Gordon equation, (A.1.62), with no damping. Even for $a > 0$, (A.2.40) is still hyperbolic, and discontinuities will propagate along the characteristics. The damping is too weak to smooth the discontinuities.

An exact solution of (A.2.40) is

$$(A.2.42) \qquad E(x, \tau) = \begin{cases} e^{-2a\tau} J_0(2\sqrt{x\tau}), & \tau > 0, \quad x > 0, \\ 0, & \tau < 0, \quad x > 0, \end{cases}$$

where $J_0(r)$ is the Bessel function of order zero. This solution is attained by abruptly turning on $E(0, \tau)$ at $\tau = 0$, then turning it off exponentially slowly. The discontinuity at $\tau = 0$ propagates to all $x > 0$. Elsewhere the solution decays as $x \to \infty$, or as $\tau \to \infty$. The induced polarization, Λ, is much more complicated. It is evident from (A.2.36) that at any fixed x, Λ is excited by the cumulative effect of E. Once excited, however, $\Lambda(x, \tau)$ remains excited as $\tau \to \infty$ even though E vanishes. Thus, E tends to zero as $\tau \to \infty$; Λ has no limit. This feature, that different parts of the solution behave much differently, is common in problems without dispersion relations.

We now return to the general case, (A.2.38). It turns out that after Λ has been eliminated, (A.2.38) has a dispersion relation for any acceptable broadening, $g(\alpha)$. To see this, take its Laplace transform in τ:

(A.2.43) $$\tilde{E}_x(x, p) = -\tilde{E}(x, p)\tilde{G}(p).$$

$\tilde{G}(p)$ is analytic for Re $(p) > 0$, and has an explicit representation:

$$G(p) = \int_{-\infty}^{\infty} \frac{g(\alpha)}{2i\alpha + p} d\alpha.$$

Note that for Re$(p) > 0$, the path of α-integration goes *under* the apparent singularity at $\alpha = ip/2$. Solving (A.2.43) and transforming back, we get

(A.2.44a) $$E(x, \tau) = \frac{1}{2\pi i} \int_C \tilde{E}(0, p) \exp\{p\tau - \tilde{G}(p)x\} dp,$$

where C denotes the Bromwich contour. $\tilde{E}(0, p)$ can have no singularities for Re $(p) > 0$, because $E(0, \tau)$ vanishes as $\tau \to \infty$. Thus, we may push the contour in (A.2.44a) to the imaginary axis and set $p = i\omega$, provided we also deform the contour of α-integration to stay *under* the singularity. The result is

(A.2.44b) $$E(x, \tau) = \frac{1}{2\pi} \int_{-\infty}^{\infty} \tilde{E}(0, -i\omega) \exp\{-i\omega\tau - \tilde{G}(-i\omega)x\} d\omega.$$

This dispersion relation is

(A.2.45) $$k(\omega) = i\tilde{G}(-i\omega) = \int_U \frac{g(\alpha)}{2\alpha - \omega} d\alpha,$$

where the contour is taken under the singularity. in particular,

(A.2.46) $$\text{Im}(k) = \frac{\pi}{2} g\left(\frac{\omega}{2}\right).$$

 (i) Because $g(\alpha)$ is nonnegative, there are no unstable modes.
 (ii) If $g(\alpha) > 0$, as in (A.2.39), all modes damp. $\int |E|^2 d\tau \to 0$ as $x \to \infty$; $E(x, \tau)$ may or may not also decay pointwise as $x \to \infty$, as we have seen in (A.2.42).
 (iii) If $g(\alpha) \geq 0$, some modes damp, while others are neutral. The asymptotic $(x \to \infty)$ behavior of E is determined by the undamped, dispersive modes.

This completes our discussion of the linearized equations of self-induced transparency. We saw in Chapter 4 that in the nonlinear problem, the part of the solution related to the continuous spectrum behaves qualitatively like this linearized solution. The behavior of the solitons, of course, is different.

There are a number of other linear problems of physical significance that have no dispersion relation. Two of these are discussed in Exercises 2 and 5. As we have mentioned, even if a linear problem has a dispersion relation, the method of Fourier transforms fails if the set of modes is not complete. We shall not discuss this aspect of the problem here. However, examples of this are examined in Exercises 6 and 7.

EXERCISES

Section A.1

1. (a) Find the Fourier transform representation of the solution of Example 1, (A.1.23–25), that evolves from

$$\Psi(x) = \begin{cases} a, & |x| < \dfrac{1}{a^2}, \\ 0, & |x| > \dfrac{1}{a^2}; \end{cases}$$

i.e., a particle is known to be somewhere in an interval of width $2/a^2$.

(b) Show that differentiation under the integral sign is permitted after the contour is rotated.

(c) As $t \to \infty$, where are the most likely regions in which to find the particle?

2. Solve Example 1 with the real initial conditions

$$\Psi(x) = a\,e^{-a^2|x|}.$$

Evaluate the Fourier integral as $t \to \infty$ by the method of steepest descents. Show that in addition to the terms given in (A.1.39), there may be a contribution from one of the poles of $\hat{\Psi}(k)$, depending on (x/t). How do these additional contributions affect the asymptotic behavior of ψ? What do they mean in the context of the probabilistic interpretation of this problem?

3. Solve

$$T_t = \kappa T_{xx}, \quad x > 0, \quad t > 0, \quad \kappa > 0,$$

$$T = 0 \quad \text{at } x = 0,$$

$$T \to 0 \quad \text{as } x \to \infty,$$

$$T(x, t = 0) = \begin{cases} 1, & x < L, \\ 0, & x > L. \end{cases}$$

For any $t > 0$, the maximum temperature in x must satisfy $T_x = 0$. Find an approximate formula for the location of the maximum, $\bar{x}(t)$, as $t \to \infty$. When does this formula become valid? Does $\bar{x}(t)$ remain within $[0, L]$? Why not? How does the maximum temperature depend on time (as $t \to \infty$)? Compare these results with the behavior on the infinite interval. Does the motion of $\bar{x}(t)$ constitute a wave?

4. Show that the solution of

$$u_t + cu_x + \alpha u_{xxx} = 0, \quad -\infty < x < \infty, \quad t > 0,$$

is equivalent to the solution of (A.1.49) by making a Galilean change of variables.

5. The first order differential equation

$$u_t + cu_x = 0$$

may be approximated by the difference equation

$$\frac{u_n^{m+1} - u_n^m}{\Delta t} + \frac{c}{4h}\left[u_{n+1}^{m+1} - u_{n-1}^{m+1}\right] + \frac{c}{4h}\left[u_{n+1}^m - u_{n-1}^m\right] = 0,$$

where the subscripts and superscripts denote indices.

(a) Multiply the difference equation by $(u_n^{m+1} + u_n^m)$ and sum over n. Show that

$$\sum_n (u_n^{m+1})^2 = \sum_n (u_n^m)^2.$$

This is the "energy integral" for the difference equation. What is the energy integral for the differential equation?

(b) What are the dispersion relations for the differential equation and the difference scheme?

(c) This approximation introduces no numerical diffusion (i.e., $|\Omega| = 1$), but is known to produce phase errors. By expanding each term of the difference equation in a Taylor series (i.e., $u_{n+1}^m = u_n^m + \partial u/\partial x|_n^m h + \cdots$) about $u_n^{m+\frac{1}{2}}$, show that the solution of the difference equation is approximated *better* by the solution of the equation in Exercise 4 than it is by the solution of the first order equation in Exercise 5. Find α in Exercise 4 in terms of $(h, \Delta t, c)$. For given initial data, estimate the time beyond which the solutions of the two equations differ significantly (this requires that you define "significantly"). For further discussion of these phase errors, see Orszag and Israeli (1974) and the references cited there.

6. (a) Find the energy integral of

$$u_{tt} = c^2 u_{xx} - \varepsilon^2 u_{xxxx}, \quad -\infty < x < \infty, \quad t > 0.$$

(b) For some appropriate choice of initial data, show that the Fourier transform solution splits into two parts, and that in terms of either the group velocity or the phase velocity, one of these is "left-going" and the other is "right-going."

(c) For small ε, show that either of these may be approximated by the problem in Exercise 4, and therefore by (A.1.49). What is required for ε to be "small"? (Note that this can be answered in a variety of ways, but all of them involve the initial conditions.) Find the range of validity in (x, t) of this approximate solution. Reconstruct the entire solution of the problem in (a) in terms of these two approximate solutions. Discuss qualitatively how the solution changes if the initial conditions contain significant short wave oscillations.

7. (a) The linearized "Boussinesq equation" is
$$u_{tt} = c^2 u_{xx} + \varepsilon^2 u_{xxxx}, \qquad -\infty < x < \infty, \quad t > 0.$$
Show that this problem is ill-posed. How does this difficulty appear when you look for an energy integral? The dispersion relation for the linearized problem of water waves is
$$\omega^2 = gk \tanh kh.$$
Show that this dispersion relation corresponds to a problem that is well-posed, and that the ill-posed problem above comes from taking only two terms in the small k-expansion of $\tanh kh$. Show that the approximation
$$u_{tt} = c^2 u_{xx} + \left(\frac{h^2}{3}\right) u_{ttxx}$$
is as accurate for small k, and is well-posed. Does it have an energy integral?

(b) $$u_{yy} = c^2 u_{xx} + \varepsilon^2 u_{xxxx}$$
is the linearized and time-independent version of the Kadomtsev–Petviashvili equation (see § 2.1). Suppose $u(x, y)$ is periodic in x with period $L < 2\pi\varepsilon/c$. Show that the problem is well-posed if we specify $u(x, 0)$ and $u(x, Y)$, $Y \neq 0$.

8. (a) Show that, if we change a sign in (A.1.60),
$$u_{TT} - u_{XX} = u,$$
the energy integral ceases to be positive definite, even though it is still conserved. Show that this equation is unstable as an initial value problem. What is the maximum growth rate?

(b) Show that
$$u(X, T) = \begin{cases} \dfrac{T}{\sqrt{T^2 - X^2}} J_1(\sqrt{T^2 - X^2}), & T^2 \geq X^2, \\ 0, & T^2 < X^2, \end{cases}$$
is an exact (similarity) solution of (A.1.60). To what initial conditions does this solution correspond? Does it have a finite energy integral?

9. Solve (A.1.72) with initial conditions corresponding to those in Exercise 1. Find explicitly the dominant terms in the asymptotic solution. Where is the wavefront? How do the two solutions compare behind the wavefront?

10. (a) Show that the energy integral for (A.1.73) is
$$\sum_{n=-\infty}^{\infty} |\psi_n^{m+1}|^2 = \sum_{n=-\infty}^{\infty} |\psi_n^m|^2.$$

(b) Find the dispersion relation for (A.1.73). Show that $|z| = 1 \Rightarrow |\Omega| = 1$.

(c) Construct the "Fourier transform" solution of this equation, analogous to (A.1.80). In what sense does it satisfy the equation?

(d) Show that the phase velocity is unbounded in this formulation but that the group velocity is bounded. What is the maximum group velocity?

(e) What is the asymptotic behavior of the solution as $m \to \infty$? How does it compare with the asymptotic solutions of (A.1.23) and (A.1.72)?

(f) Show that if (A.1.73) is replaced with the explicit scheme,

$$-i\frac{\psi_n^{m+1} - \psi_n^m}{\Delta t} = \frac{\psi_{n+1}^m + \psi_{n-1}^m - 2\psi_n^m}{h^2},$$

then both the phase and group velocities are bounded.

11. The linearized Toda lattice is

$$\frac{d^2}{dt^2}\phi_n(t) = \phi_{n+1} + \phi_{n-1} - 2\phi_n.$$

(a) Is there an energy integral?

(b) Find the dispersion relation. Is it real for $|z| = 1$?

(c) Construct the "Fourier transform" solution that evolves from appropriate initial data.

(d) What is the long time behavior of the solution? Compare this with the solution of the wave equation, $u_{tt} = u_{xx}$.

Section A.2.

1. (a) If $U(x)$ in (A.2.1) is real, show that:
   (i) $U(x) \geq 0 \Rightarrow K(y) \geq 0$;
   (ii) $K(-y) = K(y)$;
   (iii) $|K(y)| \leq K(0)$.

(Hint: use the Fourier representation of $K(y)$.)

(b) Compute $K(y)$ if

$$U(x) = A e^{-a|x|}, \quad a > 0.$$

Show that if $G(x) = 0$, then (A.2.16) is equivalent to a third order ordinary differential equation, and that all three roots of the characteristic equation are negative if α is small enough.

Find an approximate solution for f, and show that the asymptotic decay rate is a special case of (A.2.22).

(c) What is the asymptotic shape of $g(x, t)$? The fact that $g(x, t)$ is localized in x means that particles become uncorrelated when their separation distance increases beyond the effective width of g. The fact that this model has this property, even though three-body interactions were neglected, is of some importance in statistical mechanics (Kritz, Ramanathan and Sandri (1970)).

(d) What is the consequence of $G(x) \neq 0$ in (A.2.1)?

2. The Weisskopf–Wigner (1930) model for the radioactive decay of an unstable particle, when restricted to one dimension, is

$$i\frac{d\chi}{dt} = \int U(x)\psi(x,t)\,dx, \qquad t>0,$$

$$i\frac{\partial \psi}{\partial t} + \frac{\partial^2 \psi}{\partial x^2} = U(x)\chi(t), \qquad -\infty < x < \infty, \quad t>0,$$

$$\psi \to 0 \quad \text{as } |x| \to \infty,$$

$U(x)$ given, real, $\qquad U(x) \in L_2 \cap L_1,$

$\chi(0), \psi(x,0)$ given, $\qquad \psi(x,0) \in L_2.$

It has been analyzed from different viewpoints by Wellner (1960), and by Boldt and Sandri (1964). It can be analyzed along exactly the same lines as (A.2.1).

(a) Show that an energy integral of this problem is

$$|\chi(t)|^2 + \int |\psi(x,t)|^2\,dx.$$

(b) By taking a Fourier transform and eliminating $\tilde{\chi}(\omega)$, show that ω satisfies a quadratic equation with real coefficients if k is real, and that it always has two roots; (i.e., real $k \Rightarrow$ real ω).

(c) Show that this problem has no dispersion relation.

(d) Solve for $\psi(x,t)$ in terms of $\chi(t)$. Show that if $\psi(x,0) \equiv 0$, then $\chi(t)$ satisfies

$$\frac{d\chi}{dt} = -\int_0^t K(t-\tau)\chi(\tau)\,d\tau,$$

where

$$K(y) = \frac{1}{2\pi} \int \hat{U}(k)\hat{U}(-k)\,e^{-ik^2 y}\,dk.$$

(e) Solve this integrodifferential equation by Laplace transforms. Show that the only singularities in the inverse transform occur on the imaginary axis.

(f) Find the asymptotic decay rate of $\chi(t)$ as $t \to \infty$. Are there potentials $U(x)$ for which there is no decay? What is the asymptotic behavior of $\psi(x,t)$?

(g) What happens as $t \to -\infty$?

3. The linearized "Benjamin–Ono" equation is

$$\frac{\partial u}{\partial t} + \frac{\partial^2}{\partial x^2} \frac{1}{\pi} \int_{-\infty}^{\infty} \frac{1}{y-x} u(y,t)\,dy = 0.$$

(a) Find its energy integral.

LINEAR PROBLEMS

(b) Show that this integrodifferential equation has the dispersion relation
$$\omega + k|k| = 0.$$

(c) More generally, show that any integrodifferential equation of the form
$$\frac{\partial u}{\partial t} + \int_{-\infty}^{\infty} K(x-y)\frac{\partial u}{\partial y}\,dy = 0$$
has a dispersion relation (Whitham (1974, Chapt. 11)).

(d) Show that the integral term in (A.2.10) is in the form of a convolution.

4. Sketch the level curves of the E-field in (A.2.42), for $a > 0$ and for $a = 0$. What can you say about the behavior of Λ?

5. The linearized Vlasov's equations of a collisionless plasma in one dimension are
$$f_t + vf_x + \left(\frac{e}{m}\frac{dF}{dv}\right)E = 0,$$
$$E_x = -(4\pi e)\int f\,dv, \qquad -\infty < x, v < \infty, \quad t > 0,$$
$$f \to 0 \quad \text{as } |x| \to \infty.$$

(Note that all integrals are from $-\infty$ to $+\infty$.) Here $\{F(v)+f(x,v,t)\}$ is the velocity distribution function of electrons, with $|f| \ll F$. F is necessarily non-negative, but f can have either sign. $\int F\,dv = n$, the number of electrons in the equilibrium configuration. $E(x,t)$ is the averaged (perturbation) electric field strength. e and m are the charge and mass of a single electron, respectively.

(a) The linearized equations have infinitely many conservation laws. Show that
$$\left(\int f\,dv\right)_t + \left(\int vf\,dv\right)_x = 0 \qquad \text{(conservation of perturbed charge.)},$$
$$\left(\int v^2 f\,dv\right)_t + \left(\int v^3 f\,dv\right)_x = 0 \qquad \text{(conservation of perturbed kinetic energy)},$$
$$\varepsilon_{tt} + \omega_p^2 \varepsilon_{xx} = 0,$$
where $\varepsilon = \int E\,dx$, $\omega_p^2 = 4\pi ne^2/m$. (See Kruskal and Oberman (1965) and Case (1965) for the infinite set of motion constants.)

Note that none of these is positive definite. The last equation states that ε oscillates with the plasma frequency, ω_p. ε is somewhat analogous to $\int T\,dx$ in the heat equation.

(b) Look for solutions of the form
$$\binom{f}{E} \sim \left(\frac{\left(\frac{dF}{dv}\right)^{1/2}\hat{g}}{E}\right) e^{ikx - i\omega t}.$$

Show that

$$k\omega \int |\hat{g}|^2 \, dv - k^2 \int v|\hat{g}|^2 \, dv + \omega_p^2 \left| \int \left(\frac{dF}{dv}\right)^{1/2} \hat{g} \, dv \right|^2 = 0,$$

so that ω must be real if k is real.

(c) Show that there is no dispersion relation (van Kampen (1955)).

(d) Find the Fourier–Laplace transform solution of the problem. Show that the rightmost singularity in the complex p-plane is found by solving

$$k^2 - \frac{4\pi e^2}{m} \int \frac{(dF/dv)(v)}{v + \omega/k} \, dv = 0,$$

for $k > 0$ and $p = i\omega$. Landau's (1946) approximate solution of this equation is $\omega = \omega_r + i\omega_i$, where

$$k^2 - \frac{4\pi e^2}{m} \int \frac{dF}{dv}(v) \left\{ v + \frac{\omega_r}{k} \right\}^{-1} dv = 0,$$

$$\frac{\omega_i}{k} \int \frac{d^2 F}{dv^2}(v) \left\{ v + \frac{\omega_r}{k} \right\}^{-1} dv = \pi \frac{dF}{dv}\left(-\frac{\omega_r}{k}\right).$$

6. (a) Find the Fourier series solution of

$$\phi_t + c\phi_x = \mu\phi_{xx}, \quad 0 < x < L, \quad t > 0$$

$$\phi(0, t) = 1, \quad c < 0, \quad \mu > 0$$

$$\phi(L, t) = 0,$$

$\phi(x, 0)$ given and real, with $\int_0^L \phi^2 \, dx < \infty$.

(b) Find the corresponding Fourier integral solution, obtained by letting $L \to \infty$. Construct an example to show that at $t = 0$, this "solution" cannot represent arbitrary (real) initial data $\varepsilon L_2(0, \infty)$. Show that the Fourier integral does represent the true solution if the initial data satisfy both

$$\int_0^\infty |\phi| \, e^{|c|x/(2\mu)} \, dx < \infty, \quad \int_0^\infty |\phi|^2 \, e^{|c|x/\mu} \, dx < \infty.$$

(c) This is an example in which one obtains an incomplete set of modes by separating variables. Using a Laplace transform, find the general solution of the problem for initial data that satisfy both

$$\int_0^\infty |\phi| \, dx < \infty \quad \text{and} \quad \int_0^\infty |\phi|^2 \, dx < \infty.$$

7. The linear stability of two-dimensional inviscid, plane Couette flow, first analyzed correctly by Case (1960), is a famous example of a problem in which the modes obtained by separating variables are incomplete. In that problem, the unperturbed velocity field is given by

$$u_0(x, y, t) = y, \qquad 0 < y < 1,$$

$$v_0 = 0.$$

The equations for small deviations from this state are

$$\frac{\partial u}{\partial t} + y\frac{\partial u}{\partial x} + v = -\frac{\partial p}{\partial x},$$

$$\frac{\partial v}{\partial t} + y\frac{\partial v}{\partial x} = -\frac{\partial p}{\partial y},$$

$$\frac{\partial u}{\partial x} + \frac{\partial v}{\partial y} = 0,$$

subject to the boundary conditions that

$$v(y=0) = v(y=1) = 0.$$

Look for solutions of this set in the form

$$\phi(x, y, t) \sim \phi(y)e^{ikx - i\omega t},$$

where (k, ω) are constant. Reduce the differential equations to

$$(-\omega + ky)\left[\frac{\partial^2 v}{\partial y^2} - k^2 v\right] = 0.$$

Show that this equation has no nontrivial solutions that satisfy the boundary conditions at $y = 0$, $y = 1$. Thus, this set of normal modes is empty, and certainly is not complete. Case (1960) calls this the "careless treatment." He goes on to solve the complete linearized problem by taking a Fourier transform in x and a Laplace transform in t. The reader is referred to the original paper for details.

Bibliography

M. J. ABLOWITZ (1971), *Applications of slowly varying nonlinear dispersive wave theories*, Stud. Appl. Math., 50, pp. 329-344.
—— (1972), *Approximate methods for obtaining multi-phase modes in nonlinear dispersive wave problems*, Stud. Appl. Math., 51, pp. 51-56.
—— (1978), *Lectures on the inverse scattering transform*, Stud. Appl. Math., 58, pp. 17-94.
M. J. ABLOWITZ AND D. J. BENNEY (1970), *The evolution of multi-phase modes for nonlinear dispersive waves*, Stud. Appl. Math., 49, pp. 225-238.
M. J. ABLOWITZ AND H. AIRAULT (1981), *Perturbations finies et forme particulière de certaines solutions de l'équation de Korteweg-de Vries*, Comptes Rendus Acad. Sci. Paris, 292 (1981), pp. 279-281.
M. J. ABLOWITZ AND H. CORNILLE (1979), *On solutions of the Korteweg-deVries equation*, Phys. Lett., 72A, pp. 277-280.
M. J. ABLOWITZ AND R. HABERMAN (1975a), *Nonlinear evolution equations—two and three dimensions*, Phys. Rev. Lett., 35, pp. 1185-1188.
—— (1975b), *Resonantly coupled nonlinear evolution equations*, J. Math. Phys., 16, pp. 2301-2305.
M. J. ABLOWITZ, D. J. KAUP AND A. C. NEWELL (1974), *Coherent pulse propagation, a dispersive, irreversible phenomenon*, J. Math. Phys., 15, pp. 1852-1858.
M. J. ABLOWITZ, D. J. KAUP, A. C. NEWELL AND H. SEGUR (1973a), *Method for solving the sine-Gordon equation*, Phys. Rev. Lett., 30, pp. 1262-1264.
—— (1973b), *Nonlinear evolution equations of physical significance*, Phys. Rev. Lett., 31, pp. 125-127.
—— (1974), *The inverse scattering transform—Fourier analysis for nonlinear problems*, Stud. Appl. Math., 53, pp. 249-315.
M. J. ABLOWITZ AND Y. KODAMA (1979), *Transverse instability of one-dimensional transparent optical pulses in resonant media*, Phys. Lett. 70A, pp. 83-86.
—— (1980), *A note on the asymptotics of the Korteweg-deVries equation with solitons*, Stud. Appl. Math., to appear.
M. J. ABLOWITZ, M.D. KRUSKAL AND J. F. LADIK (1979), *Solitary wave collisions*, SIAM. J. Appl. Math., 36, pp. 428-437.
M. J. ABLOWITZ, M.D. KRUSKAL AND H. SEGUR (1979), *A note on Miura's transformation*, J. Math. Phys., 20, pp. 999-1003.
M. J. ABLOWITZ AND J. LADIK (1975), *Nonlinear differential-difference equations*, J. Math. Phys., 16, pp. 598-603.
—— (1976a), *Nonlinear differential-difference equations and Fourier analysis*, J. Math. Phys., 17, pp. 1011-1018.

——— (1976b), (1977), *On the solution of a class of nonlinear partial difference equations*, Stud. Appl. Math., 55, pp. 213ff; 57, pp. 1–12.
M. J. ABLOWITZ, AND Y. C. MA (1981), *The periodic cubic Schrödinger equation*, Stud. Appl. Math., to appear.
M. J. ABLOWITZ AND A. C. NEWELL (1973), *The decay of the continuous spectrum for solutions of the Korteweg-deVries equation*, J. Math. Phys., pp. 1277–1284.
M. J. ABLOWITZ, A. RAMANI AND H. SEGUR (1978), *Nonlinear evolution equations and ordinary differential equations of Painlevé type*, Lett. Nuovo Cim., 23, pp. 333–338.
——— (1980a, b,), *A connection between nonlinear evolution equations and ordinary differential equations of p-type, I, II*, J. Math. Phys., 21, pp. 715–721; pp. 1006–1015.
M. J. ABLOWITZ AND J. SATSUMA (1978), *Solitons and rational solutions of nonlinear evolution equations*, J. Math. Phys., 19, pp. 2180–2186.
M. J. ABLOWITZ AND H. SEGUR (1975), *The inverse scattering transform: semi-infinite interval*, J. Math. Phys., 16, pp. 1054–1056.
——— (1977a), *Asymptotic solutions of the Korteweg-deVries equation*, Stud. Appl. Math., 57, pp. 13–44.
——— (1977b), *Exact linearization of a Painlevé transcendent*, Phys. Rev. Lett., 38, pp. 1103–1106.
——— (1979), *On the evolution of packets of water waves*, J. Fluid Mech., 92, pp. 691–715.
——— (1980), *Long interval waves in fluids of great depth*, Stud. Appl. Math., 62, pp. 249–262.
M. J. ABLOWITZ AND A. ZEPPETELLA, (1979), *Explicit solutions of Fisher's equation for a special wave speed*, Bull. Math. Biol., 41, pp. 835–840.
M. ADLER (1979), *On a trace functional for formal pseudo-differential operators and the symplectic structure of the Korteweg-deVries equation*, Inv. Math., 50, pp. 219–248.
M. ADLER AND J. MOSER (1978), *On a class of polynomials connected with the Korteweg-deVries equation*, Comm. Math. Phys., 61, pp. 1–30.
Z. AGRANOVICH AND V. MARCHENKO (1963), *The Inverse Problem of Scattering Theory*, Gordon and Breach, New York.
M. AIKAWA AND M. TODA (1979), private communication.
H. AIRAULT (1979), *Rational solutions of Painlevé equations*, Stud. Appl. Math., 61, pp. 33–54.
H. AIRAULT, H. P. MCKEAN AND J. MOSER (1977), *Rational and elliptic solutions of the KdV equation and related many-body problems*, Comm. Pure Appl. Math., 30, pp. 95–198.
G. B. AIRY (1845) *Tides and waves,* Encyclopedia Metropolitana, vol. 5, London, pp. 241–396.
S. A. AKHMANOV AND R. V. KHOKHLOV (1972), *Problems in Nonlinar Optics,* Gordon and Breach, New York.
S. A. AKHMANOV, R. V. KHOKHLOV AND A. P. SUKHORUKOV (1972), *Self-focusing, self-defocusing and self-modulation of laser beams*, in Laser Handbook, F. T. Arecchi and E. O. Schulz-Dubois, eds., North-Holland, Amsterdam.
C. J. AMICK AND J. F. TOLAND (1979), *Finite amplitude solitary water waves*, MRC Rep. 2012, Mathematics Research Center, Univ. of Wisconsin, Madison.
R. L. ANDERSON AND N. J. IBRAGIMOV (1979), *Lie-Bäcklund Transformations in Applications*, SIAM Studies 1, Society for Industrial and Applied Mathematics, Philadelphia.
D. ANKER AND N. C. FREEMAN (1978), *On the soliton solutions of the Davey-Stewartson equation for long waves*, Proc. Roy. Soc. London A, 360, pp. 529–540.
L. ARMI (1977), *The dynamics of the bottom boundary layer of the deep ocean*, in Bottom Turbulence, J. C. J. Nihoul, ed., Elsevier, Amsterdam.
V. I. ARNOLD (1978), *Mathematical Methods in Classical Mechanics*, Springer-Verlag, New York.
V. I. ARNOLD AND A. A. AVEZ(1968), *Ergodic Problems in Classical Mechanics*, W. A. Benjamin, New York.
A. V. BÄCKLUND (1880), *Zur Theorie der partiellen Differential gleichungen erster Ordnung*, Math. Ann., 17, pp. 285–328.

F. K. BALL (1964), *Energy transfer between external and internal gravity waves*, J. Fluid Mech., 19, pp. 465-478.

E. BAROUCH, B. M. MCCOY AND T. T. WU (1973), *Zero-field susceptibility of the two-dimensional Ising Model Near T_c*, Phys. Rev. Lett., 31, pp. 1409-1411.

R. J. BAXTER (1972), *Partition function of the eight vertex lattice model*, Ann. Phys., 70, pp. 193-229.

A. BEER (1852), *Bestimmung der Absorption des rothen Lichts in farbiger Flüssigkeigen*, Ann. Physik. Chem., 26, ser. 3, pp. 78-88.

V. A. BELINSKII AND V. E. ZAKHAROV (1978), *Integration of the Einstein equations by the inverse scattering method and calculation of the exact soliton solutions*, Sov. Phys. JETP, 48, pp. 985-993.

T. B. BENJAMIN (1966), *Internal waves of finite amplitude and permanent form*, J. Fluid Mech., 25, pp. 241-270.

———— (1967), *Internal waves of permanent form in fluids of great depth*, J. Fluid Mech., 29, pp. 559-592.

T. B. BENJAMIN AND J. F. FEIR (1967), *The disintegration of wave trains on deep water*, J. Fluid Mech., 27, pp. 417-430.

D. J. BENNEY (1966), *Long nonlinear waves in fluid flows*, J. Math. Phys., (Stud. on Appl. Math.), 45, pp. 52-63.

———— (1973), *Some properties of long waves*, Stud. Appl. Math., 52, pp. 45-69.

———— (1977), *A general theory for interactions between short and long waves*, Stud. Appl. Math., 56, pp. 81-94.

D. J. BENNEY AND A. C. NEWELL (1967), *The propagation of nonlinear wave envelopes*, J. Math. Phys., (Stud. Appl. Math.), 46, pp. 133-139.

D. J. BENNEY AND G. J. ROSKES (1969), *Wave instabilities*, Stud. Appl. Math., 48, pp. 377-385.

F. A. BEREZIN AND A. M. PERELOMOV (1980), *Group theoretic interpretation of the Korteweg-deVries type equations*, Comm. Math. Phys., 74 pp. 129-140.

F. A. BEREZIN, G. P. POKHIL AND V. M. FINKEL'BERG (1964), *Schrödinger equation for one-dimensional systems of particles with point interactions*, Vestn. MGU, Ser. 1, pp. 21-28. (In Russian.)

H. BERGKNOFF AND H. B. THACKER (1979), *Structure and solution of the massive Thirring model*, Phys. Rev. D., 19, pp. 3666-3681.

H. A. BETHE (1931), *Zur Theorie der Metalle. I*, Z. Physik., 71, pp. 205-226.

L. BIANCHI (1902), *Lezioni de Geometria Differenziale*, vol. II, Pisa.

G. BIRKHOFF AND G.-C. ROTA (1969), *Ordinary Differential Equations*, Blaisdell-Ginn, Waltham, MA.

M. BLAZEK (1966), *On a method for solving the inverse problem in potential scattering*, Comm. Math. Phys., 3, pp. 282-291.

N. A. BLEISTEIN AND R. A. HANDLESMAN (1975), *Asymptotic Expansions of Integrals*, Holt, Rinehart, Winston, New York.

T. L. BOCK AND M. K. KRUSKAL (1979), *A two-parameter Miura transformation of the Benjamin-Ono equation*, Phys. Lett., 74A, pp. 173-176.

I. L. BOGOLUBSKY AND V. G. MAKHANKOV (1976), *Lifetime of pulsating solitons in certain classical models*, Sov. Phys. JETP Lett., 24, pp. 12-14.

I. L. BOGOLUBSKY, V. G. MAKHANKOV AND A. B. SHVACHKA (1977), *Dynamics of the collisions of two space-dimensional pulsons in ϕ^4 field theory*, Phys. Lett., A63, pp. 225-227.

N. N. BOGOLYUBOV (1962), *Studies in Statistical Mechanics*, North-Holland, Amsterdam.

M. BOITI AND F. PEMPINELLI (1979), *Similarity solutions of the Korteweg-deVries equation*, Il Nuovo Cimento, 51B, pp. 70-78.

———— (1980), *Similarity solutions and Bäcklund transformations of the Boussinesq equation*, Il Nuovo Cimento, 56B, ser. 11, pp. 148-156.

E. BOLDT AND G. SANDRI (1964), *Theory of unstable particles*, Phys. Rev. B., 135, pp. B1086–B1088.

J. BOUSSINESQ (1871), *Theorie de l'intumescence liquid appelée onde solitaire ou de translation, se propagente dans un canal rectangulaire*, Compte Rendus Acad. Sci. Paris, 72, pp. 755–759.

―――― (1872), *Theorie des ondes et de remous qui se propagent . . .* , J. Math. Pures Appl., Ser. 2, 17, pp. 55–108.

F. P. BRETHERTON (1964), *Resonant interactions between waves: the case of discrete oscillations*, J. Fluid Mech., 20, pp. 457–479.

R. K. BULLOUGH AND P. J. CAUDREY (1978), *The double-sine-Gordon equation: wobbling solitons?* Rocky Mountain J. Math., 8, pp. 53–70.

―――― eds., (1980), *Solitons*, Springer-Verlag, New York.

F. J. BUREAU (1972), *Integration of some nonlinear systems of ordinary differential equations*, Ann. Matematica N, 94, pp. 344–359.

J. M. BURGERS (1948), *A mathematical model illustrating the theory of turbulence*, Adv. Appl. Mech., 1, pp. 171–199.

P. B. BURT (1978), *Exact, multiple soliton solutions of the double sine-Gordon equation*, Proc. Roy. Soc. London A, 359, 479–495.

P. F. BYRD AND M. D. FRIEDMAN (1971), *Handbook of Elliptic Integrals for Scientists and Engineers*, Springer-Verlag, New York.

R. A. CAIRNS (1979), *The role of negative energy waves in some instabilities of parallel flows*, J. Fluid Mech., 92, pp. 1–14.

C. G. CALLAN AND D. J. GROSS (1975), *Quantum perturbation theory of solitons*, Nucl. Phys. B., 93, pp. 29–55.

F. CALOGERO (1971), *Solution of the one-dimensional N-body problems with quadratic and/or inversely quadratic pair potentials*, J. Math. Phys., 12, pp. 419–436.

―――― (1976), *Exactly solvable two-dimensional many-body problem*, Lett., Il Nuovo Cimento, 16, pp. 35–38.

―――― ed. (1978a), *Nonlinear Evolution Equations Solvable by the Spectral Transform*, Pitman, London.

―――― (1978b), *Nonlinear evolution equations solvable by the inverse spectral transform*, in Mathematical Problems in Theoretical Physics, F. Calogero, ed., Springer-Verlag, New York.

F. CALOGERO AND A. DEGASPERIS (1976), (1977), *Nonlinear evolution equations solvable by the inverse spectral transform, I, II*, Il Nuovo Cimento 32B, pp. 201–242; 39B, pp. 1–54.

K. M. CASE (1960), *Stability of inviscid plane Couette flow*, Phys. Fluids, 3, pp. 143–148.

―――― (1965), *Constants of the linearized motion of Vlasov plasmas*, Phys. Fluids, 8, pp. 96–101.

―――― (1973), *On discrete inverse scattering problems II*, J. Math. Phys., 14, pp. 916–920.

―――― (1978a), *Some properties of internal waves*, Phys. Fluids, 21, pp. 18–29.

―――― (1978b), *The N-soliton solution of the Benjamin-Ono equation*, Proc. Nat'l. Acad. Sci., 75, pp. 3562–3563.

―――― (1979), *Benjamin-Ono-related equations and their solutions*, Proc. Nat'l. Acad. Sci., 76, pp. 1–3.

K. CASE AND M. KAC (1973), *A discrete version of the inverse scattering problem*, J. Math. Phys., 14, pp. 594–603.

P. J. CAUDREY, R. K. DODD AND J. D. GIBBON (1976), *A new hierarchy of Korteweg-deVries equations*, Proc. Roy. Soc., London A, 351, pp. 407–422.

P. J. CAUDREY, J. D. GIBBON, J. C. EILBECK AND R. K. BULLOUGH (1973a), *Exact multisoliton solutions of the self-induced transparency and sine-Gordon equations*, Phys. Rev. Lett., 30, pp. 237–238.

―――― (1973b), *Multiple soliton and bisoliton bound state solutions of the sine-Gordon equation and related equations in nonlinear optics*, J. Phys. A, 6, pp. L112–L115.

H. H. CHEN (1974), *General derivation of Bäcklund transformations from inverse scattering problems*, Phys. Rev. Lett., 33, pp. 925–928.

——— (1976), *Relation between Bäcklund transformations and inverse scattering problems*, in Bäcklund Transformations, R. M. Miura, ed., Lecture Notes in Mathematics 515, Springer-Verlag, New York.

H. H. CHEN AND Y. C. LEE (1979), *Internal-wave solitons of fluids with finite depth.*, Phys. Rev. Lett., 43, pp. 264–266.

H. H. CHEN, Y. C. LEE AND C. S. LIU (1979), *Integrability of nonlinear Hamiltonian systems by inverse scattering method*, Physica Scripta, 20, pp. 490–492.

H. H. CHEN, Y. C. LEE AND N. R. PEREIRA (1979), *Algebraic internal wave solitons and the integrable Calogero-Moser-Sutherland N-body problem*, Phys. Fluids, 22, pp. 187–188.

I. V. CHEREDNIK (1978), *Differential equations for the Baker-Akhiezer functions of algebraic curves*, Funct. Anal. Appl. 12, pp. 195–203.

S. S. CHERN AND C-L. TERNG (1980), *Analogue of Bäcklund's theorem in affine geometry*, Rocky Mountain J. Math, 10, pp. 105–124.

C. CHESTER, B. FRIEDMAN AND F. URSELL (1957), *An extension of the method of steepest descents*, Proc. Camb. Phil. Soc., 53, pp. 599–611.

S. C. CHIU AND J. F. LADIK (1977), *Generating exactly soluble nonlinear discrete evolution equations by a generalized Wronskian technique*, J. Math. Phys., 18, pp. 690–700.

D. V. CHUDNOVSKY AND G. V. CHUDNOVSKY (1977), *Pole expansions of nonlinear partial differential equations*, Il Nuovo Cimento, 40B, pp. 339–353.

J. CLAIRIN (1903), *Sur quelques équations aux dérivées partielles du second ordre*, Ann. Fac. Sci. Univ. Toulouse, 2e Ser., 5, pp. 437–458.

A. COHEN (1979), *Existence and regularity of solutions of the Korteweg-deVries equation*, Arch. Rat. Mech. Anal., 71, pp. 143–175.

E. D. COKELET (1977), *Steep gravity waves on water of arbitrary uniform depth*, Phil. Trans. Roy. Soc. London, A, 286, pp. 183–230.

J. D. COLE (1951), *On a quasilinear parabolic equation occurring in aerodynamics*, Quart. App. Math., 9, pp. 225–236.

——— (1968), *Perturbation Methods in Applied Mathematics*, Ginn-Blaisdell, Waltham, MA.

E. T. COPSON (1965), *Asymptotic Expansions*, Cambridge University Press, London.

H. CORNILLE (1967), *Connection between the Marchenko formalism and N/D equations: regular interactions. I.*, J. Math. Phys., 8, pp. 2268–2280.

——— (1976a), *Differential equations satisfied by Fredholm determinants and application to the inversion formalism for parameter-dependent potentials*, J. Math. Phys., 17, pp. 2143–2158.

——— (1976b), *Generalization of the inversion equations and application to nonlinear partial differential equations I*, J. Math. Phys., 18, pp. 1855–1869.

——— (1979), *Solutions of the nonlinear 3-wave equations in three spatial dimensions*, J. Math. Phys., 20, pp. 1653–1666.

J. P. CORONES (1976), *Solitons and simple pseudopotentials*, J. Math. Phys., 17, pp. 756–759.

J. P. CORONES, B. L. MARKOVSKI AND V. A. RIZOV (1977), *A Lie group framework for soliton equations*, J. Math. Phys., 18, pp. 2207–2213.

J. P. CORONES AND F. J. TESTA (1976), *Pseudopotentials and their applications*, in Bäcklund Transformations, R. M. Miura, ed., Lecture Notes in Mathematics 515, Springer-Verlag, New York.

R. COURANT AND K. O. FRIEDRICHS (1948), *Supersonic Flow and Shock Waves*, Interscience, New York.

E. COURTENS (1972), *Nonlinear coherent resonant phenomena*, in Laser Handbook, F. T. Arecchi and E. O. Schultz-DuBois, eds., North-Holland, Amsterdam.

A. D. D. CRAIK AND J. A. ADAM (1979), *"Explosive" instability in a three-layer fluid flow*, J. Fluid Mech., 92, pp. 15–33.

M. M. Crum (1955), *Associated Sturm-Liouville systems*, Quart. J. Math., 6, pp. 121–127.

R. F. Dashen, B. Hasslacher and A. Neveu (1974), (1975), *Nonperturbative methods and extended-hadron models in field theory*, Phys. Rev. D., 10, pp. 4114–4138; 11, pp. 3424–3450; 12, pp. 2443–2458.

E. Date and S. Tanaka (1976a), *Analogue of inverse scattering theory for the discrete Hill's equation and exact solutions for the periodic Toda lattice*, Prog. Theoret. Phys., 55, pp. 457–465.

―――― (1976b), *Periodic multi-soliton solutions of Korteweg-deVries equation and Toda lattice*, Prog. Theoret. Phys. Suppl., 59, pp. 107–126.

A. Davey and K. Stewartson (1974), *On three-dimensional packets of surface waves*, Proc. Roy. Soc. London A, 338, pp. 101–110.

R. C. Davidson (1972), *Methods in Nonlinear Plasma Theory*, Academic Press, New York.

R. E. Davis and A. Acrivos (1967), *Solitary internal waves in deep water*, J. Fluid Mech., 29, pp. 593–608.

P. Debye (1916), *Vorträge öber die Kinetische Theorie der Materie und der Electrizität*, Leipzig, Germany.

P. Deift, F. Lund and E. Trubowitz (1980), *Nonlinear wave equations and constrained harmonic motion*, Comm. Math. Phys., 74, pp. 141–188.

P. Deift and E. Trubowitz (1979), *Inverse scattering on the line*, Comm. Pure Appl. Math., 32, pp. 121–251.

―――― (1980), *Some remarks on the Korteweg-deVries and Hill's equations*, in Nonlinear Dynamics, R. H. G. Helleman, ed., Ann. New York Academy of Science, vol. 357, pp. 55–64.

V. D. Djordjevic and L. G. Redekopp (1977), *On two-dimensional packets of capillary gravity waves*, J. Fluid Mech., 79, pp. 703–714.

R. K. Dodd, R. K. Bullough and S. Duckworth (1975), *Multisoliton solutions of nonlinear dispersive wave equations not solvable by the inverse method*, J. Phys. A, 8, pp. L64–L68.

Dorfman and I. M. Gel'fand (1979), *Hamiltonian operators and algebraic structures related to them*, Funct. Anal. Appl., 13, pp. 248–262.

V. Dryuma (1974), *Analytic solution of the two-dimensional Korteweg-deVries (KdV) equation*, Sov. Phys. JETP Lett. 19, 387–388.

B. A. Dubrovin (1975), *Periodic problems for the Korteweg-deVries equation in the class of finite band potentials*, Funct. Anal. Appl., 9, pp. 215–223.

B. A. Dubrovin, V. B. Matveev and S. P. Novikov (1976), *Nonlinear equations of Korteweg-deVries type, finite zoned linear operators, and Abelian varieties*, Russ. Math. Surveys, 31, pp. 59–146.

B. A. Dubrovin and S. P. Novikov (1975), *Periodic and conditionally periodic analogues of the many-soliton solutions of the Korteweg-deVries equation*, Sov. Phys. JETP, 40, pp. 1058–1063.

J. C. Eilbeck (1972), *Reflection of short pulses in linear optics*, J. Phys. A., 5, pp. 1355–1363.

―――― (1978), *Numerical studies of solitons*, Proc. Symposium on Nonlinear Structure and Dynamics in Condensed Matter, Bishop and Schneider, eds., Springer-Verlag, New York.

J. C. Eilbeck, J. D. Gibbon, P. J. Caudrey and R. K. Bullough (1973), *Solitons in nonlinear optics I. A more accurate description of the 2π-pulse in self-induced transparency*, J. Phys., A., 6, pp. 1337–1347.

L. P. Eisenhart (1909), *A Treatise on the Differential Geometry of Curves and Surfaces*, Ginn and Co.; reprinted by Dover, New York, 1960.

N. P. Erugin (1976), *Theory of nonstationary singularities of second order equations*, J. Differential Equations, 12, pp. 267–289; 405–420.

F. B. Estabrook (1981), *Prolongation structures of nonlinear evolution equations*, preprint.

F. ESTABROOK AND H. WAHLQUIST (1978), *Prolongation structures, connection theory and Bäcklund transformation*, in Nonlinear Evolutions Equations Solvable by the Spectral Transform, F. Calogero, ed., Pitman, London.

L. D. FADDEEV (1963), (transl. from the Russian by B. Seckler), *The inverse problem in the quantum theory of scattering*, J. Math. Phys., 4, pp. 72–104.

E. FERMI, J. PASTA AND S. ULAM (1974), *Studies of nonlinear problems, I*, Los Alamos Rep. LA1940, 1955; reprod. in Nonlinear Wave Motion, A. C. Newell, ed., American Mathematical Society, Providence, RI.

R. A. FISHER (1937), *The wave of advance of an advantageous gene*, Ann. Eugen., 7, pp. 355–369.

H. FLASCHKA (1974a), *The Toda lattice. I. Existence of integrals*, Phys. Rev. B, 9, pp. 1924–1925.

——— (1974b), *On the Toda lattice II. Inverse scattering solution*, Prog. Theoret. Phys. 51, pp. 703–706.

——— (1980), *A commutator representation of Painlevé equations*, J. Math. Phys., 21, pp. 1016–1018.

H. FLASCHKA, G. FOREST AND D. W. MCLAUGHLIN (1979), *Multiphase averaging and the inverse spectral solution of KdV*, Comm. Pure Appl. Math., 33, pp. 739–784.

H. FLASCHKA AND D. W. MCLAUGHLIN (1976a), *Canonically conjugate variables for KdV and Toda lattice under periodic boundary conditions*, Prog. Theoret Phys. 55, pp. 438–456.

——— (1976b), *Some comments on Bäcklund transformations, canonical transformations and the inverse scattering method*, in Bäcklund Transformations, R. M. Miura, ed., Lecture Notes in Mathematics 515, Springer-Verlag, New York.

——— eds., (1978), *Proceedings of Conference on Theory and Appl. of Solitons*, (Tucson, 1976) Rocky Mountain J. Math., 8, 1, 2.

H. FLASCHKA AND A. C. NEWELL (1975), *Integrable systems of nonlinear evolution equations*, in Dynamical Systems, Theory and Applications, J. Moser, ed., Lecture Notes in Physics, 38, Springer-Verlag, New York.

——— (1980), *Monodromy and spectrum preserving deformations, I*, Comm. Math. Phys., 76, pp. 65–116.

A. S. FOKAS (1980), *A symmetry approach to exactly solvable evolution equations*, J. Math. Phys., 21, pp. 1318–1326.

A. S. FOKAS AND Y. C. YORTSOS (1980), *The transformation properties of the sixth Painlevé equation and one-parameter families of solutions*, Lett. Nuovo Cimento, 30, pp. 539–540.

A. R. FORSYTH (1906), *Theory of Differential Equations*, reprinted by Dover, New York.

P. A. FRANKEN, A. E. HILL, C. W. PETERS AND G. WEINREICH (1961), *Generation of optical harmonics*, Phys. Rev. Lett., 7, pp. 118–119.

N. C. FREEMAN AND A. DAVEY (1975), *On the evolution of packets of long surface waves*, Proc. Roy. Soc. London A, 344, pp. 427–433.

K. O. FRIEDRICHS (1958), *Symmetric positive linear differential equations*, Comm. Pure App. Math., 11, pp. 333–418.

Y. C. FUNG (1965), *Foundations of Solid Mechanics*, Prentice-Hall, Englewood Cliffs, NJ.

C. S. GARDNER (1971), *The Korteweg-deVries equation and generalizations. IV. The Korteweg-deVries equation as a Hamiltonian system*, J. Math. Phys., 12, pp. 1548–1551.

C. S. GARDNER, J. M. GREENE, M.D. KRUSKAL AND R. M. MIURA (1967), *Method for solving the Korteweg-deVries equation*, Phys. Rev. Lett. 19, pp. 1095–1097.

——— (1974), *The Korteweg-deVries equation and generalizations. VI. Methods for exact solution*, Comm. Pure Appl. Math., 27, pp. 97–133.

C. S. GARDNER AND G. K. MORIKAWA (1960), *Similarity in the asymptotic behavior of collision free hydromagnetic waves and water waves*, Courant Inst. Math. Sci. Res. Rep. NYO-9082, New York University, New York.

C. S. GARDNER AND C. S. SU (1969), *The Korteweg-deVries equation and generalizations. III*, J. Math. Phys., 10, pp. 536–539.

A. G. GARGETT AND B. A. HUGHES (1972), *On the interaction of surface and internal waves*, J. Fluid Mech., 52, pp. 179-191.
C. GARRETT AND W. MUNK (1972), *Space-time scales of internal waves*, Geophys. Fluid Dyn., 2, pp. 225-264.
——— (1975), *Space-time scales of internal waves: a progress report*, J. Geophys. Res., 80, pp. 291-297.
M. G. GASYMOV AND B. M. LEVITAN (1966), *The inverse problem for a Dirac system*, Soviet Phys. Doklady, 7, pp. 495-499.
I. M. GEL'FAND (1979), comment made at Joint US-USSR Symposium on Soliton Theory, Kiev, USSR, Sept. 1979.
I. M. GEL'FAND AND L. A. DIKII (1977), *Resolvents and Hamiltonian systems*, Funct. Anal. Appl., 11, pp. 93-104.
I. M. GEL'FAND AND B. M. LEVITAN (1955), *On the determination of a differential equation from its spectral function*, Amer. Math. Soc. Transl., Ser. 2, 1, pp. 259-309.
J. D. GIBBON, P. J. CAUDREY, R. K. BULLOUGH AND J. C. EILBECK (1973), *An N-soliton solution of a nonlinear optics equation derived by a general inverse method*, Lett. Nuovo Cimento, 8, pp. 775-779.
J. D. GIBBON, N. C. FREEMAN AND A. DAVEY (1978), *Three-dimensional multiple soliton-like solutions of nonlinear Klein-Gordon equations*, J. Phys. A: Math., 11, pp. $L93-L96$.
J. D. GIBBON, I. N. JAMES AND I. M. MOROZ (1979), *An example of soliton behavior in a rotating baroclinic fluid*, Proc. Roy. Soc. London A, 367, pp. 219-237.
H. M. GIBBS AND R. E. SLUSHER (1970), *Peak amplification and breakup of a coherent optical pulse in a simple atomic absorber*, Phys. Rev. Lett., 24, pp. 638-641.
R. T. GLASSEY (1977), *On the blowing up of solutions to the Cauchy problem for nonlinear Schrödinger equations*, J. Math. Phys., 18, pp. 1794-1797.
H. GOLDSTEIN (1950), *Classical Mechanics*, Addison-Wesley, Reading, MA.
V. V. GOLUBOV (1953), *Lectures on Integration of the Equations of Motion of a Rigid Body About a Fixed Point*, State Pub. House, Moscow, transl. by J. Shorr-kon, reproduced by NTIS, Springfield, VA.
K. A. GORSHKOV AND L. A. OSTROVSKY (1981), *Interactions of solitons in nonintegrable systems; direct perturbation method and applications*, Proc. Joint US-USSR Symposium on Soliton Theory, V. E. Zakharov and S. V. Manakov, eds., North-Holland, Amsterdam, pp. 428 ff.
K. A. GORSHKOV, L. A. OSTROVSKY AND V. V. PAPKO (1976), *Interactions and coupled states of solitons as classical particles*, Sov. Phys. JETP, 44, pp. 306-311.
R. H. J. GRIMSHAW (1975), *The modulation and stability of an internal gravity wave*, Res. Rep't. School of Math. Sci., Univ. Melbourne.
R. HABERMAN (1977), *Nonlinear transition layers—the 2nd Painlevé transcendent*, Stud. Appl. Math., 57, pp. 247-270.
M. HALL, JR. (1959), *The Theory of Groups*, Macmillan, New York.
J. L. HAMMACK (1979), private communication.
J. L. HAMMACK AND H. SEGUR (1974), *The Korteweg-deVries equation and water waves part 2. Comparison with experiments*, J. Fluid Mech., 65, pp. 289-314.
——— (1978), *The Korteweg-deVries equation and water waves part 3. Oscillatory waves*, J. Fluid Mech., 84, pp. 337-358.
R. H. HARDIN AND F. D. TAPPERT (1973), *Applications of the split-step Fourier method to the numerical solution of nonlinear and variable coefficient wave equations*, SIAM-SIGNUM Fall Meeting, Austin, Texas, Oct. 1972; SIAM Rev. (Chronicle) 15, p. 423.
B. HARRISON (1978), *Bäcklund transformation for the Ernst equation of general relativity*, Phys. Rev. Lett., 41, pp. 1197-1200.

A. HASEGAWA AND F. TAPPERT (1973), *Transmission of stationary nonlinear optical pulses in dispersive dielectric fibers. II. Normal dispersion*, Appl. Phys. Lett., 23, pp. 171.
H. HASIMOTO AND H. ONO (1972), *Nonlinear modulation of gravity waves*, J. Phys. Soc. Japan, 33, pp. 805–811.
K. HASSELMANN (1962), *On the nonlinear energy transfer in a gravity-wave spectrum part I. general theory*, J. Fluid Mech., 12, pp. 481–500.
——— (1963a, b) *On the nonlinear energy transfer in a gravity-wave spectrum part II, Conservation theorem; wave-particle analogy; irreversibility*, J. Fluid Mech., 15, pp. 273–281; Part III, 15, pp. 385–398.
——— (1967), *A criterion for nonlinear wave stability*, J. Fluid Mech., 30, pp. 737–739.
K. HASSELMANN, ET AL. (1973), *Measurements of wind-wave growth and swell decay during the Joint North Sea Wave Project (JONSWAP)*, Deut. Hydrogr. Z., Suppl. A, 8.
S. P. HASTINGS AND J. B. MCLEOD (1980), *A boundary value problem associated with the second Painlevé transcendent and the Korteweg-deVries equation*, Arch. Rat. Mech. Anal., 73, pp. 31–51.
M. HÉNON (1974), *Integrals of the Toda lattice*, Phys. Rev. B9, pp. 1921–1923.
R. HERMANN (1978), *Prolongations, Bäcklund transformations, and Lie theory as algorithms for solving and understanding nonlinear differential equations*, in Solitons in Action, K. Lonngren and A. C. Scott, eds., Academic Press, New York.
E. HEWITT AND K. STROMBERG (1969), *Real and Abstract Analysis*, Springer-Verlag, Berlin.
E. HILLE (1976), *Ordinary Differential Equations in the Complex Domain*, John Wiley, New York.
R. HIROTA (1971), *Exact solution of the Korteweg-deVries equation for multiple collisions of solitons*, Phys. Rev. Lett., 27, 1192–1194.
——— (1972), *Exact solution of the sine-Gordon equation for multiple collisions of solitons*, J. Phys. Soc. Japan, 33, pp. 1459–1463.
——— (1973a), *Exact envelope-soliton solutions of a nonlinear wave equation*, J. Math. Phys., 14, pp. 805–809.
——— (1973b), *Exact N-soliton solution of a nonlinear lumped network equation*, J. Phys. Soc. Japan, 35, pp. 289–294.
——— (1973c), *Exact N-soliton solution of the wave equation of long waves in shallow water and in nonlinear lattices*, J. Math. Phys., 14, pp. 810–814.
——— (1973d), *Exact three-soliton solution of the two-dimensional sine-Gordon equation*, J. Phys. Soc. Japan, 35, p. 1566.
——— (1974), *A new form of Bäcklund transformation and its relation to the inverse scattering problem*, Prog. Theoret. Phys., 52, pp. 1498–1512.
——— (1976), *Direct methods of finding exact solutions of nonlinear evolution equations,* in Bäcklund transformations, R. M. Miura, ed., Lecture Notes in Mathematics 515, Springer-Verlag, New York.
——— (1977a), *Nonlinear partial difference equations I. A difference analogue of the Korteweg-deVries equation*, J. Phys. Soc. Japan, 43, pp. 1429–1433.
——— (1977b), *Nonlinear partial difference equations II. Discrete time Toda equation*, J. Phys. Soc. Japan, 43, pp. 2074–2078.
——— (1977c), *Nonlinear partial difference equations III. Discrete sine-Gordon equations*, J. Phys. Soc. Japan, 43, pp. 2079–2089.
——— (1978), *Nonlinear partial difference equations IV. Bäcklund transformation for the discrete Toda equation*, J. Phys. Soc. Japan, 45, pp. 321–332.
——— (1979a), Lecture delivered at Soliton Workshop/Jadwisin, Warsaw, Poland, Aug. 1979.
——— (1979b), *Nonlinear partial difference equations V. Nonlinear equations reducible to linear equations*, J. Phys. Soc. Japan, 46, pp. 312–319.

———— (1980), *Direct methods in soliton theory*, in Solitons, R. K. Bullough and P. J. Caudrey, eds., Topics of Modern Physics, Springer-Verlag, New York.
R. HIROTA AND J. SATSUMA (1976a), *A variety of nonlinear network equations generated from the Bäcklund transformation for the Toda lattice*, Prog. Theoret. Phys. Suppl., 59, pp. 64–100.
———— (1976b), *N-soliton solution of nonlinear network equations describing a Volterra system*, J. Phys. Soc. Japan, 40. pp. 891–900.
———— (1978), *A simple structure of superposition formula of the Bäcklund transformation*, J. Phys. Soc. Japan, 45, pp. 1741–1750.
R. HIROTA AND K. SUZUKI (1970), *Studies on lattice solitons by using electrical networks*, J. Phys. Soc. Japan 28, pp. 1366–1369.
———— (1973), *Theoretical and experimental studies of lattice solitons in nonlinear lumped networks*, Proc. IEEE, 61, pp. 1483–1491.
R. HIROTA AND M. WADATI (1979), *A functional integral representation of the soliton solutions*, J. Phys. Soc. Japan, 47, pp. 1385–1386.
L. V. HMURCIK AND D. J. KAUP (1979), *Solitons created by chirped initial profiles in coherent pulse propagation*, J. Opt. Sci. Amer., 69, pp. 597–604.
J. HONERKAMP, P. WEBER AND A. WIESLER (1979), *On the connection between the inverse transform method and the exact quantum states*, Nucl. Phys. B, 152, pp. 266–272.
E. HOPF (1950), *The partial differential equation $u_t + uu_x = \mu u_{xx}$*, Comm. Pure Appl. Math., 3, pp. 201–230.
F. HOPPENSTAEDT (1975), *Mathematical Theories of Populations: Demographics, Genetics and Epidemics*, CBMS Regional Conference Series in Applied Mathematics 20, Society for Industrial and Applied Mathematics, Philadelphia.
W. H. HUI AND J. HAMILTON (1979), *Exact solutions of a three-dimensional Schrödinger equation applied to gravity waves*, J. Fluid Mech., 93, pp. 117–133.
N. J. IBRAGIMOV AND A. B. SHABAT (1979), *Korteweg-deVries equation from the group standpoint*, Doklady Akad. Nauk., USSR 244, pp. 57–61.
E. L. INCE (1927), *Ordinary Differential Equations*, reprinted by Dover, New York, 1956.
M. ITO (1980), *An extension of nonlinear evolution equations of the KdV (mKdV) type to higher orders*, preprint.
A. R. ITS AND V. B. MATVEEV (1975), *The periodic Korteweg-deVries equation*, Funct. Anal. Appl. 9, pp. 67ff.
N. JACOBSON (1962), *Lie Algebras*, Wiley-Interscience, New York.
M. JAULENT (1976), *Inverse scattering problems in absorbing media*, J. Math. Phys., 17, pp. 1351–1360.
M. JIMBO, T. MIWA, Y. MŌRI AND M. SATO (1979), *Density matrix of impenetrable bose gas and the fifth Painlevé transcendent*, Pub. RIMS, 303, Kyoto Univ., Japan.
M. JIMBO, T. MIWA, Y. MŌRI AND M. SATO (1979), *Density matrix of impenetrable bose gas and the fifth Painlevé transcendant*, Pub. RIMS, 303, Kyoto Univ., Japan.
R. S. JOHNSON (1973), *On an asymptotic solution of the Korteweg-deVries equation with slowly varying coefficients*, J. Fluid Mech., 60, pp. 813–824.
R. I. JOSEPH (1977), *Solitary waves in a finite depth fluid*, J. Phys. A, 10, pp. L225–L227.
R. I. JOSEPH AND R. EGRI (1978), *Multi-soliton solutions in a finite depth fluid*, J. Phys. A. Math., 11, pp. L97–L102.
T. M. JOYCE (1974), *Nonlinear interactions among standing surface and internal gravity waves*, J. Fluid Mech., 63, pp. 801–825.
M. KAC AND P. VAN MOERBEKE (1975a), *On an explicitly soluble system of nonlinear differential equations related to certain Toda lattices*, Advances in Math., 16, pp. 160–169.
———— (1975b), *On periodic Toda lattices*, Proc. Nat'l. Acad. Sci., 72, pp. 1627–1629.
———— (1975c), *A complete solution of the periodic Toda problem*, Proc. Nat'l. Acad. Sci., 72, pp. 2879–2880.

B. B. KADOMTSEV AND V. I. PETVIASHVILI (1970), *On the stability of solitary waves in weakly dispersing media*, Sov. Phys. Doklady, 15, pp. 539-541.

C. F. F. KARNEY, A. SEN AND F. Y. F. CHU (1979), *Nonlinear evolution of lower hybrid waves*, Phys. Fluids 22, pp. 940-952.

V. I. KARPMAN (1975), *Nonlinear Waves in Dispersive Media*, Pergamon Press, Oxford.

V. I. KARPMAN AND E. M. MASLOV (1978), *Structure of tails produced under the action of perturbations on solitons*, Sov. Phys. JETP, 48, pp. 252ff.

M. KASHIWARA AND T. KAWAI (1978), *Monodromy structure of solutions of holonomic systems linear differential equations* . . . Pub. RIMS, Kyoto Univ., Japan.

D. J. KAUP (1975), *Exact quantization of the nonlinear Schrödinger equation*, J. Math. Phys., 16, pp. 2036-2041.

——— (1976a), *Closure of the squared Zakharov-Shabat eigenstates*, J. Math. Anal. Appl., 54, pp. 849-864.

——— (1976b), *The three-wave interaction—a nondispersive phenomenon*, Stud. Appl. Math., 55, pp. 9-44.

——— (1976c), *A perturbation theory for inverse scattering transforms*, SIAM J. Appl. Math., 31, pp. 121-123.

——— (1977a), *Coherent pulse propagation: a comparison of the complete solution with the McCall-Hahn theory and others*, Phys. Rev. A., 16, pp. 704-719.

——— (1977b), *Soliton particles, and the effects of perturbations* in The Significance of Nonlinearity in the Natural Sciences, B. Kursunoghu, A. Perlmutter, and L. F. Scott, eds., Plenum, New York.

——— (1978), *Simple harmonic generation: an exact method of solution*, Stud. Appl. Math, 59, pp. 25-35.

——— (1980), *The Wahlquist-Estabrook method with examples of applications*, Physica D, 1, pp. 391-411.

——— (1981), *The solution of the general initial value problem for the full three dimensional three-wave resonant interaction*, Proc. Joint US-USSR Symposium on Soliton Theory, Kiev, 1979, V. E. Zakharov and S. V. Manakov, eds., North-Holland, Amsterdam, pp. 374-395.

D. J. KAUP AND A. C. NEWELL (1978a), *An exact solution for a derivative nonlinear Schrödinger equation*, J. Math. Phys. 19, pp. 798-801.

——— (1978b), *Solitons as particles, oscillators and in slowly changing media: a singular perturbation theory*, Proc. Roy. Soc. London A, 361, pp. 413-446.

——— (1978c), *The Goursat and Cauchy problems for the sine-Gordon equation*, SIAM J. Appl. Math., 34, pp. 37-54.

D. J. KAUP AND L. R. SCACCA (1980), *Generation of O π pulses from a zero-area pulse in coherent pulse propagation*, J. Opt. Sci. Amer., 70, pp. 224-230.

I. KAY AND H. E. MOSES (1956), *Reflectionless transmission through dielectrics and scattering potentials*, J. Appl. Phys., 27, pp. 1503-1508.

D. KAZHDAN, B. KOSTANT AND S. STERNBERG (1978), *Hamiltonian group actions and dynamical systems of Calogero type*, Comm. Pure Appl. Math., 31, pp. 481-507.

J. P. KEENER AND D. W. MCLAUGHLIN (1977), *Solitons under perturbations*, Phys. Rev., 16A, pp. 777-790.

P. L. KELLEY (1965), *Self-focusing of optical beams*, Phys. Rev. Lett., 15, pp. 1005-1008.

LORD (W. THOMPSON) KELVIN (1887), *On the waves produced by a single impulse in water*, in his Math. Phys. Papers, vol. 4, pp. 303-306.

G. H. KEULEGAN (1948), *Gradual damping of solitary waves*, J. Res., N.B.S., 40, pp. 487-498.

——— (1953), *Characteristics of internal solitary waves*, J. Res., N.B.S., 51, pp. 133-140.

D. A. KLEINMAN (1972), *Optical harmonic generation in nonlinear media*, in Laser Handbook, F. T. Arecchi and E. D. Schultz-Dubois, eds., North-Holland, Amsterdam.

C. J. KNICKERBOCKER AND A. C. NEWELL (1980), *Shelves and the Korteweg–deVries equation*, J. Fluid Mech., 98, pp. 803–818.
K. KO AND H. H. KUEHL (1978), *Korteweg-deVries soliton in a slowly varying medium*, Phys. Rev. Lett. 40, pp. 233–236.
K. K. KOBAYASHI AND M. IZUTSU (1976), *Exact solution of the N-dimensional sine-Gordon equation*, J. Phys. Soc. Japan, 41, pp. 1091–1092.
Y. KODAMA (1975), *Complete integrability of nonlinear evolution equations*, Prog. Theoret. Phys., 54, pp. 669–686.
Y. KODAMA AND M. J. ABLOWITZ (1980), *Transverse instability of breathers in resonant media*, J. Math. Phys., 21, pp. 928–931.
——— (1981), *Perturbations of solitons and solitary waves*, Stud. Appl. Math., to appear.
Y. KODAMA, J. SATSUMA AND M. J. ABLOWITZ (1981), *Nonlinear intermediate long wave equation: analysis and method of solution*, Phys. Rev. Lett., 46, pp. 687–690.
Y. KODAMA AND T. TANIUTI (1978), *Higher order approximation in the reductive perturbation method I. The weakly dispersive system*, J. Phys. Soc. Japan, 45, pp. 298–310.
——— (1979), *Higher order corrections to the soliton-velocity and the linear dispersion relation*, Physica Scripta, 20, pp. 486–489.
K. KONNO, H. SANUKI AND Y. H. ICHIKAWA (1974), *Conservation laws of nonlinear evolution equations*, Prog. Theoret. Phys., 52, pp. 886–889.
K. KONNO AND M. WADATI (1975), *Simple derivation of Bäcklund transformation from the Ricatti form of the inverse method*, Prog. Theoret. Phys., 53, pp. 1652–1656.
C. G. KOOP AND G. BUTLER (1981), *An investigation of internal solitons in a two-fluid system*, J. Fluid Mech., to appear.
V. E. KOREPIN AND L. D. FADDEEV (1975), *Quantization of solitons*, Theoret. Math. Phys., 25, pp. 1039–1049.
D. J. KORTEWEG AND G. DEVRIES (1895), *On the change of form of long waves advancing in a rectangular canal, and on a new type of long stationary waves*, Philos. Mag. Ser. 5, 39, pp. 422–443.
I. M. KRICHEVER (1976), *Algebraic curves and commuting matrix differential operators*, Funct. Anal. Appl., 10, pp. 144–146.
I. M. KRICHEVER AND S. P. NOVIKOV, *Holomorphic bundles and nonlinear equations*, Proc. Joint US-USSR Symposium on Soliton Theory, Kiev, 1979, V. E. Zakharov and S. V. Manakov, eds., North-Holland, Amsterdam, pp. 267–293.
A. H. KRITZ, G. V. RAMANATHAN AND G. SANDRI (1970), *The two-particle correlation function in nonequilibrium statistical mechanics*, in Kinetic Equations, R. Liboff and N. Rostekev, eds., Gordon and Breach, New York.
M. D. KRUSKAL (1963), in *Asymptology in Mathematical Models in Physical Sciences*, S. Probot, ed., Prentice-Hall, Englewood Cliffs, NJ.
——— (1965), in Proc. IBM Scientific Computing Symposium on Large-Scale Problems in Physics, IBM Data Processing Division, White Plains, NY; Thomas J. Watson Research Center, Yorktown Heights, New York.
——— (1974), *The Korteweg-deVries equation and related evolution equations*, in Nonlinear Wave Motion, A. C. Newell, ed., AMS Lectures in Applied Mathematics, 15, American Mathematical Society, Providence, RI.
——— (1975), *Nonlinear wave equations*, in Dynamical Systems, Theory and Applications, J. Moser, ed., Lecture Notes in Physics, 38, Springer-Verlag, New York.
M. KRUSKAL AND C. OBERMAN (1965), *Some constants of the linearized motion of Vlasov plasmas*, J. Math. Phys., 6, pp. 327–335.
M. D. KRUSKAL AND N. J. ZABUSKY (1963), *Princeton Plasma Physics Laboratory annual report* MATT-Q-21, pp. 301ff, unpublished.
T. KUBOTA, D. R. S. KO AND L. DOBBS (1978a), *Propagation of weakly nonlinear internal waves in a stratified fluid of finite depth*, Report of AIAA 16th Aerospace Sciences Meeting, Huntsville, AL.

―― (1978b), *Weakly nonlinear long interval gravity waves in stratified fluids of finite depth*, AIAA J. Hydronautics, 12, pp. 157–165.
A. E. KUDRYAVASEV (1975), *Soliton-like collisions for a Higgs scalar field*, Sov. Phys. JETP Lett., 22, pp. 82–83.
P. P. KULISH, S. V. MANAKOV AND L. D. FADDEEV (1976), *Comparison of the exact quantum and quasiclassical results for a nonlinear Schrödinger equation*, Theoret. Math. Phys., 28, pp. 615–620.
B. B. KUPERSHMIDT AND YU I. MANIN (1977), (1978), *Equations of long waves with a free surface, I, II*, Funct. Anal. App., 11, pp. 188–197; 12, pp. 20–28.
E. A. KUZNETSOV AND A. V. MIKHAILOV (1975), *Stability of stationary waves in nonlinear weakly dispersive media*, Sov. Phys. JETP, 40, pp. 855–859.
J. F. LADIK AND S. C. CHIU (1977), *Solutions of nonlinear network equations by the inverse scattering method*, J. Math. Phys., 18, pp. 701–704.
B. M. LAKE, H. C. YUEN, H. RUNGALDIER AND W. E. FERGUSON (1977), *Nonlinear deep water waves: theory and experiment. Part 2*, J. Fluid Mech., 83, pp. 49–74.
G. L. LAMB (1971), *Analytical descriptions of ultrashort optical pulse propagation in a resonant medium*, Rev. Mod. Phys., 43, pp. 99–124.
―― (1973), *Phase variation in coherent-optical-pulse propagation*, Phys. Rev. Lett., 31, pp. 196–199.
―― (1974), *Bäcklund transformations for certain nonlinear evolution equations*, J. Math. Phys., 15, pp. 2157–2165.
―― (1976), *Bäcklund transformations at the turn of the century*, in Bäcklund Transformations, R. M. Miura, ed., Lecture Notes in Mathematics, 515, Springer-Verlag, New York.
―― (1980), *Elements of Soliton Theory*, John Wiley, New York.
H. LAMB (1932), *Hydrodynamics*, Dover, New York.
L. D. LANDAU (1946), *On the vibrations of electric plasmas*, J. Phys. USSR, 10, pp. 25–34.
R. LANDAUER (1967), *Sign of slow nonlinearities in nonabsorbing optical media*, Phys. Lett., 25A, pp. 416–417.
P. D. LAX (1954), *Weak solutions of nonlinear hyperbolic equations and their numerical computation*, Comm. Pure App. Math., 7, pp. 159–193.
―― (1968), *Integrals of nonlinear equations of evolution and solitary waves*, Comm. Pure Appl. Math., 21, pp. 467–490.
―― (1973), *Hyperbolic Systems of Conservation Laws and the Mathematical Theory of Shock Waves*, CBMS Regional Conference Series in Applied Mathematics 11, Society for Industrial and Applied Mathematics, Philadelphia, 1973.
―― (1975), *Periodic solutions of the KdV equation*, Comm. Pure Appl. Math., 28, pp. 141–188.
P. D. LAX AND D. LEVERMORE (1979), *The zero dispersion limit for the Korteweg-deVries equation*, Proc. Nat'l. Acad. Sci., 76, pp. 3602–3606.
LEBEDEV AND YU. I. MANIN (1978), *Hidden symmetries*, ITEP, preprint.
S. LEIBOVICH AND G. D. RANDALL (1973), *Amplification and decay of long nonlinear waves*, J. Fluid Mech., 58, pp. 481–493.
C. LEONE (1974), *Gradual damping of internal solitary waves*, M.S. Thesis, Clarkson College, New York.
C. LEONE AND H. SEGUR (1981), *Long internal waves of moderate amplitude II. Viscous decay*, preprint.
J. E. LEWIS, B. M. LAKE AND D. R. S. KO (1974), *On the interaction of internal waves and surface gravity waves*, J. Fluid Mech., 63, pp. 773–800.
E. H. LIEB AND W. LININGER (1963), *Exact analysis of an interacting Bose gas I.*, Phys. Rev., 130, pp. 1605–1616; (E. H. LIEB), *II*, 130, pp. 1616–1624.
G. LIEBBRANDT (1978), *New exact solutions of the classical sine-Gordon equation in $2+1$ and $3+1$ dimensions*, Phys. Rev. Lett., 41, pp. 435–438.

M. J. LIGHTHILL (1958), *Introduction to Fourier Analysis and Generalized Functions*, University Press, Cambridge.

A. LIU AND S. H. DAVIS (1977), *Viscous attenuation of mean drift in water waves*, J. Fluid. Mech., 81, pp. 63-84.

R. R. LONG (1964), *Solitary waves in the westerlies*, J. Atmos. Sci., 21, pp. 156-179.

K. LONNGREN AND A. C. SCOTT, eds., (1978), *Solitons in Action*, Academic Press, New York.

M. S. LONGUET-HIGGINS (1975), *Integral properties of periodic gravity waves of finite amplitude*, Proc. Roy. Soc. London A, 392, pp. 157-174.

M. S. LONGUET-HIGGINS AND M. J. H. FOX (1977), *Theory of the almost-highest wave I. The inner solution*, J. Fluid Mech., 80, pp. 721-742.

——— (1978), *Theory of the almost-highest wave II. Matching and analytic extension*, J. Fluid Mech., 85, pp. 769-786.

A. E. H. LOVE (1944), *A Treatise on the Mathematical Theory of Elasticity*, Dover, New York.

N. A. LUKASHEVICH (1971), *The second Painlevé equation*, Differential Equations, 7, pp. 853-854.

F. LUND (1977), *Example of a relativistic, completely integrable, Hamiltonian system*, Phys. Rev. Lett., 38, pp. 1175-1178.

Y. C. MA (1978), *On the long-wave/short-wave resonant interaction*, Stud. Appl. Math., 59, pp. 201-221.

Y. C. MA AND L. G. REDEKOPP (1979), *Some solutions pertaining to the resonant interactions of long and short waves*, Phys. Fluids, 22, pp. 1872-1876.

W. MAGNUS AND W. WINKLER (1966), *Hill's Equation*, Wiley-Interscience, New York.

D. MAISON (1978), *Are the stationary, axially symmetric Einstein equations completely integrable?*, Phys. Rev. Lett., 41, pp. 521-522.

——— (1979), *On the complete integrability of the stationary, axially symmetric Einstein equations*, J. Math. Phys., 20, pp. 871-877.

V. G. MAKHANKOV (1978), *Dynamics of classical solutions in non-integrable systems*, Phys. Rep., 35, pp. 1-128.

S. V. MANAKOV (1974a), *Nonlinear Fraunhofer diffraction*, Sov. Phys. JETP, 38, pp. 693-696.

——— (1974b), *On the theory of two-dimensional stationary self-focusing of electro-magnetic waves*, Sov. Phys. JETP, 38, pp. 248-253.

——— (1975), *Complete integrability and stochastization of discrete dynamical systems*, Sov. Phys. JETP, 40, pp. 269-274.

——— (1981), *The inverse scattering transform for the time-dependent Schrödinger equation and Kadomtsev-Petviaskvili equation*, Proc. Joint US-USSR Symposium on Soliton Theory, Kiev, 1979, V. E. Zakharov and S. V. Manakov, eds., North-Holland, Amsterdam, pp. 420-427.

S. V. MANAKOV, P. M. SANTINI AND L. A. TAKHTADZHYAN (1980), *Asymptotic behavior of the solutions of the Kadomtsev-Petviashvili equation*, Phys. Lett., 75A, pp. 451-454.

S. V. MANAKOV, V. E. ZAKHAROV, L. A. BORDAG, A. R. ITS AND V. B. MATVEEV (1977), *Two-dimensional solitons of the Kadomtsev-Petviashvili equation and their interaction*, Phys. Lett., 63A, pp. 205-206.

YU. I MANIN, *Hidden symmetries of long waves*, Proc. Joint US-USSR Symposium on Soliton Theory, Kiev, 1979, V. E. Zakharov and S. V. Manakov, eds., North-Holland, Amsterdam, pp. 400-409.

V. A. MARCHENKO (1974), *The periodic Korteweg-deVries problem*, Mat. Sb., 95, pp. 331-356.

D. V. MARTIN AND H. C. YUEN (1980), *Quasi-recurring energy leakage in the two-space-dimensional nonlinear Schrödinger equation*, Phys. Fluids, 23, pp. 881-883.

F. P. MATTAR AND M. C. NEWSTEIN (1977), *Transverse effects associated with the propagation of coherent optical pulses in resonant media*, IEEE J. Quantum Electronics, QE-13, pp. 507-520.

V. B. MATVEEV (1976), *Abelian functions and solitons*, Inst. Theor. Phys., Univ. Wroclaw, preprint 373.

T. Maxworthy and L. G. Redekopp (1976), *Theory of the Great Red Spot and other observed features of the Jovian atmosphere*, Icarus, 29, pp. 261-271.

T. Maxworthy, L. G. Redekopp and P. D. Wiedman (1978), *On the production and interaction of planetary waves. . .* , Icarus, 33, pp. 388-409.

S. L. McCall and E. L. Hahn (1965), cited in Bull. Am. Phys. Soc., 10, p. 1189.

——— (1967), *Self-induced transparency by pulsed coherent light*, Phys. Rev. Lett., 18, pp. 908-911.

——— (1969), *Self-induced transparency*, Phys. Rev., 183, pp. 457-485.

——— (1970), *Pulse area-pulse energy description of a traveling wave laser amplifier*, Phys. Rev. A, 2, pp. 861-870.

C. H. McComas and F. P. Bretherton (1977), *Resonant interaction of oceanic interval waves*, J. Geophys. Res., 82, pp. 1397-1412.

B. M. McCoy, C. A. Tracy and T. T. Wu (1977), *Painlevé functions of the third kind*, J. Math. Phys., 18, pp. 1058-1092.

A. D. McEwan (1971), *Degeneration of resonantly-excited standing internal gravity waves*, J. Fluid Mech., 50, pp. 431-448.

A. D. McEwan, D. W. Mander and R. K. Smith (1972), *Forced resonant second-order interaction between damped internal waves*, J. Fluid Mech., 55, pp. 589-608.

J. B. McGuire (1964), *Study of exactly soluble one-dimensional N-body problems*, J. Math. Phys., 5, pp. 622-636.

H. P. McKean (1981), *The sine-Gordon and sinh-Gordon equations on the circle*, Comm. Pure Appl. Math., 34, pp. 197-257.

H. P. McKean and E. Trubowitz (1976), *Hill's operator and hyperelliptic function theory in the presence of infinitely many branch points*, Comm. Pure Appl. Math., 29, pp. 143-226.

H. P. McKean and P. van Moerbeke (1975), *The spectrum of Hill's equation*, Invent. Math., 30, pp. 217ff.

D. W. McLaughlin (1975), *Four examples of the inverse method as a canonical transformation*, J. Math. Phys., 16, pp. 96-99, *Erratum*, 16.

D. W. McLaughlin and A. C. Scott (1973), *A restricted Bäcklund transformation*, J. Math. Phys., 14, pp. 1817-1828.

J. B. McLeod and P. J. Olver (1981), *The connection between completely integrable partial differential equations and ordinary differential equations of Painlevé type*, Rep. 2135, Mathematics Research Center, University of Wisconsin, Madison.

N. N. Meiman (1977), *The theory of one-dimensional Schrödinger operators with a periodic potential*, J. Math. Phys., 18, pp. 834-848.

A. V. Mikhailov (1981), *The reduction problem and the inverse scattering method*, Proc. Joint US-USSR Symposium on Soliton Theory, Kiev, 1979. V. E. Zakharov and S. V. Manakov, eds., North-Holland, Amsterdam, pp. 73-117.

J. W. Miles (1977a), *Obliquely interacting solitary waves*, J. Fluid Mech., 79, pp. 157-169.

——— (1977b), *Resonantly interacting solitary waves*, J. Fluid Mech., 79, pp. 171-179.

——— (1979), *The asymptotic solution of the Korteweg-deVries equation in the absence of solitons*, Stud. Appl. Math., 60, pp. 59-72.

——— (1980), *Solitary waves*, Ann. Rev. Fluid Mech., 12, pp. 11-43.

R. M. Miura (1968), *Korteweg-deVries equation and generalizations I. A remarkable explicit nonlinear transformation*, J. Math. Phys., 9, pp. 1202-1204.

——— (1974), *Conservation laws for the fully nonlinear long-wave equations*, Stud. Appl. Math., 53, pp. 45-56.

——— (1976a), *The Korteweg-deVries equation: a survey of results*, SIAM Rev. 18, pp. 412-459.

——— ed. (1976b), *Bäcklund Transformations*, Lecture Notes in Mathematics 515, Springer-Verlag, New York.

R. M. Miura, C. S. Gardner and M. D. Kruskal (1968), *Korteweg-deVries equation and generalizations. II. Existence of conservation laws and constants of motion*, J. Math. Phys., 9, pp. 1204-1209.

H. C. MORRIS (1979), *Prolongation structures and nonlinear evolution equations in two spatial dimensions*, J. Phys. A, 12, pp. 261-267.

P. M. MORSE AND H. FESHBACH (1953), *Methods of Theoretical Physics*, McGraw-Hill, New York.

J. MOSER (1975a), *Dynamical systems, finitely many mass points on the line under the influence of an exponential potential—an integrable system*, in Dynamical Systems, Theory and Applications, J. Moser, ed., Lecture Notes in Physics 38, Springer-Verlag, New York.

——— (1975b), *Integrable systems of nonlinear evolution equations*, in Dynamical Systems, Theory and Applications, J. Moser, ed., Lecture Notes in Physics 38, Springer-Verlag, New York.

H. E. MOSES (1976), *A solution of the Korteweg-deVries equation in a half-space bounded by a wall*, J. Math. Phys., 17, pp. 73-75.

G. MOTT (1973), *Elastic wave propagation in an infinite isotropic solid cylinder*, J. Acoust. Soc. Amer., 53, pp. 1129-1135.

A. C. MURRAY (1978), *Solutions of the Korteweg-deVries equation evolving from a "box" and other irregular initial functions*, Duke Math. J., 45, pp. 149-181.

R. NAKACH (1977), *Tech Rep't #2*, School of Electrical Engineering, Chalmers Univ. Techn., Göteborg, Sweden.

A. NAKAMURA (1979a), *A direct method of calculating periodic wave solutions to nonlinear evolution equations I. Exact two-periodic wave solution*, J. Phys. Soc. Japan, 47, pp. 1701-1705.

——— (1979b), *Bäcklund transform and conservative laws of the Banjamin-Ono equation*, J. Phys. Soc. Japan 47, pp. 1335-1340.

V. V. NEMYTSKII AND V. V. STEPANOV (1960), *Qualitative Behavior of Differential Equations*, Princeton Univ. Press, Princeton, NJ.

G. NEUGEBAUER (1979), *Bäcklund transformations of axially symmetric stationary gravitational fields*, J. Phys. A, 12, pp. 167-170.

A. C. NEWELL (1974), *Nonlinear Wave Motion*, American Mathematical Society, Providence, RI

A. C. NEWELL AND L. REDEKOPP (1977), *Breakdown of Zakahrov-Shabat theory and soliton creation*, Phys. Rev. Lett., 38, pp. 377-380.

R. G. NEWTON (1966), *Scattering Theory of Waves and Particles*, McGraw-Hill, New York.

——— (1979), *A new result on the inverse scattering problem in three dimensions*, Phys. Rev. Lett., 43, pp. 541-542.

L. P. NIZHIK (1973), *Inverse Nonstationary Problem of Scattering Theory*, Naukova Dumka, Kiev, USSR.

S. P. NOVIKOV (1974), *The periodic problem for the Korteweg-deVries equation*, Funct. Anal. Appl., 8, pp. 236-246.

S. P. NOVIKOV AND I. M. KRICHEVER (1981), *Algebraic geometry and mathematical physics*, cited in Proc. Joint US-USSR Symposium on Soliton Theory, Kiev, 1979, V. E. Zakharov and S. V. Manakov, eds., North-Holland, Amsterdam.

S. OISHI (1979), *Relationship between Hirota's method and the inverse spectral method—the Korteweg-deVries equation's case*, J. Phys. Soc. Japan, 47, pp. 1037-1038.

D. J. OLBERS AND K. HERTERICH (1979), *The spectral energy transfer from surface waves to internal waves*, J. Fluid Mech., 92, pp. 349-379.

M. OLSHANETSKY AND A. PERELOMOV (1976a), *Completely integrable classical systems connected with semisimple Lie algebras*, Lett. Math. Phys., 1, pp. 187-193.

——— (1976b), *Explicit solutions of some completely integrable systems*, Lett. Nuovo Cimento, 17, pp. 97-101.

——— (1976c), *Completely integrable Hamiltonian systems connected with semisimple Lie algebras*, Invent. Math., 37, pp. 93-108.

F. W. OLVER (1974), *Asymptotics and Special Functions*, Academic Press, New York.

H. Ono (1975), *Algebraic solitary waves in stratified fluids*, J. Phys. Soc., Japan, 39, pp. 1082-1091.
――― (1976), *Algebraic soliton of the modified Korteweg-deVries equation*, J. Phys. Soc. Japan, 41, pp. 1817-1818.
L. Onsager (1949), *Statistical hydrodynamics*, Il Nuovo Cimento, Ser. 9, 6, Suppl. No. 2, pp. 279-287.
S. A. Orszag and M. Israeli (1974), *Numerical simulation of viscous incompressible flows*, Ann. Rev. Fluid Mech., 6, pp. 281-318.
A. R. Osborne and T. L. Burch (1980), *Interval solitons in the Andaman Sea*, Science, 258, pp. 451-460.
E. Ott and R. N. Sudan (1970), *Damping of solitary waves*, Phys. Fluids, 13, pp. 1432-1434.
S. Oxford (1979), *The Hamiltonian of the quantized nonlinear Schrödinger equation*, Ph.D. thesis, Univ. of California, Los Angeles.
C. K. N. Patel (1970), *Investigation of pulse delay in self-induced transparency*, Phys. Rev. A., Ser. 3, 1, pp. 979-982.
J. Pedlosky (1971), *Geophysical fluid dynamics*, Lectures in Applied Mathematics 13, American Mathematics Society, Providence, RI.
B. M. Peek (1958), *The Planet Jupiter*, Faber and Faber, London.
A. S. Peters and J. J. Stoker (1960), *Solitary waves in liquids having nonconstant density*, Comm. Pure Appl. Math., 13, pp. 115-164.
O. M. Phillips (1960), *On the dynamics of unsteady gravity waves of finite amplitude I. The elementary interactions*, J. Fluid Mech., 9, pp. 193-217.
――― (1974), *Nonlinear dispersive waves*, Ann. Rev. Fluid Mech., 6, pp. 93-110.
――― (1977), *The Dynamics of the Upper Ocean*, 2nd ed., Cambridge Univ. Press, London.
F. A. E. Pirani (1979), *Local Jet Bundle Formulation of Bäcklund Transformations*, Reidel, Boston, 1979.
W. Pogorzelski (1966), *Integral Equations and Their Applications*, Pergamon Press, London.
G. V. Ramanathan and G. Sandri (1969), *Model for the derivation of kinetic theory*, J. Math. Phys., 10, pp. 1763-1773.
Lord (J. W. Strutt) Rayleigh (1876), *On waves*, Philos. Mag., Ser. 5, 1, pp. 257-279.
C. Rebbi (1979), *Solitons*, Scientific American, February, pp. 92-116.
L. G. Redekopp (1977), *On the theory of solitary Rossby waves*, J. Fluid Mech., 82, pp. 225-745.
――― (1980), *Similarity solutions of two-dimensional wave equations*, Stud. Appl. Math., 63, pp. 185-207.
L. G. Redekopp and P. D. Weidman (1978), *Solitary Rossby waves in zonal shear flows and their interactions*, J. Atmos. Sci., 35, pp. 790-804.
R. D. Richtmeyer and K. W. Morton (1967), *Difference Methods for Initial-Value Problems*, Interscience, New York.
M. H. Rizk and D. R. S. Ko (1978), *Interaction between small-scale surface waves and large scale internal waves*, Phys. Fluids, 21, pp. 1900-1905.
R. Rosales (1978), *Exact solutions of some nonlinear evolution equations*, Stud. Appl. Math., 59, pp. 117-151.
G. G. Rossby (1939), *Relation between variations in the intensity of the zonal circulation of the atmosphere*, J. Marine Res., 2, pp. 38-55.
S. N. M. Ruijsenaars (1980), *On one-dimensional integrable quantum systems with infinitely many degrees of freedom*, Ann. Phys., 128, pp. 335-362.
W. Rund (1976), *Variational problems and Bäcklund transformations*, in Bäcklund Transformations, R. M. Miura, ed., Lecture Notes in Mathematics 515, Springer-Verlag, New York.
J. S. Russell (1838), *Report of the committee on waves*, Report of the 7th Meeting of British Association for the Advancement of Science, Liverpool, pp. 417-496.
――― (1844), *Report on waves*, Report of the 14th Meeting of the British Association for the Advancement of Science, John Murray, London, pp. 311-390.

G. J. SALAMO, H. M. GIBBS AND G. C. CHURCHILL (1974), *Effects of degeneracy on self-induced transparency*, Phys. Rev. Lett., 33, pp. 273-276.

S. SALIHOGLU (1980), *Two-dimensional O(N) nonlinear σ-model and the fifth Painlevé transcendent*, Phys. Lett., 89B, p. 367.

H. SAMELSON (1969), *Notes on Lie Algebra*, Van Nostrand-Reinhold, New York.

R. SASAKI (1979a), *Geometrization of soliton equations*, Phys. Lett., 71A, pp. 390ff.

——— (1979b), *Soliton equations and pseudo-spherical surfaces*, Nucl. Phys. B., 154, pp. 343-357.

M. SATO, T. MIWA AND M. JIMBO (1977), (1978), *Studies on holonomic quantum fields*, Proc. Japan Acad. Ser. A. Math. Sci., *I*, 53, pp. 6-10; *II*, 53, pp. 147-152; *III*, 53, pp. 153-158; *IV*, 53, pp. 183-185; *V*, 53, pp. 219-224; *VI*, 54, pp. 1-5; *VII*, 54, pp. 36-41 (series continues). See also *Holononic quantum fields*, Pub. RIMS, *I*, 14, pp. 223-267 (series continues).

J. SATSUMA (1976), *N-soliton solution of the two-dimensional Korteweg-deVries equation*, J. Phys. Soc. Japan 40, pp. 286-290.

——— (1979), private communication.

J. SATSUMA AND M. J. ABLOWITZ (1979), *Two-dimensional lumps in nonlinear dispersive systems*, J. Math. Phys., 20, pp. 1496ff.

——— (1980), *Solutions of an internal wave equation describing a stratified fluid with finite depth*, in Nonlinear Partial Differential Equations in Engineering and Applied Science, R. L. Sternberg, A. J. Kalinowki and J. S. Papadakis, eds., Marcel Dekker, New York.

J. SATSUMA, M. J. ABLOWITZ AND Y. KODAMA (1979), *On an internal wave equation describing a stratified fluid with finite depth*, Phys. Lett., 73A, pp. 283-286.

J. SATSUMA AND D. J. KAUP (1977), *A Bäcklund transformation for a higher order Korteweg-deVries equation*, J. Phys. Soc. Japan, 43, pp. 692-697.

J. SATSUMA AND N. YAJIMA (1974), *Initial value problems of one-dimensional self modulation of nonlinear waves in dispersive media*, Supp. Prog. Theoret. Phys., 55, pp. 284-306.

A. C. SCOTT, F. Y. F. CHU AND D. W. MCLAUGHLIN (1973), *The soliton—A new concept in applied science*, Proc. IEEE, 61, pp. 1443-1483.

H. SEGUR (1973), *The Korteweg-deVries equation and water waves, solutions of the equation, I*, J. Fluid Mech., 59, pp. 721-736.

——— (1976), *Asymptotic solutions and conservation laws for the nonlinear Schrödinger equation, Part II.*, J. Math. Phys., 17, pp. 714-716.

——— (1980), *Resonant interactions of surface and internal gravity waves*, Phys. Fluids, 23, pp. 2556-2557.

H. SEGUR AND M. J. ABLOWITZ (1976), *Asymptotic solutions and conservation laws for the nonlinear Schrödinger equation, Part I.*, J. Math. Phys., 17, pp. 710-713.

——— (1981), *Asymptotic solutions of nonlinear evolution equations and a Painlevé transcendent*, Proc. Joint US-USSR Symposium on Soliton Theory, Kiev, 1979, V. E. Zakharov and S. V. Manakov, eds., North-Holland, Amsterdam, pp. 165-184.

H. SEGUR AND J. L. HAMMACK (1981), *Long internal waves of moderate amplitude I. Solitons*, preprint.

A. B. SHABAT (1973), *On the Korteweg-deVries equation*, Sov. Math. Doklady, 14 (1973), pp. 1266-1269.

——— (1975), *Inverse-scattering problem for a system of differential equations*, Funct. Anal. Appl. 9, pp. 244-247.

——— (1981), cited in Proc. Joint US-USSR Symposium on Soliton Theory, Kiev, V. E. Zakharov and S. V. Manakov, eds., North-Holland, Amsterdam

W. F. SIMMONS (1969), *A variational method for weak resonant wave interactions*, Proc. Roy. Soc. London, *A*, 309, pp. 551-575.

E. SKLYANIN (1979), *Method of the inverse scattering problem and the nonlinear quantum Schrödinger equation*, Sov. Phys. Doklady, 24, pp. 107-109.

E. SKLYANIN AND F. D. FADDEEV (1978), *Quantum mechanical approach to a completely integrable field*, Sov. Phys. Doklady, 23, pp. 902–904.
R. E. SLUSHER AND H. M. GIBBS (1972), *Self-induced transparency in atomic rubidium*, Phys. Rev. A, 5, pp. 1634–1659; *Erratum*, 6, p. 1255.
J. J. STOKER (1957), *Water Waves*, Interscience, New York.
G. G. STOKES (1847), *On the theory of oscillatory waves*, Trans. Camb. Philos. Soc., 8, pp. 441–455.
R. SUGAYA, M. SUGAWA AND H. NOMOTO (1977), *Experimental observation of explosive instability due to a helical electron beam*, Phys. Rev. Lett., 39, pp. 27–31.
B. SUTHERLAND (1972), *Exact results for a quantum many-body problem in one-dimension, II*, Phys. Rev., 5A, pp. 1372–1376.
T. TAHA AND M. J. ABLOWITZ (1981), *Numerical simulations*, Clarkson College, Potsdam, NY.
L. A. TAKHTADZHYAN (1972), Diploma Thesis, Leningrad State University.
―――― (1981), *The quantum inverse problem method and the XYZ Heisenberg model*, Proc. Joint US-USSR Symposium on Soliton Theory, Kiev, 1979, V. E. Zakharov and S. V. Manakov, eds., North-Holland, Amsterdam, pp. 231–245.
V. I. TALANOV (1965), *Self focusing of wave beams in nonlinear media*, Sov. Phys. JETP Lett., 2, pp. 138–141.
S. TANAKA (1972a), *On the N-tuple wave solutions of the Korteweg-deVries equation*, Publ. RIMS, Kyoto Univ., 8, pp. 419–428.
―――― (1972b), *Some remarks on the modified Korteweg-deVries equation*, Publ. RIMS, Kyoto Univ., 8, pp. 429–437.
―――― (1974), *Korteweg-deVries equation: construction of solutions in terms of scattering data*, Osaka J. Math., 11, pp. 49–59.
―――― (1975), *Korteweg-deVries equation; asymptotic behavior of solutions*, Publ. RIMS, Kyoto Univ., 10, pp. 367–379.
T. TANIUTI AND C. C. WEI (1968), *Reductive perturbation method in nonlinear wave propagation I*, J. Phys. Soc. Japan, 24, pp. 941–946.
G. I. TAYLOR (1959), *Waves on thin sheets of water II.*, Proc. Roy. Soc. London A, 253, pp. 296–312.
H. B. THACKER (1978), *Polynomial conservation laws in (1 + 1)-dimensional classical and quantum field theory*, Phys. Rev. D, 17, pp. 1031–1040.
H. B. THACKER AND D. WILKINSON (1979), *Inverse scattering transform as an operator method in quantum field theory*, Phys. Rev. D, 19, pp. 3660–3665.
W. THICKSTUN (1976), *A system of particles equivalent to solitons*, J. Math. Anal. Appl., 55, pp. 335–346.
S. A. THORPE (1966), *On standing internal gravity waves of finite amplitude*, J. Fluid Mech., 24, pp. 737–751.
M. TODA (1967a), *Vibration of a chain with nonlinear interaction*, J. Phys. Soc. Japan, 22, pp. 431–436.
―――― (1967b), *Wave propagation in anharmonic lattices*, J. Phys. Soc. Japan, 23, pp. 501–506.
―――― (1970), *Waves in nonlinear lattices*, Prog. Theoret. Phys., Suppl. 45, pp. 174–200.
―――― (1971), private communication.
K. UENO (1981), *Monodromy preserving deformation and its application to soliton theory*, RIMS Kyoto Univ., preprint.
N. G. VAN KAMPEN (1955), *On the theory of stationary waves in plasmas*, Physica, 21, pp. 949–963.
A. VLASOV (1945), *On the kinetic theory of an assembly of particles with a collective interaction*, J. Phys. USSR, 19, pp. 25–40.
J. VON NEUMANN (1944), *Proposal and analysis of a numerical model for the treatment of hydrodynamical shock problems*, Nat. Def. Res. Com. Rep. AM-551.
M. WADATI (1972), *The modified Korteweg-deVries equation*, J. Phys. Soc. Japan, 32, pp. 1681ff.

M. WADATI, K. KONNO AND Y. H. ICHIKAWA (1979), *New integrable nonlinear evolution equations*, J. Phys. Soc. Japan, 47, pp. 1698ff.

M. WADATI, H. SANUKI AND K. KONNO (1975), *Relationships among inverse method, Bäcklund transformation and an infinite number of conservation laws*, Prog. Theoret. Phys., 53, pp. 419–436.

M. WADATI AND M. TODA (1972), *The exact N-soliton solution of the Korteweg-deVries equation*, J. Phys. Soc. Japan 32, pp. 1403–1411.

H. D. WAHLQUIST AND F. B. ESTABROOK (1973), *Bäcklund transformation for solutions of the Korteweg-deVries equation*, Phys. Rev. Lett., 23, pp. 1386–1389.

——— (1975), (1976), *Prolongation structures and nonlinear evolution equations*, J. Math. Phys., 16, pp. 1–7; 17, pp. 1293–1297.

O. D. WATERS (1967), *Oceanographic Atlas of the North Atlantic Ocean, II, Physical Properties*, U.S.N. Oceanographic Office, Washington, DC.

K. M. WATSON, B. J. WEST AND B. I. COHEN (1976), *Coupling of surface and internal gravity waves: a mode coupling model*, J. Fluid Mech., 77, pp. 185–208.

J. V. WEHAUSEN AND E. V. LAITONE (1960), *Surface waves*. Handbuch der Physik, vol. 9, Springer-Verlag, Berlin.

P. D. WEIDMAN AND T. MAXWORTHY (1978), *Experiments on strong interactions between solitary waves*, J. Fluid Mech., 85, pp. 417–431.

V. WEISSKOPF AND E. WIGNER (1930), *Berechnung der Naturlichen Linienbriete auf Grund*, Z. Physik, 63, pp. 54–73.

M. WELLNER (1960), *Energy renormalization in ordinary wave mechanics*, Phys. Rev., 118, pp. 875–877.

B. J. WEST, J. A. THOMSON AND K. M. WATSON (1974), *Statistical mechanics of ocean waves*, J. Hydronaut., 9, pp. 25–31.

G. B. WHITHAM (1974), *Linear and Nonlinear Waves*, Wiley-Interscience, New York.

——— (1979), *Comments on some recent multi-soliton solutions*, J. Phys. A. Math., 12, pp. L1–L3.

J. WILLEBRAND (1975), *Energy transport in a nonlinear and inhomogeneous random gravity wave field*, J. Fluid Mech., 70, pp. 113–126.

T. T. WU, B. M. MCCOY, C. A. TRACY AND E. BAROUCH (1976), *Spin-spin correlation functions for the two-dimensional Ising model-exact theory in the scaling region*, Phys. Rev. B, 13, pp. 316–374.

N. YAJIMA (1974), *Stability of envelope solitons*, Prog. Theoret. Phys., 52, pp. 1066ff.

N. YAJIMA AND Y. H. ICHIKAWA, eds. (1979), *Soliton Phenomena in Plasmas*, Selected Papers in Physics 73, Physical Society of Japan.

N. YAJIMA AND N. KAKUTANI, eds. (1975), *Nonlinear Dispersive Wave Motion*, Selected Papers in Physics 59, Physical Society of Japan.

N. YAJIMA AND M. OIKAWA (1975), *A class of exactly solvable nonlinear evolution equations*, Prog. Theoret Phys., 54, pp. 1576–1578.

——— (1976), *Formation and interaction of sonic-Langmuir solitons—inverse scattering method*, Prog. Theoret. Phys., 56, pp. 1719–1739.

N. YAJIMA, M. OIKAWA AND J. SATSUMA (1978), *Interaction of ion-acoustic solitons in three-dimensional space*, J. Phys. Soc. Japan, 44, pp. 1711–1714.

A. YARIV (1975), *Quantum Electronics*, 2nd ed., John Wiley, New York.

H. C. YUEN AND W. FERGUSON (1978), *Fermi-Pasta-Ulam recurrence in the two-space dimensional nonlinear Schrödinger equation*, Phys. Fluids, 21, pp. 2116–2118.

H. C. YUEN AND B. M. LAKE (1975), *Nonlinear deep water waves: theory and experiment*, Phys. Fluids, 18, pp. 956–960.

——— (1980), *Instabilities of waves on deep water*, Ann. Rev. Fluid Mech., 12, pp. 303–334.

N. J. ZABUSKY (1962), *Phenomena associated with the oscillations of a nonlinear model string*, Mathematical Models in the Physical Sciences, S. Drobat, ed., Prentice-Hall, Englewood Cliffs, NJ.

—— (1967), *A synergetic approach to problems of nonlinear dispersive wave propagation and interaction*, in Proc. Symposium on Nonlinear Partial Differential Equations, W. F. Ames, ed., Academic Press, New York.

—— (1969), *Nonlinear lattice dynamics and energy sharing*, J. Phys. Soc. Japan, 26, Suppl., pp. 196ff.

N. J. ZABUSKY AND C. J. GALVIN (1971), *Shallow water waves, the Korteweg-deVries equation and solitons*, J. Fluid Mech., 47, pp. 811–824.

N. J. ZABUSKY AND M. D. KRUSKAL (1965), *Interaction of solitons in a collisionless plasma and the recurrence of initial states*, Phys. Rev. Lett., 15, pp. 240–243.

V. E. ZAKHAROV (1968), *Stability of periodic waves of finite amplitude on the surface of a deep fluid*, Sov. Phys. J. Appl. Mech. Tech. Phys., 4, pp. 190–194.

—— (1971), *Kinetic equation for solitons*, Sov. Phys. JETP, 33, pp. 538–541.

—— (1972), *Collapse of Langmuir waves*, Sov. Phys. JETP, 35, pp. 908–914.

—— (1974), *On stochastization of one-dimensional chains of nonlinear oscillations*, Sov. Phys. JETP, 38, pp. 108–110.

—— (1975), *Instability and nonlinear oscillations of solitons*, Sov. Phys. JETP Lett., 22, pp. 172–173.

—— (1976), *Exact solutions to the problem of the parametric interaction of three-dimensional wave packets*, Sov. Phys. Doklady, 21, pp. 322–323.

—— (1981), *On the Benney equations*, Proc. US-USSR Symposium on Soliton Theory, Kiev, 1979, V. E. Zakharov and S. V. Manakov, eds., North-Holland, Amsterdam, pp. 193–202.

V. E. ZAKHAROV AND V. A. BELINSKII (1978), *Integration of the Einstein equations by the inverse scattering method and calculation of the exact soliton solutions*, Sov. Phys. JETP, 48, pp. 985–994.

V. E. ZAKHAROV AND L. D. FADDEEV (1971), *Korteweg-deVries equation, a completely integrable Hamiltonian system*, Funct. Anal. Appl., 5, pp. 280–287.

V. E. ZAKHAROV AND S. V. MANAKOV (1973), *Resonant interaction of wave packets in nonlinear media*, Sov. Phys. JETP Lett., 18, pp. 243–245.

—— (1974), *On the complete integrability of the nonlinear Schrödinger equation*, Theoret. Math. Phys., 19, pp. 551–559.

—— (1976a), *The theory of resonance interaction of wave packets in nonlinear media*, Sov. Phys., JETP, 42, pp. 842–850.

—— (1976b), *Asymptotic behavior of nonlinear wave systems integrated by the inverse scattering method*, Sov. Phys. JETP, 44, pp. 106–112.

—— (1979), *Soliton theory*, Phys. Rev. (Sov. Scient. Rev.) 1, pp. 133–190.

—— eds. (1981), *Proceedings of the Joint US-USSR Symposium on Soliton Theory*, Kiev, 1979, North-Holland, Amsterdam.

V. E. ZAKHAROV, S. V. MANAKOV, S. P. NOVIKOV AND L. P. PITAYEVSKY (1980), *Theory of Solitons. The Method of the Inverse Scattering Problem*, Nauka, Moscow. (In Russian).

V. E. ZAKHAROV AND A. V. MIKHAILOV (1978a), *Example of nontrivial interaction of solitons in two-dimensional classical field theory*, Sov. Phys. JETP Lett., 27, pp. 42–46.

—— (1978b), *Relativistically invariant two-dimensional models of field theory which are integrable by means of the inverse scattering problem method*, Sov. Phys. JETP, 47, pp. 1017–1027.

V. E. ZAKHAROV, S. L. MUSHER AND A. M. RUBENCHIK (1974), *Nonlinear stage of parametric wave excitation in a plasma*, Sov. Phys. JETP Lett., 19, pp. 151–152.

V. E. ZAKHAROV AND A. M. RUBENCHIK (1974), *Instability of wave guides and solitons in nonlinear media*, Sov. Phys. JETP, 38, pp. 494–500.

V. E. ZAKHAROV AND P. B. SHABAT (1972), *Exact theory of two-dimensional self-focusing and one-dimensional self-modulation of waves in nonlinear media*, Sov. Phys. JETP, 34, pp. 62–69.

——— (1973), *Interaction between solitons in a stable medium*, Sov. Phys. JETP, 37, pp. 823–828.

——— (1974), *A scheme for integrating the nonlinear equations of mathematical physics by the method of the inverse scattering problem. I*, Funct. Anal. Appl., 8, pp. 226–235.

——— (1979), *Integration of the nonlinear equations of mathematical physics by the method of the inverse scattering problem. II*, Funct. Anal. Appl., 13, pp. 166–174.

V. E. ZAKHAROV AND E. I. SHULMAN (1980), *Degenerate dispersion laws, motion invariance and kinetic equations*, Physica D, 1, pp. 192–202.

V. E. ZAKHAROV AND V. S. SYNAKH (1976), *The nature of the self-focusing singularity*, Sov. Phys. JETP, 41, pp. 465–468.

J. M. ZIMAN (1960), *Electrons and Phonons*, Clarendon Press, Oxford.

Index

AUTHOR INDEX

Ablowitz, M. J., 1, 9, 19, 28, 39, 40, 41, 42, 45, 46, 47, 48, 68, 71, 78, 80, 81, 82, 85, 89, 93, 97, 110, 111, 113, 114, 115, 121, 122, 124, 134, 148, 159, 172, 191, 212, 215, 216, 217, 218, 219, 228, 231, 233, 236, 238, 239, 240, 243, 244, 245, 248, 250, 251, 252, 260, 268, 294, 316, 318, 320, 323, 324, 325, 334, 335, 337, 347
Acrivos, A., 293
Adam, J. A., 312
Adler, M., 153, 191, 196
Agranovich, Z., 15
Aikawa, M., 148
Airault, H., 191, 203, 209, 233, 246, 248
Airy, G. B., 277
Akhmanov, S. A., 302, 307, 308, 314, 316, 317
Amick, C. J., 283
Anderson, R. L., 169, 171
Anker, D., 113, 324
Armi, L., 348
Arnold, V. I., 58, 64, 355
Avez, A. A., 64

Bäcklund, A. V., 155
Ball, F. K., 301, 304, 307, 309
Barouch, E., 248
Baxter, R. J., 342
Beer, A., 330
Belinskii, V. A., 337, 338, 339
Benjamin, T. B., 203, 282, 293, 294, 311, 325
Benney, D. J., 89, 148, 277, 297, 301, 312, 313, 318, 320

Berezin, F. A., 341
Bergknoff, H., 342
Bethe, H. A., 341
Bianchi, L., 159
Birkhoff, G., 355
Blazek, M., 217
Bleistein, N. A., 356
Bock, T. L., 216
Bogolyubov, N. N., 378
Boiti, M., 191, 246, 248
Boldt, E., 388
Bordag, L. A., 191, 199, 232
Boussinesq, J., 3
Bretherton, F. P., 301, 309
Bullough, R. K., 181, 333, 337
Burch, T. L., 296
Bureau, F. J., 236
Burgers, J. M., 154
Burt, P. B., 190
Butler, G., 296
Byrd, P. F., 345

Cairns, R. A., 312
Callan, C. G., 342
Calogero, F., 52, 159, 203, 206
Case, K. M., 114, 115, 133, 204, 206, 389, 391
Caudrey, P. J., 181, 190, 333, 337
Chen, H. H., 152, 157, 161, 204, 206, 211
Cherednik, I. V., 147
Chern, S. S., 330
Chester, C., 365
Chiu, S. C., 52, 122
Chu, F. Y. F., 327
Chudnovsky, D. V., 203, 210

415

Chudnovsky, G. V., 203, 210
Churchill, G. C., 337
Clairin, J., 161, 164
Cohen, A., 309, 363
Cokelet, E. D., 323
Cole, J. D., 154, 280
Copson, E. T., 356, 359, 360, 364
Cornille, H., 99, 217, 233
Corones, J. P., 152, 161, 163, 265
Courant, R., 366
Courtens, E., 330
Craik, A. D. D., 312
Crum, M. M., 160

Dashen, R. P., 341
Date, E., 134
Davey, A., 190, 318, 320
Davidson, R. C., 301, 312
Davis, R. E., 293
Davis, S. H., 348
Debye, P., 3
Degasperis, F., 52
Deift, P., 10, 15, 26, 31, 52, 153, 159, 160, 161, 217
deVries, G., 3, 4, 277
Dikii, L. A., 153
Djordjevic, V. D., 97, 296, 318, 320, 322
Dobbs, L., 211, 294
Dodd, R. K., 190, 337
Dorfman, I., 153
Dryuma, V., 114, 231
Dubrovin, B. A., 134, 147, 146
Duckworth, S., 337

Egri, R., 211
Eilbeck, J. C., 38, 181, 332, 333
Eisenhart, L. P., 328, 329
Erugin, N. P., 197, 248
Estabrook, F. B., 153, 158, 161, 162, 163, 166, 263

Faddeev, L., 26, 55, 56, 57, 58, 67, 138, 217, 341, 342
Feir, J. F., 311, 325
Ferguson, W. E., 323, 325
Fermi, E., 3, 4
Feshbach, H., 366
Finkel'berg, V. M., 341
Fisher, R. A., 166

Flaschka, H., 56, 58, 64, 67, 114, 115, 132, 147, 148, 239, 240, 248
Fokas, A. S., 156, 197
Forest, G., 148
Forsyth, A. R., 156, 169, 171
Fox, M. J. H., 323
Franken, P. A., 307
Freeman, N. C., 113, 190, 318, 324
Friedman, B., 345, 365
Friedrichs, K. O., 352, 366
Fung, Y. C., 342

Galvin, C. J., 282
Gardner, C. S., 1, 3, 6, 7, 8, 38, 41, 47, 52, 55, 159
Gargett, A. G., 312
Garrett, C., 309
Gasymov, M. G., 10
Gel'fand, I. M., 21, 153
Gibbon, J. D., 181, 190, 292, 327, 333
Gibbs, H. M., 330, 331, 336, 337
Glassey, R. T., 317
Goldstein, H., 58
Golubov, V. V., 235
Greene, J. M., 1, 7, 8
Grimshaw, R. H. J., 296
Gross, D. J., 342

Haberman, R., 54, 93, 97, 105, 110, 111, 231, 244, 324, 347
Hahn, E. L., 330, 332, 335, 336, 337, 350
Hall, 357
Hamilton, J., 324
Hammack, J. L., 282, 283, 284, 285, 286, 287, 288, 291, 326
Handlesman, R. A., 356
Hardin, R. H., 38
Harrison, R., 337, 339
Hasegawa, A., 28
Hasimoto, H., 318
Hasselmann, K., 313
Hasslacher, B., 341
Hastings, S. P., 244
Hénon, M., 149
Hermann, R., 152
Herterich, K., 309
Hewitt, E., 358
Hill, A. E., 307
Hille, E., 235

INDEX

Hirota, R., 40, 114, 115, 151, 161, 171, 172, 175, 179, 181, 182, 183, 185, 187, 190, 200
Honerkamp, J., 342
Hopf, E., 154
Hoppenstaedt, F., 166
Hughes, B. A., 312
Hui, W. H., 324

Ibragimov, N. J., 156, 169, 171
Ichikawa, Y. H., 14, 54, 275, 277
Ince, E. L., 234, 235, 240, 241, 266
Israeli, M., 385
Its, A. R., 134, 147, 191, 199, 232
Izutsu, M., 190

Jacobson, N., 167, 169
James, I. N., 292, 327
Jaulent, M., 28
Jimbo, M., 238, 248
Johnson, R. S., 250
Joseph, R. I., 211, 294
Joyce, T. M., 310, 311

Kac, M., 114, 115, 116, 133, 134
Kadomtsev, B. B., 250, 289
Kakutani, N., 275, 277
Karpman, V. I., 250, 252, 253, 254, 275, 277, 314
Kashiwara, M., 248
Kaup, D. J., 1, 8, 9, 14, 19, 42, 44, 45, 46, 47, 48, 51, 85, 93, 95, 99, 101, 107, 110, 159, 161, 163, 172, 233, 238, 250, 252, 253, 255, 301, 306, 330, 334, 335, 336, 341, 369
Kawai, T., 248
Kay, I., 28, 40
Kazhdan, D., 152
Keener, J. P., 250
Kelley, P. L., 315, 316
Keulegan, G. H., 285, 295, 296
Khokhlov, R. V., 302, 307, 308, 314, 316, 317
Kleinman, D. A., 302, 307
Knickerbocker, C. J., 255
Ko, D. R. S., 211, 294, 311, 312
Ko, K., 250
Kobayashi, K. K., 190
Kodama, Y., 41, 51, 56, 58, 172, 215, 216, 250, 252, 260, 337

Konno, K., 14, 54, 159
Koop, C. G., 296
Korepin, V. E., 341
Korteweg, D. J., 3, 4, 277
Kostant, B., 152
Krichever, I. M., 134, 147, 153
Kritz, A. H., 387
Kruskal, M. D., 1, 4, 5, 6, 7, 8, 15, 38, 41, 47, 52, 55, 68, 80, 149, 159, 168, 170, 203, 216, 279, 282, 337, 339, 389
Kubota, T., 211, 294
Kudryavesev, A. E., 38
Kuehl, H. H., 250
Kulish, P. P., 341
Kuperschmidt, B. B., 89
Kuznetsov, E. A., 233

Ladik, J. F., 38, 52, 114, 115, 121, 122, 124, 337
Laitone, E. V., 277
Lake, B. M., 311, 325
Lamb, G. L., 159, 164, 169, 171, 330, 331, 334, 335, 337
Lamb, H., 1, 2, 277, 293, 294
Landau, L. D., 378, 390
Landauer, R., 314
Lax, P. D., 4, 8, 9, 14, 239, 359
Lebedev, 153
Lee, Y. C., 152, 204, 206, 211
Leibovich, S., 252
Leone, C., 296
Levitan, B. M., 10, 21
Lewis, J. E., 311
Lieb, E. H., 341
Liebbrandt, G., 190
Lighthill, M. J., 375
Lininger, W., 341
Liu, C. S., 152
Long, R. R., 297
Longuet-Higgins, M. S., 323
Lonngren, K., 1
Love, A. E. H., 342
Lui, A., 348
Lukashevich, N. A., 246
Lund, F., 52

Ma, Y. C., 134, 296, 312
Magnus, W., 136, 137
Maison, D., 337, 338
Makhankov, V. G., 37, 275, 277

418 INDEX

Manakov, S. V., 21, 56, 58, 68, 71, 87, 90, 93, 97, 99, 110, 114, 116, 132, 191, 199, 232, 290, 341
Mander, D. W., 311
Manin, Yu I., 89, 153
Marchenko, V., 15, 147
Markovski, B. L., 152
Martin, D. V., 323
Maslov, E. M., 250, 252, 253, 254, 275, 277, 314
Matsubara, T., 277
Mattar, F. P., 337
Matveev, V. B., 134, 147, 191, 199, 232
Maxworthy, T., 282, 297, 299, 300
McCall, S. L., 330, 332, 335, 336, 337, 350
McComas, C. H., 309
McCoy, B. M., 248
McEwan, A. D., 311
McGuire, J. B., 341
McKean, H. P., 134, 137, 153
McLaughlin, D. W., 58, 88, 147, 148, 168, 170, 250, 327
McLeod, J. B., 236, 238, 239, 244
Meiman, N. N., 147
Mikhailov, A. V., 99, 168, 233, 267
Miles, J. W., 68, 81, 168, 189, 232, 266, 267, 277, 289, 297
Mirva, T., 248
Miura, R. M., 1, 3, 6, 7, 52, 55, 79, 89, 159, 266, 363
Miwa, T., 248
Mōri, Y., 248
Morikawa, G. K., 3
Moroz, I. M., 292, 327
Morris, H. C., 161
Morse, P. M., 366
Morton, K. W., 352
Moser, J., 132, 191, 196, 203, 206, 209
Moses, H. E., 28, 40
Mott, G., 342, 344
Munk, W., 309
Musher, S. L., 114

Nakamura, A., 172, 190, 216
Nemytskii, V. V., 134
Neugebauer, G., 337
Neveu, A., 341
Newell, A. C., 1, 8, 9, 19, 42, 45, 46, 47, 48, 56, 58, 64, 67, 68, 85, 110, 159, 232, 239, 240, 248, 255, 301, 313, 334, 335

Newstein, M. C., 337
Newton, R. G., 15, 114
Nizhik, L. P., 110
Novikov, S., 134, 146, 147, 153

Oberman, C., 389
Oishi, S., 190
Oikawa, M., 190
Olbers, D. J., 309
Olshanetsky, M., 206
Olver, F. W., 356
Olver, P. J., 236, 238, 239
Ono, H., 191, 202, 203, 294, 318
Onsager, L., 203
Orszag, S. A., 385
Osborne, A. R., 296
Ott, E., 255
Oxford, S., 341

Pasta, J., 3, 4
Patel, C. K. N., 335
Pedlosky, J., 297
Peek, B. M., 300
Pempinelli, F., 191, 246, 248
Pereira, N. R., 204, 206
Perelomov, M., 206
Peters, C. W., 307
Petviashvili, V. I., 250, 289
Phillips, O. M., 290, 296, 301, 309, 310, 312, 348
Pirani, F. A. E., 156
Pitayevsky, L. P., 1
Pogorzelski, W., 17, 24
Pokhil, G. P., 341

Ramanathan, G. V., 374, 387
Ramani, A., 40, 217, 218, 219, 236, 238, 239, 240, 243, 244, 248, 268
Randall, G. D., 252
Rayleigh, Lord (J. W. Strutt), 279
Rebbi, C., 342
Redekopp, L. G., 97, 232, 268, 296, 297, 299, 300, 312, 318, 320, 322
Richtmyer, R. D., 352
Rizk, M. H., 312
Rizov, V. A., 152
Rosales, R., 40
Roskes, G. J., 318, 320
Rossby, G. G., 297
Rota, G. C., 355

Rubenchik, A. M., 114, 250, 260
Rund, W., 2, 168, 170, 261
Rungaldier, H., 325
Russell, J. S., 277, 282
Ruijsenaars, S. N. M., 342

Salamo, G. J., 337
Salihoglu, S., 238
Samelson, H., 167
Sandri, G., 374, 387, 388
Santini, P. M., 290
Sanuki, H., 54, 159
Sasaki, R., 330
Sato, M., 248
Satsuma, J., 84, 85, 113, 114, 152, 172, 183, 187, 190, 191, 200, 211, 212, 215, 232
Scott, A. C., 168, 170, 327
Segur, H., 1, 8, 9, 28, 40, 42, 68, 71, 75, 78, 80, 81, 82, 85, 89, 113, 217, 218, 219, 228, 236, 238, 239, 240, 243, 244, 245, 248, 260, 268, 294, 296, 309, 316, 318, 320, 323, 324, 325
Shabat, A. B., 1, 8, 10, 11, 28, 34, 40, 53, 68, 80, 99, 110, 114, 156, 217, 228, 229, 233
Shulman, E. I., 152
Simmons, W. F., 310
Sklyanin, E., 342
Slusher, R. E., 330, 331, 336, 337
Smith, R. K., 311
Sternberg, S., 152
Stewartson, K., 318, 320
Stoker, J. J., 277
Stokes, G. G., 348
Stromberg, K., 358
Su, C. S., 277
Sudan, R. N., 255
Sugawa, M., 312
Sugaya, R., 312
Sutherland, B., 203, 206
Suzuki, K., 114, 183
Synakh, V. S., 256, 258, 273, 316, 317

Taha, T., 124
Takhtadzhyan, L. A., 58, 290, 342
Talanov, V. I., 258, 315
Tanaka, S., 41, 68, 134
Taniuti, T., 51, 250, 277
Tappert, F. D., 28, 38

Taylor, G. I., 290
Terng, C.-L., 330
Testa, F. J., 163, 265
Thacker, H. B., 341, 342
Thickstun, N., 203
Thomson, J. A., 313
Thorpe, S. A., 310
Toda, M., 114, 148
Toland, 283
Tracy, C. A., 248
Trubowitz, E., 10, 15, 26, 31, 52, 134, 137, 153, 159, 160, 161, 217

Ueno, K., 248
Ulam, S., 3, 4
Ursell, F., 365

van Kampen, N. G., 376, 390
van Moerbeke, P., 116, 134, 137, 153
Vlasov, A., 378
von Neumann, J., 353

Wadati, M., 8, 14, 34, 40, 41, 54, 159, 190
Wahlquist, H. D., 158, 161, 162, 163, 166, 263
Watson, K. M., 309, 313
Weber, P., 342
Wehausen, J. V., 277
Wei, C. C., 250, 277
Weidman, P. D., 282, 297, 300
Weinreich, G., 307
Weisskopf, V., 388
Wellner, M., 388
West, B. J., 309, 313
Whitham, G. B., 148, 190, 250, 251, 267, 277, 302, 351, 389
Wiesler, A., 342
Wigner, E., 388
Wilkinson, D., 341, 342
Willebrand, J., 313
Winkler, W., 136, 137
Wu, T. T., 248

Yajima, N., 84, 85, 97, 190, 250, 275, 277
Yariv, A., 302, 307, 308, 314, 316, 347
Yortsos, Y. C., 197
Yuen, H. C., 318, 323, 325, 326

Zabusky, N. J., 1, 4, 5, 6, 38, 149, 282, 339
Zakharov, V. E., 1, 8, 10, 11, 28, 34, 40,

41, 53, 55, 56, 57, 58, 67, 68, 71, 87, 89, 93, 97, 98, 99, 110, 114, 138, 152, 191, 199, 217, 228, 229, 232, 233, 237, 250, 256, 258, 260, 262, 267, 273, 290, 316, 317, 323, 337, 338, 339
Zeppetella, A., 266

SUBJECT INDEX

Abel's transformation, 144, 145
Ablowitz, Kaup, Newell and Segur method (AKNS), 9
Action-angle variable, 53, 58, 61, 66, 67, 69, 88, 149, 151, 157, 206, 207, 316, 340, 346, 355
Adjoint eigenfunction, 101
Adjoint operator, 251
Adjoint problem, 44, 251, 253, 256, 259
Ado's theorem, 168, 169
Airy function, 76, 228, 244, 255, 364
Algebraic geometry, 134
Algebraic soliton, 197, 199, 202, 203
   See also Lump solution.
Almost periodic function, 134, 147
Amplification factor, 122, 353
Amplifier, 359
Analytic, 58, 124, 132, 138, 234, 235, 242, 279
Angle variable, 206
Anharmonic lattice, 3
Anisotropic material, 303, 305
Antihermitian, 208
Antikink, 36, 348
Asymptotic line, 328, 348
Asymptotic solution, 68-83, 245, 282, 286, 287, 290, 355, 356, 359-384, 387, 388
Attenuator, 329, 337
Auxiliary spectrum, 136, 137, 142, 143

Bäcklund transformation (BT), 151, 153-161, 163, 167, 170, 171, 184-188, 198, 237, 246, 248, 250, 261-265
   in Painlevé or bilinear form, 184-188, 191, 194-197, 201, 215, 265
Baroclinic instability, 326
Benjamin–Feir instability, 310, 324
Benjamin–Ono equation, 203, 206, 209, 211, 212, 216, 268, 269, 276, 292, 294, 295, 388
Bessel function, 382

β-plane, 297
Bethe ansatz, 340
Bilinear equation, 159, 172, 179, 180, 183, 191, 200, 201, 212, 265
Bilinear theory, 152, 171-191, 211
Bloch eigenfunction, 135, 136
Bound states, 18-20, 28, 29, 30, 31, 32, 34, 45, 106, 109, 133, 148
   See also Discrete eigenvalues.
Boussinesq equation, 97, 98, 117, 191, 197, 201, 232, 237, 262, 265-268, 282, 283, 386
Branch points, 153, 235, 241, 267
   movable 240-243, 267-269
Breather, 36, 37, 38, 85, 250, 251, 260, 335, 349
Brunt–Väisälä frequency, 298

Canonical transformation, 53, 58, 61, 66, 67, 86, 88, 157
Capillary wave. *See* Surface tension.
Carrier wave, 316, 317, 323, 324, 332, 348, 349
Characteristic, 366-368, 382
Cole–Hopf transformation, 154, 166, 167, 170, 263, 266
Collision-free hydromagnetic wave, 3
Compact support, 100
Complete integrability, 52, 53, 58, 65, 67, 157, 276, 340, 341
Conditionally periodic, 134, 146, 147
Conductivity, thermal, 3
Conjugate variables, 59-64, 340
Connection problem, 245
Conservation law and conserved quantities, 6, 52-56, 61, 67, 69, 71, 75, 77, 88, 89, 105, 148, 151, 161, 166-170, 255, 258, 261, 263, 296, 335, 348, 349, 352, 357, 361, 389
   discrete case, 131, 132
Constant of the motion, 65, 336, 341

INDEX

Constrained harmonic motion, 52, 153
Contact transformation, 156
Continuous eigenvalue, 161
Continuous spectrum, 32, 41, 42, 160, 190, 334
Coordinates, generalized, 59, 62, 66
Coriolis parameter, 297
Coupled nonlinear Schrödinger equations, 97
Coupled oscillators, 153
Cramer's rule, 39
Crank–Nicolson scheme, 123, 369
Critical point, movable, 235, 238, 240, 267
Crystal lattice, 149
Curvature:
  Gaussian, 329
  negative, 328, 329, 330, 349

Dark pulse. *See* envelope hole soliton
Davey–Stewartson equation (multidimensional nonlinear Schrödinger equation), 113
Density, conserved, 55, 56, 88
Dependent variable transformation, 172, 179, 180, 183, 190
Determinant, 172
Dielectric medium, 301-304, 313-315, 318, 329-331, 346
Difference scheme, 153, 353, 355, 369, 385
Differential-difference equation, 115-122, 151, 211, 353
Differential-difference nonlinear Schrödinger equation, 121-131
Differential form, 161
Differential geometry, 153, 328
Diffraction, 90
Dirac delta function, 28
Direct method, 171-191, 197
Direct scattering, 134
Discrete eigenfunction, 85
Discrete eigenvalue, 35, 41, 52, 57, 65, 67, 75, 84, 85, 159-161, 169, 285, 288, 334, 335
  See also Bound state.
Discrete Gel'fand–Levitan–Marchenko integral equation (GLM), 127, 128, 133
Discrete inverse scattering, 114-133
Discrete KdV, 122
Discrete modified KdV, 121

Discrete problem (discrete nonlinear evolution equation), 114-133
Discrete Schrödinger scattering problem, 115, 124, 132
Discrete soliton, 130, 131, 133
Dispersion relation, 31, 45, 46-49, 61-64, 66, 68, 74-78, 88, 119, 122, 129, 130, 131, 152, 156, 157, 170, 179, 261, 278, 286, 293, 300-304, 307, 308, 311, 314, 335, 352-355, 358, 362, 363, 367, 370, 373-376, 381-386, 389
Dispersive, 300, 356
Dissipation, 250, 252, 256, 270, 274, 282, 295, 354
Dissipative perturbation, 250-253
Double pole solution, 36, 37
Double sine-Gordon equation, 38
Dym's equation, 15
Dynamical system. *See* N-body problem.

Eigenfunction, 160, 163, 265, 341
Eigenvalue, 164, 169-171, 249, 264, 283, 285, 298, 338
  See also Continuous eigenvalue, Discrete eigenvalue.
Einstein's equation, 336, 337
Electron displacement, 302
Electrostriction, 313, 314
Elliptic function, 237, 301, 307, 344
Elliptic function solution, 144, 203, 205, 210
Energy integral, 352-355, 362, 363, 367-370, 374, 379, 385, 386, 389
Entire function, 62, 64, 139
Envelope hole soliton, 28, 315, 323, 325
Equipartition of energy, 3, 4
Ernst equation, 338
Euler–Lagrange hypothesis, 342
Exact reduction of a PDE, 234, 236, 238, 268, 315
Exploding soliton, 34, 85
Explosive instability, 107, 109, 112
Exponential lattice. *See* Toda lattice.
Extrinsic curvature tensor, 329

Factor ordering, 341
Fast variable, 251
Fermi–Pasta–Ulam problem, 3
Fifth order KdV, 86
Finite band potential, 134

Finite perturbation, 137, 233
Focusing singularity, 256-258, 273, 316, 317
Fourier analysis, 48
Fourier integral solution of linear problem, 30, 353, 358, 362, 363, 367, 371
Fourier transform, 25, 26, 46, 48, 91, 103, 157, 211, 351, 354, 358, 363, 367-377, 382-387, 391
Fourier–Laplace transform, 351, 373, 376, 390
Fredholm alternative, 24
Fredholm determinant, 190
Fredholm integral equation, 24
Frequency-up conversion, 307
Functional (Fréchet) derivative, 58

Galilean invariant, 282, 339
Gauss–Bonnet theorem, 348
Gel'fand–Levitan–Marchenko equation (GLF), 22, 23, 26, 32, 38, 99, 110, 190, 217- 232, 236, 250, 266
General evolution equation, 42-52, 86
General evolution operator, 42-51
Generalized function, 275
Goursat problem, 20
Great primary wave of translation, 2
Great Red Spot of Jupiter, 297-300
Green's function, 250
Green's identity, 251, 260
Green's theorem, 257
Group theory, 152, 357
Group velocity, 75, 78, 356, 360, 371, 387
Guitar string, 344-346

Hamiltonian equation, 114
Hamiltonian operator, 152, 153
Hamiltonian system, 53, 58-65, 88, 89, 157, 205, 340, 355, 356
Hasselmann's theorem, 309
Heat equation, 263, 154, 361-363, 375
Hermitian operator, 12, 20, 32, 44
Higher nonlinear KdV, 252-261
Higher NLS, 252-261
Higher order KdV, 86, 138, 217
Higher order mKdV, 12, 13, 217
Higher order NLS, 12, 217
Hilbert space (L_2) 339, 354, 357, 359, 363, 374, 375, 390
Hilbert transform, 26, 204, 211, 277

Hill's equation, 136
Hirota's method, 34, 40, 171-191
Hyperelliptic function, 134, 139, 153

Ill-posed problem, 354, 386
Index of refraction, 303, 313, 329
Inhomogeneous broadening, 330-333, 336, 379, 380, 382
Initial data, 68, 69, 75, 76, 79, 80, 90, 353, 358, 368, 369, 373, 390
Integrability 58, 65, 89, 157, 162, 164, 171, 296, 338
Integral representation for eigenfunctions, 20
Intermediate long wave equation, 203, 211, 212
Internal wave, 282, 290, 292-297, 308-312, 346
Invariant of the motion, 206
  See also conserved quantity.
Inverse scattering, 15-28, 48, 49, 52, 151-157, 161, 163, 170, 172, 188, 234-239, 243, 248-250, 259, 265-268, 294, 306-311, 315-317, 322-324
  See also Periodic inverse scattering.
Inverse scattering transform (IST), 1, 7, 8, 9, 36, 46, 48, 151, 152
  and bilinear forms, 188
Involution, 64, 65, 89
Ion-acoustic waves, 97, 190
Ising model, 248
Isospectral flow, 200, 206
Isotropic dielectric material, 303-314

Jacobian elliptic function. See elliptic function.
Josephson junction, 327
Jost function, 15, 99, 124, 132, 217
Kadomtsev–Petviashvili equation (K–P), 113, 134, 188, 191, 197, 199, 200, 228, 229, 231, 259, 266, 268, 277, 289, 290, 386
Kerr effect, 314
Kerr solution in relativity, 339
Kinetic theory in one dimension, 374
Kink, 35, 36, 349
Korteweg-deVries equation (KdV), 3-14, 30, 34, 38-43, 46, 47, 52, 56, 58, 64, 67, 68, 79-86, 91, 117, 134, 138, 142, 144, 146, 148, 149, 153-163, 166, 172-174,

179, 184, 185, 188, 191, 194, 196, 199, 203, 209, 211-217, 223, 233, 246, 249-262, 265-271, 276, 278, 281-289, 292-299, 318, 344, 363
   with perturbations, 250-255
Kovalevskaya, 240

L_2. See Hilbert space.
Landau damping, 334, 373, 374, 377
Langmuir wave, 97, 322
Laser, 302, 305-308, 316, 317
Laurent series, 241, 243, 247
Lax's method, 9, 14, 239
Legendre equation, 269
Level degeneracy, 329, 334, 336
Lie algebra, 163, 167-171
   Abelian, 168-170, 263, 264
   non-Abelian, 168
Lie–Bäcklund transformation, 156
Light cone, 91
Linearized dispersion relation, 179
Liouville's theorem, 135, 139, 155
Local conservation, 105
Long wave, 279, 287-289, 292-297, 343-345
Long wave–short wave interaction, 77, 311
Lump solution, 113, 191, 197, 198, 232, 289
Linear pseudopotential, 165, 169-171, 263

Main spectrum, 13, 136, 137
Manley–Rowe relation, 306, 307
Mapping of functions, 66, 67, 86
Mass of a wave, 348
Mathematical induction, 172, 178
Maxwell–Bloch equation, 331, 341
Maxwell's equation, 302, 330
Metric tensor of a surface, 320, 338
Miura's transformation, 6, 7, 52, 79, 84, 86, 153, 160, 215, 266, 363
Modified Korteweg-deVries equation (mKdV), 6, 7, 8, 9, 13, 21, 52, 75-79, 84-86, 121, 153, 155, 163, 175, 179, 181, 191, 197, 201, 215, 217, 219, 227, 262, 265, 276, 292-296, 299, 329, 344, 363
Modulated multiperiodic solution, 148
Moment of inertia, 258
Momenta, generalized, 59, 66
Monochromatic wave, 90, 318
Monodromy, 135, 240, 248, 249

Monodromy coefficient, 40, 41, 42
Monodromy matrix, 135
Motion of poles. See N-body problem.
Multidimensional scattering problem, 110-114
Multidimensional Schrödinger scattering (eigenvalue) problem, 114
Multiple scales method, 250-261

N-band potential, 134, 146, 148, 150, 152
N-body Hamiltonian system, 205, 210
N-body problem, 203-217
N-soliton solution, 34, 38, 40, 41, 42, 151, 159, 172-175, 179-192, 195, 197, 211, 212, 239, 265, 289, 323
Natural stability, 123
Neumann series, 17, 100
Non-Abelian pseudopotential, 151, 163, 170, 171, 265
Nondispersive problem, 107
Nonlinear Fourier analysis, 45-51
Nonlinear instability, 109
Nonlinear ladder network, 183
Nonlinear optics, 301-308, 311-318, 329-336
Nonlinear Schrödinger equation (NLS), 8, 9, 12, 28, 32, 33, 46, 134, 181, 182, 217, 238, 242, 243, 256-260, 265, 270-273, 312-326, 338, 340
   differential-difference form, 120, 131
   in two dimensions, 258, 260, 271
   with dissipation, 256, 270, 274
   See also Higher NLS, Higher order NLS, Schrödinger equation.
Normalizing coefficient, 27-31, 106, 133

Orthogonality condition, 43-45, 48-52, 271, 273, 274
   See also secular condition.

Painlevé conjecture, 152, 236, 243, 266, 268, 293
Painlevé function, 203
Painlevé property, 151, 234-239, 296, 315-317, 322
Painlevé transcendent, 152, 197, 228, 233, 236, 238, 240, 243, 248
Parametric oscillation, 307
Parseval's relation, 25

Partial difference equation, 115, 122, 124, 151
  nonlinear Schrödinger, 123, 131
Partial differential equation (PDE), 151, 153, 157, 161-163, 179, 233-239, 243, 250, 269, 352, 356
Periodic boundary condition, 134-148, 324
Periodic inverse scattering, 138-142
Periodic multiple wave solution, 190
Permutability theorem, 159
Permutation relation, 184, 187, 265
Perturbation, 250-261, 327, 341, 342
  See also Finite perturbation.
Phase shift, 5, 40, 41, 75, 198, 199, 265, 299, 300
  of solitons, 40-42, 86
Piston, 283, 285, 288
Poincaré recurrence. *See* Recurrence.
Poisson bracket, 60, 64, 86, 87, 339
Polarization, 301, 331, 332, 347, 379, 382
Pole, 239, 244, 247, 248
  movable, 241, 243
Pole expansion, 203-216
  See also N-body problem.
Potential, 66, 158, 166, 170, 277, 292, 341
  See also Finite band potential, Resonance potential.
Projection operator, 26
Prolongation structure, 161, 162
Pseudopotential, 152, 161-171, 262, 263, 265
  See also Non-Abelian pseudopotential, Nonlinear pseudopotential.
Pseudospherical surface, 328, 329, 338
P-type, 235-243, 250, 266-268, 316
Pulse:
  2π, 334-336, 349, 350
  0π, 334-336
  dark, *See* Envelope hole soliton.

Quadratic forms, fundamental, 328, 329
Quadratic interaction, 301
  See also Three-wave interaction equation, Resonant triad.
Quantization, 149, 340, 341
Quantum field theory, 248, 338
Quartic resonance, 96

Radiation, 68, 75, 83, 337
Random wave, 312
Rational exponential soliton, 233

Rational solution, 191-202, 265
Recurrence, 4, 150, 316, 355, 356
Recurrence time, 150
Recursion relation, 246
Reflection coefficient, 27-30, 38, 51, 52, 79, 83, 91, 109, 132, 133, 286
Reflectionless potential, 38
Relativity, 337
Resonance, 241, 242, 300, 301
  See also Quartic resonance.
Resonant atomic frequency, 303, 329
Resonant interaction, 96, 190
Resonant quartet, 302, 303, 311-313
Resonant triad, 152, 190, 300-314, 346, 347
Reversible shock, 6
Riccati equation, 7, 235, 266
Ricci tensor, 337
Riemann–Hilbert problem, 25, 26, 99, 266, 338
  differential, 216, 266
Riemann–Lebesgue lemma, 358, 368
Riemann surface, 142, 144, 147
Riemann theta function, 147
Rossby wave, 297-300
Ruby, 305, 307, 330

Scattering data, 16, 20, 25, 28-32, 48, 49, 53, 57, 61, 65, 66, 70-75, 85, 88, 342
Scattering matrix, 99
Scattering theory, 15
Schrödinger equation, 7, 9-13, 15, 26-28, 31, 33, 38, 46, 51-53, 58, 59, 68, 75, 89, 115, 134, 139, 216, 221, 357
  See also Nonlinear Schrödinger equation.
Second harmonic generation, 306-308
Secular term, 72, 251, 275, 280, 281, 299, 305, 314, 319, 320, 326, 332, 337, 343
Secularity condition, 251, 252
Self-adjoint, 272
Self-dual network equation, 121, 131
Self-focusing singularity. *See* Focusing singularity.
Self-induced transparency (SIT), 260, 329-337, 348, 349, 379, 383
  mechanical analogue of, 349
Self-modal interaction, 96
Self-self interaction, 96
Self-similar equation, 217, 227, 228, 238, 240, 245, 246, 249, 265, 267, 268, 286, 336, 357, 363, 365, 366

Self-similar solution, 152, 232-250
Semi-infinite problem, 28
Shelf, 252-256, 284
Similarity equation, 40, 196, 203
Similarity solution, 69, 197, 233, 236, 238, 266, 356, 357, 365, 386
Sine-Gordon equation, 8, 9, 13, 34, 35, 38, 90, 134, 149, 155, 159, 170, 181, 190, 217, 237, 238, 262, 265, 267, 326-338, 348, 366
Singular equation, 26, 203
Singular-point analysis, 237, 240-243
Singularities for solitons, 33, 34
Singularities of an ODE:
    fixed, 234, 235, 239
    movable, 234, 235, 239, 240
Slinky, 345
Slow variable, 251
Solitary wave, 2, 5, 6, 238, 250, 251, 254-259, 282, 297, 336, 341, 379
Soliton, 67, 68, 74, 75, 79, 80, 83, 86, 89, 90, 91, 151, 152, 156, 159, 160, 161, 167, 171, 172, 174, 175, 179-191, 238, 250-255, 259-262, 265, 266, 269, 270, 275, 276, 281-286, 290, 295, 296, 315, 317, 318, 323-326, 334, 338, 341, 344, 349, 383
    definition, 6
Soliton perturbation, 250-261
Soliton phase shift formula, 174, 175
Soliton resonance, 189, 232, 266
Soliton stability, 259
    See also Transverse stability.
Soliton superposition (permutation) formula, 187, 188, 265
Soliton wave, 36, 38, 41
Southern Tropical Disturbance, 299, 300
Spectral band, 146
    See also Unstable band.
Spectrum, 159, 161, 307, 308, 317
    See also Auxiliary spectrum, Continuous spectrum, Eigenvalue, Main spectrum.
Square well, 28
Squared eigenfunction, 42-52
Stationary phase, 356, 357, 359, 364, 368, 371
Steepest descents, 356, 364, 384
Stochastic, 67, 276
Stokes multiplier, 249
Stratified fluid, 203, 211

Surface tension, 260, 268, 277, 280, 289-292, 319, 322
Susceptibility, 301, 302

Thermal conductivity, 3
Three-wave interaction equation, 94, 95, 99, 105, 110, 111, 233, 300, 308
Time dependence, 159, 334
Time scale, 275, 276, 294-299, 309, 313, 316, 319-323, 348
Toda lattice (exponential lattice), 114-117, 121, 132, 133, 148, 149, 183, 209, 387
Torus, 147
Trace formula, 56, 57, 70
Transverse perturbation, 250, 259, 260, 273, 289, 299, 315-318, 325, 336
Transverse stability, 250-261
Triad resonance, 96, 190
Truncation error, 123
Turning point, 244
Two-dimensional sine-Gordon equation, 190
Two-dimensional water wave, 260
Two-wave interaction equation, 95

Unstable band, 136, 137, 139, 142, 144, 145-147, 197
Unstable solution, 355

Volterra integral equation, 17, 100, 139
Volterra operator, 229

Water wave, 260, 267, 268, 277, 281-289, 314, 315, 317, 321, 322, 325, 347
Wave equation, 277, 319
    linear, 280, 281
Wave front, 287, 372, 386
Wave guide, 317, 318
Wave measurement, 282, 283
Wave packet, 90, 260, 297, 305, 310-314, 317, 319, 321, 324, 325, 327, 347-349
Wave tank, 283, 295, 315, 317, 323, 326
Weierstrass P function, 210
    See also Elliptic function.
WKB, WKBJ, 71, 101, 255
Wronskian, 65, 87

Zakharov–Shabat scattering problem, 8-14, 28, 32, 42-52, 117, 124, 163, 255
Zero of $a(k)$, 126, 148